Lignin Chemistry

Lignin Chemistry

Characterization, Isolation, and Valorization

Edited by Yuhe Liao and Bert F. Sels

WILEY-VCH

Authors

Prof. Yuhe Liao
Guangzhou Institute of Energy Conversion
Chinese Academy of Sciences
Guangzhou, 510640
China

Prof. Bert F. Sels
KU Leuven
Faculty of Bioscience Engineering
Leuven, 3001
Belgium

Library of Congress Card No.: applied for

British Library Cataloguing-in-Publication Data
A catalogue record for this book is available from the British Library.

Bibliographic information published by the Deutsche Nationalbibliothek
The Deutsche Nationalbibliothek lists this publication in the Deutsche Nationalbibliografie; detailed bibliographic data are available on the Internet at <http://dnb.d-nb.de>.

Print ISBN: 978-3-527-35107-7
ePDF ISBN: 978-3-527-83984-1
ePub ISBN: 978-3-527-83985-8
oBook ISBN: 978-3-527-83986-5

Typesetting Straive, Chennai, India
Printing and Binding CPI Group (UK) Ltd, Croydon, CR0 4YY

C9783527351077_300524

Contents

Preface

Utilization of non-renewable fossil resources in the transportation and chemical industry is responsible for the large amounts of carbon dioxide in the atmosphere. Mitigation of such emissions requires disruptive solutions in the area of renewable energy (such as solar and wind) exploitation and utilization, while replacement of fossils with renewable carbon resources will be indispensable. Given its CO_2 conversion capabilities, biomass is such a sustainable feedstock for the production of fuels and chemicals. Processing biomass in biorefineries is a challenging, perhaps underexplored, research activity but also a necessary step to advance biomass conversion field. Besides the carbohydrates, lignin (an aromatic biopolymer) represents the major fraction of the lignocellulosic biomass. Its structure is diverse, hampering quick and easy valorization, and therefore burnt for power and heat today. However, new ideas and advances in lignocellulose biorefinery are focusing on value-added utilization by generating higher-quality primary lignin products for more promising upgrades.

Lignin is a promising carbon feedstock for the production of (oxygen-free) fuels such as bio-jet fuels due to its higher energy density and original cyclic molecular structure, compared to other plant components such as sugars and vegetable oils. Even more, the unique chemical structure of lignin, with its oxygen-containing functional groups, offers opportunities for functional aromatic and cyclic chemicals, as well as for diverse functional materials. Effective and selective conversion of lignin to such high-value products is challenging. This book therefore covers the chemistry of lignin utilization. It starts with an introduction of the state-of-the-art structural characterization and understanding of lignin and its refined lignin fractions, in relation to the lignin's properties. Fractionation of lignin from plant biomass by different (catalytic) biorefinery approaches (including acid and base treatments) to obtain distinct lignin properties is then discussed. The main part of the book discloses a myriad of catalytic conversion of lignin toward high-value chemicals, including biocatalysis, oxidation, photocatalysis, and electrocatalysis, followed by lignin valorization toward functional materials such as thermoset and thermoplastic polymeric materials, nanoparticles, carbon materials, and hydrogels.

Given that it covers analytical chemistry, catalysis, biology, material chemistry, and engineering in the field of biomass conversion, thanks to the interdisciplinary expertise of the contributors; this book may be a valuable and inspiring reference for students, scientists, engineers, and professionals to motivate them to discover and advance new biorefinery and valorization technologies, with a view on commercial lignin utilization.

We would like to thank Wiley-VCH and its staff for their great support in this project. Finally, we repeat our appreciation and gratitude to all our co-authors for their invaluable work and time they have spent during the intense editing process.

Yuhe Liao
Guangzhou Institute of Energy Conversion (GIEC),
Chinese Academy of Sciences (CAS)

Bert F. Sels
KU Leuven
February 2024

Acknowledgments and Conflict of Interest

No potential conflict of interest is declared by the author. YL thanks the National Natural Science Foundation of China (52176213). BS thanks the KU Leuven, Flemish (Vlaamse Veerkracht, FWO, VLAIO, Catalysti, Province Brabant, iBOF), Belgian (ETF FOD Economy), and European (Horizon 2020 CBE, EU RRF) governmental support for biomass-related topics.

1

A Brief Introduction of Lignin

Yuhe Liao

Guangzhou Institute of Energy Conversion, Chinese Academy of Sciences, No. 2 Nengyuan Road, Tianhe, Guangzhou, 510640, China

1.1 Introduction

To reduce greenhouse gas (GHG) emissions, society should use renewable carbon resources in a sustainable way to produce chemicals, materials, and fuels next to circular use of currently available carbon feedstock. Lignocellulosic biomass has been considered as an abundant, carbon dioxide neutral, and renewable carbon resource [1]. The major compositions in lignocellulosic biomass are three oxygen-containing biopolymers: cellulose (40–60%), hemicellulose (10–40%), and lignin (15–30%), and the minor compositions include proteins, fats, pectins, inorganic matter, and others [2]. Cellulose and hemicellulose are carbohydrate-based biopolymers, whereas lignin is a complex aromatic biopolymer with a high carbon content (Figure 1.1) [3, 4]. Typically, cellulose determines the structure of cell walls in the form of microfibrils. The cross-linked lignin and hemicellulose wrap around the cellulose microfibrils (Figure 1.1). The lignin can provide additional rigidity and cause the cell walls to be hydrophobic and water impermeable. Therefore, these three main fractions are intertwined to yield the complex structure, contributing to biomass recalcitrance, which hampers the effective valorization of lignocellulosic biomass toward high-value products such as chemical and liquid fuels [5].

Different approaches have been developed to overcome the recalcitrance and valorize lignocellulosic biomass over the past decades [2, 6]. Classically, the lignocellulosic biomass is utilized to produce high-quality pulps for paper production. Emerging approaches are conversion of lignocellulose via thermal cracking (e.g. pyrolysis), biocatalysis, chemocatalysis, and integration of them toward chemicals and fuels such as bioethanol, furfurals, and levulinic acid [1, 3, 4, 7–13]. It is clear

Lignin Chemistry: Characterization, Isolation, and Valorization, First Edition.
Edited by Yuhe Liao and Bert F. Sels.

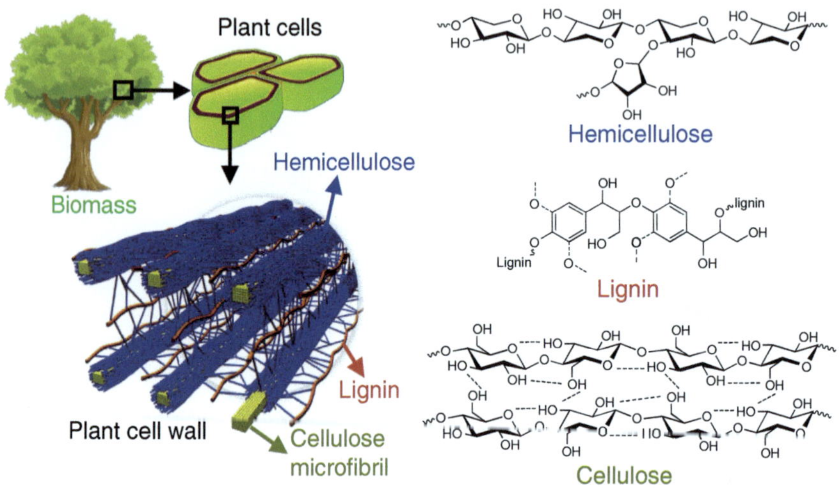

Figure 1.1 General structure of lignocellulose in plant and representative structure of cellulose, hemicellulose, and lignin. Source: From Liao et al. [2].

that these products are usually derived from cellulose and hemicellulose, whereas lignin, either left as a solid residue after conversion of cellulose and hemicellulose (such as via hydrolysis) or extracted from lignocellulose with cellulose and hemicellulose as solid residue, is considered as a waste or a low-value product for energy use. Currently, the pulping and biorefinery processes such as bioethanol production generate more than 50 million tons of lignin annually with ca. 95% used as fuel for heat and power generation due to the recalcitrance of lignin [14, 15]. The utilized 5% of lignin have several applications such as additives, surfactants, and adhesives. Although delignification (i.e. removal of lignin) of lignocellulosic biomass can facilitate the utilization of cellulose and hemicellulose to improve the economics of biorefineries, the value of biomass may not be maximized without utilization of lignin toward high-value products. Techno-economic analysis (TEA) and life-cycle assessment (LCA) have shown that valorize lignin can improve both economics and sustainability of biorefineries [16]. Hence, it is paramount to valorization of lignin with novel strategies to explore the potential of all carbon constituents of lignocellulose.

Over the past years, some progresses were achieved in the aspects of lignin characterizations to reveal the structure of lignin, isolation to obtain lignin with different properties (even native like), and valorization toward chemicals, fuels, and materials with novel approaches. Hence, this book aims to introduce the most recent advancements in these aspects, particularly the valorization methods such as oxidation, photocatalysis, electrocatalysis, and valorization of native lignin. As a preface to the following chapters, this introductory chapter will briefly introduce the structure of lignin from the point of view of monomeric units, inter-unit linkages, and biosynthesis.

1.2 The Building Blocks of Lignin

Lignin, a phenolic biopolymer, is derived primarily from three kinds of 4-hydroxyphenylpropanoids (i.e. monolignols, e.g. *p*-coumaryl alcohol, coniferyl alcohol, and sinapyl alcohol, Figure 1.2). These monolignols differ in substitution degree of methoxylation on the aromatic ring (i.e. 2 and 6 positions) and incorporate into lignin chain to produce corresponding *p*-hydroxyphenyl (H), guaiacyl (G), and syringyl (S) units, respectively [17]. Besides, it is found that lignin is derived from numerous other building blocks, such as *p*-coumarates, ferulates, caffeyl alcohol, *p*-hydroxybenzoates, hydroxycinnamaldehdyes, tricin, and hydroxystilbenes [18, 19]. These building blocks are shown in Figure 1.2.

Currently, the only known route to form these building blocks is the phenylpropanoid pathway starting from phenylalanine via multiple steps in all plants with different enzymes [20, 21]. Whereas tyrosine can be an additional starting substrate for grasses. Figure 1.3 overviews the complete pathway of phenylpropanoid pathway for these building blocks. The phenylalanine was first deaminated in the presence of phenylalanine ammonia-lyase (PAL) toward cinnamic acid, which is then hydroxylated toward *p*-coumaric acid in the presence of cinnamate 4-hydroxylase (C4H). While in the case of tyrosine, *p*-coumaric acid can be produced from a shortcut pathway, direct deamination with tyrosine ammonia-lyase (TAL) or PAL [14, 23].

Then, *p*-coumaric acid is enzymatically converted toward either *p*-coumaroyl-CoA via 4-coumarate: CoA ligase (4CL) or caffeic acid through hydroxylation via *p*-coumarate 3-hydroxylase (C3H). *p*-Coumaryl alcohol is produced from reduction of *p*-coumaroyl-CoA via cinnamoyl-CoA reductase (CCR) and cinnamyl alcohol dehydrogenase (CAD). For caffeic acid, methylation via caffeic acid *O*-methyltransferase (COMT) forms ferulic acid, which can be transformed into feruloyl-CoA. Meanwhile, caffeic acid can be converted toward cafferoyl-CoA, which can be further methylated to feruloyl-CoA via caffeoyl-CoA *O*-methyltransferase (CCoAOMT). In addition, conversion of *p*-coumaroyl-CoA in the presence of shikimate/quinate hydroxycinnamoyl transferase (HCT) yields *p*-coumaroyl shikimate. Hydroxylation of *p*-coumaroyl shikimate via C3H can produce *p*-caffeoyl shikimate, which can be transformed toward caffeic acid and caffeoyl-CoA in the presence of caffeoyl shikimate esterase (CSE) and HCT, respectively. The shikimate intermediates are currently recognized as the favored substrates for hydroxylation [22].

Reduction of feruloyl-CoA via CCR yields coniferaldehyde, which can be transformed to coniferyl alcohol and 5-hydroxyconiferaldehyde through reduction (via CAD) and hydroxylation (via ferulate 5-hydroxylase, F5H), respectively (Figure 1.3). The main pathway to produce sinapyl alcohol is hydroxylation (via F5H) of coniferyl aldehyde followed by tandem methylation (via COMT) and reduction. Hydroxylation (via F5H) of coniferyl alcohol followed by methylation can form sinapyl alcohol as well. Oxidation of coniferylaldehyde by hydroxycinnamaldehyde dehydrogenase (HCALDH) forms ferulic acid. These synthesized monolignols are transported to the cell wall and integrated into a growing lignin chain.

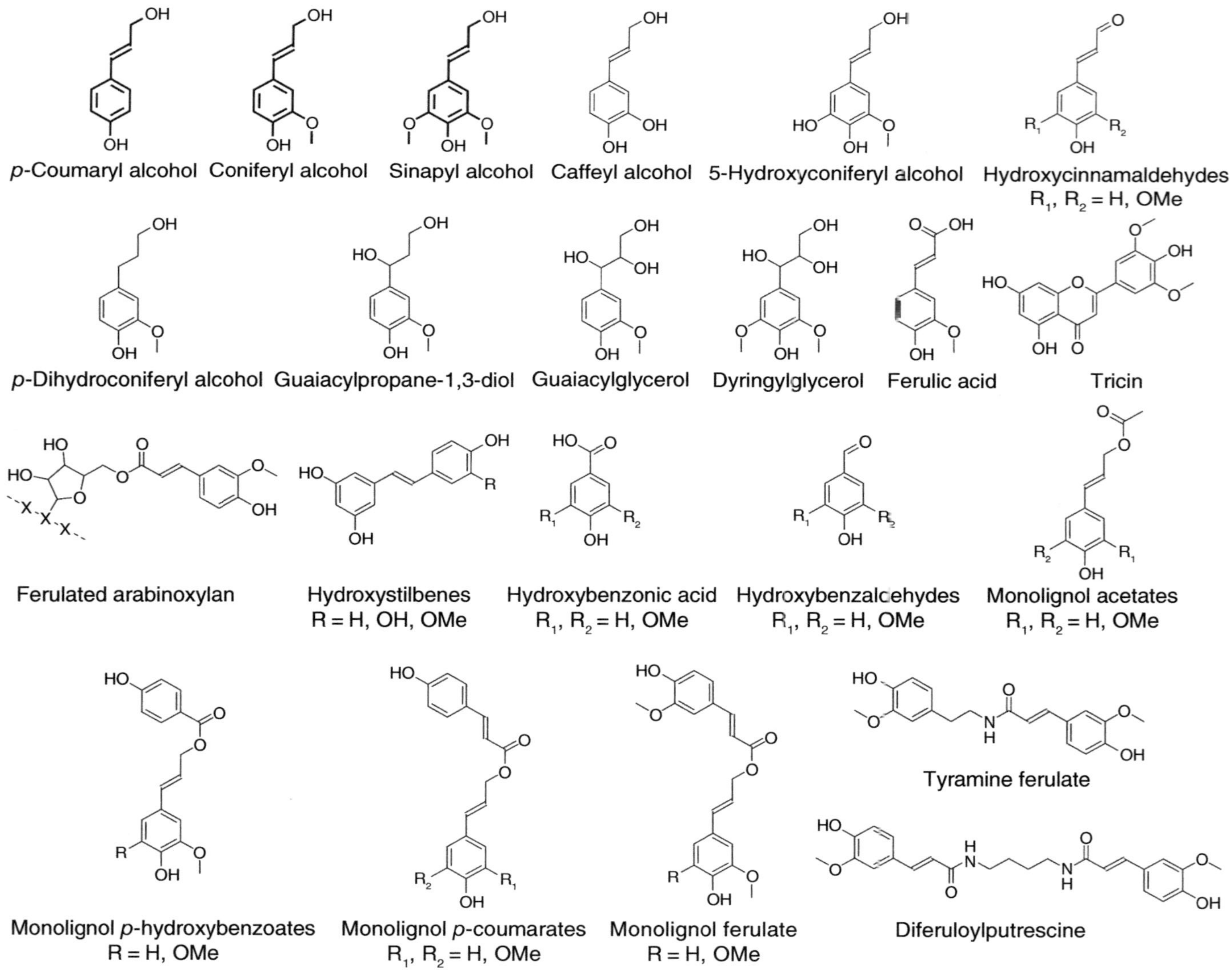

Figure 1.2 Lignin building blocks. Three main monolignols (i.e. *p*-coumaryl alcohol, coniferyl alcohol, and sinapyl alcohol) are shown in bold.

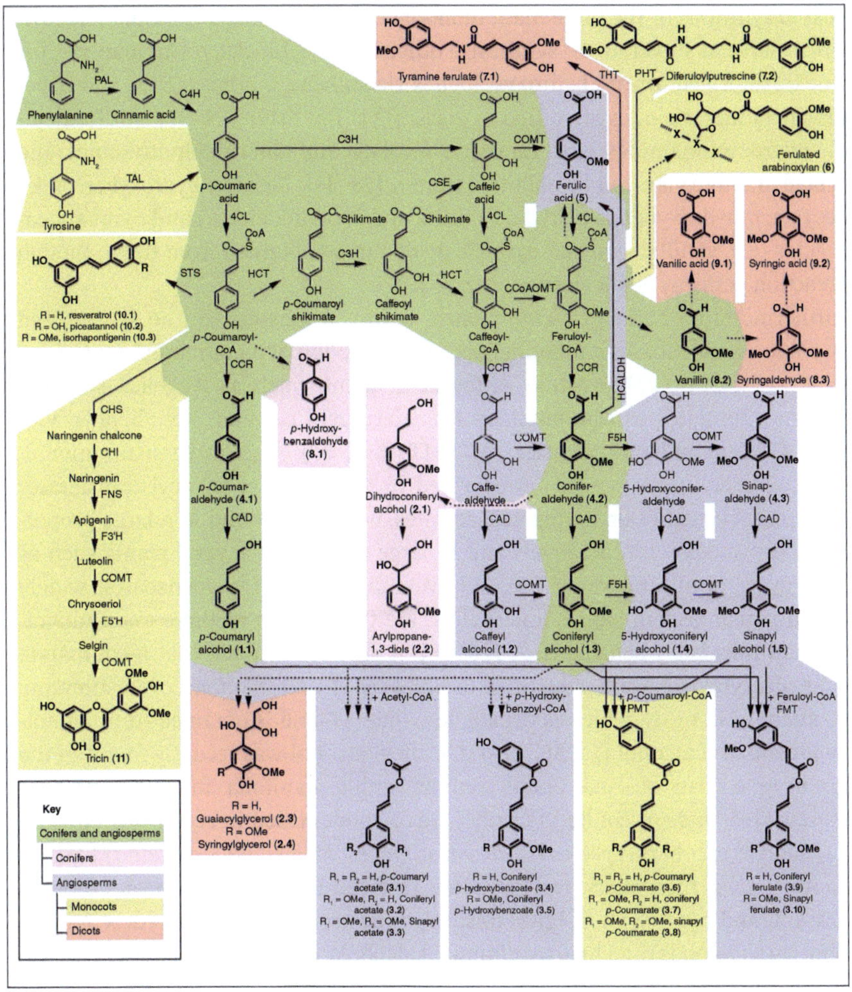

Figure 1.3 Summary of the metabolic pathway toward lignin building blocks. PAL, phenylalanine ammonia-lyase (PAL); cinnamate 4-hydroxylase (C4H); tyrosine ammonia-lyase (TAL); *p*-coumarate 3-hydroxylase (C3H); caffeic acid *O*-methyltransferase (COMT); 4-coumarate:CoA ligase (4CL); cinnamoyl-CoA reductase (CCR); cinnamyl alcohol dehydrogenase (CAD); *p*-hydroxycinnamoyl-CoA: quinate/shikimate *p*-hydroxycinnamoyltransferase (HCT); caffeoyl-CoA *O*-methyltransferase (CCoAOMT); caffeoyl shikimate esterase (CSE); ferulate 5-hydroxylase (F5H); *p*-coumaroyl-CoA monolignol transferase (PMT); feruloyl-CoA monolignol transferase (FMT); chalcone synthase (CHS); chalcone isomerase (CHI); flavone synthase (FNS); flavonoid 3'-hydroxylase (F3'H); flavonoid 5'-hydroxylase (F5'H); stilbene synthase (STS); hydroxycinnamaldehyde dehydrogenase (HCALDH); hydroxycinnamoyl-CoA: putrescine hydroxycinnamoyl transferase (PHT); hydroxycinnamoyl-CoA: tyramine *N*-hydroxycinnamoyltransferase (THT). Source: Adapted from Vanholme et al. [22].

Next to synthesis of the three main monolignols, *p*-coumaroyl-CoA and feruloyl-CoA can serve as substrates to synthesize other building blocks. *p*-Coumaroyl-CoA can be transformed toward hydroxystilbenes via stilbene synthase (STS) and flavonoids (including tricin, as identified in grasses) via multistep [24–27]. For feruloyl-CoA, hydroxycinnamides (e.g. tyramine ferulate and diferuloylputrescine) and ferulated arabinoxylan can be obtained from it [28–31]. Besides, hydroxybenzaldehydes (e.g. *p*-hydroxybenzaldehyde, vanillin, and syringaldehyde) and hydroxybenzoic acids (e.g. vanillic acid and syringic acid) can be obtained from them, though the reaction pathway is not clear yet [32, 33].

Furthermore, monolignol *p*-coumarates, found in grasses, can be synthesized from *p*-coumaroyl-CoA and monolignols with the action of *p*-coumaroyl-CoA monolignol transferase (PMT), while as found in plants such as Angelica sinensis, monolignol ferulates can be obtained from feruloyl-CoA and monolignols with feruloyl-CoA monolignol transferase (FMT) [34, 35]. The three main monolignols and acetyl-CoA can be transformed in the presence of acyl transferases toward γ-O-acylated ester conjugates, acetates, which are found in a large proportion of hardwoods with low levels and in large amount in several plants such as palms, abaca, sisal, and kenaf [36, 37]. Monolignol *p*-hydroxybenzoates, widely found in poplar, aspen, palms, and willow, are the product of three main monolignols and *p*-hydroxy-benzoyl-CoA [36, 38]. It should be noted that *p*-coumarate and *p*-hydroxybenzoate are often misinterpreted as H-unit fractions, overestimating the content of H units, which is usually less than 5% and is obtained from monolignol *p*-coumaryl alcohol [14, 36, 39–41]. Caffeyl alcohol was recently found in the lignin of seed coats of some plants such as vanilla planifolia and cactaceae species [42–44]. 5-Hydroxyconiferyl alcohol was also identified in seed coats of some plants such as escobaria species (*E. dasyacantha*, *E. lloydii*, and *E. zilziana*) [45].

Another two monomers, arylpropane-1,3-diols (guaiacylpropane-1,3-diol in Figure 1.3) and hydroxyphenylglycerols, might be produced from dihydroconiferyl alcohol (in softwoods) and hydroxycinnamyl alcohols, respectively, likely via radical oxidation instead of enzymatical reaction [22, 46, 47].

1.2.1 Interlinkages in Lignin

Lignin is obtained from polymerization of these building blocks after transportation to the cell wall. In the polymerization process, the phenols are first oxidatively converted via enzymes to form phenolic radicals, which are then subjected to coupling to form a racemic polymer. Peroxidases and laccases are involved in the oxidation step [48]. Hydrogen peroxide is the substrate of peroxidases, while laccases oxidize using oxygen. Lignification is a purely chemical process and not enzymatically determined once the radicals are formed [49].

Generally, the monolignol is first dehydrogenated (i.e. oxidation) to form a phenol radical, which is relatively stable due to the unpaired electron delocalization [40]. Afterward, two monolignol radicals couple to yield a (dehydro)dimer, thereby forming a covalent bond. These monolignol radicals are preferred to couple at their β position, leading to the formation of β-β (resinol), β-O-4 (β-aryl ether), and β-5 (phenylcoumaran) (dehydro)dimers (Figure 1.4) [50]. These are the most common interunit linkages in lignin. These dimers are further oxidized to form radicals, which then couple with another monolignol radical to increase polymer chain. Usually, the new monolignol radical is coupled exclusively at the β position of the growing oligomer, forming β-O-4 interunit linkage. Therefore, β-O-4 interunit linkage is the most abundant linkage in lignin, and it accounts for 50–80% of total linkages in native lignin [19, 50], while isolation of lignin can decrease the amount of β-O-4 interunit linkage in isolated lignin due to degradation of lignin, and the reduced amount depends on the process conditions [14]. This type of polymerization is termed as endwise coupling, where a monolignol radical is added to the growing lignin polymer (also radical) in one reaction. As the polymerization progresses, two lignin oligomeric phenolic end units are coupled to form 5-5 (biphenyl) and 4-O-5 (diphenyl ether) interlinkages, which are not from the coupling between monomer and oligomer or between monomer and monomer [19, 50]. This is rare for S-unit rich lignin due to the presence of the two methoxy groups, while it is common in G-unit rich lignin as only one methoxy group in the monolignol. Besides, β-1 (spirodienone) linkage is formed by coupling a β-aryl ether dimeric end unit with a monolignol. The possibility of different linkages depends on the type of monolignols and the conditions in the cell wall [50].

The lignin obtained from combinatorial radical coupling is very complex and difficult to valorize and characterize. The composition (i.e. the relative abundance of units), structure, and amount of lignin differ between plant species and within the plant species [51, 52]. Besides, the lignin can vary at different growth stages of plants.

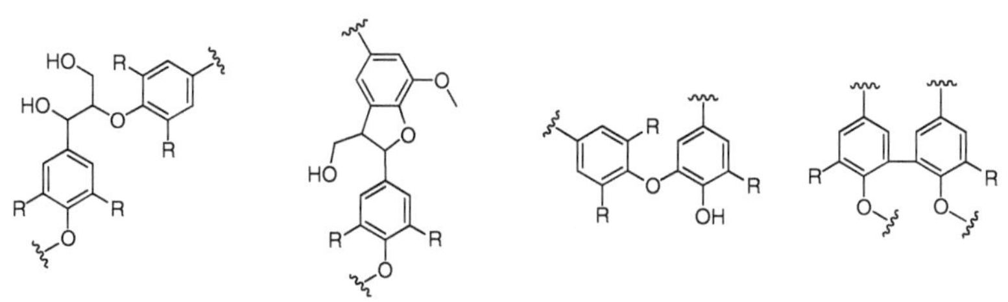

β-O-4 (β-aryl ether) β-5/α-O-4 (phenylcoumaran) 4-O-5 (biphenyl ether) (free phenolic) 5-5 (biphenyl)

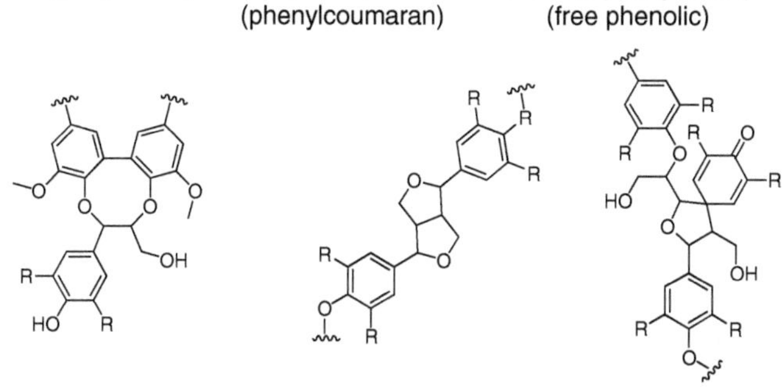

R = H, OMe

5-5/4-O-β (dibenzodioxocin) (free phenolic) β-β (resinol) β-1 (spirodienone)

Figure 1.4 Representative interlinkages in lignin.

Generally, the lignin in a gymnosperm/softwood is composed of mostly G units, and lignin in angiosperm/dicot/hardwood contains both G and S units with trace amount of H units [19]. The lignin in (commelinid) monocot/grass contains H, G, and S units [19, 53]. The model lignin structure of these three types of plants is displayed in Figure 1.5 [39]. Additional building blocks that have been found in the past are also integrated into the chain of lignin to form new linkages. Indeed, most of the lignin has a very complex structure, but there is an exception for C-lignin, which is obtained from endwise β-O-4 coupling of caffeyl alcohol [42–44]. C-lignin is a linear homopolymer linked with benzodioxane (Figure 1.6) [43]. Therefore, selective depolymerization of C-lignin is easier than other types of lignin as they contain different C–O and C–C interlinkages. For instance, C-lignin can be nearly quantitatively converted into monomeric catechol via hydrogenolysis [54–56]. Besides, 5H-lignin, obtained from coupling of 5-hydroxyconiferyl alcohol, has the same benzodioxane units [43, 45].

Lignin has been hypothesized as a highly branched polymer for a long time, but this is now doubted. It is assumed that lignin chains connected via 5-5 (biphenyl) or 4-O-5 (diphenyl ether) interlinkages can form a Y-branch polymer. However, the yielded structures are mostly free phenolic (which is traditionally hypothesized to be branching points), as identified by 2D NMR recently, and thus the native lignin must be less branched than usually assumed [45, 57]. This was also observed for β-1 (spirodienone) linkage (i.e. free phenolic instead of etherified) [58].

1.2.2 Bioengineering of Lignin

As discussed in the previous section, the monolignols are obtained from phenylalanine and tyrosine with the action of different enzymes. Upon genetic modifications of the phenylpropanoid pathway can influence the lignin composition, structure, and amount. For instance, downregulation of the genes of PAL, C4H, and 4CL, which are the enzymes involved in the early stages of phenylpropanoid biosynthesis, can lead to decreased flux, and thus reduce the lignin content in plant [59]. Although low content of recalcitrant lignin can be formed in the plant through this strategy, detrimental agronomic effect can occur as plants need a certain amount of lignin [60].

The favored approach is downregulation or upregulation of enzymes to yield plant lignin, which can facilitate the valorization of biomass (including lignin). For instance, the S/G ratio in angiosperms can be regulated via overexpression or underexpression of the F5H gene [59]. Overexpression of the F5H, when forced by a powerful lignin promoter, can yield linear lignin with extremely high S unit content (98.3%) [61]. The interlinkages in this type of high S lignin only involve β-O-4 and β-β, where the β-O-4 linkage accounts for 88.8% (i.e. low content of recalcitrant C–C interlinkage). Depolymerization of this type of linear lignin can produce high monomer yield, such as 78% of monomer was obtained from hydrogenolysis [61, 62]. In contrast, downregulation of F5H can obtain plant lignin with solely guaiacyl unit. Similarly, underexpression of the genes of C3H, HCT, and CSE enzymes, which are related to 3-hydroxylation steps, can produce plant lignin rich in H units [59, 63–67]. However, in the typical wild-type lignin H units are a minor

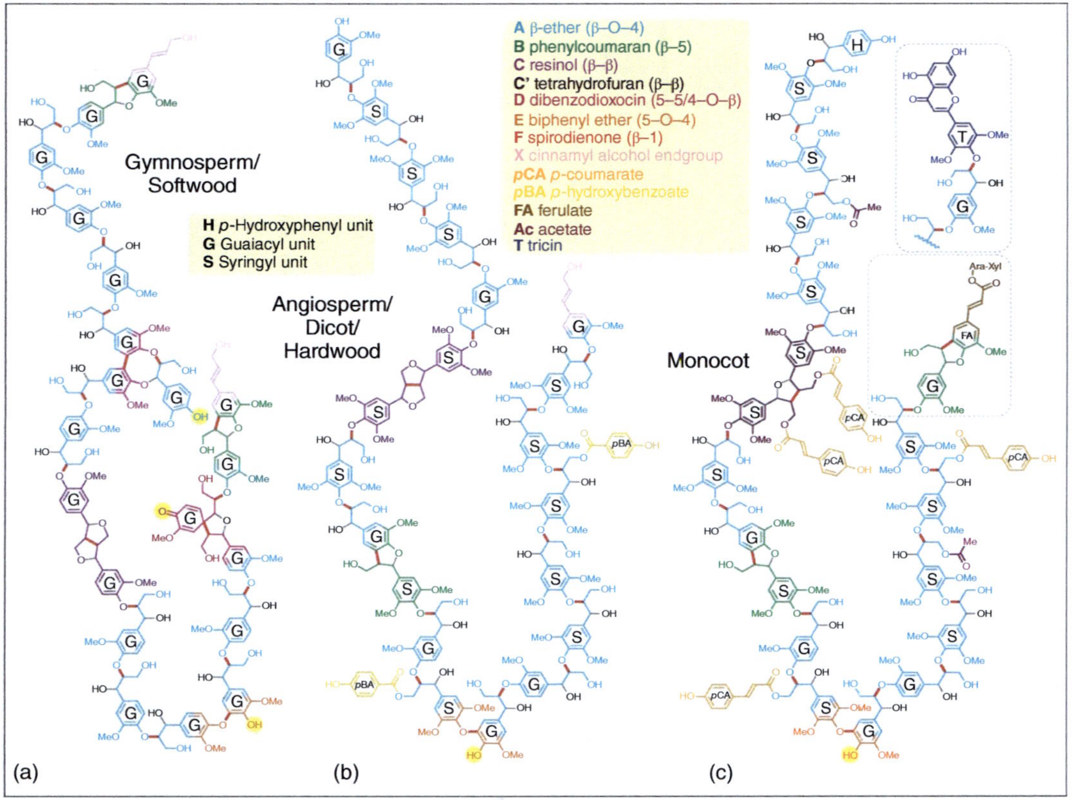

Figure 1.5 Lignin model structures containing 20 units. (a) A gymnosperm/softwood, (b) an angiosperm/dicot/hardwood, and (c) a (commelinid) monocot/grass. Source: Adapted from Ralph et al. [39].

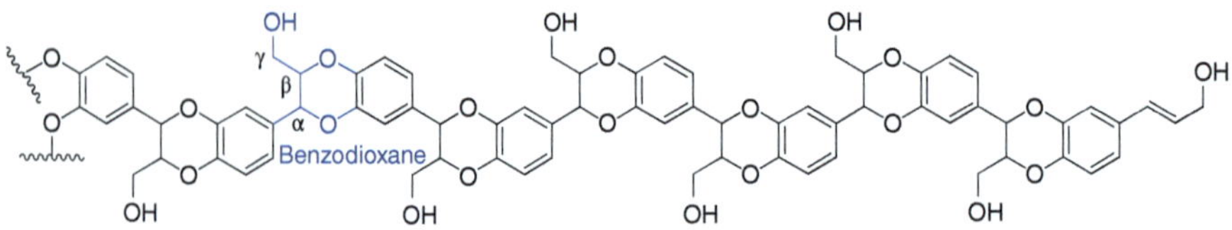

Figure 1.6 Representative chemical structural models of C-lignin. It is based on Ref. [43].

fraction. Besides, if chemically labile bonds are introduced into the lignin polymer, it will be easier to degrade lignin [68]. "Zip-lignin" is a successful example in this aspect. It introduces ester bonds, which are more chemically reactive than ether bonds, into lignin chain by acyl transferases [69]. More details about the bioengineering of lignin can be found in the decent reviews [14, 39, 59, 70].

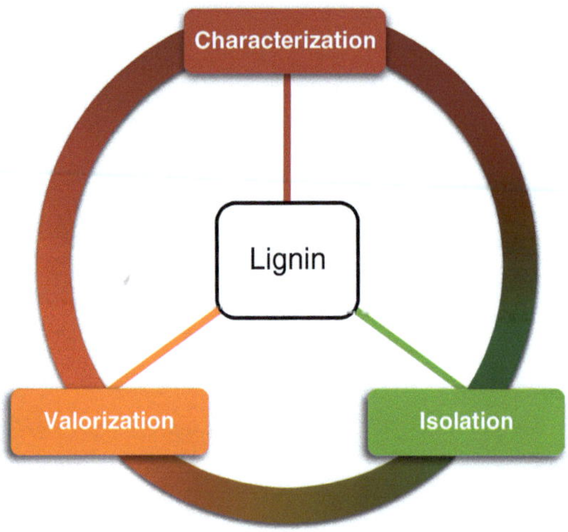

Figure 1.7 The scope of this book.

1.3 Scope of this Book

Besides the inherent complexity as introduced in this chapter, the structure of lignin is modified as the result of depolymerization (i.e. cleavage of linkages) and repolymerization (i.e. formation of new linkages) in most of the biomass processing technologies. Depending on processing technologies and the severity of the process, the structure of the yielded lignin can be varied. From a chemistry and technology point of view, rational utilization of lignin for different applications is determined by three important aspects: (i) characterization of lignin to understand the structure (including linkages), (ii) isolation to obtain lignin with desired properties, and (iii) on purpose valorization of lignin based on the inherent properties (Figure 1.7). Therefore, this interdisciplinary book includes chapters focused on introducing the latest progress in these aspects, and experts from analytic chemistry, organic chemistry, chemocatalysis, biocatalysis, chemical engineering, and material chemistry will contribute to this book. The main focus is valorization of lignin toward fuels, chemicals, and materials.

References

1 Schutyser, W., Renders, A.T., Van den Bosch, S. et al. (2018). *Chemical Society Reviews* 47: 852–908.

2 Liao, Y., de Beeck, B.O., Thielemans, K. et al. (2020). *Molecular Catalysis* 487: 110883.

3 Sun, Z., Fridrich, B., De Santi, A. et al. (2018). *Chemical Reviews* 118: 614–678.

4 Li, C., Zhao, X., Wang, A. et al. (2015). *Chemical Reviews* 115: 11559–11624.

5 Himmel, M.E., Ding, S.-Y., Johnson, D.K. et al. (2007). *Science* 315: 804–807.

6 Agbor, V.B., Cicek, N., Sparling, R. et al. (2011). *Biotechnology Advances* 29: 675–685.

7 Zhang, B., Biswal, B.K., Zhang, J., and Balasubramanian, R. (2023). *Chemical Reviews*.

8 Huber, G.W., Iborra, S., and Corma, A. (2006). *Chemical Reviews* 106: 4044–4098.

9 Zhou, C.-H., Xia, X., Lin, C.-X. et al. (2011). *Chemical Society Reviews* 40: 5588–5617.

10 Alonso, D.M., Wettstein, S.G., and Dumesic, J.A. (2012). *Chemical Society Reviews* 41: 8075–8098.

11 Zhang, Z., Song, J., and Han, B. (2017). *Chemical Reviews* 117: 6834–6880.

12 Wong, S.S., Shu, R., Zhang, J. et al. (2020). *Chemical Society Reviews* 49: 5510–5560.

13 Van de Vyver, S., Geboers, J., Jacobs, P.A., and Sels, B.F. (2011). *ChemCatChem* 3: 82–94.

14 Rinaldi, R., Jastrzebski, R., Clough, M.T. et al. (2016). *Angewandte Chemie International Edition* 55: 8164–8215.

15 Zhang, C., Shen, X., Jin, Y. et al. (2023). *Chemical Reviews* 123: 4510–4601.

16 Ragauskas, A.J., Beckham, G.T., Biddy, M.J. et al. (2014). *Science* 344: 1246843.

17 Boerjan, W., Ralph, J., and Baucher, M. (2003). *Annual Review of Plant Biology* 54: 519–546.

18 Lu, F. (2019). Lignin: Biosynthesis, Functions and Economic Significance. https://novapublishers.com/shop/lignin-biosynthesis-functions-and-economic-significance/.

19 Sun, R.C. (2020). *ChemSusChem* 13: 4385–4393.

20 Barros, J., Serrani-Yarce, J.C., Chen, F. et al. (2016). *Nature Plants* 2: 1–9.

21 Jansen, F., Gillessen, B., Mueller, F. et al. (2014). *Biotechnology and Applied Biochemistry* 61: 646–654.

22 Vanholme, R., De Meester, B., Ralph, J., and Boerjan, W. (2019). *Current Opinion in Biotechnology* 56: 230–239.

23 Higuchi, T., Ito, Y., and Kawamura, I. (1967). *Phytochemistry* 6: 875–881.

24 Tropf, S., Lanz, T., Rensing, S. et al. (1994). *Journal of Molecular Evolution* 38: 610–618.

25 Eudes, A., Dutta, T., Deng, K. et al. (2017). *PLoS One* 12: e0178160.

26 Ralph, J., Kim, H., Lu, F. et al. (2023). *Recent Advances in Polyphenol Research*, vol. 8, 1–50. Wiley.

27 del Río, J.C., Rencoret, J., Gutiérrez, A. et al. (2021). *Recent Advances in Polyphenol Research*, vol. 7, 177–206. Wiley.

28 Onkokesung, N., Gaquerel, E., Kotkar, H. et al. (2012). *Plant Physiology* 158: 389–407.

29 Farmer, M.J., Czernic, P., Michael, A., and Negrel, J. (1999). *European Journal of Biochemistry* 263: 686–694.
30 del Río, J.C., Rencoret, J., Gutiérrez, A. et al. (2022). *Advances in Botanical Research* 104: 1–39.
31 Balk, M., Sofia, P., Neffe, A.T., and Tirelli, N. (2023). *International Journal of Molecular Sciences* 24: 11668.
32 Qualley, A.V., Widhalm, J.R., Adebesin, F. et al. (2012). *Proceedings of the National Academy of Sciences* 109: 16383–16388.
33 Widhalm, J.R. and Dudareva, N. (2015). *Molecular Plant* 8: 83–97.
34 Petrik, D.L., Karlen, S.D., Cass, C.L. et al. (2014). *The Plant Journal* 77: 713–726.
35 Karlen, S.D., Zhang, C., Peck, M.L. et al. (2016). *Science Advances* 2: e1600393.
36 Ralph, J. (2010). *Phytochemistry Reviews* 9: 65–83.
37 Del Río, J.C., Marques, G., Rencoret, J. et al. (2007). *Journal of Agricultural and Food Chemistry* 55: 5461–5468.
38 Lu, F., Karlen, S.D., Regner, M. et al. (2015). *Bioenergy Research* 8: 934–952.
39 Ralph, J., Lapierre, C., and Boerjan, W. (2019). *Current Opinion in Biotechnology* 56: 240–249.
40 del Río, J.C., Rencoret, J., Gutiérrez, A. et al. (2020). *ACS Sustainable Chemistry & Engineering* 8: 4997–5012.
41 Vanholme, R., Morreel, K., Darrah, C. et al. (2012). *New Phytologist* 196: 978–1000.
42 Chen, F., Tobimatsu, Y., Havkin-Frenkel, D. et al. (2012). *Proceedings of the National Academy of Sciences* 109: 1772–1777.
43 Wang, S., Shen, Q., Su, S. et al. (2022). *Trends in Chemistry*.
44 Li, Y., Meng, X., Meng, R. et al. (2023). *RSC Advances* 13: 12750–12759.
45 Chen, F., Tobimatsu, Y., Jackson, L. et al. (2013). *The Plant Journal* 73: 201–211.
46 Ralph, J., MacKay, J.J., Hatfield, R.D. et al. (1997). *Science* 277: 235–239.
47 Ralph, J., Kim, H., Peng, J., and Lu, F. (1999). *Organic Letters* 1: 323–326.
48 Barrière, Y., Ralph, J., Méchin, V. et al. (2004). *Comptes Rendus Biologies* 327: 847–860.
49 Ralph, J., Brunow, G., Harris, P.J. et al. (2009). *Recent Advances in Polyphenol Research* 1: 36–66.
50 Ralph, J., Lundquist, K., Brunow, G. et al. (2004). *Phytochemistry Reviews* 3: 29–60.
51 Lourenco, A., Rencoret, J., Chemetova, C. et al. (2016). *Frontiers in Plant Science* 7: 1612.
52 Rencoret, J., Marques, G., Gutierrez, A. et al. (2008). Structural characterization of milled wood lignins from different eucalypt species. *Holzforschung* 62: 514–526. https://www.degruyter.com/document/doi/10.1515/HF.2008.096/html
53 Buranov, A.U. and Mazza, G. (2008). *Industrial Crops and Products* 28: 237–259.
54 Li, Y., Shuai, L., Kim, H. et al. (2018). *Science Advances* 4: eaau2968.
55 Wang, S., Zhang, K., Li, H. et al. (2021). *Nature Communications* 12: 416.
56 Stone, M.L., Anderson, E.M., Meek, K.M. et al. (2018). *ACS Sustainable Chemistry & Engineering* 6: 11211–11218.
57 Li, Y., Akiyama, T., Yokoyama, T., and Matsumoto, Y. (2016). *Biomacromolecules* 17: 1921–1929.

58 Zhang, L., Gellerstedt, G., Ralph, J., and Lu, F. (2006). *Journal of Wood Chemistry and Technology* 26: 65–79.

59 Vanholme, R., Morreel, K., Ralph, J., and Boerjan, W. (2008). *Current Opinion in Plant Biology* 11: 278–285.

60 Voelker, S.L., Lachenbruch, B., Meinzer, F.C. et al. (2011). *Plant, Cell & Environment* 34: 655–668.

61 Shuai, L., Amiri, M.T., Questell-Santiago, Y.M. et al. (2016). *Science* 354: 329–333.

62 Lan, W., Amiri, M.T., Hunston, C.M., and Luterbacher, J.S. (2018). *Angewandte Chemie International Edition* 57: 1356–1360.

63 Vanholme, R., Cesarino, I., Rataj, K. et al. (2013). *Science* 341: 1103–1106.

64 Ralph, J., Akiyama, T., Coleman, H.D., and Mansfield, S.D. (2012). *Bioenergy Research* 5: 1009–1019.

65 Wagner, A., Ralph, J., Akiyama, T. et al. (2007). *Proceedings of the National Academy of Sciences* 104: 11856–11861.

66 Franke, R., Hemm, M.R., Denault, J.W. et al. (2002). *The Plant Journal* 30: 47–59.

67 Coleman, H.D., Park, J.-Y., Nair, R. et al. (2008). *Proceedings of the National Academy of Sciences* 105: 4501–4506.

68 Zhou, S., Runge, T., Karlen, S.D. et al. (2017). *ChemSusChem* 10: 3565–3573.

69 Wilkerson, C., Mansfield, S., Lu, F. et al. (2014). *Science* 344: 90–93.

70 De Meester, B., Vanholme, R., Mota, T., and Boerjan, W. (2022). *Plant Communications* 3: 100465.

2

NMR Characterization of Lignin

Cheng-Ye Ma, Ying Xu, Ling-Hua Xu, Chen Zhang, Ji-Yun Qi and Jia-Long Wen

Beijing Forestry University, Beijing Key Laboratory of Lignocellulosic Chemistry, No. 35 Tsinghua East Road, Haidian District, Beijing, 100083, China

2.1 Introduction

The lignin macromolecule is primarily connected through carbon–carbon and carbon–oxygen ether bonds among the building blocks of phenylpropane monomers. The structure of lignin is complex, irregular, and highly heterogeneous, with no regular extended repeating unit structures observed [1, 2]. Compared to other biopolymers, the structural determination of lignin is more challenging due to its complexity and the difficulty in isolation of a highly representative and structurally unchanged lignin sample from plant species [3–5]. Currently, lignin is a major component of lignocellulosic biomass and an underutilized feedstock in the growing pulping and biofuel industries. At present, industrial lignin exceeds 50 million tons produced annually by pulp and paper industry around the world. About 95% of industrial lignin is used as fuel for thermal power plants because of its complex, destructive, and condensed structures that hinder most industrial use. Only 5% of industrial lignin is modified for potential applications such as additives, adhesives, dispersants, and surfactants [6, 7]. With the upgrading of lignin separation and purification technology, more research studies about lignin-based energy, materials, and chemicals have emerged. At the same time, a large amount of lignin with different structural characteristics and properties can be produced by different biorefinery processes. Various fractionation processes cause condensation of lignin to different degrees, and this poses many difficulties for lignin valorization. Particularly, depolymerization of condensed lignin leads to low yields of monomers [8, 9]. Therefore, other than developing viable valorization methods for different sources of lignin, structural characterization of lignin fractions could open the way toward new fractionation and valorization methods. In other words, the development of technologies for its structural characterization is currently

Lignin Chemistry: Characterization, Isolation, and Valorization, First Edition.
Edited by Yuhe Liao and Bert F. Sels.

desirable for lignin valorization, which will improve the techno-economics of applications of lignin in industries.

As mentioned above, structural characterization of lignin is important not only for revealing the inherent structure of lignin in different feedstocks but also for structural characterization and chemical reactivity determination of lignin isolated from different industrial and pretreatment processes. Destructive and nondestructive methods for lignin characterization have been successively developed in the past decades. The destructive methods include acidolysis, hydrogenolysis, alkaline nitrobenzene oxidation, copper(II) oxidation, permanganate oxidation, ozonation, thioacidolysis, and derivatization followed by reductive cleavage. These methodologies have been reviewed previously [10, 11]. Through careful analyses of degraded lignin monomers and dimers, S/G ratios of lignin and other information can be obtained by these methods. However, almost all these methods mainly focus on selective cleavage of β-*O*-4 linkage, and thus only a part of lignin fragments can be used for structural analysis, since lignin generally contains β-*O*-4 and carbon–carbon (C–C) linkages. In fact, many lignin isolation methods (i.e. harsh isolation conditions) lead to cleavage of β-*O*-4 linkages in native lignin and formation of C–C linkages during the isolation process, and thus the applicability of these destructive characterization methods is greatly reduced.

Nuclear magnetic resonance spectroscopy (NMR) spectroscopy has enormously facilitated investigations into structural aspects of complex lignin polymers. Although the exact structure of native lignin in a plant is still not fully understood, advances in spectroscopic methods (especially NMR techniques) and computational modeling have enabled scientists to elucidate the predominant structural features of lignin, such as inter-unit linkages and their relative abundances in plant biomass. Some decades ago, proton NMR (^{1}H-NMR) was mainly used for qualitative lignin analysis. Later, the ^{1}H-NMR spectrum of acetylated lignin is used to determine the quantity of different hydroxyl groups. Prof. Lundquist and co-authors have published many works about NMR characterization of lignin, which have promoted the utilization of NMR techniques in lignin characterization [12, 13]. Principally due to its high resolution and the high dispersion of chemical shifts, ^{13}C-NMR has played an important role in the qualitative and quantitative understanding of lignin structure [14–17]. Now modern two-dimensional correlative methods (e.g. 2D-HSQC) are powerful tools for identifying structural units of lignin based on previous studies on lignin model. At the same time, there were many researchers using NMR technique to study the structure of lignin, which contributed to the understanding of the structure of lignin at that time. In the year of 2010, Prof. Ralph and Landucci contributed a chapter on "NMR of Lignins" in the book *Lignin and Lignans: Advances in Chemistry*, in which they gave a comprehensive and in-depth review on NMR characterization of polymeric lignin models and natural lignins [18]. Regarding to the 2D-HSQC technique, Prof. Ralph John and his groups at University of Wisconsin, Madison have made the outstanding contributions to the 2D-HSQC NMR characterization of lignin, especially the synthesis and characterization of lignin models and the uncovering of unconventional lignin structures [19–22]. In addition to 2D-HSQC, ^{31}P NMR is also a very important lignin characterization method, mainly used to quantify the content of hydroxyl and carboxyl groups in the lignin fractions [23, 24]. In this respect, Prof. Dimitris S Argyropoulos at North Carolina State University developed the application of ^{31}P-NMR spectroscopy in lignin characterization more than two decades ago and improved the method over a long period of practice [25–27]. Meanwhile, Prof. Arthur J. Ragauskas in Georgia Institute of Technology further expanded the application of NMR technique (^{1}H-NMR, ^{13}C-NMR, 2D-HSQC NMR, and ^{31}P-NMR) in the lignin characterization, especially the structural analysis of lignin obtained after the different mainstream pretreatment (Dilute acid pretreatment, organosolv pretreatment, and other pretreatments) of biomass [6, 28, 29]. Despite the contributions of the above-mentioned groups, there are also many excellent research teams that have made a series of progress in the field of lignin characterizations via 2D-HSQC techniques. Alternatively, ^{19}F-NMR was also introduced in this chapter because the content of hydroxyl and carbonyl groups can be distinguished and quantified via this technique [17, 30]. In short, the main objective of this chapter is to introduce the application of NMR technique in characterizing the structural and chemical activity of lignin from different sources, such as native lignin, biorefinery lignin, and industrial lignin fractions, aiming to extend the lignin valorization and improve the techno-economics of applications of lignins in industries (see Figure 2.1).

2.2 ^{1}H NMR

^{1}H NMR spectroscopy is a valuable technique for the characterization and classification of lignin structural features. Earlier work on the characterization of lignin by NMR mainly relied on ^{1}H NMR spectroscopy, and extensive databases of ^{1}H NMR chemical shifts were established for lignin model compounds and functional groups [31, 32]. Since the signal intensity of the hydrogen spectrum is proportional to the number of protons, theoretically, according to the integration curves of different signals of the lignin ^{1}H NMR spectrum, the ratio of the area of each integration region can be derived, so as to calculate the corresponding percentage of protons in each region. Combined with the elemental composition, methoxylation (hydrogen spectroscopy), and other analyses, the chemical formula of lignin can be preliminarily deduced. Therefore, ^{1}H NMR spectroscopy can be used to analyze the structural features of lignin: carboxylic acid, aldehyde, unsubstituted/substituted phenolic, aromatic, aliphatic, methoxyl, and others.

2.2.1 Sample Preparation and Program

2.2.1.1 Sample Preparation of Non-Acetylated Lignin

For the non-acetylated lignin, which can be directly dissolved in 0.5 ml of DMSO-d_6 (dimethylsulfoxide-d_6), the concentration has little effect on the sample detection, about 10–50 mg lignin samples.

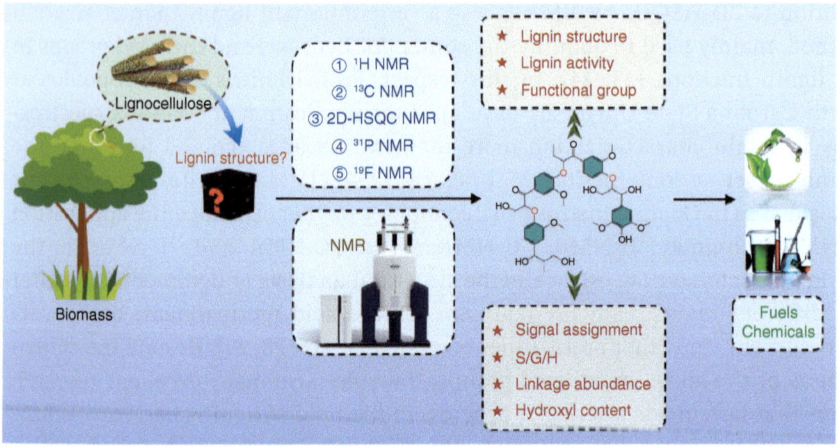

Figure 2.1 Scheme for NMR characterization of lignin.

2.2.1.2 Sample Preparation of Acetylated Lignin

In order to dissolve lignin in the solvent and increase the signal of H protons, pyridine and acetic anhydride reagents are commonly used to acetylate lignin. Compared with ^{1}H NMR spectra of unacetylated lignin, ^{1}H NMR spectra of acetylated lignin can quantitatively analyze the functional groups of lignin. The method of lignin acetylation can be referred to as follows [11]: 200 mg lignin was dissolved in 8 ml of a solution of acetic anhydride/pyridine (1 : 1). After stirring for 24 hours at room temperature in the dark, the mixture was dropped slowly into 200 ml ice water (pH 2.0) to induce precipitation, and the precipitate was washed with deionized water. The precipitates were centrifuged and freeze-dried leading to acetylated lignin.

For the quantitative analysis of the acetylated lignin, when the sample is dissolved in DMSO-d_6, internal standard (IS) *p*-nitrobenzaldehyde (NBA) should be added. First, 1 mg/ml of NBA IS chloroform solution is prepared. Second, 10 mg of acylated lignin is dissolved in 0.5 ml of DMSO-d_6, then add 1 ml of NBA IS chloroform solution for quantitative determination.

2.2.1.3 ^{1}H Program

The ^{1}H NMR spectra can be acquired on a Bruker Avance 400 MHz spectrometer fitted with a 5 mm gradient probe with inverse geometry (proton coils closest to the sample). The chemical shifts were calibrated relative to the signals from DMSO-d_6, used as an IS at 2.49 ppm for the ^{1}H NMR spectra. There are many publications that described the specific testing procedures, such as the acquiring time (AQ) can adopt 3.98 seconds, and relaxation time is 1.0 second [33, 34].

2.2.2 ^{1}H-NMR Spectra Analysis

2.2.2.1 ^{1}H-NMR Spectra of Non-Acetylated Lignin

Although ^{1}H-NMR Spectra of the acetylation lignin can provide quantitative information, the acetylation procedure may cause some unwanted chemical

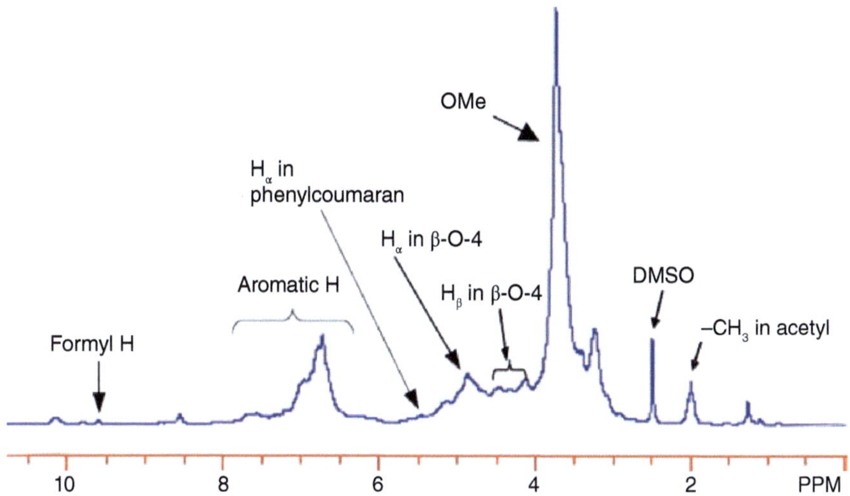

Figure 2.2 An example of ^{1}H NMR spectrum of a poplar milled wood lignin using DMSO as solvent.

modifications to the sample. ^{1}H NMR of underivatized lignin is informative about some key lignin functionalities. Figure 2.2 provides an example of ^{1}H NMR spectrum of underivatized milled wood lignin (MWL) isolated from the poplar biomass [35].

2.2.2.2 ^{1}H-NMR Spectra of Acetylated Lignin

For the acetylated lignin, the methoxy group, phenolic hydroxyl group, and other groups can be quantitatively analyzed by ^{1}H-NMR spectra [36]. Table 2.1 lists assignments and chemical shifts of typical structural features of acetylated spruce MWL in a ^{1}H NMR spectrum [37]. The functional groups (phenolic hydroxyl, aliphatic hydroxyl, and methoxy content) were all calculated using the following formula:

$$F = \frac{\frac{I_F}{3} \times \frac{4}{I_{NBA}} \times \frac{W_{NBA}}{151} \times 1000}{W_L - \frac{I_{AC}}{3} \times \frac{4}{I_{NBA}} \times \frac{W_{NBA}}{151} \times 42}$$

where F represents the number of functional groups (mmol) contained in 1 g of lignin, and I_F is the integral interval for a specific functional group (methoxy, 4.10–3.10 ppm; phenolic hydroxyl, 2.50–2.17 ppm; aliphatic hydroxyl, 2.17–1.70 ppm). 3 represents the three protons on the methoxy and acetyl groups; 4 represents the four on the benzene ring of *p*-nitrobenzaldehyde protons; I_{NBA} represents the integral of the NBA molecule; 151 represents the molecular weight of NBA; 42 represents the molecular weight of the acetyl group minus the molecular weight of one proton; I_{Ac} represents the integral interval of the acetyl group (2.50–1.70 ppm); and W_L represents the mass of the acetylated lignin (mg).

Table 2.1 Signal assignments in the ^{1}H NMR spectrum of acetylated milled wood lignin.

δ (ppm)	Assignment
2.01	Aliphatic acetate
2.29	Aromatic acetate
3.08	H_β in β–β structures
3.76	Proton in the methoxyl group
4.18–4.43	H_γ in several types of structures
4.60	H_β in β-*O*-4 structures
5.44	H_α in β-5 structures
6.01	H_α in β-*O*-4 and β-1 structures
6.60	Aromatic protons in syringyl units
6.94	Aromatic protons in guaiacyl units
7.50	Aromatic protons located ortho to carbonyl groups
9.67	Formyl protons in cinnamaldehyde units
9.86	Formyl protons in benzaldehyde units

Due to the complexity of lignin molecular structure, the resolution between different signal peaks is low in ^{1}H NMR analysis, and it is difficult to accurately determine the chemical shift of proton signal peaks. Therefore, the different groups in lignin can only be roughly analyzed, and the quantitative information is limited. In addition, different signal peaks in ^{1}H NMR spectra of lignin are easy to overlap, which is not conducive to comprehensive characterization of complex structures, so precise analysis needs to be combined with other NMR spectroscopy techniques. At present, the quantitative analysis of lignin using ^{1}H NMR spectroscopy is rarely used. Especially for non-acetylated lignin samples, ^{1}H NMR spectroscopy is generally only used for preliminary determination of sample structure (purity and structural unit prediction). In addition, ^{1}H NMR spectra of lignin can also obtain structural information that is difficult to obtain by other analytical methods, such as the content of condensed and non-condensed aromatic groups, and the content of methoxy groups.

2.3 ^{13}C-NMR

^{13}C-NMR spectroscopy is one of the most reliable and frequently used techniques for lignin characterization [17], which can determine the type of carbon atoms in a compound and also provide information, such as the carbon skeleton of the compound, thus providing more information about the molecular structure. Since the 1980s, ^{13}C-NMR has been used to aid in the elucidation of pulping or delignification mechanism (soda pulping, kraft pulping, and oxygen/peracid treatments), as well as pretreatments, which are discussed in detail in a recent book by Ralph and

Landucci [18]. In lignin structural studies, ^{13}C-NMR spectra are generally classified into nonquantitative and quantitative modes, which can be used for both qualitative structural characterization of lignin and semiquantitative quantitative studies of specific structures. The use of quantitative ^{13}C-NMR allows for the determination of information on the basic composition of lignin (S/G ratio), aryl ether bonding, carbon–carbon bonding, methoxylation, and condensed and non-condensed lignin structural units [38]. Therefore, ^{13}C-NMR can be used to determine the different chemical changes of lignin, the position of chemical bonding, and to reveal the reaction mechanism in the studies of lignin structure elucidation and chemical processing [39–41].

2.3.1 Sample Preparation and Program

2.3.1.1 Sample Preparation

Quantitative ^{13}C-NMR analysis requires a number of conditions to be fulfilled. First, the lignin sample must be free of contaminants, such as carbohydrates and extractives. Also, the lignin-solvent solution must be made as concentrated as possible to maximize the signal-to-noise ratio and minimize baseline and phase distortions. Typically, the solvent used is DMSO-d_6. Acetylated lignins can provide more ideal samples. They are quite soluble in common organic solvents and solute interactions are less severe. The methods of lignin acetylation have been discussed in the Section 2.2.1.2.

2.3.1.2 ^{13}C-NMR Program

A typical quantitative ^{13}C-NMR experiment is illustrated as follows: the standard program "C13IG" is selected from the program library, the inverse-gated (IG) decoupling sequence is used, which involves turning off the proton decoupling during the recovery between pulses so that the nuclear Overhauser effect (NOE) is avoided [17]. If the concentration of lignin is more than 20% (more than 100 mg lignin in 0.5 ml DMSO-d_6), then a pulse sequence: A 90° pulse width, with a 1.4 seconds acquisition time, and a 1.7 seconds relaxation delay is used [11]. It is worth noting that the higher concentration of lignin has a better spectral signal. In addition, to achieve sufficient relaxation in a feasible time, 20 ml of chromium(III) acetylacetonate (0.01 M) was added as a relaxation agent for the quantitative ^{13}C spectrum to reduce the relaxation delay according to a previous report [42]. Generally, a total of 20 000–30 000 scans are collected at 400–600 MHz NMR instruments.

2.3.2 Quantitative ^{13}C-NMR

2.3.2.1 Quantitative ^{13}C-NMR of Non-Acetylated Lignin

Generally, both qualitative (signal assignments) and quantitative information (the relative abundance of substructures per aromatic ring) can be obtained by the quantitative ^{13}C-NMR spectrum of non-acetylated lignin. Table 2.2 lists an extensive compilation of structural assignments of a typical lignin (^{13}C-NMR spectra of non-acetylated lignin) that have been derived from the previous study [11].

Table 2.2 The chemical shift value (δ, ppm) of ^{13}C-NMR spectrum of non-acetylated lignin.

δ (ppm)	Assignment	δ (ppm)	Assignment
166.5	C-9 in *p*-coumarate (PCE)	123.0	C-6, FE ester
160.0	C-4 in PCE	122.6	C-1 and C-6 in Ar–C(=O)C–C units
144.7	C-7 in PCE	119.4	C-6 in G units
130.3	C-2/C-6 in PCE	118.4	C-6 in G units
125.1	C-1 in PCE	115.1	C-5 in G units
116.0	C-3/C-5 in PCE	114.7	C-5 in G units
115.0	C8 in PCE	111.1	C-2 in G units
152.5	C-3/C-5, etherified S units	110.4	C-2 in G units
149.7	C-3, etherified G units	106.8	C-2/C-6, S with α-CO
148.4	C-3, G units	104.3	C-2/C-6, S units
148.0	C-3, G units	86.6	C-α in G type β-5 units
146.8	C-4, etherified G	84.6	C-β in G type β-*O*-4 units (*threo*)
145.8	C-4, non-etherified G	83.8	C-β in G type β-*O*-4units (*erythro*)
145.0	C-4, etherified 5–5	72.4	C-γ in β-β; C-γ, β-aryl ether
143.3	C-4, non-etherified 5–5	71.8	C-α in G type β-*O*-4 units (*erythro*)
138.2	C-4, S etherified	71.2	C-α in G type β-*O*-4 units (*threo*)
134.6	C-1, S etherified C-1, G etherified	63.2	C-γ in G type β-*O*-4 units with α–C=O
133.4	C-1, S non-etherified; C-1, G non-etherified	62.8	C-γ in G type β-5, β-1 units
132.4	C-5, etherified 5–5	60.2	C-γ in G type β-*O*-4 units
131.1	C-1, non-etherified 5–5	55.6	C in Ar-OCH_3
129.3	C-β in Ar-CH=CH–CHO	53.9	C-β in β-β units
128.0	C-α and C-β in Ar–CH=CH–CH_2OH	53.4	C-β in β-5 units
128.1	C-2/C-6, in H units	36.8	CH_3 group, ketones (conj) or in aliphatic
125.9	C-5/C-5′ in non-etherified 5–5	29.2	CH_2 in aliphatic side chain
122.6	C-1 and C-6 in Ar–C(=O) C–C	26.7	CH_3 or CH_2 group in saturated side chains
125.9	C-5, non-etherified 5–5	14.0	γ-CH_3 in *n*-propyl side chain

Generally, quantitative ^{13}C-NMR is used only for the estimation of some specific moieties. The most current practices in the use of quantitative ^{13}C-NMR spectroscopy for the investigation of lignin are confined to using the aromatic and methoxyl signals as ISs in expressing the various functional groups per C9. Such a practice is applicable to native lignin, but it is inapplicable for industrial lignin or modified lignin [43]. To overcome the above defects, Xia et al. [43] suggested a novel protocol for acquiring quantitative ^{13}C-NMR spectra of lignins by using the internal reference compounds 1,3,5-trioxane and pentafluorobenzene. The ISs could be used for the quantification of ^{13}C signals in lignins, expressed in absolute units of millimoles (mmol) per gram of sample. In addition, such quantification becomes particularly important for severely altered lignin samples. Under the inspiration of these works, quantitative ^{13}C-NMR spectroscopy was used to illustrate the structural changes of lignin, isolated from various pretreated biomass, aiming to clarify the mechanism of lignin transformation during these pretreatments.

2.3.2.2 Quantitative ^{13}C-NMR of Acetylated Lignin

The ^{13}C-NMR spectra of acetylated lignin can be used to determine the different structures of different lignin sources. For instance, the ^{13}C-NMR spectra of acetylated lignin can be used to compare the structural differences between lignins of hardwoods, softwoods, and grasses. In addition, quantitative ^{13}C-NMR spectra of acetylated lignin samples were also used to determine the amount of primary, secondary, and phenolic hydroxyl groups. The chemical shifts between 102 and 160 ppm were first set as 600, the corresponding values in the range of between δ_C 170 and 169, 169 and168, and 168 and167 ppm were then obtained, which represent the number of hydroxyl groups in the lignin (results were expressed by per/Ar) (Figure 2.3) [44].

2.3.3 Applications of the ^{13}C-NMR

Quantitative ^{13}C-NMR is frequently used to determine the changes of the linkages in lignin during the pretreatments and delignification process. Hallac et al. [45] used quantitative ^{13}C-NMR to characterize the structure of lignin isolated from shrub *Buddleja davidii*, they reported that the guaiacyl/syringyl (G/S) ratio of the

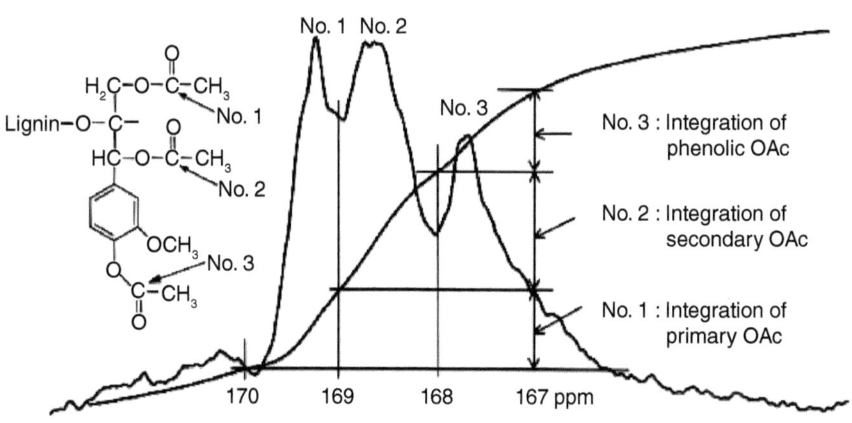

Figure 2.3 Quantitative evaluation of hydroxyl groups by ^{13}C-NMR spectra of acetylated lignin.

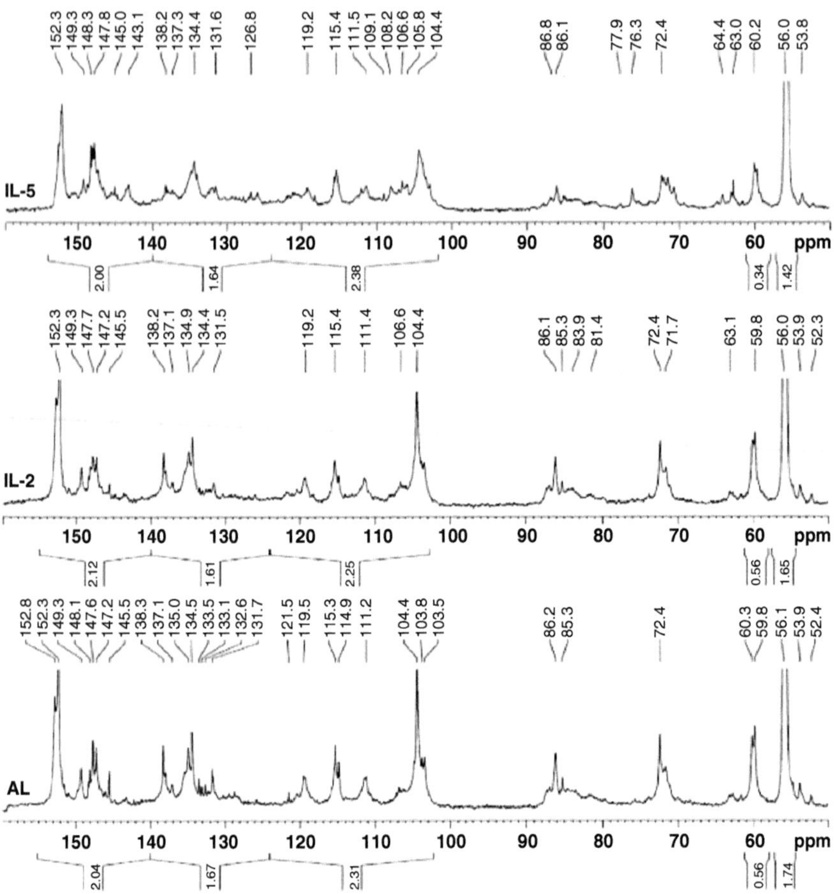

Figure 2.4 Quantitative ^{13}C-NMR spectra of the lignins.

lignin was about 81 : 19, and H-unit was not observed. Using ^{13}C-NMR and other NMR techniques, they further investigated the chemical transformation of *B. davidii* lignin during pretreatment with ethanol organic solvents and demonstrated quantitative changes in lignin structural characteristics after pretreatment, such as a substantial reduction in β-*O*-4 bonds [39]. Using ^{13}C-NMR, Samuel et al. [46] observed a 36% decrease in β-*O*-4 linkages, and an S/G ratio decreases from 0.80 to 0.53 for ball-milled lignin isolated from Alamo switchgrass after dilute acid pretreatment. Furthermore, Wen et al. [47] used quantitative ^{13}C-NMR to analyze the lignin samples during ionic liquid pretreatment and compared them with that of the alkaline lignin (Figure 2.4). The newly appeared signals at 145.0 and 126.8 ppm in the pretreated lignin samples were probably assigned to the C-4 in etherified 5–5′ and C-5/C-5′ in non-etherified 5–5′ linkages, which suggested that the 5–5′ linkages were formed during the ionic liquid pretreatment. They observed that the pretreatment under mild temperature has a mild effect on the content of β-*O*-4′

linkages, while higher temperature has a significant effect on the content of β-*O*-4′ linkages. They also reported that demethoxylation occurred in the ionic liquid pretreatment process and preferentially took place in G units, especially at higher temperatures.

Some drawbacks remain concerning the efficient use of ^{13}C-NMR, such as precise signal assignments and true quantification based on ^{13}C-NMR spectra of lignins, which are difficult due to signal overlap and other factors. The ^{13}C nucleus is not sensitive enough (1/5800 compared to the proton), and it is only 1.1% naturally abundant. Consequently, relatively large sample sizes and long acquisition times are essential. There are other options, but all are imperfect to some degree. For the signal assignments, it could be improved if advanced two-dimensional spectra were to be jointly applied. In addition, the best quantification method remains the relatively tedious IG technique, along with an IS substance. Fortunately, most of the semi-quantitative methods that have been described are adequate when the researcher wants to follow changes in structure during a particular treatment provided that the desired precision is maintained.

2.4 2D-HSQC (Heteronuclear Single-Quantum Correlation)

As a reliable method to study the structure of lignin, two-dimensional heteronuclear single quantum correlation (2D-HSQC) NMR technique is crucial for the identification and analysis of the molecular structure of lignin. The 2D-HSQC NMR experiment can not only improve the sensitivity of ^{13}C nuclei through polarization transfer but also separate overlapping signals that often appear in 1D spectra, making the ^{1}H–^{13}C correlation method very effective in lignin structure analysis. 2D-HSQC can provide the relative content of different connecting bonds and structural units in lignin and display the structural characteristics and changes of lignin macromolecules to the greatest extent. In the 2D-HSQC spectra, different signal peaks in lignin could be distinguished individually. By combining the signals of ^{1}H spectrum and ^{13}C spectrum, the aromatic ring region and side chain region (including aliphatic region) of lignin can be displayed in 2D-HSQC spectrum, respectively. In the aromatic ring region of lignin, the types of structural units in lignin can be clearly identified. The relevant signals of the aliphatic region mainly come from CH_2 and CH_3 of lignin side chains and the signals of acetyl groups, which provide less information about the composition and structure of lignin. The lignin side chain region mainly reflects the bond type of lignin (such as β-*O*-4, β-β, β-5, and β-1), functional groups (such as methoxy groups), and carbohydrates connected to lignin. Furthermore, the use of 2D NMR technique helps to advance the analysis of lignin structures, especially to discover the existence of new lignin subunits and lignin-carbohydrate complexes, such as dibenzodioxocin and spirodienone structures [48–52]. In addition, the side chain region of lignin can also be integrated to quantitatively calculate the content of each linking bond in lignin, such as β-*O*-4, β-β, β-5, and β-1 linkages. In this section, the test method, signal assignments, and quantitative method and strategy of 2D-HSQC NMR analysis of lignin will be introduced, respectively.

2.4.1 Sample Preparation and Program

In 2D-HSQC NMR characterization of isolated lignin, it is only necessary to dissolve the lignin with a suitable solvent, the most common solvent being deuterated dimethyl sulfoxide (DMSO-d_6). About 20–100 mg of lignin was dissolved in 0.5 ml of solvent to obtain a uniformly dissolved lignin solution. Transfer the lignin solution to NMR tubes for testing. In addition, although early lignin NMR studies mainly focused on acetylated lignin, lignin without acetylation can better reflect the true structure of lignin.

Although the Q-HSQC NMR pulse sequence can be used to suppress the $^1J_{C\text{-}H}$ dependence of the HSQC NMR signal, however, the Q-HSQC sequence is still unable to quantify structures in polymer samples because it does not resolve the error [53]. For the 2D-HSQC NMR characterization of lignin, the adiabatic pulse sequence named "hsqcetgpsisp.2" (Bruker standard pulse sequence) is the most popular procedure. Phase-sensitive 2D-HSQC using echo-de-echo and adiabatic pulses for inversion and refocusing, selectable from the Bruker pulse program library. This program has the advantage of J-independence and offset insensitivity over an essentially unlimited active bandwidth. The program is less sensitive to differences in one-bond ^{13}C–^1H coupling constants, and the response over the entire spectral range is more uniform, suggesting that improved quantification should result [54, 55]. A typical method for measuring 2D-HSQC NMR spectra of lignin using the Bruker standard pulse sequence "hsqcetgpsisp.2" is summarized in Section 2.4.2 [56]. Take the Bruker AVANCE 600 MHz NMR spectrometer as an example. Experiments were performed at room temperature (25 °C). Chemical shifts are referenced to the central DMSO solvent peak ($\delta^{13}C$ 39.5, δ^1H 2.49). The spectral width of the F2 (^1H) dimension is 16 ppm with 2048 data points (TD1); the spectral width of the F1 (^{13}C) dimension is 240 ppm with 256 data points (TD2); the number of scans (SN) is 128; delay between scans (D1) was 1 second; and acquisition time was 10 hours. The following parameters were used: TD1, 1024 data points; TD2, 128 data points; NS, 96; and DW, 62.4 μs. The NMR data processing and analysis were performed using Bruker's Topspin software.

2.4.2 Semi-Quantification of Lignin in 2D-HSQC NMR Spectra

In the structural analysis of lignin, the assignment and confirmation of lignin structure is very important, but the quantification of specific functional groups is more important, especially in the study of pretreatment mechanism. It is worth noting that HSQC is not absolutely quantitative. The simple proportionality of the volume integral versus molarity of individual CHn groups in lignin can be greatly disturbed by a number of factors inherent in the co-pulse program, including the aforementioned mismatch in the $^1J_{CH}$-related delay parameters as well as several others. The

C_2–H_2 correlation of the guaiacyl (G) and the C_2–H_2/C_6–H_6 correlation of the symmetric syringyl (S) and *p*-hydroxyphenyl (H) units are often used to obtain the important H/G/S estimation of lignin composition ratio. In addition, the relative proportions of various bonds in lignin, the relative proportions of methoxy groups, etc. can be determined by semiquantitative calculations. At present, there are three NMR semiquantitative methods: (i) Relative quantitative method (without IS); (ii) Relative quantitative method (IS is aromatic ring); and (iii) Based on the combination of ^{13}C and 2D-HSQC Quantitative method. When studying the structural changes of lignin during pretreatment, most of them use relative quantitative method (IS is aromatic ring). This quantitative approach assumes that the aromatic rings of lignin are relatively unchanged.

2.4.2.1 Relative Quantitative Method (Without Internal Standard)

Relative quantification without an IS is the most commonly used method for diagnosing the relative abundance of lignin structures based on 2D-HSQC spectra. It should be noted that this method uses a "normalized" quantitative method, and the obtained calculation results cannot be used to compare the absolute differences between different lignin components. By integrating the volume of the HSQC crossover signal corresponding to the key structure, a semiquantitative analysis was performed based on the integrated value. Note that all integrations should be done on the same contour level. Estimate the relative percentage of lignin basic structural units (Figure 2.6) by the following formula:

$$I_X\% = I_X/(I_A + I_B + I_C + I_D) \times 100\%$$

where I_A, I_B, I_C, and I_D are β-*O*-4 (**A**), β–β (**B**), β-5 (**C**), and β-1 (**D**), respectively.

2.4.2.2 Relative Quantitative Method (Aromatic Ring as IS)

This quantification strategy selects "aromatic unit" as IS. Specifically, this method uses signal clusters representing all C_9 units (i.e. IS). Select G_2, $0.5IS_{2,6} + G_2$, and $0.5IS_{2,6} + IG_2 + 0.5IH_{2,6}$ signals as IS for softwood, hardwood lignin, and grass lignin, respectively. Results are expressed as how many linkages per aromatic ring (<1.0). This integration process should also be maintained at the same contour level.

$$IC_9 \text{ units} = IG_2 \left(\text{Softwood Lignin}\right)$$

$$IC_9 \text{ units} = 0.5IS_{2,6} + IG_2 \left(\text{Hardwood Lignin}\right)$$

$$IC_9 \text{ units} = 0.5IS_{2,6} + IG_2 + 0.5IH_{2,6} \left(\text{Grass Lignin}\right)$$

$$I_X\% = I_X/I_{C9} \times 100\%$$

where $IS_{2,6}$ is the integral of $S_{2,6}$ including S and S′, and IG_2 is the integral value of $G_2$2. IH_2 is the integral value of $H_{2,6}$. IC_9 represents the integral value of the aromatic ring. According to the IS (IC_9), the amount of I_X% can be obtained by the following formula, I_X is **A** (β-*O*-4), **B** (β-β), **C** (β-5), and **D** (β-1) α bit integral value.

2.4.2.3 Based on the Combination of ^{13}C and 2D-HSQC Technique

This method, proposed by Zhang and Gellerstedt [57], combines HSQC and quantitative ^{13}C-NMR techniques, which can provide more reliable data about lignin bonds in lignin samples. The specific calculation method is as follows:

$$I_X\% = 2I_X/2I(90-78/6.0-2.5)\times{}^{13}\text{C}(90.0-78.0)/{}^{13}\text{C}(163.0-103.6)\times 600$$

where I_X is the integral of A_β (A_β(G/H) and A_β(S)), B_α, C_α, and D_α in the region of 90–78/6.0–2.5, and I (90–78/6.0–2.5) is the total integral of the area (should have the same profile level as **A**, **B**, **C** and **D** cells). The key principle of the quantification method is based on "absolute quantification of ^{13}C-NMR spectra, relative quantification by 2D-HSQC spectra (distinguishing lignin units from signals overlapping with carbohydrates)." In general, total aromatic carbon (δ_C/δ_H = 163.0–103.6) integral value is set to 600.0.

2.4.3 Assignments of Lignin in 2D-HSQC NMR spectra

Assignments of lignin-related signals were obtained from an extensive database of lignin model compounds. The representative works published by Ralph's groups have played an important role in the development of the field through 2D-HSQC characterization of lignin and synthetic model compounds from different species of plants to identify unknown structures [58–61]. In general, a typical lignin 2D-HSQC profile consists of two regions: side chains (δ_C/δ_H = 50–90/2.5–6.0) and aromatics (δ_C/δ_H = 100–150/5.5–8.5) regions. The side chain region of the 2D-HSQC NMR spectrum provides useful information about the different bonds (β-*O*-4, β–β, β-5, β-1, etc.) present in lignin. With this information, the main structures of natural lignin from different sources can be easily determined. Likewise, lignin fractions isolated from different industrial and pretreatment processes can also be characterized by the 2D-HSQC NMR technique. Table 2.3 shows the signal attribution of 2D-HSQC of bamboo MWL. In fact, these signal peaks also include most of the common lignin signal peaks, which can be used to quickly identify the structural characteristics of the original lignin separated and the lignin separated during different pretreatment processes.

2.4.3.1 Gramineous Lignin

Grasses have the most complex lignin structure. In addition to the conventional lignin units and linkages, gramineous lignin also has the cross-linking structure of hydroxycinnamic acid. Bamboo is a representative gramineous plant. In recent years, with the gradual deepening of the research on the chemical components of bamboo, bamboo has been widely developed in the fields of pulp and paper,

Table 2.3 Assignments of 2D-HSQC spectra of MWL from bamboo.

Label	δ_C/δ_H	Assignments
B'_β	49.8/2.56	C_β–H_β in β-β tetrahydrofuran (B′)
C_β	53.1/3.46	C_β–H_β in phenylcoumaran (C)
B_β	53.5/3.07	C_β–H_β in β-β (resinol) (B)
D_β	59.8/2.75	C_β–H_β in spirodienones (D)
OCH_3	56.4/3.70	C–H in methoxyls
A_γ	59.9/3.35–3.80	C_γ–H_γ in β–*O*–4 substructures (A)
A'_γ	63.0/4.36	C_γ–H_γ in γ-acylated β–*O*–4 (A′)
C_γ	62.2/3.76	C_γ–H_γ in phenylcoumaran (C)
Iγ	61.2/4.09	Cγ–Hγ in cinnamyl alcohol end-groups (I)
I'_γ	64.0/4.80	C_γ–H_γ in γ-acylated cinnamyl alcohol (I′)
B_γ	71.2/3.82–4.18	C_γ–H_γ in β-β resinol (B)
A_α	71.8/4.86	C_α–H_α in β-*O*-4 unit (A) (*Erythro*)
A_α	71.8/4.86	C_α–H_α in β-*O*-4 unit (A) (*Thero*)
$A_{\beta(G)}$	83.4/4.38	C_β–H_β in β-*O*-4 linked to G (A)
B_α	84.8/4.66	C_α–H_α in β–β resinol (B)
B'_α	83.2/4.94	C_α–H_α in β–β (B′, tetrahydrofuran)
A'_β	82.8/5.23	C_β–H_β in β-*O*-4 substructures (A)
$A'_{\beta(G)}$	80.8/4.62	C_β–H_β in β-*O*-4 linked to G (A)
$A_{\beta(S)}$	85.8/4.12	C_β–H_β in β-*O*-4 linked to S (A, *Erythro*)
$A_{\beta(S)}$	86.7/4.00	C_β–H_β in β-*O*-4 linked to S (A, *Thero*)
D_α	81.0/5.10	C_α–H_α in spirodienones (D)
D'_α	79.4/4.10	C'_α–H'_α in spirodienones (D)
E_α	79.6/5.60	C_α–H_α in α,β-diaryl ethers (E)
C_α	86.8/5.45	C_α–H_α in phenylcoumaran (C)
$T'_{2,6}$	103.9/7.34	$C'_{2,6}$–$H'_{2,6}$ in tricin (T)
T_6	98.9/6.23	$C_{2,6}$–$H_{2,6}$ in tricin (T)
T_8	94.2/6.60	C_8–H_8 in tricin (T)
T_3	106.2/7.07	C_3–H_3 in tricin (T)
$S_{2,6}$	103.9/6.70	$C_{2,6}$–$H_{2,6}$ in syringyl units (S)
$S'_{2,6}$	106.3/7.32	$C_{2,6}$–$H_{2,6}$ in oxidized S units (S′)
G_2	110.8/6.97	C_2–H_2 in guaiacyl units (G)
G_5	114.5/6.70	C_5–H_5 in guaiacyl units (G)
G_{5e}	115.1/6.95	C_5–H_5 in etherified guaiacyl units (G)
G_6	119.0/6.78	C_6–H_6 in guaiacyl units (G)
J_β	126.1/6.76	C_β–H_β in cinnamyl aldehyde end-groups (J)

(Continued)

Table 2.3 (Continued)

Label	δ_C/δ_H	Assignments
$H_{2,6}$	127.7/7.17	$C_{2,6}$–$H_{2,6}$ in H units (H)
$PCE_{3,5}$	115.6/6.77	$C_{3,5}$–$H_{3,5}$ in *p*-coumarate (PCE)
$PCE_{2,6}$	130.2/7.48	$C_{2,6}$–$H_{2,6}$ in *p*-coumarate (PCE)
PCE_7	144.8/7.51	C_7–H_7 in *p*-coumarate (PCE)
PCE_8	113.7/6.24	C_8–H_8 in *p*-coumarate (PCE)
FA_2	110.7/7.35	C_2–H_2 in ferulate (FA)
FA_6	123.1/7.20	C_6–H_6 in ferulate (FA)
FA_7	144.8/7.51	C_7–H_7 in ferulate (FA)
J_α	153.4/7.59	C_α–H_α in cinnamyl aldehyde end-groups (J)

bioenergy, and biorefinery. Bamboo lignin not only contains the three basic units of lignin (G, S and H) but also cross-linked hydroxycinnamic acids (*p*-coumaric acid and ferulic acid). Research on the structure of bamboo lignin will be beneficial to the rational utilization of bamboo biomass resources. Figure 2.5 shows the 2D-HSQC NMR spectra of MWL from bamboo pole (MWL_P) and bamboo stick (MWL_S), respectively [62]. The attribution of the main relevant signals in the spectrum is shown in Table 2.3, and the main basic connection structures and structural units are shown in Figure 2.6. In Figure 2.5, the (A_γ') signals for lignin γ-acylation can be observed at 63.0 and 4.36 ppm In addition, a signal at 80.8/4.52 ppm (A′β(*G*)) was found in MWLS due to acylation of the β-*O*-4 linkage, but this signal was not present in MWL_P, it may be caused by the low G content in MWL_P (high S/G ratio). In addition to the β-*O*-4 ether bond structural unit, these maps also show a large number of resin alcohol β–β structures (B), respectively, through their α (δ_C/δ_H 85.1/4.66), β (53.7/3.05) and γ (71.3/4.18 and 3.82) position related signals are identified. At the same time, another β–β connection form (B′) was also found in the spectra (δ_C/δ_H = 49.8/2.56 and 83.2/4.94), which is the tetrahydrofuran structure, which is a lignin monomer through γ-acetylation formed by coupling reaction. Therefore, it can be speculated that the appearance of the tetrahydrofuran structure is related to the high acylation of the γ-position. Likewise, the presence of β-5 and β-1 was found based on the signal. In addition, another structural unit α,β-diaryl ether bond (**E**) was also determined by the chemical shift of its α-position (δ_C/δ_H = 79.6/5.60). In general, the content of α,β-diaryl ether linkages is rather low or even undetectable, but such units are found in large quantities in tobacco lignin. Other units such as cinnamyl end groups and acetylated cinnamyl end groups are also found. In the aromatic ring region of lignin, it is mainly composed of lignin basic units (G, S and H) and hydroxycinnamic acid (*p*-coumaric acid, ferulic acid). Compared with the amount of G and S units, the content of H units is very low (1–2%).

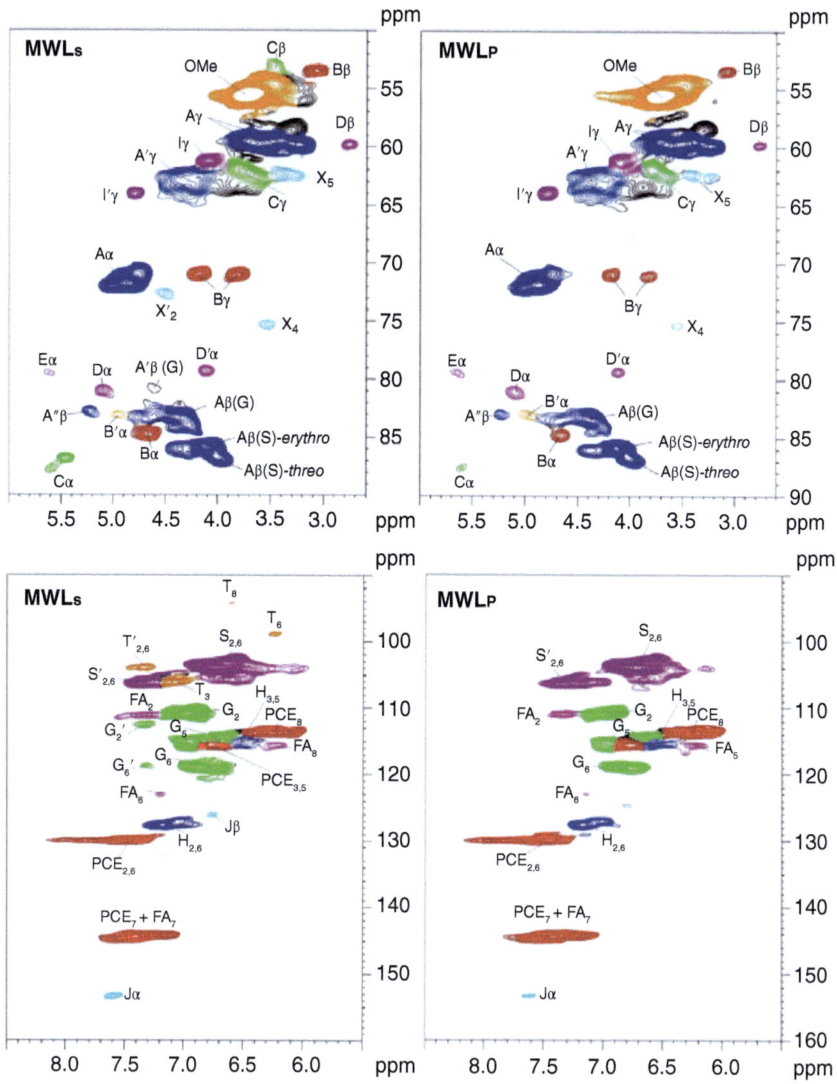

Figure 2.5 The 2D-HSQC NMR spectra of milled wood lignin from bamboo pole (MWL_P) and bamboo stick (MWL_S, respectively [62].

2.4.3.2 Hardwood Lignin

As one of the common broad-leaf trees in northern China, poplar is widely used in pulp and paper and biorefinery industries due to its characteristics of fast growth, high yield, and strong climate adaptability. Figure 2.7 shows the 2D-HSQC spectra of double-enzymatic lignin (DEL) from *Populus tomentosa* at different growth stages, as well as the main connecting bonds and basic structural units. Please refer to Table 2.4 for signal attribution. By observing the side chain region of the

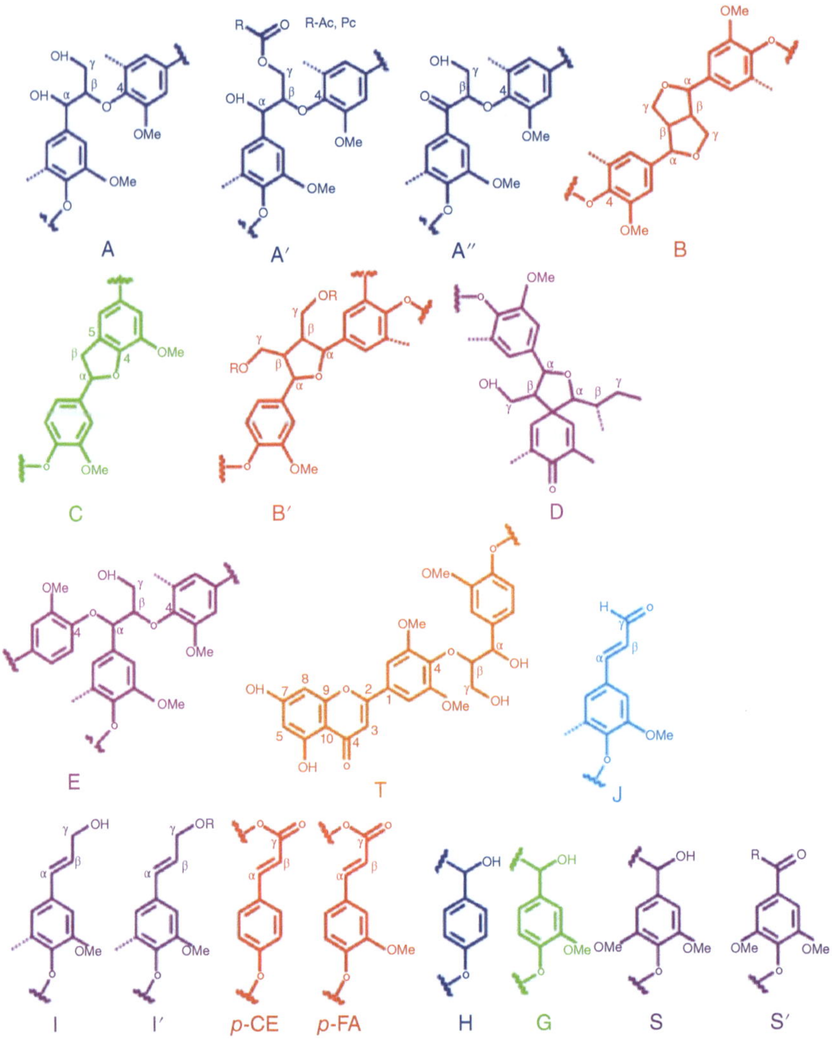

Figure 2.6 Main structures present in bamboo lignin: (**A**) β-*O*-4 alkyl-aryl ethers; (**A′**) β-*O*-4 alkyl-aryl ethers with acylated γ-OH with *p*-coumaric acid; (**A″**) Cα-oxidized β-*O*-4′ structures; (**B**) resinols; (**B′**) tetrahydrofuran; (**C**) phenylcoumarans; (**D**) spirodienones; (**E**) α,β-diaryl ethers; (**T**) a likely incorporation of tricin into the lignin polymer through a G-type β-*O*-4 linkage; (**I**) *p*-hydroxycinnamyl alcohol end-groups; (**I′**) *p*-hydroxycinnamyl alcohol end-groups acylated at the γ-OH; (**J**) cinnamyl aldehyde end-groups; (***p*CA**) *p*-coumarates; (**FA**) ferulates; (**H**) *p*-hydroxyphenyl units; (**G**) guaiacyl units; (**S**) syringyl units; (**S′**) oxidized syringyl units bearing a carbonyl at C_{α}.

2D-HSQC spectrum of lignin, it can be seen that all spectra show obvious β-*O*-4 (**A**) connecting bond signals, including C_{α}–H_{α} (A_{α}), C_{β}–H_{β} (A_{β}, linked to S, G, or H building blocks), and C_{γ}–H_{γ} (A_{γ}). In addition, it can be observed that the content of β–β linkages in lignin also gradually increases with age. Small amounts of β-5 and *p*-hydroxycinnamyl end group (**I**) signals can also be observed in the spectrum. At

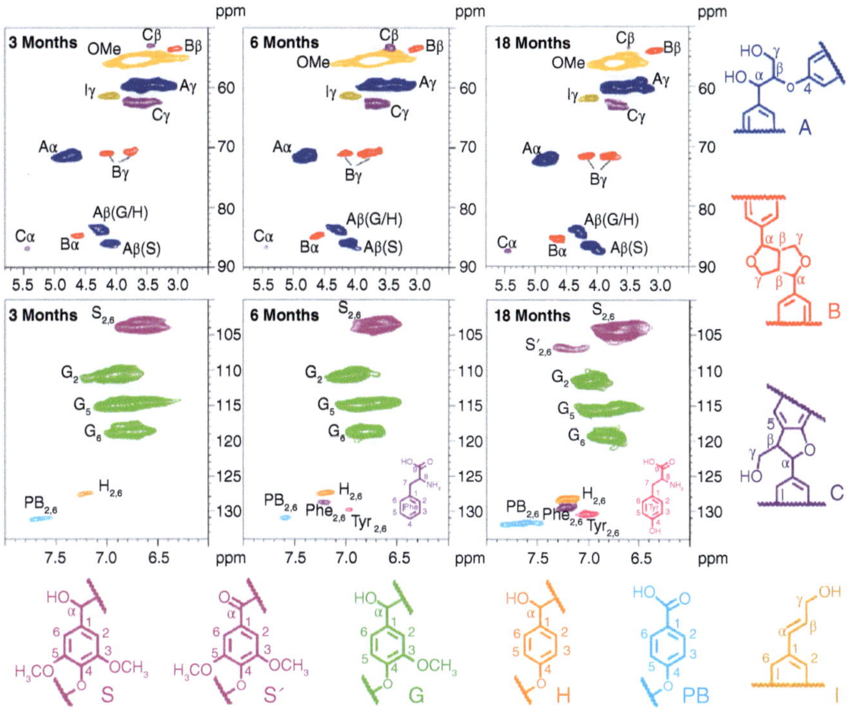

Figure 2.7 The 2D-HSQC spectra and main linkages and structural units of poplar DEL components in different growth stages.

Table 2.4 Quantification of 2D-HSQC spectra of HL, SL, and BL lignin (results are calculated per 100 benzene rings).

Lignin sample	β-*O*-4	β–β	β-5	β-1	S/G
HL	62.1	11.6	0.2	0.3	3.45
SL	62.4	7.7	0.4	0.1	2.74
BL	69.0	2.9	1.0	0.0	2.91

the same time, in the aromatic ring area of the 2D-HSQC spectrum, the signals of the main units of lignin (G, S, H) can be clearly seen. The signal intensity of these structural units increases with the growth of wood, which reflects the growth of the plant cell wall. Natural lignification process. In addition, as a unique structure in poplar wood, *p*-hydroxybenzyl (PB) content gradually increases with the extension of the growth cycle.

Extracting lignin from the heartwood, sapwood and bark of *Eucalyptus grandis* and characterizing its structure will provide a theoretical basis for the efficient utilization of lignin in different parts of the eucalyptus tree. The side chain (δ_C/δ_H = 95–48/5.2–2.5) and aromatic (δ_C/δ_H = 132–100/7.6–6.2) regions in the 2D-HSQC NMR spectrum are shown in Figure 2.8. Its main connecting bonds and lignin structural units are shown in Figure 2.9. The side chain regions of heartwood lignin (HL), sapwood lignin (SL), and bark lignin (BL) show obvious signal peaks such as methoxy group to lignin linkage, including β-*O*-4 aryl ether bond (**A**), resin alcohol unit (**B**), phenyl aromatic Legumin unit (**C**), spirodiene substructure (**D**), and *p*-hydroxycinnamyl alcohol end group (**I**). The 2D spectra revealed differences in the type and strength of each connecting bond between HL, SL, and BL. For all samples analyzed, β-*O*-4 units are the major linkages in lignin. The α position of the β-*O*-4 aryl ether bond appears at δ_C/δ_H = 71.6/4.84, the β position appears at δ_C/δ_H = 83.5/4.28 and 86.7/4.08, and the γ position appears at δ_C/δ_H = 59.9/3.25–3.68, and the signal with xylan was partially overlapping. The α, β, and γ positions of the resin alcohol unit (β–β′, **B**) appear at δ_C/δ_H = 85.0/4.63, 53.5/3.03, 70.9/3.79, and 4.16 ppm, respectively. The signal of the phenylcoumarin unit (β-5, **C**) is observed in all spectra, but with lower intensity. Trace amounts of spirodienone substructure (β-1, **D**) and *p*-hydroxycinnamyl alcohol end group (**I**) were detected in both HL and SL, whereas their signals were not detected in the spectrum of BL. The aromatic region of the 2D-HSQC spectrum shows obvious

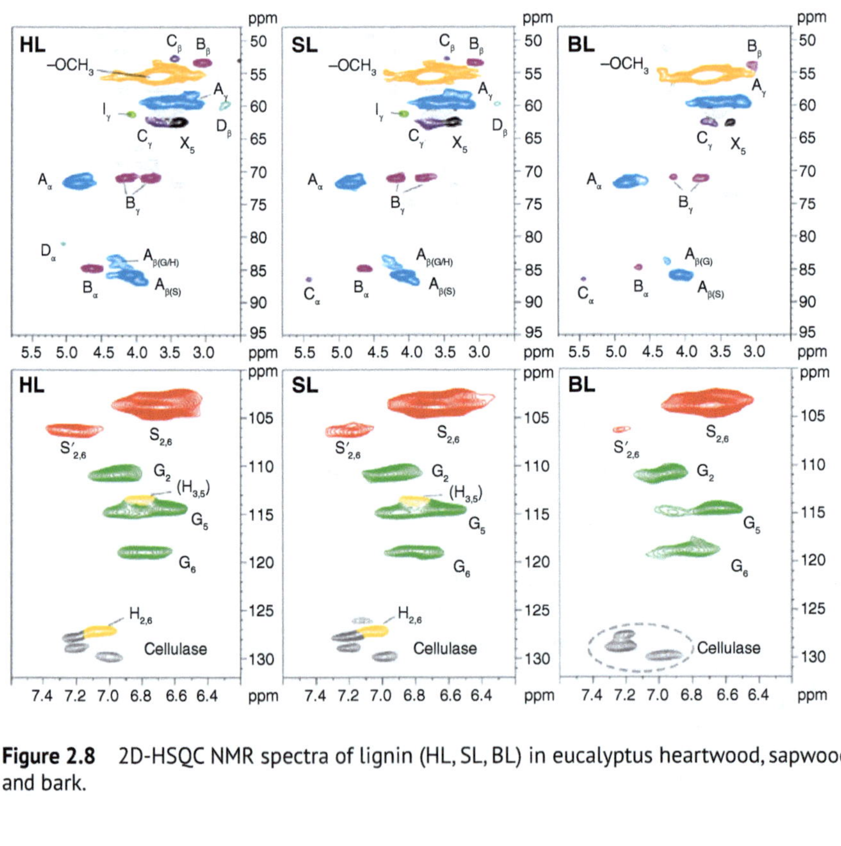

Figure 2.8 2D-HSQC NMR spectra of lignin (HL, SL, BL) in eucalyptus heartwood, sapwood, and bark.

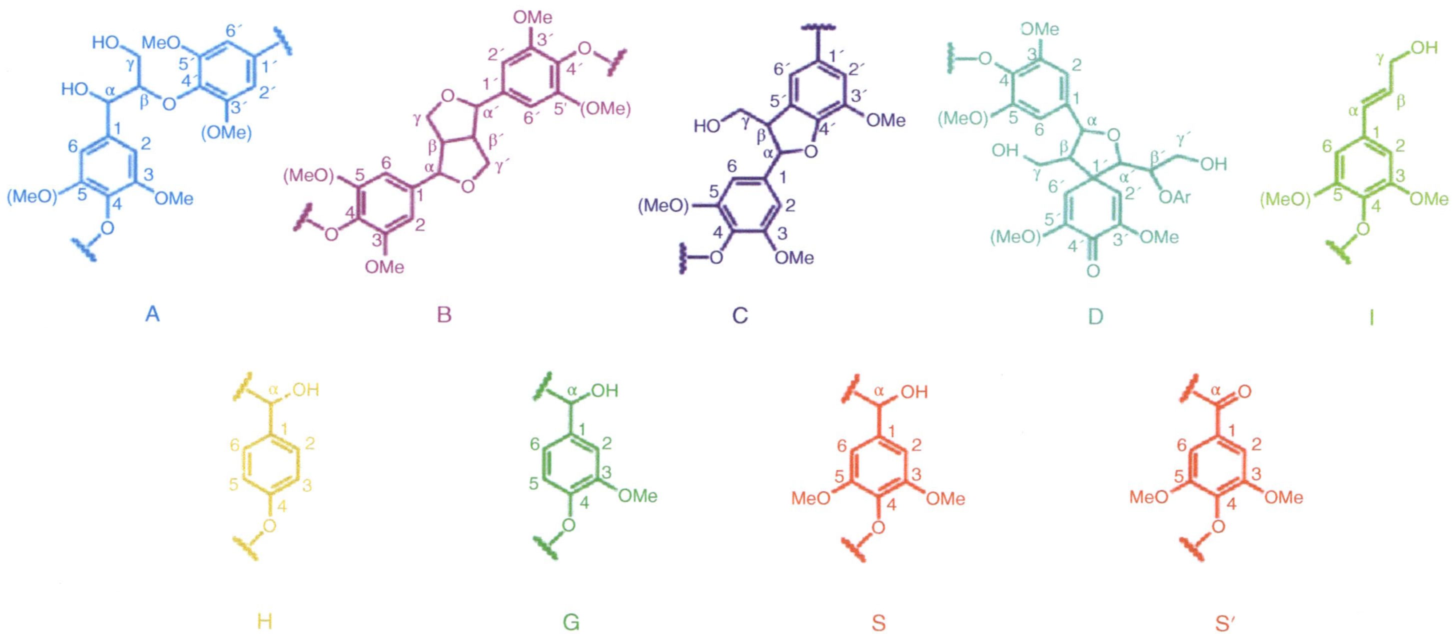

Figure 2.9 Main linkages and structural units contained in eucalyptus lignin.

signals of S and G units. Positions 2 and 6 in the S unit are observed at $\delta_C/\delta_H = 103.8/6.68$ and 106.3/7.17. Positions 2, 5, and 6 of the G unit are observed at $\delta_C/\delta_H = 110.7/6.95$, 114.7/6.68–6.94, and 118.8/6.78, respectively. Obvious *p*-hydroxyphenyl (H) units were observed in the spectra of HL and SL with 3,5 positions at $\delta_C/\delta_H = 114.5/6.79$ and 2,6 positions at $\delta_C/\delta_H = 127.3/7.08$, but in the BL spectrum No corresponding signal could be found.

Based on the semiquantitative analysis results of lignin substructure and lignin units based on the two-dimensional HSQC NMR spectrum (Table 2.5), it is obvious that β-*O*-4 aryl ether accounts for the majority, and its content ranges from HL to BL increasing in sequence (62.1%, 62.4%, and 69.0% for HL, SL, and BL, respectively). The HL also has the highest relative content of β–β substructure (11.6%), which indicates that HL contains more C—C bond condensation structures through SL and BL. Although the contents of β-5 and β-1 linkage are quite low, they still show opposite proportions in HL and SL. The results of S/G ratio showed that lignin varied greatly in different tissues. HL has the highest S-type lignin content (S/G ratio of 3.45), while SL has a slightly higher S-type lignin content (a S/G ratio of 2.74). This may indicate increased deposition of S-type lignin as wood matures. The S/G ratio of BL is 2.91, which is lower than that of HL. Considering that BL also has relatively high β-5 bonds, lignin in bark must contain more G units, and a possible explanation is that the rapid lignification process in the cell wall of bark tissue leads to the enrichment of G units. Through a comprehensive

Table 2.5 Assignments of ^{13}C-^{1}H cross signals in 2D-HSQC NMR spectra of lignin fractions.

Label	δ_C/δ_H	Assignments
C_β	53.1/3.46	C_β–H_β in phenylcoumaran substructures (C)
B_β	53.5/3.05	C_β–H_β in β–β (resinol) substructures (B)
OMe	56.4/3.70	C–H in methoxyls (OMe)
A_γ	59.9/3.35–3.80	C_γ–H_γ in γ-hydroxylated β–O–4 substructures (A)
C_γ	62.2/3.76	C_γ–H_γ in β-β phenylcoumaran l substructures (C)
I_γ	61.2/4.10	C_γ–H_γ in cinnamyl alcohol end-groups (I)
B_γ	71.0/3.79–4.16	C_γ–H_γ in β–β resinol substructures (B)
A_α	71.8/4.86	C_α–H_α in β–O–4 substructures (A)
C_α	86.8/5.45	C_α–H_α in phenylcoumaran (C)
B_α	84.8/4.66	C_α–H_α in β–β resinol (B)
A_β(G)	83.4/4.38	C_β–H_β in β–O–4 substructures linked to a G unit (A)
G_2	110.8/6.97	C_2–H_2 in guaiacyl units (G)
G_5	114.5/6.70	C_5–H_5 in guaiacyl units (G)
G_6	119.0/6.78	C_6–H_6 in guaiacyl units (G)
$H_{2,6}$	127.7/7.17	$C_{2,6}$–$H_{2,6}$ in H units (H)

comparison of the lignin structures isolated from different eucalyptus species, it can be seen that all eucalyptus lignin is dominated by β-*O*-4 aryl ether bonds, followed by β–β resin alcohol units. The β-5 and β-1 substructures are less abundant.

2.4.3.3 Softwood

Unlike hardwood, the xylem of pine usually exhibits a more uniform woody structure (mainly composed of tracheids and wood rays) and a higher lignin content. During the lignification process, the coupling of lignin monomers tends to be at the β position due to electron delocalization (generated by the conjugation structure of the benzene ring and the C=C bond of the side chain). Figure 2.10 compares the interunit connections of lignin macromolecules in different years (2, 5, 14, and 35 years) of Chinese pine (*P. tabuliformis*). The assignment of signal peaks assigned to each substructure is shown in Table 2.5. Clearly, the signals identified by the different correlations of β-*O*-4 aryl ethers (**A**) are fully shown in the spectra. In addition, the signals of β–β resinol (**B**) and β-5 (**C**) also appeared in the spectra, and the intensity of Bα and Cα signals gradually increased with the growth of pine trees. At the same time, the HSQC spectrum also revealed the presence of *p*-hydroxycinnamyl alcohol terminal group (I). The overlapping signals of H units in L5, L14, and L35 are a combination of H-unit and enzyme-specific signals [33]. The increased number of H unit signals in the adult stage may not only be related to gene expression but also contribute to other cell wall phenolics such as *p*-coumarate and paraben (H units) [64]. During the dehydrodimerization process of lignin monomers, the phenolic groups produced by dehydrogenation of lignin monomers are relatively stable due to electron delocalization, providing single electron density around the aromatic ring and at the β position, resulting in higher β-*O*-4 scale [65]. Pine lignin is a highly branched polymer that may contain more β–β linkages [66]. In addition, the amount of β-*O*-4 was significantly enhanced from juvenile (2 years) to adult (35 years) pine trees, which correlated with changes in lignin accumulation. It has been reported that β–β linkages may serve as nucleation sites during lignification because β–β is formed from two lignin monomers linked together [64]. This increasing trend is consistent with the changing trend of H-unit content with lignification in the process of pine lignification from juvenile to adult, showing the difference of lignin accumulation.

2.4.4 Application of 2D-HSQC NMR Technique in the Field of Lignin

The ultimate goal of characterizing lignin is to promote lignin applications in different fields such as chemicals and materials. The detailed information of lignin structure can be accurately identified through 2D-HSQC NMR technique. By comparing the separated lignin sample with the original lignin, we can understand the structural changes of lignin during plant growth and the structural changes during lignin separation. This detailed structural information will provide an important reference for the application of lignin, such as the production of chemicals and the preparation of composite materials.

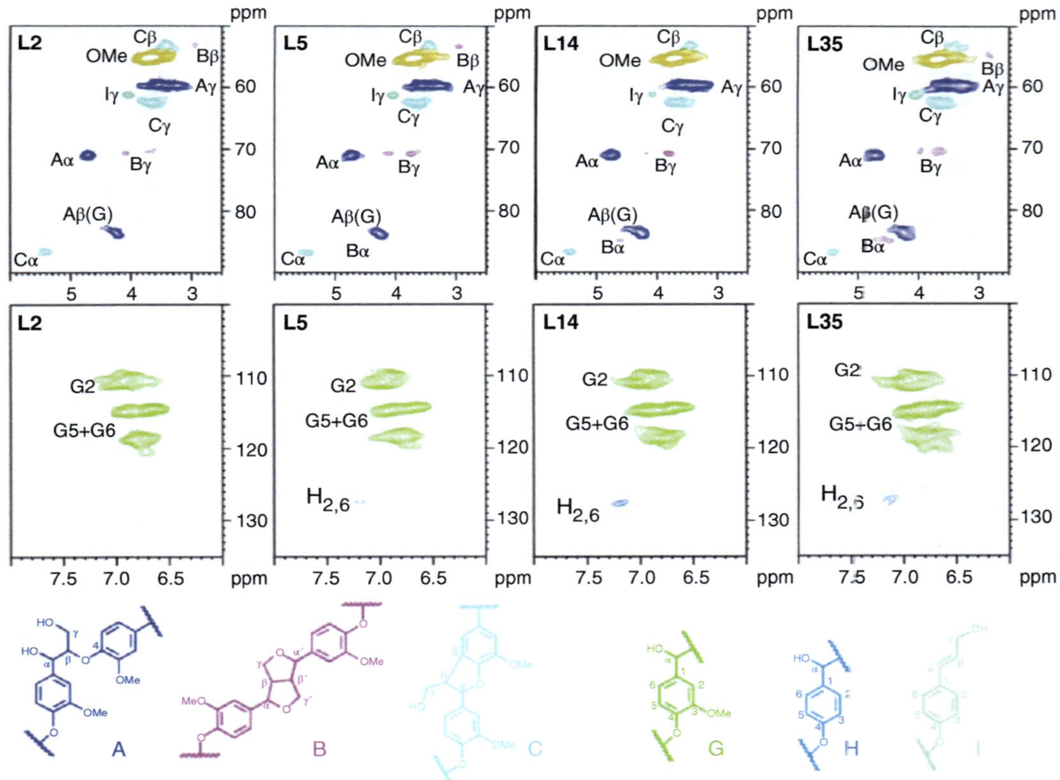

Figure 2.10 2D-HSQC NMR spectra of DEL from *Pinus tabuliformis* in different years and identified substructures: (A) β-*O*-4; (B) β–β; (C) β-5 [63]. Source: Reprinted with permission from Elsevier.

2.4.4.1 Structure Characterization of Native Lignin from Different Plant Sources

Detailed lignin structure information can be obtained by isolating original lignin from plants of different species and growing years for structural comparison. This information will provide a reference for determining the source species of plants, plant growth conditions, and changes in lignin structure caused by plant gene regulation. The original lignin separation methods that have been developed include MWL, CEL, and DEL, and detailed introductions can be found in the previous chapters. Validation by comparing 2D-HSQC spectra between lignin model compounds and lignin samples allows accurate identification of specific lignin bonds present in natural lignin [61, 67]. In the side chain region of the 2D-HSQC NMR spectrum ($\delta_C/\delta_H = 50–90/2.5–6.0$), there is useful information about the different bonds of lignin (β-*O*-4, β–β, β-5, β-1, etc.). Crossing signals from H, S, and G units are readily observed in the aromatic region of the 2D-HSQC NMR spectrum. Furthermore, significant signals corresponding to *p*-coumarate (*p*CA) and ferulate (FA) were observed in the 2D-HSQC NMR spectra (side chain regions) of the gramineous lignin samples. Likewise, paraben cross-over signals can be found in 2D-HSQC NMR spectra (side chain regions) of lignin from specific hardwood species such as poplar and willow. Based on the aromatic zone information, the S/G (S/G/H) ratio of the entire lignin can be easily obtained. In addition, by comparing with the 2D-HSQC NMR spectra of the discovered substructures of lignin, information on the basic substructures in lignin can be obtained. Several studies have made progress in elucidating the native substructure of lignin through NMR methods [68, 69]. In summary, 2D-HSQC NMR technique is very valuable for determining specific changes in lignin from different sources, growth years, and structural information.

2.4.4.2 Structure Changes of Lignin During Delignification and Pretreatment Process

By comparing the 2D-HSQC NMR spectra of lignin under different separation conditions with the original lignin, the detailed structural changes of lignin during the separation process can be obtained, such as ionic liquid pretreatment [33, 70, 71], deep eutectic solvent pretreatment [72–74], and γ-valerolactone pretreatment [75, 76]. Detailed signal assignment of lignin according to the type of feedstock can be found in the above tables and in the literature given. The detailed structural information of lignin during the separation process is analyzed through 2D-HSQC NMR technique, thereby analyzing the structural changes of lignin and guiding subsequent applications. In addition to the qualitative assignment of signals in 2D-HSQC NMR spectra, the quantitative determination of lignin structure in comparable models is another important study of lignin structure, which has important implications for the development of value-added chemicals from lignin. For example, lignin fractions containing abundant β-*O*-4 bonds and fewer C—C bonds are suitable for the production of chemicals through catalytic hydrogenolysis.

2.5 ^{31}P NMR

^{31}P NMR spectroscopy is a promising detection technique for determining the type and content of different hydroxyl groups in different lignin fractions [77]. ^{31}P NMR analysis of lignin was first proposed by Argyropoulos et al. and has been further popularized by many lignin research groups [78, 79]. In general, the content of phenolic hydroxyl, aliphatic hydroxyl, and carboxyl groups in lignin has a great influence on its antioxidant activity, thermal stability, and the properties of the target materials, and these functional groups are also important for the chemical modification of lignin [23, 35]. It is difficult to accurately identify or quantify aliphatic hydroxyl and phenolic hydroxyl groups in lignin by conventional ^{1}H and ^{13}C-NMR techniques. However, the type and quantity of functional groups in lignin is not only particularly critical for its high-value conversion and utilization but also an important basis for its own functionality. Compared to conventional methods for the determination of hydroxyl content, ^{31}P NMR spectroscopy presents unique advantages for the determination of hydroxyl groups in lignin with high signal resolution. The method requires a small sample volume (approximately 20 mg) and a short experimental time (30–120 minutes) [23, 35, 77–79]. By contrast, the conventional methods can only provide general information on all phenolic hydroxyl groups or a specific hydroxyl group, ^{31}P NMR can distinguish the contents of different kinds of aliphatic hydroxyl groups, phenolic hydroxyl groups, and carboxyl groups, and can also be used to track the change trends of different hydroxyl groups in lignin samples, providing an important technical support for chemical modification and structural monitoring of lignin in different pretreatment/delignification process.

2.5.1 Factors Affecting ^{31}P NMR

Dissolution system. Complete dissolution of the lignin sample in a suitable solvent is a necessary prerequisite for accurate quantitative measurements by solution-state NMR. According to the principles established by previous ^{31}P NMR analysis, a mixture of deuterated chloroform and anhydrous pyridine is recommended as solvent [24, 26, 80]. Nevertheless, the solvents used could be adjusted according to the characteristics or solubility of the analyzed lignin substrates. For lignin samples with limited solubility in chloroform and pyridine, *N,N*-dimethylformamide (DMF) can be introduced as a third solvent to help facilitate dissolution [81]. Stücker et al. reported that a novel solvent system containing DMF, deuterated DMF, and pyridine (4.5 : 1 : 1 (v/v/v)) was able to completely solubilize sulfonated lignin [82]. The preswelling of lignosulfonates in dimethyl sulfoxide has also been proven to be an effective method for overcoming the solubilization barrier of lignosulfonates in conventional mixed solvents of deuterated chloroform and pyridine [83]. The same study also showed that while ultrasonication increased the solubility of lignin, it also led to a dramatic increase in aliphatic and phenolic hydroxyl groups, suggesting that depolymerization of lignin during ultrasonication pretreatment occurred. Therefore, the use of ultrasonic treatment is not recommended. When employing novel solvents to assist in ^{31}P NMR analysis, it should always be ensured that these solvents are inert to the phosphorylated reagents or the chemical shifts will be altered. Derivatization of lignin samples is usually carried out in a pyridine/deuterated chloroform (1.6 : 1, v/v) dissolution system, which has three advantages. First, it facilitates sample dissolution. Second, the pyridine can neutralize hydrochloric acid (HCl), a byproduct of the phosphorylation process. Third, deuterated chloroform has a unique single-line signal that can be used to lock the NMR spectrometer during the acquisition process.

Internal standard. The derivatization of lignin samples was performed in the presence of an IS. After derivatization of the lignin sample and IS, the amount of each hydroxyl group can be quantified by comparing the peak integrals of the lignin sample and the standard. There are a variety of compounds that can be considered ISs. These compounds are characterized by a single hydroxyl group per molecule and produce a sharp signal in the ^{31}P NMR spectra after derivatization. ISs must be carefully selected to ensure that their signals do not overlap with those of the derivatized sample. In addition, it is critical to ensure that the sample is completely dry, because the phosphorylation reagent will react almost immediately with water. All solvents used in the ^{31}P NMR analysis should be of high purity and anhydrous. The cyclohexanol and cholesterol were widely used in the early years. However, the employment of cholesterol has been limited because of partial overlap of signals generated by cholesterol and aliphatic hydroxyl groups. In addition, the Ragauskas' group investigated a range of *N*-hydroxyl compounds that could be used as ISs for lignin ^{31}P NMR analysis [23, 24, 35]. The results showed that N-hydroxy compounds were suitable as ISs for lignin ^{31}P NMR analysis, and the chemical shifts of phosphorylated *N*-hydroxy compounds (i.e. 150.7–153.6 ppm) were at a greater distance from those of the lignin-derived fractions, which could avoid the interference of some specific lignin hydroxyl signals and improve the accuracy of the determination. A variety of ISs are available for ^{31}P NMR analysis of lignin, but cyclohexanol is the most widely used from a practical point of view.

Phosphorylation reagent. The most common phosphorylation reagent used in the determination of ^{31}P NMR is 2-chloro-4,4,5,5-tetramethyl-1,3,2-dioxaphospholane (TMDP), which is quantified mainly by detecting the signal generated by the phosphorylation reaction of lignin groups with phosphorus-containing reagents. Previous researchers have studied trivalent and pentavalent phosphorus reagents and found that trivalent phosphorus reagents have the greatest difference in chemical shifts [23, 24, 26]. Wroblewski et al. investigated the derivatization of organic model compounds by five trivalent phosphorylation reagents, including phenols, fatty alcohols, aromatic acids, fatty acids, amines, and thiols [84]. The results showed that 2-chloro-4,4,5,5-tetramethyl-1,3,2-dioxaphospholane (TMDP) was considered as the best reagent with chemical shifts that usually do not overlap each other. In the presence of organic bases (e.g. pyridine), TMDP reacts with various hydroxyl groups (aliphatic, phenolic, and carboxylic acid groups) in lignin, as well as with hydroxyl groups in ISs, to produce phosphorylated products (Figure 2.11). Figure 2.11 displays a typical ^{31}P NMR spectra of lignin derivatized by TMDP using cyclohexanol as an IS. In addition, it has been reported that 2-chloro-1,3,2-dioxaphospholane (CDP) can be used to quantify carboxyl and guaiacol hydroxyl groups in lignin, as well as to differentiate

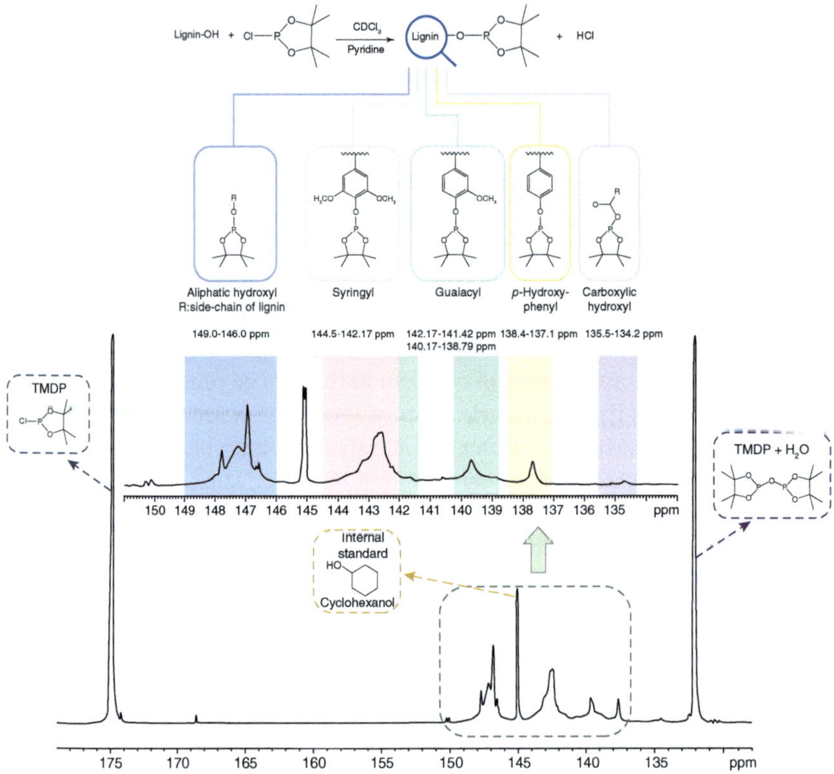

Figure 2.11 Phosphorylation products and ^{31}P NMR spectra of various hydroxyl groups (e.g. aliphatic, phenolic, and carboxylic acids) in lignin reacting with TMDP in the presence of a solvent mixture of pyridine, deuterated chloroform, and cyclohexane.

between primary and secondary hydroxyl groups in the *erythro*- and *threo*-conformations of β-*O*-4 structures [78, 85]. However, signal overlap between eugenol and C_5-substituted guaiacol can be observed when using this reagent, and this reagent has not been routinely applied to native lignin, especially hardwood lignin. Compared to CDP, TMDP has higher signal resolution for various phenolic hydroxyl groups and better stability of phosphorylation products. Furthermore, ^{31}P NMR data can be obtained more quickly even with smaller sample volumes. Dissolving lignin in the pyridine/deuterated chloroform system could facilitate the full reaction of each functional group in lignin with the phosphorylation reagent (TMDP), ensuring the accuracy of the final determination.

^{31}P NMR program. In NMR acquisitions, the routine use of reverse selective decoupling pulses is required to minimize the enhancement of the nuclear Overholser effect for quantitative measurements. In addition, a relatively long inter-pulse delay (e.g. 10 seconds) is required to ensure that the spin relaxation is completed before applying subsequent radiofrequency pulses [23, 77]. Spectral quality should be judged by such characteristics as whether the baseline is flat, the peaks are sharp, the phase is consistent, and the presence of spectral artifacts. Chemical shifts are usually internally calibrated using the sharp peak (around 132.2 ppm) of the phosphorylation product of TMDP with water (Figure 2.11). It is worth noting that ^{31}P NMR analysis is much more sensitive than ^{13}C-NMR, but not as sensitive as ^{1}H NMR. This is due to the 100% abundance, relatively high magnetic ratio, and 1/2 nuclear spin of the ^{31}P isotopes. ^{31}P spectroscopy typically has a wide range of chemical shifts, so that different classes of hydroxyl peaks can be separated.

2.5.2 ^{31}P NMR Operation

The solvent used for ^{31}P NMR analysis of lignin is typically an anhydrous pyridine and deuterated chloroform (1.6 : 1, v/v) system containing a relaxation agent (chromium(III) acetylacetonate) and an IS (cyclohexanol). Pyridine is used as a base to capture the hydrogen chloride released in the phosphorus-benzoylation reaction. Deuterated chloroform provides a locked deuterium signal in NMR experiments, solubilizes derivatized lignin samples, and prevents precipitation of pyridine-HCl salts. Accurately weigh 54.25 mg of cyclohexanol and dilute to 5 ml with pyridine/deuterated chloroform solution to obtain the desired IS solution. 25.0 mg of chromium(III) acetylacetonate was weighed and stoichiometrically quantified to 5 ml with pyridine/deuterated chloroform solution to obtain a 5 mg/ml solution of relaxation reagent. The lignin phosphorylation reaction requires that the lignin sample be dried in advance, usually in a vacuum oven at ~35 °C for 12 hours. An accurately weighed sample of lignin (approximately 20 mg) was dissolved in pyridine/deuterated chloroform solution (0.50 ml), and 0.10 ml of the IS solution and 0.10 ml of the relaxation reagent solution were added after sufficient dissolution. The above solutions were mixed thoroughly and then 0.1 ml of 2-chloro-4,4,5,5-tetramethyl-1,3,2-dioxaphosphorane (TMDP) phosphorylation reagent was added and stirred for a few moments at room temperature. Finally, the reaction mixture was transferred to a 5 mm NMR tube for ^{31}P NMR analysis. Since phosphorylation reagents are sensitive to humidity, there is a need to minimize contact with water. Quantitative ^{31}P NMR spectroscopy is typically recorded with a long pulse delay of at least five times the longest spin–lattice relaxation time of the ^{31}P nucleus to ensure that the phosphorus nucleus reaches thermal equilibrium before subsequent pulses. Chromium(III) acetylacetonate in the solvent system can be used as a relaxation reagent to shorten the spin–lattice relaxation time of phosphorus nuclei [23, 77]. In general, a pulse delay of 10 seconds is considered suitable for quantitative ^{31}P NMR analysis of lignin. ^{31}P NMR spectra can be obtained by performing the appropriate number of acquisitions (approximately 128 ~ 256 acquisitions for about 1 ~ 2 hours) at room temperature. The phosphorylated water exhibits a clear and stable signal at 132.2 ppm in the pyridine/deuterated chloroform solvent. The quantitative information on lignin functional groups can be obtained by first attributing the ^{31}P NMR spectra to different functional groups and then comparing the integral of the attributed signals with the IS signals.

Water-soluble lignin is inhomogeneous and has low solubility in common organic solvents, thus its ^{31}P NMR analysis procedure still remains a great challenge. The most common water-soluble lignins (lignosulfonates) are the second most produced type of industrial lignin, which are obtained through the sulfite pulping process [86].

The solubility is essential to obtain the reliable ^{31}P NMR spectra. In order to overcome this problem, the lignosulfonate can be converted to the corresponding lignosulfonic acid prior to derivatization. The treatment of lignosulfonate solutions with a strong acid (e.g. sulfuric acid) or an acidic exchange resin (e.g. Dowex ^{1}H, a strongly acidic cation exchanger) induces the conversion of all sulfonic acid groups to the acidic form. Using a selective adsorbent resin (XAD-7, a polar adsorbent) can remove the product from the acidic solution [87]. In this field, relevant researchers have already conducted investigations [83].

2.5.2.1 Detailed Procedures

Sample preparation. The lignin samples were dried overnight in a vacuum oven at 35 °C. Special attention needs to be paid to the choice of temperature because temperatures above 35 °C might alter the structure of the lignin sample. After drying, the samples were quickly transferred to a desiccator until they reached room temperature. This step is necessary to avoid the sample absorbing water from the environment.

Preparation of dissolution solutions. The dissolution solution was prepared by mixing anhydrous pyridine and deuterated chloroform at a ratio of 1.6/1 (v/v) in a 20 ml sample bottle. The preparation and usage of dissolution solutions should be performed with appropriate gloves and in a well-ventilated fume hood.

Preparation of internal standard solution. Accurately weigh 54.25 mg of cyclohexanol and dilute to 5 ml with pyridine/deuterated chloroform solution to obtain the desired IS solution. This step must be weighed accurately.

Preparation of ^{31}P NMR sample solution. Accurately weigh approximately 20 mg of lignin sample into a 2 ml bottle. Seal the reagent bottle with a septum cap. Add 0.5 ml of dissolution solutions to the sample bottle. After sufficient dissolution, add 0.10 ml of the IS solution and 0.10 ml of the relaxation reagent solution. Stir the solution until all the lignin is dissolved and a clear solution is formed (make sure the sample is completely dissolved). Add 0.1 ml of TMDP to the sample solution. The sample solution should be kept sealed. Use TMDP in a well-ventilated fume hood while wearing appropriate gloves, because TMDP and its vapors are corrosive, harmful to humans, and react quickly with water. Finally, the sample solution was transferred to an NMR tube.

^{31}P NMR programs. Setting up the experiment parameters (Pulse program: Inverse gated decoupling pulse; Nucleus: ^{31}P; Spectral width: 100 ppm; Acquisition time: ~0.8 seconds; Relaxation delay: ≥10 seconds; Number of scans: 64 or more; Center of spectrum: 140 ppm). The resonance frequency of deuterated chloroform was used to set the spectrometer frequency. A standard Bruker reverse-gated decoupling pulse sequence was used to acquire the signal.

Processing and analysis of ^{31}P spectra. Perform calibration phase and baseline manually. The signal of phosphorylated water is set to a chemical shift value of 132.2 ppm, and the spectrum can be amplified in the spectral range from 132 to about 150. ppm Set the IS solution to 1.0, and integrate the different hydroxyl groups (Aliphatic OH: 149.0–146.0 ppm; Phenolic OH: 144.0–137.4 ppm; Syringyl OH: 141.7–140.2 ppm; Guaiacyl OH: 140.2–138.8 ppm; *p*-hydroxyphenyl OH: 138.8–137.4 ppm; COOH: 136.0–133.6 ppm). If the lignin sample is 20 mg and the unit of expression for the final hydroxyl group is the amount of hydroxyl contained in 1 g of lignin, the hydroxyl content of the IS is calculated according to the following formula:

$$x = \frac{n}{m} = \frac{m/M}{m} = \frac{\frac{1}{50} \times 54.25 \times 10^{-3} / \left(100.158 \times 10^{-3}\right)}{20} \times 1000 = 0.54\,\text{mmol/g}$$

where x is the amount of substance of the IS OH corresponding to 1 g of lignin, n is the amount of substance corresponding to 54.25 mg of the IS, m is the mass of 20 mg of lignin sample weight, and 1/50 is the ratio of 20 mg of lignin to 1 g of lignin. Ultimately, the formula for calculating the hydroxyl content in lignin can be simplified as follows.

$$I\left(\text{different types of hydroxyls}\right) = \text{Integral value} \times 0.54$$

2.5.3 Applications of the ^{31}P NMR

The quantitative ^{31}P NMR technique is widely used to characterize the variation of lignin hydroxyl content during biorefining processes in the present time [23, 24]. Biomass pretreatment strategies are an important step in breaking the "biomass recalcitrance", which usually requires acidic or alkaline conditions [6, 7, 88]. Under these conditions, the lignin structure usually undergoes various depolymerization or condensation reactions. There are recent studies emphasizing the quantitative ^{31}P NMR analysis of lignin isolated from various biomass pretreatment strategies [89–91]. Since lignins have become a key genetic engineering target for improving wood quality and biofuel production, the native lignins in transgenic biomass were also quantitatively analyzed by ^{31}P NMR techniques, which in turn provided a clear view of the effect of transgenic technology on the structural characteristics of lignin in biomass [89, 92–94]. In addition, a comprehensive study of the structural transformation of transgenic biomass during pretreatment was conducted, providing a theoretical basis for downstream high-value-added applications of transgenic biomass [89, 94]. Other researchers have used ^{31}P NMR technology to monitor the dynamic lignin synthesis process in plants during different growth years, providing support for data in the field of biology to investigate the changes in plant growth [95, 96].

2.6 ^{19}F NMR

As well as ^{31}P NMR spectroscopy to determine the content of different hydroxyl groups in lignin, ^{19}F NMR can also determine the content of hydroxyl groups in lignin. In contrast to ^{31}P NMR spectroscopy, the ^{19}F NMR technique has the advantage to determine the carbonyl group in lignin [30, 97]. Two decades ago, Barrelle et al. developed ^{19}F NMR spectroscopy for the quantitative analysis of phenolic hydroxyls, aliphatic hydroxyls, and carbonyl groups using fluorinated compounds following lignin derivatization [30, 97, 98]. Among them, fluorobenzylation and fluorobenzoylation methods can be used to quantify phenolic and aliphatic hydroxyl groups, whereas trifluoromethylphenylhydrazine derivatization can be used to determine carbonyl groups. Detailed procedures can be found in related studies. ^{19}F NMR has many advantages for the study of lignin structure: (A) the fluorinated derivatives of lignin are very stable; (B) The position of fluorine atoms on the aromatic ring is very sensitive to chemical shift, so it is easy to obtain signals various for hydroxyl and carbonyl groups [98]. The ^{19}F-NMR acquisition time is quite short (less than 30 minutes for the lignin samples), and only a small amount of lignin derivatives (15 mg) is required for quantitative analysis [30]. The commonly used fluorine reagents include 4-fluorobenzyl chloride, 4-fluorophenacyl bromide, 4-fluorobenzoic acid anhydride, and trifluoromethylphenylhydrazine.

Constant et al. employed ^{19}F NMR to quantify and classify the cations of ketone and aldehyde carbonyl functional groups contained in lignin. They improved and simplified the known method of derivatizing the carbonyl groups to the corresponding hydrazones with 4-(trifluoromethyl)phenylhydrazine, which can be performed directly in the NMR tube without additional work prior to ^{19}F NMR quantification [99]. Esakkimuthu et al. quantified the hydroxyl groups by fluorobenzylation of the lignin followed by ^{19}F NMR spectroscopy [100]. They tested the assignments proposed in previous studies and examined the reactivity of various types of lignin hydroxyl groups with derivatizing agents. They successfully determined the phenolic group content in commercial organic lignins by ^{19}F NMR analysis. In addition, Kenny et al. reported a complementary method that employed pentafluoropyridine (PFP) as a derivatization reagent followed by ^{19}F NMR spectroscopy to accurately identify and quantify phenolic molecules in lignin samples [101] (see Figure 2.12).

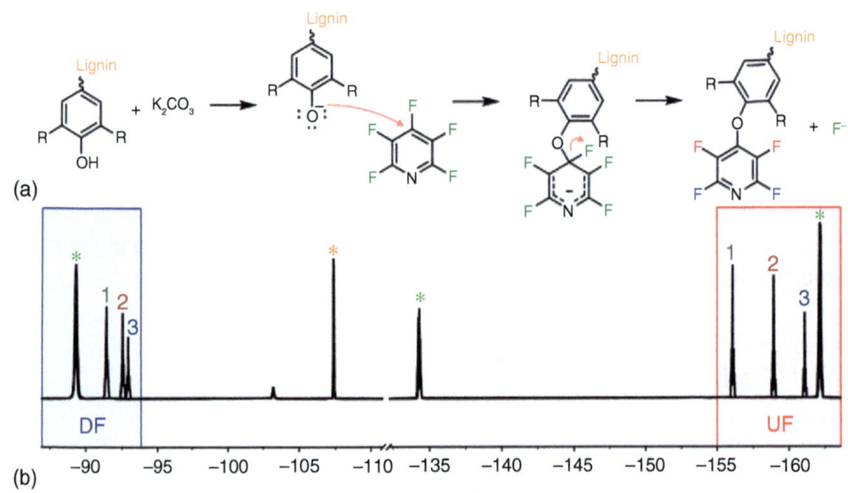

Figure 2.12 (a) Reaction of pentafluoropyridine with a phenolic compound to form a tetrafluoropyridyl ether. (b) The blue and red fluorine atoms in the tetrafluoropyridyl product correspond to the ^{19}F NMR resonance regions from 89 to 95 ppm (downfield) and 155 to 162 ppm (upfield), respectively. Resonances marked with a green asterisk belong to pentafluoropyridine, and the resonance marked with an orange asterisk is for the internal standard, 4,4-difluorobenzophenone. Numbered resonances correspond to 4-propylphenol (1), 4-propylguaiacol (2), and 4-propylsyringol (3) derivatized in 40% H_2O/DMSO.

2.7 Solution-State NMR of Whole Plant Cell Wall

Solution-state NMR techniques, coupled with appropriate dissolution or gelling solvents, have been proven to be powerful tools for characterizing lignocellulosic biomass without the need for isolation and purification of individual components. The development of this technology has made it possible to rapidly screen plant material and discern information throughout the cell wall. Therefore, it becomes possible that the structural changes of lignin could be investigated by an *in situ* 2D-HSQC NMR approach (without lignin isolation). Generally, prior to *in situ* 2D-HSQC NMR characterization, ball-milling dissolution of plant cell wall materials (e.g. chemical or biological treated cell wall) was needed. Up to now, there have been several approaches to achieve the goal of direct NMR characterization of biomass. (i) solution-state 2D-HSQC NMR of acetylated plant cell walls; (ii) solution-state 2D-HSQC NMR of non-acetylated plant cell walls in DMSO-d_6.

2.7.1 Solution-State 2D-HSQC NMR of Acetylated Plant Cell Walls

The concept of whole plant cell wall characterization by solution-state 2D-HSQC NMR was first proposed by Lu and Ralph, who developed a bisolvent system of *N*-methylimidazole (NMI) and dimethyl sulfoxide (DMSO) to dissolve and acetylate finely ball-milled plant cell walls and performed detailed structural studies on the acetylated plant cell-wall solutions using high-resolution NMR spectroscopy [60]. In the result of 2D-HSQC spectra, most of lignin structural features are evident and correlations from the aromatic region and provide very useful information about the syringyl/guaiacyl (S/G) composition of entire lignin. Although the initial results cannot meet the elaborate requirements of characterizing the lignin polymer in the cell wall, the idea realizes the lignin characterization of the ratio of S/G. Besides, under the enlightenment of the bisolvent system, Holtman et al. applied the DMSO-NMI solvent system to prepare several acetylated lignin samples, ball-milled wood, MWL, and residual lignin (RL), from the sapwood of Loblolly pine (*Pinus taeda*) [102]. All samples were treated with industrial cellulase to remove the carbohydrates before dissolution in the DMSO/NMI system and then acetylation. Results from this work suggested that the residual enzymatic lignin (REL) is closer in structure to the original lignin than the MWL and the residues after cellulase treatment are readily acetylated in the DMSO/NMI solvent system, which may be the most suitable preparation for lignin characterization by NMR techniques [102]. Other successful examples were monitoring the lignin changes during the biodegradation process via *in situ* HSQC NMR by using this solvent system (DMSO/NMI) [103, 104]. The application of DMSO/NMI solvent system could also predict biomass processing efficiency as well as for optimizing pretreatment methods in various biorefinery processes. For example, *in situ* 2D-HSQC NMR characterization of the structural heterogeneity of lignin polymers during successive treatment of bamboo was emphatically performed in a recent study [105] (see Figure 2.13).

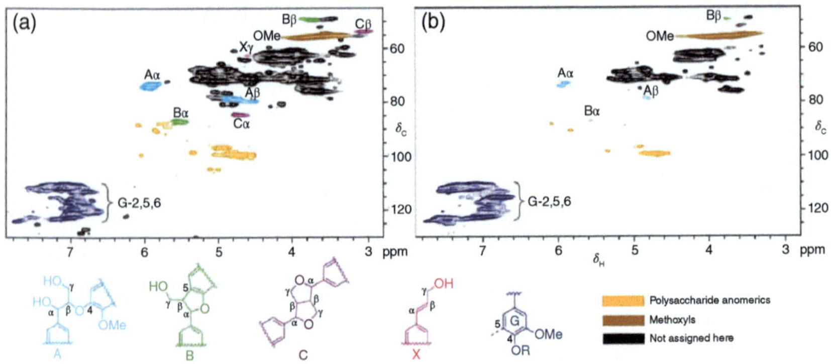

Figure 2.13 HSQC NMR spectra of acetylated sound spruce wood (A) and acetylated, brown-rotted spruce wood (B) in DMSO-d_6. Most of the unassigned structures are attributable to non-anomeric carbon–hydrogen bonds in polysaccharides. OMe: methoxyl. R = lignin polymer or H. Source: Reprinted from [[103]]. Copyright 2008 Wiley.

In addition to the DMSO/NMI solvent system, ionic liquid ([Bmim]Cl) was also used as reaction medium to acetylate cell walls for *in situ* 2D-HSQC NMR characterization. The *in situ* HSQC NMR technique of acetylated cell wall was also applied to detect the lignin changes between archaeological wood and fresh wood [106, 107]. Afterward, Qu et al. identified the whole cell-wall components (including lignin, cellulose, and hemicelluloses) by the 2D-HSQC NMR technique, with the aid of isolated lignin and commercial cellulose and hemicelluloses (arabinoxylan, galactomannan, and glucomannan) [108]. However, the above-mentioned chemical modification of the plant cell wall, with even simple derivatization, led to the loss of some structural information. For example, natural acetylation in the plant cell wall was masked when the sample was per-acetylated.

2.7.2 Solution-State 2D-HSQC NMR of Non-Acetylated Plant Cell Walls

It would be ideal to have a way to examine "native" whole plant cell walls by solution-state NMR. It was noted that Yelle et al. were the first to introduce a mixture of deuterated solvents (DMSO-d_6/NMI-d_6) to investigate the entire wood cell wall without acetylation. Fortunately, through this approach, native acylations were revealed [109].

Thanks to the different whole cell wall dissolved systems found by researchers, the *in situ* 2D-HSQC NMR characterization of the plant cell wall could be realized with satisfactory results via advanced NMR equipment (400 MHz or higher). Structural characterization of lignin could be directly investigated by NMR investigations via non-derivatization solvent systems, such as DMSO-d_6/NMI-d_6 [109], DMSO-d_6 [21], and DMSO-d_6/pyridine-d_5 [20]. The development of DMSO-d_6/NMI-d_6 was to enhance the peak dispersion of the polysaccharides and to enable visualization of natural acylations in woody plants since NMI facilitates dissolution of cellulose and xylan/mannan [109]. However, some overlapping solvent signals still mask portions of the lignin aromatic region. Because of this, the DMSO-d_6 method has been developed subsequent to the DMSO-d_6/NMI-d_6 work to use NMR as a rapid screening process for lignin chemistry. The rapid NMR characterization method provides what appears to be the best tool for the detailed structural study of the complex cell wall polymers. For example, the correlations for *p*-coumarate units and simple ferulates in the corn sample are readily seen and well resolved [20]. Aromatic regions of the 2D-HSQC spectra highlight the differences in the *p*-hydroxyphenyl: guaiacyl: syringyl (H : G : S) distributions in the lignin. Figure 2.14 shows the significant variations in the compositions of the lignin polymers and other aromatic constituents in the wall [10].

The literature from Ralph's team have developed and reported in detail on such methods [10, 21]. However, when it comes to lignin structure characterization, the separation of representative native lignin for NMR characterization remains a more accurate approach.

2.7.3 Comparison of the Dissolution Methods and Applications of Whole Plant Cell Wall Solution State NMR

At present, there have been numerous reports on whole plant cell wall solution state NMR analysis. Table 2.6 is a summary of biomass dissolution methods used for NMR analysis, including their advantages and disadvantages.

Currently, whole plant cell wall solution state NMR technologies have been applied to the fields, such as (i) characterization of biomass; (ii) estimation of changes occurring during pretreatment; (iii) analysis of isotopic enrichment in biomass; (iv) rapid chemical metrology and S/G ratio estimation of plant cell walls.

2.8 Conclusion

For lignin, there are few guidelines and there are no absolutes due to structural uncertainty and complexity. In terms of the polymer properties of lignin, the higher molecular weight, greater viscosity, and lower solubility of polymers compared to small molecules make for low mobility, leading to short relaxation times and broadening of signals and limiting a detailed characterization of polymeric components. Additionally, lignin fractions from different sources and different methods have different structural features and molecular weights. Absolute quantification in lignin fractions is difficult due to signal overlap and other factors. Therefore, any so-called quantitative values obtained from a nonquantitative technique must be compared to corresponding values obtained from an IG experiment to determine its quantitative validity. The best quantification method remains the relatively tedious IG technique, along with an IS substance. The detailed chemical composition and structural features of the isolated lignin should be illustrated by a reliable and powerful method, such as advanced 2D-HSQC technique, which can provide the qualitative and quantitative information of lignin. The well-characterized composition and linkages in lignin can screen suitable raw materials for catalytic

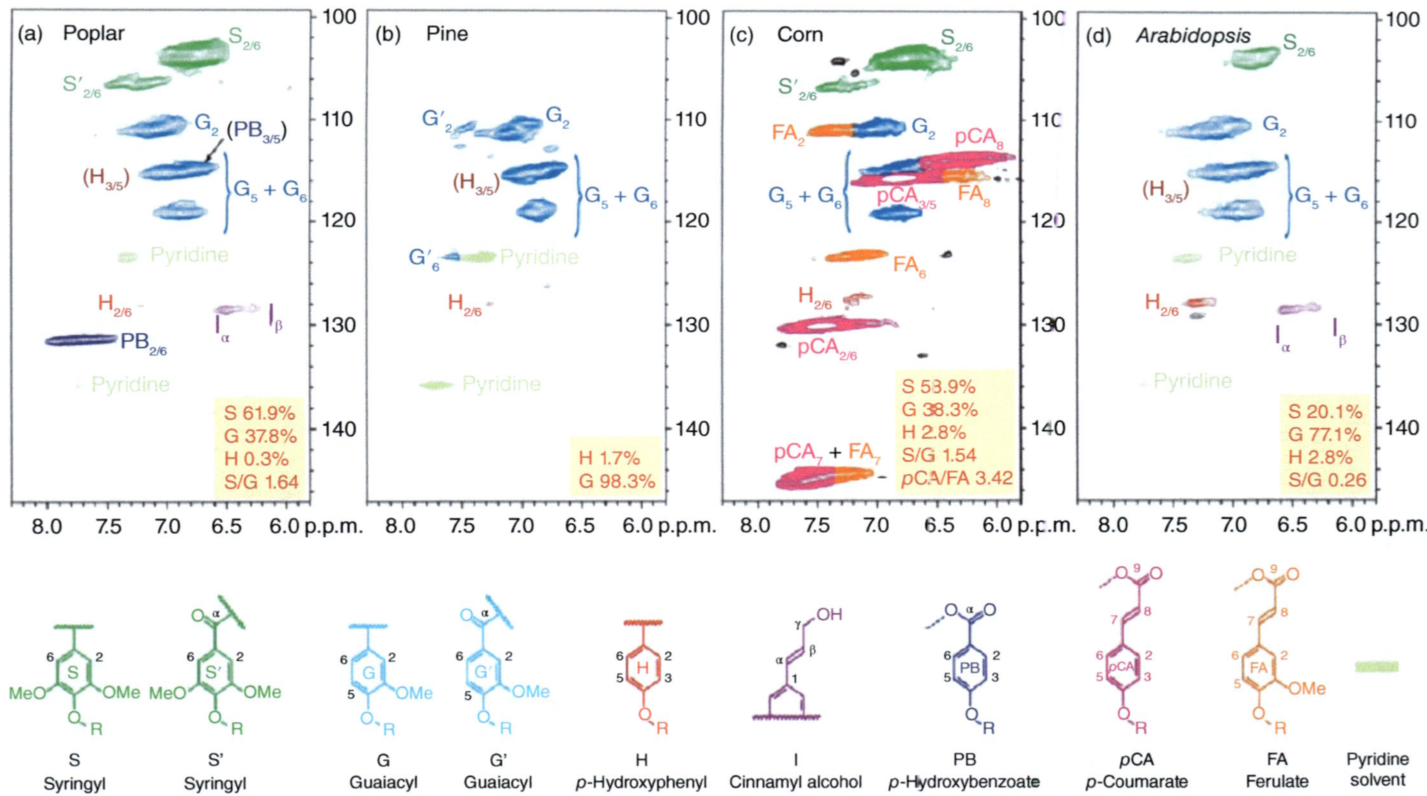

Figure 2.14 2D-HSQC NMR spectra revealing lignin unit compositions. 2D-HSQC spectra (aromatic regions only) of cell wall gels in 4:1 (vol/vol) DMSO-d_6/pyridine-d_5 from (a) two-year-old greenhouse-grown poplar wood; (b) mature pine wood; (c) senesced corn stalks and (d) senesced Arabidopsis inflorescence stems. Contours in this region are used to measure S/G/H ratios, as well as relative p-hydroxybenzoate (PB, in poplar), p-coumarate (PCA in corn), and ferulate (FA in corn) levels. Source: Reprinted from [10]. Copyright 2012 Nature.

Table 2.6 Summary of the dissolution methods and their advantages and disadvantages.

Solvent system	Advantages	Disadvantages	References
1. Derivatizing-biomass acetylation:			
Derivatization in DMSO/TBAF or DMSO/NMI	Simple and cheap solvent system that facilitates derivatization of the entire biomass for NMR analysis in common deuterated solvents	Requires significant milling requirement and harsh conditions for derivatization	[21, 60, 105, 110]
Derivatization in ILs	Higher solubilities at less harsh sample preparation and derivatization conditions	ILs are costly and may have reactive characteristics that alter biomass structure	[108, 111]
2. Non-derivatizing:			
DMSO-d_6	Cheap and simple sample preparation	Requires significant milling, has high sample viscosity, and does not solubilize crystalline cellulose	[10, 21, 22]
DMSO-d_6/pyridine-d_5 or DMSO-d_6/NMI-d_6	Enhanced swelling compared to pure DMSO with a lower sample viscosity, better biomass solubilization, and greater polymer chain mobility		[20, 109, 112, 113]
ILs	Higher solubilities with less harsh sample preparation, reduced size-reduction requirements, and solubilizes crystalline cellulose	Expensive and may have reactive characteristics that alter biomass structure	[102, 114]
Organic electrolyte	Reduced sample viscosity and reduced use of ILs, and thus, less costly than using pure ILs. Solubilizes crystalline cellulose		[51, 56, 115–122]

degradation of lignin. Additionally, ^{31}P-NMR technique was also used to quantify the amounts of different hydroxyl groups in lignins, which will be beneficial to the chemical modification of lignin in developing lignin-based materials. Fortunately, most of the semiquantitative methods that have been described are adequate when the researcher wants to follow structural changes during different pretreatments provided that the desired precision is maintained. In fact, the application of this aspect is also the most-used semiquantitative application of lignin chemistry and its structural changes. Further elucidation of the supramolecular structures and inherent properties of lignin nanoparticles by more advanced techniques in the future will stimulate more opportunities and contribute to extensive and advanced applications of these green nanomaterials in value-added applications of lignin. In short, it is believed that lignin characterization by NMR methodologies for elucidating the structural characteristics and chemical reactivity of lignin polymers will maximize the ultimate utilization of lignin in the biorefinery process, such as developing lignin-based materials and chemicals.

References

1 Calvo-Flores, F.G. and Dobado, J.A. (2010). Lignin as renewable raw material. *ChemSusChem* 3 (11): 1227–1235. https://doi.org/10.1002/cssc.201000157.

2 Lora, J.H. and Glasser, W.G. (2002). Recent industrial applications of lignin: a sustainable alternative to nonrenewable materials. *Journal of Polymers and the Environment* 10: 39–48. https://doi.org/10.1023/A:1021070006895.

3 Bujanovic, B., Ralph, S., Reiner, R. et al. (2010). Polyoxometalates in oxidative delignification of chemical pulps: effect on lignin. *Materials* 3 (3): 1888–1903. https://doi.org/10.3390/ma3031888.

4 Higuchi, T. (1990). Lignin biochemistry: biosynthesis and biodegradation. *Wood Science and Technology* 24: 23–63. https://doi.org/10.1007/bf00225306.

5 Ralph, J., Lundquist, K., Brunow, G. et al. (2004). Lignins: natural polymers from oxidative coupling of 4-hydroxyphenyl-propanoids. *Phytochemistry Reviews* 3: 29–60. https://doi.org/10.1023/B:PHYT.0000047809.65444.a4.

6 Ragauskas, A.J., Beckham, G.T., Biddy, M.J. et al. (2014). Lignin valorization: improving lignin processing in the biorefinery. *Science* 344 (6185): 1246843. https://doi.org/10.1126/science.1246843.

7 Schutyser, W., Renders, A.T., Van den Bosch, S. et al. (2018). Chemicals from lignin: an interplay of lignocellulose fractionation, depolymerisation, and upgrading. *Chemical Society Reviews* 47 (3): 852–908. https://doi.org/10.1039/c7cs00566k.

8 Lu, F. and Ralph, J. (2010). *Chapter 6-Lignin, Cereal Straw as a Resource for Sustainable Biomaterials and Biofuels*, 169–207. Amsterdam: Elsevier.

9 Su, S., Wang, S., and Song, G. (2021). Disassembling catechyl and guaiacyl/syringyl lignins coexisting in Euphorbiaceae seed coats. *Green Chemistry* 23 (18): 7235–7242. https://doi.org/10.1039/d1gc02131a.

10 Mansfield, S.D., Kim, H., Lu, F., and Ralph, J. (2012). Whole plant cell wall characterization using solution-state 2D NMR. *Nature Protocols* 7 (9): 1579–1589. https://doi.org/10.1038/nprot.2012.064.

11 Wen, J.-L., Sun, S.-L., Xue, B.-L., and Sun, R.-C. (2013b). Recent advances in characterization of lignin polymer by solution-state nuclear magnetic resonance (NMR) methodology. *Materials* 6 (1): 359–391. https://doi.org/10.3390/ma6010359.

12 Lundquist, K. (1980). NMR studies of lignins. 4. Investigation of spruce lignin by ^{1}H NMR spectroscopy. *Acta Chemica Scandinavica. Series B* 34: 21–26.

13 Lundquist, K. (1981). NMR studies of lignins. 5. Investigation of non-derivatized spruce and birch lignin by ^{1}H NMR spectroscopy. *Acta Chemica Scandinavica. Series B* 35: 497–501.

14 Gellerstedt, G., Robert, D., Parker, V. et al. (1987). Quantitative 13C NMR analysis of kraft lignins. *Acta Chemica Scandinavica. Series B* 41 (7): 541–546.

15 Hawkes, G.E., Smith, C.Z., Utley, J.H., et al. (1993). A comparison of solution and solid state 13C NMR spectra of lignins and lignin model compounds. https://doi.org/10.1515/hfsg.1993.47.4.302

16 Pan, D.-R., Tai, D.-S., Chen, C.-L., and Robert, D. (1990). Comparative studies on chemical composition of wood components in recent and ancient woods of *Bischofia polycarpa*. *Holzforschung* 44: 7–16. https://doi.org/10.1515/hfsg.1990.44.1.7.

17 Robert, D. (1992). Carbon-13 nuclear magnetic resonance spectrometry. In: *Methods in Lignin Chemistry*, 250–273. Springer.

18 Ralph, J., Landucci, L., Heitner, C. et al. (2010). *NMR of Lignins, in Lignin and Lignans: Advances in Chemistry*, 137–244. CRC Press.

19 Hu, W.-J., Harding, S.A., Lung, J. et al. (1999). Repression of lignin biosynthesis promotes cellulose accumulation and growth in transgenic trees. *Nature Biotechnology* 17 (8): 808–812.

20 Kim, H. and Ralph, J. (2010). Solution-state 2D NMR of ball-milled plant cell wall gels in DMSO-d_6/pyridine-d_5. *Organic & Biomolecular Chemistry* 8 (3): 576–591. https://doi.org/10.1039/b916070a.

21 Kim, H., Ralph, J., and Akiyama, T. (2008). Solution-state 2D NMR of ball-milled plant cell wall gels in DMSO-d_6. *Bioenergy Research* 1: 56–66. https://doi.org/10.1007/s12155-008-9004-z.

22 Lu, F. and Ralph, J. (2011). Solution-state NMR of lignocellulosic biomass. *Journal of Biobased Materials and Bioenergy* 5 (2): 169–180. https://doi.org/10.1166/jbmb.2011.1131.

23 Meng, X., Crestini, C., Ben, H. et al. (2019). Determination of hydroxyl groups in biorefinery resources via quantitative ^{31}P NMR spectroscopy. *Nature Protocols* 14 (9): 2627–2647. https://doi.org/10.1038/s41596-019-0191-1.

24 Pu, Y., Cao, S., and Ragauskas, A.J. (2011). Application of quantitative ^{31}P NMR in biomass lignin and biofuel precursors characterization. *Energy & Environmental Science* 4 (9): 3154–3166. https://doi.org/10.1039/c1ee01201k.

25 Argyropoulos, D. (1995). 31 P NMR in wood chemistry: a review of recent progress. *Research on Chemical Intermediates* 21: 373–395.

26 Granata, A. and Argyropoulos, D.S. (1995). 2-Chloro-4, 4, 5, 5-tetramethyl-1, 3, 2-dioxaphospholane, a reagent for the accurate determination of the uncondensed and condensed phenolic moieties in lignins. *Journal of Agricultural and Food Chemistry* 43 (6): 1538–1544. https://doi.org/10.1021/jf00054a023.

27 Vvu, S. and Argyropoulos, D. (2003). An improved method for lsolating lignin in high yield and purity. *Journal of Pulp and Paper Science* 29 (7): 235–240.

28 Foston, M. and Ragauskas, A.J. (2012). Biomass characterization: recent progress in understanding biomass recalcitrance. *Industrial Biotechnology* 8 (4): 191–208. https://doi.org/10.1089/ind.2012.0015.

29 Meng, X. and Ragauskas, A.J. (2014). Recent advances in understanding the role of cellulose accessibility in enzymatic hydrolysis of lignocellulosic substrates. *Current Opinion in Biotechnology* 27: 150–158. https://doi.org/10.1016/j.copbio.2014.01.014.

30 Barrelle, M. (1993). A new method for the quantitative ^{19}F NMR spectroscopic analysis of hydroxyl groups in lignins. https://doi.org/10.1515/hfsg.1993.47.3.261

31 Ferguson, R.C. and Phillips, W.D. (1967). High-resolution nuclear magnetic resonance spectroscopy. *Science* 157 (3786): 257–267. https://doi.org/doi:10.1126/science.157.3786.257.

32 Lundquist, K. (1992). Proton (^{1}H) NMR spectroscopy. In: *Methods in Lignin Chemistry* (ed. S.Y. Lin and C.W. Dence), 242–249. Berlin, Heidelberg: Springer.

33 Wen, J.-L., Sun, S.-L., Yuan, T.-Q., and Sun, R.-C. (2015). Structural elucidation of whole lignin from *Eucalyptus* based on preswelling and enzymatic hydrolysis. *Green Chemistry* 17 (3): 1589–1596. https://doi.org/10.1039/c4gc01889c.

34 Wen, J.L., Sun, Z.J., Sun, Y.C. et al. (2010). Structural characterization of alkali-extractable lignin fractions from bamboo. *Journal of Biobased Materials and Bioenergy* 4 (4): 408–425. https://doi.org/10.1166/jbmb.2010.1111.

35 Pu, Y., Hallac, B., and Ragauskas, A.J. (2013a). Plant biomass characterization: application of solution- and solid-state NMR spectroscopy. In: *Aqueous Pretreatment of Plant Biomass for Biological and Chemical Conversion to Fuels and Chemicals*, 369–390. Wiley.

36 Abreu, H. and Freire, M. (1995). Methoxyl content determination of lignin by ^{1}H NMR. *Anais da Academia Brasileira de Ciências* 67: 379–382.

37 Lundquist, K., Sjöholm, R.E., Teien, G. et al. (1979). NMR studies of lignins. 2. Interpretation of the ^{1}H NMR spectrum of acetylated birch lignin. *Acta Chemica Scandinavica* 33: 27–30.

38 Balakshin, M.Y., Capanema, E.A., Santos, R.B. et al. (2016). Structural analysis of hardwood native lignins by quantitative 13C NMR spectroscopy. *Holzforschung* 70 (2): 95–108. https://doi.org/doi:10.1515/hf-2014-0328.

39 Hallac, B.B., Pu, Y., and Ragauskas, A.J. (2010). Chemical transformations of *Buddleja davidii* lignin during ethanol organosolv pretreatment. *Energy & Fuels* 24 (4): 2723–2732. https://doi.org/10.1021/ef901556u.

40 Sannigrahi, P., Ragauskas, A.J., and Miller, S.J. (2008). Effects of two-stage dilute acid pretreatment on the structure and composition of lignin and cellulose in loblolly pine. *Bioenergy Research* 1 (3): 205–214. https://doi.org/10.1007/s12155-008-9021-y.

41 Sannigrahi, P., Ragauskas, A.J., and Miller, S.J. (2010). Lignin structural modifications resulting from ethanol organosolv treatment of loblolly pine. *Energy & Fuels* 24 (1): 683–689. https://doi.org/10.1021/ef900845t.

42 Holtman, K.M., Chang, H.M., Jameel, H., and Kadla, J.F. (2006). Quantitative 13C NMR characterization of milled wood lignins isolated by different milling techniques. *Journal of Wood Chemistry and Technology* 26 (1): 21–34. https://doi.org/10.1080/02773810600582152.

43 Xia, Z., Akim, L.G., and Argyropoulos, D.S. (2001). Quantitative (13)C NMR analysis of lignins with internal standards. *Journal of Agricultural and Food Chemistry* 49 (8): 3573–3578. https://doi.org/10.1021/jf010333v.

44 Choi, J.W. and Faix, O. (2011). NMR study on residual lignins isolated from chemical pulps of beech wood by enzymatic hydrolysis. *Journal of Industrial and Engineering Chemistry* 17 (1): 25–28. https://doi.org/doi:10.1016/j.jiec.2010.10.004.

45 Hallac, B.B., Sannigrahi, P., Pu, Y. et al. (2009). Biomass characterization of *Buddleja davidii*: a potential feedstock for biofuel production. *Journal of Agricultural and Food Chemistry* 57 (4): 1275–1281. https://doi.org/10.1021/jf8030277.

46 Samuel, R., Pu, Y., Raman, B., and Ragauskas, A.J. (2010). Structural characterization and comparison of switchgrass ball-milled lignin before and after dilute acid pretreatment. *Applied Biochemistry and Biotechnology* 162 (1): 62–74. https://doi.org/10.1007/s12010-009-8749-y.

47 Wen, J.-L., Yuan, T.-Q., Sun, S.-L. et al. (2014). Understanding the chemical transformations of lignin during ionic liquid pretreatment. *Green Chemistry* 16 (1): 181–190. https://doi.org/10.1039/c3gc41752b.

48 Du, X., Pérez-Boada, M., Fernández, C. et al. (2014). Analysis of lignin–carbohydrate and lignin–lignin linkages after hydrolase treatment of xylan–lignin, glucomannan–lignin and glucan–lignin complexes from spruce wood. *Planta* 239 (5): 1079–1090. https://doi.org/10.1007/s00425-014-2037-y.

49 Liitiä, T.M., Maunu, S.L., Hortling, B. et al. (2003). Analysis of technical lignins by two- and three-dimensional NMR spectroscopy. *Journal of Agricultural and Food Chemistry* 51 (8): 2136–2143. https://doi.org/10.1021/jf0204349.

50 Rencoret, J., Prinsen, P., Gutiérrez, A., and Martínez, Á.T., del Río, J.C. (2015). Isolation and structural characterization of the milled wood lignin, dioxane lignin, and cellulolytic lignin preparations from brewer's spent grain. *Journal of Agricultural and Food Chemistry* 63 (2): 603–613. https://doi.org/10.1021/jf505808c.

51 Samuel, R., Foston, M., Jiang, N. et al. (2011). Structural changes in switchgrass lignin and hemicelluloses during pretreatments by NMR analysis. *Polymer Degradation and Stability* 96 (11): 2002–2009. https://doi.org/doi:10.1016/j.polymdegradstab.2011.08.015.

52 Zhang, L. and Gellerstedt, G. (2001). NMR observation of a new lignin structure, a spiro-dienone. *Chemical Communications* 24: 2744–2745. https://doi.org/10.1039/B108285J.

53 Zhang, L. and Gellerstedt, G. (2007). Quantitative 2D HSQC NMR determination of polymer structures by selecting suitable internal standard references. *Magnetic Resonance in Chemistry* 45 (1): 37–45. https://doi.org/doi:10.1002/mrc.1914.

54 Long, Y.-Y., Lv, J., Li, B.-X., and Liu, Y.-G. (2021). Speedy quantitative microstructure determination of poly(ethylene-*co*-1-hexene) at triads by ^{1}H–^{13}C two-dimensional NMR. *Polymer* 229: 123993. https://doi.org/doi:10.1016/j.polymer.2021.123993.

55 Wagner, A., Donaldson, L., Kim, H. et al. (2008). Suppression of 4-coumarate-CoA ligase in the coniferous gymnosperm pinus radiata. *Plant Physiology* 149 (1): 370–383. https://doi.org/10.1104/pp.108.125765.

56 Cheng, K., Sorek, H., Zimmermann, H. et al. (2013). Solution-state 2D NMR spectroscopy of plant cell walls enabled by a dimethylsulfoxide-d 6/1-ethyl-3-methylimidazolium acetate solvent. *Analytical Chemistry* 85 (6): 3213–3221. https://doi.org/10.1021/ac303529v.

57 Heikkinen, S., Toikka, M.M., Karhunen, P.T., and Kilpeläinen, I.A. (2003). Quantitative 2D HSQC (Q-HSQC) via suppression of J-dependence of polarization

transfer in NMR spectroscopy: application to wood lignin. *Journal of the American Chemical Society* 125 (14): 4362–4367. https://doi.org/10.1021/ja029035k.

58 del Río, J.C., Rencoret, J., Gutiérrez, A. et al. (2018). Structural characterization of lignin from maize (*Zea mays* L.) fibers: evidence for diferuloylputrescine incorporated into the lignin polymer in maize kernels. *Journal of Agricultural and Food Chemistry* 66 (17): 4402–4413. https://doi.org/10.1021/acs.jafc.8b00880.

59 del Río, J.C., Rencoret, J., Prinsen, P. et al. (2012). Structural characterization of wheat straw lignin as revealed by analytical pyrolysis, 2D-NMR, and reductive cleavage methods. *Journal of Agricultural and Food Chemistry* 60 (23): 5922–5935. https://doi.org/10.1021/jf301002n.

60 Lu, F. and Ralph, J. (2003). Non-degradative dissolution and acetylation of ball-milled plant cell walls: high-resolution solution-state NMR. *The Plant Journal* 35 (4): 535–544. https://doi.org/doi:10.1046/j.1365-313X.2003.01817.x.

61 Yue, F., Lu, F., Ralph, S., and Ralph, J. (2016). Identification of 4–O–5-units in softwood lignins via definitive lignin models and NMR. *Biomacromolecules* 17 (6): 1909–1920. https://doi.org/10.1021/acs.biomac.6b00256.

62 Wen, J.-L., Sun, S.-L., Xue, B.-L., and Sun, R.-C. (2013a). Quantitative structural characterization of the lignins from the stem and pith of bamboo (*Phyllostachys pubescens*). *Holzforschung* 67 (6): 613–627. https://doi.org/doi:10.1515/hf-2012-0162.

63 Sun, Q., Wang, H.-M., Ma, C.-Y. et al. (2023). Dynamic structural evolution of lignin macromolecules and hemicelluloses during Chinese pine growth. *International Journal of Biological Macromolecules* 235: 123688. https://doi.org/doi:10.1016/j.ijbiomac.2023.123688.

64 Ralph, J., Lapierre, C., and Boerjan, W. (2019). Lignin structure and its engineering. *Current Opinion in Biotechnology* 56: 240–249. https://doi.org/doi:10.1016/j.copbio.2019.02.019.

65 del Río, J.C., Rencoret, J., Gutiérrez, A. et al. (2020). Lignin monomers from beyond the canonical monolignol biosynthetic pathway: another brick in the wall. *ACS Sustainable Chemistry & Engineering* 8 (13): 4997–5012. https://doi.org/10.1021/acssuschemeng.0c01109.

66 Belyy, V.A., Karmanov, A.P., Kocheva, L.S. et al. (2019). Comparative study of chemical and topological structure of macromolecules of lignins of birch (*Betula verrucosa*) and apple (*Malus domestica*) wood. *International Journal of Biological Macromolecules* 128: 40–48. https://doi.org/doi:10.1016/j.ijbiomac.2019.01.095.

67 Lahive, C.W., Kamer, P.C., Lancefield, C.S., and Deuss, P.J. (2020). An introduction to model compounds of lignin linking motifs; synthesis and selection considerations for reactivity studies. *ChemSusChem* 13 (17): 4238–4265. https://doi.org/10.1002/cssc.202000989.

68 Rinaldi, R., Jastrzebski, R., Clough, M.T. et al. (2016). Paving the way for lignin valorisation: recent advances in bioengineering, biorefining and catalysis. *Angewandte Chemie International Edition* 55 (29): 8164–8215. https://doi.org/10.1002/anie.201510351.

69 Stark, N.M., Yelle, D.J., and Agarwal, U.P. (2016). Techniques for characterizing lignin. *Lignin in Polymer Composites* 49–66. https://doi.org/10.1016/B978-0-323-35565-0.00004-7.

70 Dutta, T., Isern, N.G., Sun, J. et al. (2017). Survey of lignin-structure changes and depolymerization during ionic liquid pretreatment. *ACS Sustainable Chemistry & Engineering* 5 (11): 10116–10127. https://doi.org/10.1021/acssuschemeng.7b02123.

71 Sathitsuksanoh, N., Holtman, K.M., Yelle, D.J. et al. (2014). Lignin fate and characterization during ionic liquid biomass pretreatment for renewable chemicals and fuels production. *Green Chemistry* 16 (3): 1236–1247. https://doi.org/10.1039/c3gc42295j.

72 Hong, S., Li, H.Y., Shen, X.J. et al. (2022). Unveiling the migration and transformation mechanism of lignin in eucalyptus during deep eutectic solvent pretreatment. *ChemSusChem* 15 (15): e202200553. https://doi.org/10.1002/cssc.202200553.

73 Hong, S., Shen, X.-J., Pang, B. et al. (2020). In-depth interpretation of the structural changes of lignin and formation of diketones during acidic deep eutectic solvent pretreatment. *Green Chemistry* 22 (6): 1851–1858. https://doi.org/10.1039/d0gc00006j.

74 Shen, X.-J., Chen, T., Wang, H.-M. et al. (2019). Structural and morphological transformations of lignin macromolecules during bio-based deep eutectic solvent (DES) pretreatment. *ACS Sustainable Chemistry & Engineering* 8 (5): 2130–2137. https://doi.org/10.1021/acssuschemeng.9b05106.

75 Cheng, F., Liu, S., Karlen, S.D. et al. (2023). Poplar lignin structural changes during extraction in γ-valerolactone (GVL). *Green Chemistry* 25 (1): 336–347. https://doi.org/10.1039/d2gc03446h.

76 Luterbacher, J.S., Azarpira, A., Motagamwala, A.H. et al. (2015). Lignin monomer production integrated into the γ-valerolactone sugar platform. *Energy & Environmental Science* 8 (9): 2657–2663. https://doi.org/10.1039/c5ee01322d.

77 Argyropoulos, D.S., Pajer, N., and Crestini, C. (2021). Quantitative ^{31}P NMR analysis of lignins and tannins. *JoVE (Journal of Visualized Experiments)* 174: e62696. https://doi.org/10.3791/62696.

78 Archipov, Y., Argyropoulos, D., Bolker, H., and Heitner, C. (1991a). ^{31}P NMR spectroscopy in wood chemistry. I. Model compounds. *Journal of Wood Chemistry and Technology* 11 (2): 137–157. https://doi.org/0.1080/02773819108050267.

79 Archipov, Y., Argyropoulos, D.S., Bolker, H., and Heitner, C. (1991b). ^{31}P-NMR spectroscopy in wood chemistry. Phosphite derivatives of carbohydrates. *Carbohydrate Research* 220: 49–61. https://doi.org/10.1016/0008-6215(91)80005-8.

80 Faruk, O. and Sain, M. (2015). *Lignin in Polymer Composites*. William Andrew.

81 Hoareau, W., Trindade, W.G., Siegmund, B. et al. (2004). Sugar cane bagasse and curaua lignins oxidized by chlorine dioxide and reacted with furfuryl alcohol: characterization and stability. *Polymer Degradation and Stability* 86 (3): 567–576. https://doi.org/10.1016/j.polymdegradstab.2004.07.005.

82 Stücker, A., Podschun, J., Saake, B., and Lehnen, R. (2018). A novel quantitative 31 P NMR spectroscopic analysis of hydroxyl groups in lignosulfonic acids. *Analytical Methods* 10 (28): 3481–3488. https://doi.org/10.1039/c8ay01272e.

83 Wurzer, G.K., Bacher, M., Hettegger, H. et al. (2021). A general solvent system for the analysis of lignosulfonates by 31 P NMR. *Analytical Methods* 13 (45): 5502–5508. https://doi.org/10.1039/d1ay01241j.

84 Wroblewski, A., Lensink, C., Markuszewski, R., and Verkade, J.G. (1988). Phosphorus-31 NMR spectroscopic analysis of coal pyrolysis condensates and extracts for heteroatom functionalities possessing labile hydrogen. *Energy & Fuels* 2 (6): 765–774. https://doi.org/10.1021/ef00012a008.

85 Jiang, Z.-H. and Argyropoulos, D. (1994). The stereoselective degradation of arylglycerol-beta-aryl ethers during kraft pulping. *Journal of Pulp and Paper Science* 20 (7): J183–J188.

86 Calvo-Flores, F.G., Dobado, J.A., Isac-García, J., and Martín-Martínez, F.J. (2015). *Lignin and Lignans as Renewable Raw Materials: Chemistry, Technology and Applications*. Wiley.

87 Sumerskii, I., Korntner, P., Zinovyev, G. et al. (2015). Fast track for quantitative isolation of lignosulfonates from spent sulfite liquors. *RSC Advances* 5 (112): 92732–92742. https://doi.org/10.1039/c5ra14080c.

88 Ding, S.-Y., Liu, Y.-S., Zeng, Y. et al. (2012). How does plant cell wall nanoscale architecture correlate with enzymatic digestibility? *Science* 338 (6110): 1055–1060. https://doi.org/10.1126/science.1227491.

89 Ma, C.-Y., Gao, X., Peng, X.-P. et al. (2021). Microwave-assisted deep eutectic solvents (DES) pretreatment of control and transgenic poplars for boosting the lignin valorization and cellulose bioconversion. *Industrial Crops and Products* 164: 113415. https://doi.org/10.1016/j.indcrop.2021.113415.

90 Ma, C.-Y., Xu, L.-H., Sun, Q. et al. (2022a). Tailored one-pot lignocellulose fractionation to maximize biorefinery toward controllable producing lignin nanoparticles and facilitating enzymatic hydrolysis. *Chemical Engineering Journal* 450: 138315. https://doi.org/10.1016/j.cej.2022.138315.

91 Ma, C.-Y., Xu, L.-H., Sun, Q. et al. (2022b). Ultrafast alkaline deep eutectic solvent pretreatment for enhancing enzymatic saccharification and lignin fractionation from industrial xylose residue. *Bioresource Technology* 352: 127065. https://doi.org/10.1016/j.biortech.2022.127065.

92 Akim, L.G., Argyropoulos, D.S., Jouanin, L. et al. (2001). Quantitative ^{31}P NMR spectroscopy of lignins from transgenic poplars. *Holzforschung* 55: 386–390. https://doi.org/10.1515/HF.2001.064.

93 Peng, X.-P., Bian, J., Yao, S.-Q. et al. (2021). Effects of P-coumarate 3-hydroxylase downregulation on the compositional and structural characteristics of lignin and hemicelluloses in poplar wood (*Populus alba* × *Populus glandulosa*). *Frontiers in Bioengineering and Biotechnology* 9: 790539. https://doi.org/10.3389/fbioe.2021.790539.

94 Xu, L.-H., Ma, C.-Y., Zhang, C. et al. (2022). Ultrafast fractionation of wild-type and CSE down-regulated poplars by microwave-assisted deep eutectic solvents (DES) for cellulose bioconversion enhancement and lignin nanoparticles fabrication. *Industrial Crops and Products* 176: 114275. https://doi.org/10.1016/j.indcrop.2021.114275.

95 Chen, T.-Y., Wang, B., Wu, Y.-Y. et al. (2017). Structural variations of lignin macromolecule from different growth years of Triploid of *Populus tomentosa* Carr. *International Journal of Biological Macromolecules* 101: 747–757. https://doi.org/10.1016/j.ijbiomac.2017.03.146.

96 Wang, H.-M., Ma, C.-Y., Li, H.-Y. et al. (2019). Structural variations of lignin macromolecules from early growth stages of poplar cell walls. *ACS Sustainable Chemistry & Engineering* 8 (4): 1813–1822. https://doi.org/10.1021/acssuschemeng.9b05845.

97 Barrelle, M., Fernandes, J.C., Froment, P., Lachenal, D. 1992. An approach to the determination of functional groups in oxidized lignins by ^{19}F NMR. 413–424. https://doi.org/10.1080/02773819208545789

98 Sevillano, R., Mortha, G., Barrelle, M., and Lachenal, D. (2001). ^{19}F NMR spectroscopy for the quantitative analysis of carbonyl groups in lignins. *Holzforschung* https://doi.org/10.1515/hf.2001.048.

99 Constant, S., Lancefield, C.S., Weckhuysen, B.M., and Bruijnincx, P.C. (2017). Quantification and classification of carbonyls in industrial humins and lignins by ^{19}F NMR. *ACS Sustainable Chemistry & Engineering* 5 (1): 965–972. https://doi.org/10.1021/acssuschemeng.6b02292.

100 Esakkimuthu, E.S., Marlin, N., Brochier-Salon, M.-C., and Mortha, G. (2020). Study of the reactivity of lignin model compounds to fluorobenzylation using ^{13}C and ^{19}F NMR: application to lignin phenolic hydroxyl group quantification by ^{19}F NMR. *Molecules* 25 (14): 3211. https://doi.org/10.3390/molecules25143211.

101 Kenny, J.K., Medlin, J.W., and Beckham, G.T. (2023). Quantification of phenolic hydroxyl groups in lignin via ^{19}F NMR spectroscopy. *ACS Sustainable Chemistry & Engineering* 11 (14): 5644–5655. https://doi.org/10.1021/acssuschemeng.3c00115.

102 Holtman, K.M., Chang, H.M., and Kadla, J.F. (2007). An NMR comparison of the whole lignin from milled wood, MWL, and REL dissolved by the DMSO/NMI procedure. *Journal of Wood Chemistry and Technology* 27 (3–4): 179–200. https://doi.org/10.1080/02773810701700828.

103 Yelle, D.J., Ralph, J., Lu, F., and Hammel, K.E. (2008b). Evidence for cleavage of lignin by a brown rot basidiomycete. *Environmental Microbiology* 10 (7): 1844–1849. https://doi.org/10.1111/j.1462-2920.2008.01605.x.

104 Yelle, D.J., Wei, D., Ralph, J., and Hammel, K.E. (2011). Multidimensional NMR analysis reveals truncated lignin structures in wood decayed by the brown rot basidiomycete *Postia placenta*. *Environmental Microbiology* 13 (4): 1091–1100. https://doi.org/10.1111/j.1462-2920.2010.02417.x.

105 Wen, J.-L., Xue, B.-L., Xu, F., and Sun, R.-C. (2012). Unveiling the structural heterogeneity of bamboo lignin by in situ HSQC NMR technique. *Bioenergy Research* 5: 886–903. https://doi.org/10.1007/s12155-012-9203-5.

106 Colombini, M.P., Lucejko, J.J., Modugno, F. et al. (2009). A multi-analytical study of degradation of lignin in archaeological waterlogged wood. *Talanta* 80 (1): 61–70. https://doi.org/10.1016/j.talanta.2009.06.024.

107 Salanti, A., Zoia, L., Tolppa, E.-L. et al. (2010). Characterization of waterlogged wood by NMR and GPC techniques. *Microchemical Journal* 95 (2): 345–352. https://doi.org/10.1016/j.microc.2010.02.009.

108 Qu, C., Kishimoto, T., Kishino, M. et al. (2011). Heteronuclear single-quantum coherence nuclear magnetic resonance (HSQC NMR) characterization of acetylated fir (*Abies sachallnensis* MAST) wood regenerated from ionic liquid. *Journal of Agricultural and Food Chemistry* 59 (10): 5382–5389. https://doi.org/10.1021/jf200498n.

109 Yelle, D.J., Ralph, J., and Frihart, C.R. (2008a). Characterization of nonderivatized plant cell walls using high-resolution solution-state NMR spectroscopy. *Magnetic Resonance in Chemistry* 46 (6): 508–517. https://doi.org/10.1002/mrc.2201.

110 Ostlund, Å., Lundberg, D., Nordstierna, L. et al. (2009). Dissolution and gelation of cellulose in TBAF/DMSO solutions: the roles of fluoride ions and water. *Biomacromolecules* 10 (9): 2401–2407. https://doi.org/10.1021/bm900667q.

111 Qu, C., Kishimoto, T., Ogita, S. et al. (2012). Dissolution and acetylation of ball-milled birch (*Betula platyphylla*) and bamboo (*Phyllostachys nigra*) in the ionic liquid [Bmim] Cl for HSQC NMR analysis. *Holzforschung* 66 (5): 607–614. https://doi.org/10.1515/hf.2011.186.

112 Chen, F., Tobimatsu, Y., Havkin-Frenkel, D. et al. (2012). A polymer of caffeyl alcohol in plant seeds. *Proceedings of the National Academy of Sciences* 109 (5): 1772–1777. https://doi.org/10.1073/pnas.1120992109.

113 Kim, H. and Ralph, J. (2014). A gel-state 2D-NMR method for plant cell wall profiling and analysis: a model study with the amorphous cellulose and xylan from ball-milled cotton linters. *RSC Advances* 4 (15): 7549–7560. https://doi.org/10.1039/c3ra46338a.

114 Glas, D., Van Doorslaer, C., Depuydt, D. et al. (2014). Lignin solubility in non-imidazolium ionic liquids. *Journal of Chemical Technology & Biotechnology* 90 (10): 1821–1826. https://doi.org/10.1002/jctb.4492.

115 Bali, G., Foston, M.B., O'Neill, H.M. et al. (2013). The effect of deuteration on the structure of bacterial cellulose. *Carbohydrate Research* 374: 82–88. https://doi.org/10.1016/j.carres.2013.04.009.

116 Foston, M., Hubbell, C.A., Samuel, R. et al. (2011). Chemical, ultrastructural and supramolecular analysis of tension wood in *Populus tremula* × *alba* as a model substrate for reduced recalcitrance. *Energy & Environmental Science* 4 (12): 4962–4971. https://doi.org/10.1039/c1ee02073k.

117 Foston, M., Samuel, R., and Ragauskas, A.J. (2012a). 13 C cell wall enrichment and ionic liquid NMR analysis: progress towards a high-throughput detailed chemical analysis of the whole plant cell wall. *Analyst* 137 (17): 3904–3909. https://doi.org/10.1039/c2an35344j.

118 Foston, M.B., McGaughey, J., O'Neill, H. et al. (2012b). Deuterium incorporation in biomass cell wall components by NMR analysis. *Analyst* 137 (5): 1090–1093. https://doi.org/10.1039/c2an16025k.

119 Jiang, N., Pu, Y., and Ragauskas, A.J. (2010). Rapid determination of lignin content via direct dissolution and ^{1}H NMR analysis of plant cell walls. *ChemSusChem* 3 (11): 1285–1289. https://doi.org/10.1002/cssc.201000120.

120 Jiang, N., Pu, Y., Samuel, R., and Ragauskas, A.J. (2009). Perdeuterated pyridinium molten salt (ionic liquid) for direct dissolution and NMR analysis of plant cell walls. *Green Chemistry* 11 (11): 1762–1766. https://doi.org/10.1039/b913609f.

121 Jiang, N. and Ragauskas, A.J. (2011). Perdeuterated pyridinium ionic liquids for direct biomass dissolution and characterization. In: *Ionic Liquids: Applications and Perspectives*. IntechOpen.

122 Li, M., Foster, C., Kelkar, S. et al. (2012). Structural characterization of alkaline hydrogen peroxide pretreated grasses exhibiting diverse lignin phenotypes. *Biotechnology for Biofuels* 5 (1): 38. https://doi.org/10.1186/1754-6834-5-38.

3

Advances in the Molar Mass and Functionality Analysis of Lignin

Oliver Musl and Antje Potthast

Department of Chemistry, Institute of Chemistry of Renewable Resources, University of Natural Resources and Life Sciences, Vienna, Konrad-Lorenz-Strasse 24 A-3430, Tulln, Austria

3.1 Introduction

The term *lignin* comprises actually a rich and diverse class of biopolymers that show a tremendous variability in their chemical composition. Lignins are major cell wall constituents in plants that perform many different physiological functions, from water barrier to resistance against microbial decay. Naturally, the monomeric composition and thus the molecular shape and structure of the lignin are strongly dependent on the botanical origin, i.e. plant genus, tissue, cell wall layer, development stage, and so on. So far, a real plethora of possible monomers has been identified to take part in lignin biosynthesis to form the irregular bonding pattern characteristic of native lignin [1–4]. Consequently, lignins are considered one of nature's most intricate polymers due to their peculiar chemical structure.

Industrial exploitation of lignocellulosic biomass usually aims at the removal of lignin from the plant feedstock to obtain polysaccharide-rich fractions that are further utilized as materials or in fermentation. The extraction of lignin during biomass processing leads to degradation and rearrangement of the native structure and to the introduction of new structural motifs [5]. Hence, more or less pronounced changes in monomeric composition, bonding pattern, branching structure, molar mass, and functionality are to be expected depending on the process technology and its parameters. Concurrently, structural changes crucially affect the polymer properties of lignin, i.e. solubility, compatibility, accessibility, reactivity, adsorptive capacity, chemical resistance, thermal stability, mechanical strength, and many more. For this reason, accurate characterization of the molecular structure of (technical) lignins is of vital importance for their valorization in a future bioeconomy. Currently, molar mass and functionality are the most prominently analyzed structural properties of lignins because of their immediate effect on many properties and the readily accessible analytical infrastructure to many researchers. In contrast, analysis of the monomeric composition, bonding, and branching patterns struggles

Lignin Chemistry: Characterization, Isolation, and Valorization, First Edition.
Edited by Yuhe Liao and Bert F. Sels.

with costly equipment, required personal with specialized knowledge, and a lack of fundamental knowledge on technical lignin structure and composition in general. Consequently, far more structure–property relationships are hypothesized and theorized for molar mass and functionality, emphasizing their importance in guiding current processes and material developments.

In this chapter, we focus on current practice and recent advances in molar mass and functionality analysis of (technical) lignins. General instructions for proper lignin analysis are also provided to help researchers in their quest for accurate measurement data. Section 3.6 is dedicated to the interrelationship between molar mass and functionality; a fact that still receives too little attention in lignin analytics.

3.2 General Aspects

Proficient lignin analysis starts with a comprehensive review of the sample's history. Lignin samples can come in many different shapes and forms. Since their chemical and structural nature is strongly influenced by their botanical origin, the applied extraction (pulping) process, and additional purification, fractionation, or modification steps, it is essential to know where the samples come from and what has been done to them. The nature of the native lignin structure is studied either *in-situ* (i.e. by FTIR, Raman spectroscopy, etc.) or after mild extraction (i.e. milling, cellulolytic enzymes, etc.) from plant biomass, which aims to preserve the original structure as much as possible [6–11]. However, the vast majority of samples originate from industrial pulping processes, as the huge quantities of technical lignin they produce are of economic and technical interest. In this sense, lignin samples can be divided into close-to-native and technical lignins.

Industrial pulping is a global affair and to produce the rich portfolio of cellulosic products we consume today, pulp mills make use of different biomass feedstock, pulping chemistry, and changes in many other process variables. The paramount motive is to efficiently remove lignin from the cellulosic polysaccharide fractions, rendering technical lignins quite diverse and structurally different from their native counterpart. In addition, technical lignin samples are collected from industrial process streams such as spent pulping liquor or other process effluents (e.g. from oxygen delignification, bleaching, fermentation, etc.). Hence, their consistency (i.e. solids content) and degree of purity as well as their chemical nature can vary wildly depending on the sample collection site. For all these reasons, to know the origin of a technical lignin is of particular importance for further analyses. Once collected, (liquid) samples should be stored cool or even frozen to prevent degradation or fouling. While lignin is not particularly susceptible to microbial attack, carbohydrate-based impurities are.

An important prerequisite for accurate analysis is to provide the lignin sample in the required purity, hence, adequate sample pretreatment and preparation is usually a must. Although there is a strong desire to save time and effort by avoiding sample preparation at all, direct analysis of components in the liquor/effluent is quite challenging due to their complex matrix. In addition to lignin, spent pulping liquors (or other process effluents) contain a vast mixture of components, including inorganic salts (i.e. from pulping/bleaching chemicals or biomass itself), residual poly- and oligosaccharides, and small organic compounds (i.e. extractives, monomeric sugars, degradation products, etc.). While the content of inorganics or dry solids is easily accessible by combustion or simple drying, analysis of the organic by-products is more difficult. In most cases, the success of a direct analysis method relies on an integrated separation step to isolate the compounds of interest from the liquor/effluent. For instance, high-performance thin-layer chromatography can be used for separation prior to the direct quantification of lignin and other components in spent liquors using densitometry and multivariate calibration [12]. In a fast and reliable way, important information on the liquor composition of multiple samples is gathered in a single measurement, while avoiding conventional conversion or isolation processes. The detection and quantification of small organic compounds (i.e. degradation products from pulping or modification processes) directly from liquor or effluent samples is a difficult task as they are highly diverse and occur only in traces. However, the implementation of a microscale precipitation step provides a purified supernatant available for derivatization and subsequent GC–MS analysis [13]. For analysis of volatile compounds, sample preparation can be limited to a minimum using multiple headspace sampling in solid-phase microextraction GC–MS analysis [14]. In asymmetric flow field-flow fractionation (AF4), the cross-flow can be used to purify the liquor sample prior to the molar mass determination of the lignin [15]. Recently, simultaneous quantification and molar mass determination of lignin contained in black liquors was demonstrated using an aqueous SEC system [16]. Despite progress, direct analysis of liquors and effluents remains challenging and limited to certain applications. However, *in situ* analysis of liquors/effluents by spectroscopic techniques still poses a hot topic for the use in industrial plants for real-time process control.

For the mentioned reasons, liquor samples usually undergo some form of selective fractionation as a means of purification. In general, the method of choice for lignin isolation is highly dependent on the nature of the technical lignin. On industrial scale, acid precipitation (i.e. Lignoboost® or LignoForce System™) and membrane filtration are the only technologies currently in use for the recovery of technical lignins [17–20]. On lab scale, acid precipitation using mineral acids is used for lignins from alkaline processes (i.e. soda or kraft lignins), while ultrafiltration or resin-assisted extraction is used for the water-soluble lignosulfonates [21]. However, isolation methods may be adapted to recover specific lignin fractions or to meet specific sample requirements (e.g. for organosolv, biorefinery, or modified lignins). In addition, ion exchange or protonation is often used to increase purity or to render the lignin sample more soluble in organic solvents. Independent of the isolation method, the purity and composition of the recovered lignin must be scrutinized prior to further analysis, since a certain amount of impurities must always be expected, especially in the case of industrially recovered lignins. In this respect, attention must be paid to lignin carbohydrate complexes (LCCs) where hemicelluloses are covalently linked to lignin, preventing their removal from the sample. Sample purity is often determined gravimetrically by drying (e.g. water/solids content), combustion (e.g. content of inorganics, elemental analysis), or digestion (e.g. KLASON lignin). The content of extractives can be determined by extraction with apolar solvents (e.g. hexane). Determination of the carbohydrate content and composition is more sophisticated. Typically, the lignin sample is subjected to chemical degradation (e.g. total hydrolysis or acidic methanolysis) and the products are then analyzed by ion exchange LC or after derivatization by GC–MS [22]. Extensive sample purification may be time-consuming and costly, but it is in most cases worthwhile, because impurities do not only distort the lignin content but can also have a negative impact on sample solubility or even directly affect certain measurements. This is especially true when derivatization is used in sample preparation. As an example, contamination with carbohydrates can severely influence the determination of hydroxy group contents by ^{31}P NMR, resulting in significant overestimation of lignin functionality. Thus, comprehensive analysis of the sample composition allows identifying (the causes of) measurement failures, and moreover, more accurate conclusions can be drawn from further analyses. Finally, the comprehensive analysis of liquor/effluent composition may provide valuable information to identify potentials for their successful economic exploitation.

3.3 Dissolution of Lignin Samples

For a number of measurements, complete dissolution of the lignin sample is a requirement for proper analysis. As a result of the applied pulping chemistry, technical lignins show tremendous differences in solubility due to differences in their chemical structure. Technical lignins show an amphiphilic behavior in solutions due to the various polar groups located on the otherwise hydrophobic backbone. However, the nature and amounts of functional groups present have a large impact on the actual behavior in solution. For example, the pK_a of phenolic hydroxy groups (around 10.5) renders alkali lignins (i.e. soda or kraft lignins) susceptible to partial precipitation in aqueous solutions below pH 10. In contrast, lignosulfonates are much less affected by pH due to the presence of sulfonic acid groups with a very low pK_a (around 1.5) [23]. The amphiphilic lignin structure is also the reason why binary or tertiary solvent systems often achieve greater dissolution than single solvents, as they respond more appropriately to the different polarities within the lignin structure (i.e. preferential solvation) [24, 25]. Recently, various ionic liquids, deep eutectic solvents, and (acid) hydrotropes have been proposed for lignin dissolution, used alone or proportionally as dissolution enhancer [26–35]. Whether a sample is perfectly dissolved is sometimes difficult to judge by the naked eye. Hence, turbidity or dynamic light scattering (DLS) measurements are commonly applied for this purpose.

Despite all efforts, partial sample insolubility due to self-aggregation or agglomeration are frequently observed phenomenon when working with lignins. Self-aggregation occurs due to various intermolecular interactions (e.g. non-covalent π–π interactions or hydrogen bonding), which are influenced by lignin concentration, pH, temperature, and the presence of impurities [36–43]. Moreover, ionic strength has a particular impact on the solubility of lignins, as it directly influences electrostatic and ionic interactions within the solution (e.g. screening of charges). High ionic strength may promote hydrophobic interactions, resulting eventually in lignin precipitation (i.e. salting out) [44, 45]. The magnitude of this antichaotropic effect is determined by the nature of the salt ions and can be described by the lyotropic or Hofmeister series [46, 47]. Finally, lignin dissolution is also a time-dependent process, thus complete sample dissolution may require several days on a lab shaker. Extensive sonication should be avoided to prevent sample degradation and thus changes to the lignin structure and properties [48].

In addition to the specific solubility of different lignins, solvent systems must meet the requirements of the particular measurement. For this reason, developing new solvent systems is a very demanding task. Frequently used solvent systems in lignin analysis are NaOH solutions, ethanol or acetone or dioxane/water, DMSO/LiBr, THF/LiBr, $CHCl_3$/pyridine, DMF/pyridine, etc. In some cases, derivatization is used to enhance lignin solubility; most prominently acetobromination or acetylation prior to dissolution in THF [49]. Derivatization may have an influence on the measurement, which has to be considered in data evaluation [50–52]. For LC techniques that employ high salinity to facilitate adsorption/retention of lignins (i.e. hydrophobic interaction chromatography, HIC), eluent composition, and sample dissolution are delicate tasks, as lignin precipitation prior to injection must be avoided [53].

3.4 Molar Mass Analysis

Molar mass is an important polymer property that affects the processability, reactivity, solubility, and mechanical properties of lignin materials. In addition, most structural properties are interrelated with molar mass due to degradation and transformation processes during pulping (i.e. functionality, molecular architecture, chemical composition, and bonding pattern). For polymers, molar mass is a distributed property, as they are composed of macromolecules with different molar mass. The molar mass distribution (MMD) of a polymer is typically determined by analytical separations based on molecular size, e.g. size-exclusion chromatography (SEC) or field-flow fractionation (FFF). Mean values calculated from the MMD such as the number-average molar mass (M_n) and the weight-average molar mass (M_w) are commonly used as a reference for comparison (see Formula (3.1)). Evidently, mean values do not suffice to adequately characterize dispersed polymer properties. Hence, the dispersity $Đ_M$ as a measure of the width of the distribution (i.e. heterogeneity in molar mass) is often also calculated from the ratio of M_w to M_n. Lignins – like many other natural polymers – show a broad distribution and thus $Đ_M$ in molar mass. Chemically altered (technical) lignins can even exhibit complex multimodal MMDs, although this could also be a result of inadequate separations (e.g. due to column interactions in SEC).

$$M_n = \frac{\sum n_i M_i}{\sum n_i}$$

$$M_w = \frac{\sum w_i M_i}{\sum w_i} = \frac{\sum n_i M_i^2}{\sum n_i M_i}$$

$$Đ_M = \frac{M_w}{M_n} \tag{3.1}$$

3.4.1 Size-Exclusion Chromatography (SEC)

In the last decades, SEC has matured into the reference technique for the molar mass analysis of lignins. However, a standardized method has not yet been established due to the difficulty of designing a universal SEC system to analyze the wide variety of lignins [54]. Hence, a number of reliable SEC systems emerged to allow analysis of virtually any (technical) lignin. In SEC, macromolecules are separated according to their hydrodynamic volume in the mobile phase and the accessible pore size distribution of the stationary phase. The separation mechanism is based on the fact that small molecules can penetrate deeper into the pores of the stationary phase than large molecules, causing a shift in elution time due to molecular size. Ideally, the separation is exclusively governed by entropic effects, which impose high demands on the eluent strength and the polarity of the stationary phase to avoid undesired secondary interactions.

In SEC, mobile phases must be thermodynamically good solvents to allow optimal dissolution of the analyte. Moreover, salts (e.g. LiBr, LiCl, etc.) or modifiers (e.g. alcohols, surfactants, etc.) are often added to the eluent to effectively suppress undesired electrostatic or ionic interactions between the analyte and the stationary phase. It is important to note that the eluent composition must be kept absolutely constant to maintain consistent separation conditions throughout the analysis, thus isocratic elution is the preferred mode in SEC. Otherwise, undesired shifts in elution time are inevitable. Frequently used eluents in lignin SEC are aqueous/alkaline solutions (e.g. NaOH, NH_4OH, and $NaNO_3$), or organic solvents (e.g. DMSO/LiBr, THF/LiBr, and DMF/LiBr) [54]. As mentioned before, certain sample pretreatments may be required to achieve complete dissolution of the lignins in the eluent, for instance, protonation of lignosulfonates for DMSO/LiBr, or lignin derivatization by acetobromination or acetylation for THF/LiBr [49, 55]. In recent years, DMSO/LiBr has emerged as the preferred eluent in SEC, as DMSO appears to be an optimal single-component solvent for lignins [25]. Hence, the majority of lignins are readily soluble in DMSO without the need for time-consuming pretreatments or sample derivatization.

In SEC, the polarity of the stationary phase must correspond with the polarity of the analyte and the eluent to avoid undesired adsorption phenomena. For aqueous and alkaline eluents, stationary phases based on polystyrene sulfonate (PSS) are highly suitable and can be recommended. In case of organic solvents, apolar polystyrene-divinylbenzene (PS-DVB) copolymers are the preferred choice for stationary phases when using THF as eluent, whereas hydroxylated (medium polarity) PS-DVB stationary phases pose a better fit for polar solvents such as DMSO or DMF. In addition to the compatibility in polarity, the pore size distribution of the stationary phase must match the size distribution of the targeted lignin to avoid a loss in resolution due to the exclusion of large molecules or total permeation of small molecules. Therefore, combination of columns with specific pore sizes (100–100 000 Å) or mixed bed columns are used to effectively cover the size distribution of the lignin sample. Note that the specified molar mass range for stationary phases is often determined using synthetic standards (e.g. polystyrene, PS), thus deviations are to be expected for lignins.

A drawback of conventional SEC is certainly the long run time of up to one hour and more per sample. For this reason, various high-speed methods have been developed to effectively reduce analysis time with limited loss in resolution. Designated highspeed columns take advantage of a larger column diameter to allow higher flow rates and thus shorter analysis time at the expense of resolution, hence they are mainly used in 2D applications [57]. In addition, novel stationary phases based on rigid ethylene bridged hybrid (BEH) silica particles pose a viable alternative to conventional gel-like stationary phases [56, 58]. In SEC, the separation efficiency or resolution is typically determined by the accessible pore volume of the stationary phase and thus effectively by column length (or number of columns). However, BEH columns contain particles of very small size (<3 μm) compared to conventional SEC materials (5–10 μm), leading to a tremendous increase in separation efficiency due to various effects [59]. Consequently, BEH columns are able to sacrifice column length while maintaining the required resolution, thereby significantly shortening the analysis time (see Figure 3.1). However, the high back pressure (up to 800 bar) of BEH columns requires custom LC equipment and changes in eluent composition (e.g. addition of salt to avoid freezing of the eluent) or analysis conditions (e.g. high temperature to reduce the eluent viscosity) [56].

Another way to reduce analysis time is to skip sample preparation as much as possible. Typically, SEC systems are sensitive to impurities and complex sample mixtures, which are often a cause of short column lifetimes and compromised separations. However, it is still possible to devise a viable SEC system for the direct analysis of unpurified samples (e.g. black liquor), albeit at a loss in precision and accuracy (see Figure 3.1) [16]. Above all, a high elution strength of the mobile phase for all constituents is paramount to avoid undesired interactions and irreversible sample adsorption. In addition, selective concentration detection (i.e. UV absorbance) must be employed to allow accurate quantification and molar mass determination without interference from other sample components (e.g. sugars, salts, etc.).

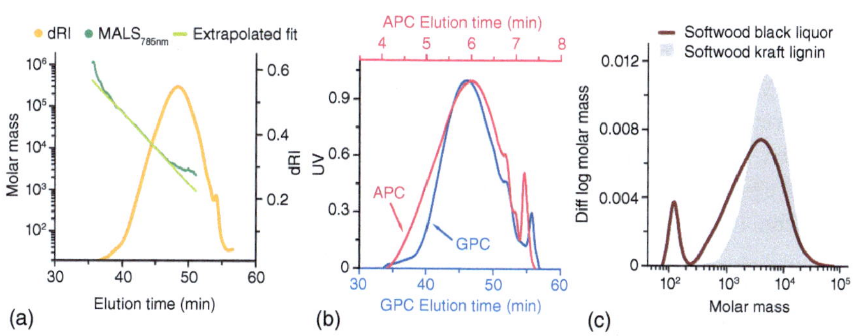

Figure 3.1 Advances in molar mass analysis. (a) Absolute molar mass determination by $MALS_{785nm}$ with correction by signal extrapolation [55]. (b) Advanced polymer chromatography (APC) using rigid BEH particles to significantly reduce analysis time compared to regular gel permeation chromatography (GPC) columns [56]. (c) Direct analysis of black liquor avoids losses during sample preparation for a more authentic MMD [16].

In this way, a powerful high-throughput method can be devised that allows fast screening of complex samples (for quality control purposes). This approach could also improve characterization of samples that suffer from heavy losses during sample preparation (e.g. acid- or water-soluble fractions).

Finally, a few application tips to secure robust and reproducible SEC analysis, especially with regard to lignins. For the sake of convenience, SEC is typically performed at room temperature (20–35 °C). Although their impact on the separation is often negligible, fluctuations in temperature have a tremendous influence on detection, as analyte properties are often very sensitive to small temperature changes (e.g. refractive index [RI]). Therefore, minute temperature control of the system (e.g. column oven, heatable flow cells in detectors, etc.) is mandatory for good SEC analysis. Some lignins tend to form aggregates when dissolved in the eluents. If sample impurities are not the cause, an increase in ionic strength may mitigate aggregation phenomena during dissolution. Intense sonication should be avoided due to the possible degradation of the sample [48]. In any case, it is advisable to filter samples through 0.45 μm PTFE or regenerated cellulose filters (depending on the polarity of the lignin) prior to injection to protect the SEC system from possible aggregates. In addition, proper filters and pre-columns should be installed in the SEC system to protect the separation columns from blockage. If possible, it is also recommended to determine the sample recovery rate when measuring new lignins to detect any undesired adsorption to the column or losses in the filtration step, which would compromise the whole analysis. It is very important to keep injection parameters (i.e. injected volume or mass) constant, as they directly affect the separation and detection [59]. For this reason, sample concentration and purity are critical to ensure consistent injection conditions. Moreover, contaminants may lead to a shift in elution time (e.g. high salt contents) or clogging of pores [57].

3.4.2 Asymmetric Flow Field-Flow Fractionation (AF4)

AF4 is a relatively new separation technique for lignins and lignin particles. It is a channel-based fractionation method that separates molecules or particles based on differences in their mobilities under the force of a perpendicular eluent flow (i.e. cross-flow). In the narrow separation channel of AF4, the elution of an analyte is primarily governed by the parabolic flow profile of the laminar flow in the axial direction with analytes closer to the center moving faster than those at the bottom. However, the relative position of a certain analyte in this flow profile (and thus their mobility) derives from two opposing forces: the cross-flow and the counteracting diffusion rate of the analyte. Therefore, the order of elution in AF4 is reversed compared to SEC, as diffusion coefficients decrease with the hydrodynamic size of the analyte. As a result, the elution behavior in AF4 is determined by the interplay of axial and cross-flow, thus fluctuations in their flow rates critically affect the elution pattern of the analyte.

In contrast to SEC, omitting the stationary phase minimizes the risk of undesired analyte interactions and shear degradation of large molecules and particles. However, the semipermeable membrane at the bottom of the channel must be selected carefully as well. Unfortunately, the available membranes for AF4 are only suited for moderate eluent conditions (e.g. aqueous buffer solutions, mild organic solvents), which limits the application of AF4 to the water-soluble lignins and lignosulfonates. A low molar mass cut off of the membrane prevents the loss of the low molar mass fraction but increases the backpressure in the system. In addition, the membrane material must be of medium polarity to avoid electrostatic repulsion or irreversible adsorption (i.e. membrane fouling) of the lignin. The addition of salts to the eluent can help to mitigate electrostatic effects. Regenerated cellulose membranes with a low molar mass cut off (3kDa) proved to be suitable and can be recommended for AF4 of lignosulfonates [15]. Polyether sulfone (PES) membranes show electrostatic repulsion and are therefore only of limited use. For instance, electrostatic repulsion can prevent the loss of sample material when the cross-flow in AF4 is used for sample focusing and purification prior to the separation in molecular size [15].

3.4.3 Detection Systems

Typically, SEC systems employ concentration-sensitive detection systems based on RI or UV absorptivity (UV) due to their ease of application and inexpensiveness. Although RI is more common due to its universal detection, UV has the advantage of high sensitivity and selectivity for lignin. Consequently, sample impurities (e.g. carbohydrates in LCC) can be detected by comparing the RI and UV traces.

Concentration sensitive detectors in this case only provide a detector signal vs. elution time plot, which must yet be translated into the respective MMD. The actual concentration of the polymer is not needed at this point. Typically, calibration with synthetic polymer standards is used to establish a relationship between elution time and molar mass. Polystyrene (PS) or PSS standards are frequently used for this purpose, due to their structural similarities to lignins compared to other SEC standards (e.g. pullulan, dextran). However, the linear structure and uniform sulfonation of PS or PSS do not match the structural heterogeneity present in lignins (i.e. branching, heterogeneous composition, etc.). This leads to significant differences in their hydrodynamic size in solution and thus a systematic bias in the molar mass determination [55, 60]. Therefore, conventional calibration is only suitable for the relative comparison of structurally similar lignins. Absolute molar mass values remain inaccessible by conventional calibration, except correction factors are provided [55]. In response, the concept of universal calibration was developed to resolve the problem of the structural mismatch between polymer standards and the target analyte. It is based on the principle that the intrinsic viscosity of a polymer is directly correlated with the hydrodynamic size and the molar mass of the analyte [59, 61]. Therefore, the additional implementation of an online viscosity detector is required to establish a universal calibration. Universal calibration is regarded as an excellent alternative to conventional calibration for the analysis of copolymers with unknown chemical composition or unknown degree of branching; typical properties of lignins. However, universal calibration still remains an exotic topic in lignin SEC, partly due to the necessity of an online viscometer in combination with salt-containing organic solvents [62, 63]. The comparably low effect of lignin on viscosity, which even decreases with branching, are additional challenges for universal calibration [64–66].

Nevertheless, universal calibration is a valuable tool for the analysis of fluorescent polymers such as lignin when scattering techniques become unreliable [66].

For many research labs, the calibration approach will remain popular simply due to its cost-effectiveness and ease of application. However, calibrations are only valid for a particular set of columns, eluent, and temperature; any change in the system requires recalibration. Moreover, calibrations are also extremely sensitive to fluctuations in flow rate, temperature, and eluent composition, because of their influence on the elution time of the analyte. The structural heterogeneity, polyfunctionality, and irregular branching of lignins generate a complex relationship between molar mass and hydrodynamic size, partially invalidating established calibrations. This is especially important when comparing lignins of vastly different chemical nature. For this reason, modern SEC or AF4 analysis of lignins strives for direct detection of the molar mass to provide more robust information on the MMD. A popular approach to study the molar mass, molecular size, and shape of polymers in solution is the use of light scattering techniques, i.e., static light scattering (SLS) or DLS. In those experiments, the scattering intensity of a monochromatic light beam (i.e. laser) is measured when it passes through a solution containing a macromolecular analyte. In DLS, the time-dependent fluctuations in scattering intensity are studied, which allows the determination of the diffusion coefficient of macromolecules and consequently their hydrodynamic size, since they are inversely related. In contrast, SLS determines the average scattering intensity, which can be directly related to the molar mass when the concentration of the analyte is accurately known. For this reason, SLS is highly relevant as a molar mass-sensitive detection method in SEC and AF4, especially in its advanced variant of multi-angle light scattering (MALS).

For absolute molar mass analysis, concentration sensitive detectors are required in tandem with MALS. As mentioned before, the calculation of molar mass via MALS is very sensitive to the concentration of the analyte. Hence, specific proportionality factors must be determined to accurately correlate the detector response with the analyte concentration. After measuring a dilution series of the analyte, the specific RI increment $\partial n/\partial c$ or the molar absorptivity (ε) can be derived from the slope of the concentration vs. response plot for RI or UV detection, respectively. For the now widely used differential refractometers, detector constants are only required in the online approach. The $\partial n/\partial c$ and the ε are generally constant across the MMD of a polymer, however, changes are to be expected in the oligomeric region due to the increasing influence of end groups. This may pose a critical issue for lignins of low molar mass. Furthermore, every technical lignin has a slightly different $\partial n/\partial c$, due to their differences in chemical structure [67, 68]. The analysis of the $\partial n/\partial c$ requires a very pure, salt-free lignin sample. Note that the $\partial n/\partial c$ is also affected by dissolution time and aggregation state of the analyte [67]. Hence, which approach is best chosen for the determination of the $\partial n/\partial c$ depends also on the eluent used. In the case of polyelectrolytic analytes (i.e. lignin) and salt-containing eluents, establishing the osmotic equilibrium directly with the SEC column(s) can be considered the easiest way [69, 70]. Still, the user has to ensure that the sample concentration or injected mass is correct, also having in mind possible inaccuracies of the autosampler metering, incomplete solubility of the lignin sample, or removal of sample portions by filtering prior to injection, just to name a few.

In the recent past, important progress has been made to incorporate MALS in the molar mass analysis of lignins [15, 55, 60, 62, 63, 67, 71, 72]. However, the fluorescent and absorptive properties of the lignin structure pose major analytical challenges in MALS. Technical lignins tend to be strongly light-absorbing and highly fluorescent analytes due to the abundance of aromatic and chromophoric structures. In addition, drastic transformations in chemical pulping often lead to the formation of additional fluorescent structures, especially in the lower molar mass range. Elimination or suppression of fluorescent and absorption effects is essential for accurate molar mass determination with MALS [72, 73]. Otherwise, the absorption and re-emittance of light in form of fluorescence contributes to the scattering intensity, usually resulting in a significant overestimation of the molar mass. In the first step, fluorescence can be attenuated by lowering the excitation energy of the used laser light, thus the use of a near-IR MALS laser (785 nm) is strongly recommended [55]. In addition, narrow bandwidth fluorescence filters (±10 nm) should be employed in addition to mitigate the residual fluorescence of the analyte. Note that later in data processing, only the input of photodiodes with filters must be included in the molar mass calculation. However, despite all efforts, the interference from fluorescence cannot be completely eliminated because the low molar mass range exhibits a relatively weak light scattering intensity compared to the emitted fluorescence. This is especially true for the highly fluorescent kraft lignins. For such cases, correction procedures involving fluorometry may offer a solution in the future [72, 73]. As for light absorption, the scattering intensity must be based on the laser beam intensity after the flow cell to correct for the loss in incident energy due to the absorption of light by the analyte [55]. Otherwise, the scattering intensity and consequently the molar mass are severely underestimated, especially in the higher molar mass range. If all analytical measures are exhausted and distortions in the observed detector signal still persist, then fitting of the scattering intensity is a good approach to correct excessive fluorescence [55]. Here, the high molar mass range proves to be a reliable starting point for extrapolation due to the high ratio of scattering intensity to fluorescence (see Figure 3.1).

An analytically very intricate structural property of lignins is their degree of branching or cross-linking. The last decades have seen much debate between proponents of a branched and a linear lignin structure [1, 74]. Both camps based their arguments mainly on data from characterizations by chemical degradation or NMR experiments. The integration of molar mass data in this debate has always been difficult, mainly due to the limited availability of reliable molar mass data. The advent of MALS, and thus the accurate determination of the absolute molar mass of lignins, offered a great opportunity to reconsider the current perspective on lignin structure, concluding with a reaffirmation of the branched and cross-linked lignin model, at least in the case of softwood lignin [74]. MALS is also frequently used in polymer analysis to determine the degree of branching directly via the angular dependence of the scattered light [59, 61]. However, the molar mass range of lignins is typically too low to exhibit this angular dependence; lignins are therefore considered isotropic light scatterers.

MALS is slowly becoming the standard for absolute molar mass determination in lignin analysis. However, even MALS in combination with SEC or AF4 is prone to a number of errors that the user should be aware of. Above all, MALS requires a correct $\partial n/\partial c$ which should be measured in the system used rather than taken from literature. In addition, SEC measurements of lignins are also known to have an error in reproducibility up to 15% in intra- and interlaboratory comparisons based on M_n and M_w values [54]. Differences in the setup of the SEC system (i.e. choice of eluent or columns), the detector sensitivity (i.e. UV or RI), and user dependency in terms of peak assignment and curve fitting are plenty of sources to introduce variability in the obtained molar mass values.

In general, molar mass determination in combination with LC can be conducted by several other techniques as well, for instance, analytical ultracentrifugation (i.e. sedimentation velocity or equilibrium method) or mass spectroscopy with a number of restrictions (i.e. MALDI TOF) [34, 75–77]. But for now, in terms of molar mass determination, they serve only as reference methods to validate findings from MALS.

3.5 Functional Group Analysis

Lignins are considered polyfunctional polymers that show a complex functional group composition, as the nature and content of their functional groups can vary widely. The native lignin structure boasts a variety of functionalities, including aliphatic, and phenolic hydroxy groups, carboxylic acid groups, aldehydes, ketones, and methoxy groups. However, the native lignin structure undergoes severe transformations during chemical pulping, leading to changes in the functional group contents and in some cases to the introduction of new functional groups (e.g. sulfonic acid groups, ethoxy groups, etc.). Moreover, modification or fractionation processes can be used to alter the functional group composition of a technical lignin to fit one's needs.

Since functionality has a profound influence on process and material qualities, the accurate determination of functional group contents is of key importance. Therefore, a wide range of different approaches has been developed to address many different functional groups of lignins. Spectroscopic techniques enjoy great popularity (e.g. FTIR, NIR, ^{1}H-NMR, ^{13}C-NMR), as they preserve the sample (nondestructive) and provide a high depth of information. In principle, the acquisition of a single spectrum may be sufficient to collect all the information on the functional group contents. However, the interpretation of lignin spectra poses a quite challenging task (e.g. peak assignment, signal overlap, etc.), which often requires chemometric modeling to obtain reliable values. Moreover, the validity of these chemometric models often depends on accurate reference values that are provided by alternative approaches such as wet chemistry (e.g. derivatization, elimination reactions, acid-base reactions, etc.) or elemental analysis (e.g. sulfur content). For this reason, wet chemical techniques are still commonplace in lignin analysis, despite the effort and skill they require. In recent years, a lot of effort has been made to standardize functional group analysis and to establish certain methods as good practice due to their superior accuracy and precision. The hydroxy group determination via ^{31}P NMR is a prime example in this regard [50, 78, 79]. Since the growing demand for analytical data is putting pressure on existing analytical facilities, the development of high-throughput methods has also attracted a lot of attention. High-throughput methods must be capable of processing hundreds of samples in an efficient and reliable manner. In order to achieve this goal, high-end optimization of analytical conditions must be accompanied by rigorous streamlining of every process step in the analysis procedure. In many cases, the automatization of sample handling, pretreatment, digestion, and measurement is pushed to the extreme by the use of robotics. In general, elimination reactions are well suited for analytical upscaling to high throughput, as the volatile reaction products can be easily separated from the sample and subsequently analyzed by headspace GC–MS (see Figure 3.2). High-throughput determination of methoxy, ethoxy, and sulfonic acid groups is already feasible with this approach [80, 81]. Unfortunately, not every method is yet scalable for high-throughput routine analysis. In most cases, the diversity in lignin structure often necessitates the development of specialized analysis methods. In addition, special lab equipment and certain resources may not be readily available to every researcher. Therefore, some diversity in available methods is also necessary for analytical flexibility.

3.5.1 Heterogeneity in Functionality

In lignin literature, the term *heterogeneity* is often used in a very general context to emphasize lignin variability and to express the diversity in structure present: as a synonym for the irregularity of lignin structure per se. It is usually not considered a quantifiable parameter. However, in polymer science, the term *heterogeneity* is used to describe the correlation between two structural properties, e.g. the change in functionality with molar mass [82, 83]. For lignins, it is well-established that the functional group contents are strongly interrelated with molar mass. Typically, the lower molar mass range is associated with high functionality (i.e. high contents of phenolic hydroxy, carboxylic acid, or sulfonic acid groups) due to the conjunction of degradation and functionalization processes in industrial pulping. However, there is no standardized procedure for determining the heterogeneity in functionality of a lignin, i.e. the shift in functionality with molar mass. So far, lab-scale fractionation processes based on ultrafiltration or gradual precipitation/extraction (i.e. dependent on pH or solvent solubility) have been used to study the composition of lignins [84–88]. However, these fractionations are constrained by the limited number of available membrane cut-offs or suitable solvents, so they typically lack the required separation efficiency to provide accurate functional group profiles. One solution to this problem involves the use of preparative chromatography (see Figure 3.2). Recently, preparative SEC in combination with offline functional group determination was used to generate high-resolution profiles of the heterogeneity in functionality of various technical lignins [68]. Alternatively, derivatization for the

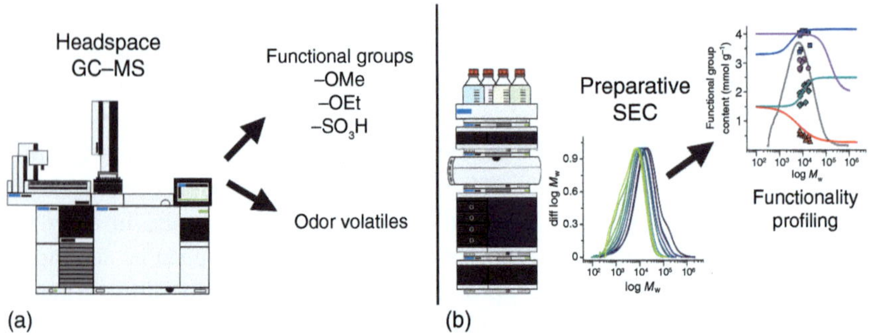

Figure 3.2 Advances in functional group analysis. (a) High-throughput analysis by automized sample digestion for GC–MS analysis [14, 80, 81]. (b) Preparative SEC with offline characterizations for functionality profiling of lignins [68].

selective detection of functional groups (e.g. fluorescent labeling) can also be used to determine the heterogeneity of a lignin in tandem with a regular SEC measurement [51, 52]. Information on the heterogeneity of lignins is useful for many reasons. Material applications often require lignins with defined properties for proper process control, thus lignins with high heterogeneity must be excluded or refined prior to use. Modification processes could be monitored and controlled at a very fundamental level when the change in heterogeneity is known. Finally, reliable information on the heterogeneity can be used to develop simulation tools that are vital for the design of efficient fractionation and purification strategies [68].

3.5.2 Dispersity in Functionality

Heterogeneity – in the aforementioned definition – conceptualizes functionality only as a function of molar mass. However, functionality is an independent structural property of lignin with its own distribution of values. The functionality (type) distribution (FTD) of lignins can be characterized by mean values (i.e. the number-averaged functionality, F_n; or the weight-averaged functionality, F_w) and the dispersity in functionality $Đ_F$, which are calculated analogous to their molar mass counterparts (see Formula (3.2)) [83]. The dispersity $Đ_F$ is a measure of the heterogeneity in functionality independent of molar mass. In contrast, local dispersity refers to the variance in functionality at a particular molar mass.

$$F_n = \frac{\sum n_i F_i}{\sum n_i}$$

$$F_w = \frac{\sum w_i F_i}{\sum w_i} = \frac{\sum n_i F_i^2}{\sum n_i F_i}$$

$$Đ_F = \frac{F_W}{F_n} \tag{3.2}$$

The current state-of-the-art in lignin analysis relies mainly on wet chemical and spectroscopic methods, which determine functionality as the average content of a sample; information on the underlying distribution is typically not accessible. To resolve the distribution of a structural property (e.g. molar mass, functionality), separation of the polymer constituents is necessary to adequately quantify them. In the first step, the heterogeneity in functionality can be determined by separations based on molar mass (see above), which already provides a higher analytical depth than conventional methods. However, information on the local dispersity in functionality is still inaccessible by such approaches. Therefore, separations based on functionality are required to address the dispersity in functionality in its full depth. In contrast to SEC, separations based on functionality make use of specific interactions between the analyte and the stationary phase (i.e. interaction chromatography). In addition, changes in eluent strength are used to promote or impede those interactions to influcnce the retention behavior of an analyte. Currently, interaction chromatography is not common in lignin analysis. In fact, only a few techniques have been pioneered, including (ion-pair) reversed-phase chromatography (RPC) and HIC [89–95]. In many cases, the separation focused on the chemical composition (i.e. lignin content, monomer composition, sugar composition, etc.) rather than the functionality of the lignin.

The goal of separation in functionality is always to produce fractions that are either enriched in or depleted of functionality. In the case of complex polymers (i.e. lignins), molar mass and functionality are often interrelated (i.e. heterogeneity in functionality), thus the evaluation of the separation sensitivity toward functionality can be very difficult. For instance, although homogenous in hydrophobicity, lignin fractions recovered by HIC still show a distribution in molar mass and functionality [96]. To verify a separation in functionality, the evaluation must therefore show a separation effect that exceeds the heterogeneity in functionality; the enrichment or depletion of a fraction in functionality independent of molar mass. For this purpose, a reliable simulation tool has been developed to estimate the functional group contents of lignin fractions based on their MMD [68]. Deviations between the estimated and the actual functional group content of a fraction could then indicate enrichment or depletion of functionality and thus separation based on functionality. So far, lignin separation based on functionality has only been demonstrated for HIC [96]. HIC was originally developed for protein purification [53]. However, it has also been used to study the hydrophobic composition of lignosulfonates, as hydrophobicity is a crucial parameter for their performance as surfactants and dispersing agents [95, 97]. In HIC, the high salt content of the mobile phase promotes interactions between hydrophobic ligands of the stationary phase and hydrophobic domains of the analyte, causing adsorption/elution based on the hydrophobicity of the analyte [53]. Consequently, the elution behavior of an analyte is governed by the opposing influence of functionality and molar mass. In the case of lignosulfonates, the elution in HIC is promoted by sulfonic acid group content and impeded by molar mass [96].

3.6 Analysis of Lignin Composition by Two-Dimensional Liquid Chromatography

In interaction chromatography, the heterogeneity of complex polymers (i.e. lignins) causes interferences in the separation mechanism, hence multidimensional approaches are often required to fully unravel their distributed properties. In two-dimensional LC, the fractions obtained from one separation (1st dimension) are subjected to another separation (2nd dimension) to achieve the effective separation of a complex analyte into its constituents. Offline combination of two LC techniques allows great flexibility due to optional concentration or purification steps, albeit at the expense of time and effort. Online coupling is a more elegant approach to combine the determination of two properties (e.g. FTD and MMD) in a single measurement. However, method development in online two-dimensional LC is a tedious process that must address the compatibility of the mobile phases, the sampling mode for the 2nd dimension (i.e. heart-cut or comprehensive transfer), the detection limits in the 2nd dimension, and the computational processing of the obtained measurement data. In addition, high-speed separations are required in 2nd dimension to maintain a reasonable total analysis time per sample. The key indicator for the success of a multidimensional separation is the orthogonality in separation. Orthogonality can be defined as the spread of the analyte across the separation plane and the extent of its occupied area. It is an expression of how well the composition of an analyte can be resolved by a particular setup. Orthogonal LC systems combine techniques with independent selectivity to achieve exclusive separations with respect to the targeted properties. However, note that the orthogonality in separation is closely correlated to the nature of the analyte; ultimately, it can only reproduce the analyte's inherent heterogeneity in composition.

In lignin analysis, two-dimensional LC is still a very young technique, and so far, only two systems have been proposed: ion-pair RPC × SEC and HIC × SEC [57, 90, 96]. In ion-pair RPC × SEC, orthogonality in separation remained unsatisfactory because of the strong influence of molar mass on the separation in ion-pair RPC [90]. Although the separation in HIC is also prone to molar mass interference, the HIC × SEC system proved to be more sensitive toward functionality and thus showed superior orthogonality compared to ion-pair RPC × SEC. The corresponding 2D plot of a HIC × SEC separation contains therefore a great depth of information about the concurrent FTD and MMD, providing a characteristic fingerprint for a particular lignin (see Figure 3.3). In this way, 2D-LC poses an exciting counterpart to multidimensional NMR (e.g. HSQC), which provides a characteristic fingerprint of the monomer composition and bonding pattern. In synthetic polymer analysis, hyphenation with information-rich detection systems (e.g. FTIR, NMR, etc.) provides information on functional group contents or chemical composition along with the 2D distribution plot [98]. Unfortunately, hyphenation is not readily available for lignin analysis due to the complications in spectra interpretation as mentioned earlier.

Although direct determination of functionality is not yet possible in 2D-LC, information on the functional composition can be already deduced from the shape of the

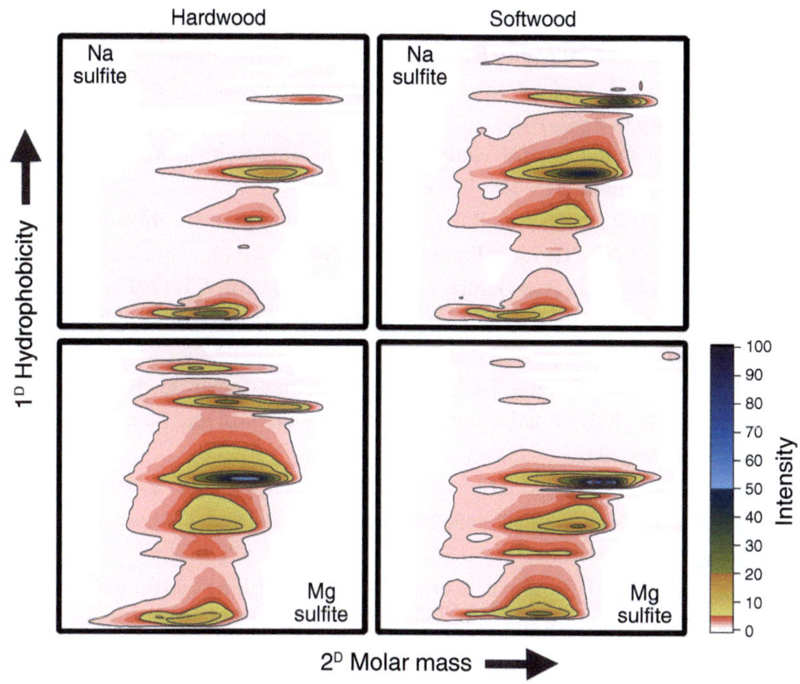

Figure 3.3 Characteristic fingerprints of the hydrophobic composition of four lignosulfonates as determined by two-dimensional HIC×SEC LC [57]. A large spread and an inclination toward the left (see Hardwood, Mg sulfite) indicates high dispersity in functionality $Đ_F$.

2D plot, as a large spread indicates higher local dispersity and consequently a higher $Đ_F$ (see Figure 3.3). Note that the separation in HIC is notably influenced by molar mass with higher molar mass fractions shifting toward later elution in HIC. Calibration with synthetic PSS standards can help to visualize this influence by constructing a reference line of constant functionality-to-molar mass ratio in the 2D plot [57]. The deviation of the sample pattern from this reference line can then be used as a measure of the heterogeneity in the functionality-to-molar mass ratio or $Đ_F$. In this way, dispersity parameters can be derived from the 2D plot, and semiquantitative analysis of the FTD becomes finally accessible for lignosulfonates. However, it remains a challenge to make this methodology accessible for other types of lignins.

References

1 Ralph, J., Lapierre, C., and Boerjan, W. (2019). Lignin structure and its engineering. *Current Opinion in Biotechnology* 56: 240–249.

2 Vanholme, R., De Meester, B., Ralph, J., and Boerjan, W. (2019). Lignin biosynthesis and its integration into metabolism. *Current Opinion in Biotechnology* 56: 230–239.

3 del Río, J.C., Rencoret, J., Gutiérrez, A. et al. (2022). *Unconventional Lignin Monomers – Extension of the Lignin Paradigm*, 1–39. Academic Press.

4 Ralph, J., Kim, H., Lu, F. et al. (2023). Lignins and lignification. In: *Recent Advances in Polyphenol Research*, 1–50. Wiley.

5 Berlin, A. and Balakshin, M. (2014). Industrial lignins. In: *Bioenergy Research: Advances and Applications*, 315–336. Elsevier.

6 Heitner, C., Dimmel, D., and Schmidt, J. (2010). *Lignin and Lignans: Advances in chemistry*. Boca Raton, FL: Taylor & Francis Group, LCC.

7 Calvo-Flores, F., Dobado, J., Isac-Garcia, J., and Martin-Martinez, F. (2015). *Lignin and Lignans as Renewable Raw Materials – Chemistry, Technology and Applications*. Singapore: Wiley.

8 Perera, P.N., Schmidt, M., Chiang, V.L. et al. (2012). Raman-spectroscopy-based noninvasive microanalysis of native lignin structure. *Analytical and Bioanalytical Chemistry* 402 (2): 983–987.

9 Capanema, E., Balakshin, M., Katahira, R. et al. (2015). How well do MWL and CEL preparations represent the whole hardwood lignin? *Journal of Wood Chemistry and Technology* 35 (1): 17–26.

10 Zinovyev, G., Sumerskii, I., Rosenau, T. et al. (2018). Ball Milling's effect on pine milled wood lignin's structure and molar mass. *Molecules* 23 (9): 2223.

11 Du, X., Gellerstedt, G., and Li, J. (2013). Universal fractionation of lignin-carbohydrate complexes (LCCs) from lignocellulosic biomass: an example using spruce wood. *Plant Journal* 74 (2): 328–338.

12 Khaliliyan, H., Schuster, C., Sumerskii, I. et al. (2020). Direct quantification of lignin in liquors by high performance thin layer chromatography-densitometry and multivariate calibration. *ACS Sustainable Chemistry & Engineering* 8 (45): 16766–16774.

13 Musl, O., Holzlechner, M., Winklehner, S. et al. (2019). Changing the molecular structure of kraft lignins—ozone treatment at alkaline conditions. *ACS Sustainable Chemistry & Engineering* 7 (18): 15163–15172.

14 Guggenberger, M., Potthast, A., Rosenau, T., and Böhmdorfer, S. (2019). Quantification of volatiles from technical lignins by multiple headspace sampling-solid-phase microextraction-gas chromatography-mass spectrometry. *ACS Sustainable Chemistry & Engineering* 7 (11): 9896–9903.

15 Sulaeva, I., Vejdovszky, P., Henniges, U. et al. (2019). Molar mass characterization of crude lignosulfonates by asymmetric flow field-flow fractionation. *ACS Sustainable Chemistry & Engineering* 7 (1): 216–223.

16 Kohlhuber, N., Musl, O., Sulaeva, I. et al. (2023). Adding speed to lignin analysis straight from black liquors. *ACS Sustainable Chemistry & Engineering* 11 (42): 15294–15301.

17 Tomani, P. (2010). The lignoboost process. *Cellulose Chemistry and Technology* 44 (1–3): 53–58.

18 Kouisni, L., Gagné, A., Maki, K. et al. (2016). LignoForce system for the recovery of lignin from black liquor: feedstock options, odor profile, and product characterization. *ACS Sustainable Chemistry & Engineering* 4 (10): 5152–5159.

19 Hubbe, M.A., Alén, R., Paleologou, M. et al. (2019). Lignin recovery from spent alkaline pulping liquors using acidification, membrane separation, and related processing steps: a review. *BioResources* 14 (1): 2300–2351.

20 Humpert, D., Ebrahimi, M., and Czermak, P. (2016). Membrane technology for the recovery of lignin: a review. *Membranes* 6 (3): 1–13.

21 Sumerskii, I., Korntner, P., Zinovyev, G. et al. (2015). Fast track for quantitative isolation of lignosulfonates from spent sulfite liquors. *RSC Advances* 5 (112): 92732–92742.

22 Sundberg, A., Sundberg, K., Lillandt, C., and Holmhom, B. (1996). Determination of hemicelluloses and pectins in wood and pulp fibres by acid methanolysis and gas chromatography. *Nordic Pulp & Paper Research Journal* 11 (4): 216–219.

23 Yan, M., Yang, D., Deng, Y. et al. (2010). Influence of pH on the behavior of lignosulfonate macromolecules in aqueous solution. *Colloids and Surfaces A: Physicochemical and Engineering Aspects* 371 (1–3): 50–58.

24 Myrvold, B.O. (2014). The hansen solubility parameters of some lignosulfonates. *World Academy of Science, Engineering and Technology International Journal of Energy and Power Engineering* 1 (October): 261.

25 Vermaas, J.V., Crowley, M.F., and Beckham, G.T. (2020). Molecular lignin solubility and structure in organic solvents. *ACS Sustainable Chemistry & Engineering* 8 (48): 17839–17850.

26 Melro, E., Alves, L., Antunes, F.E., and Medronho, B. (2018). A brief overview on lignin dissolution. *Journal of Molecular Liquids* 265 (2017): 578–584.

27 Wurzer, G.K., Bacher, M., Hettegger, H. et al. (2021). A general solvent system for the analysis of lignosulfonates by 31PNMR. *Analytical Methods* 13 (45): 5502–5508.

28 Wang, Z., Liu, Y., Barta, K., and Deuss, P.J. (2022). The effect of acidic ternary deep eutectic solvent treatment on native lignin. *ACS Sustainable Chemistry & Engineering* 10 (38): 12569–12579.

29 Wang, Z., Zhang, Z., Wang, H., and Deuss, P.J. (2022). Tuning lignin properties by mild ionic-liquid-mediated selective alcohol incorporation. *Chem Catalysis* 2 (6): 1407–1427.

30 Liu, Y., Deak, N., Wang, Z. et al. (2021). Tunable and functional deep eutectic solvents for lignocellulose valorization. *Nature Communications* 12 (1): 5424.

31 Chen, L., Dou, J., Ma, Q. et al. (2017). Rapid and near-complete dissolution of wood lignin at ≤80°C by a recyclable acid hydrotrope. *Science Advances* 3 (9).

32 Sosa, F.H.B., Abranches, D.O., da Costa Lopes, A.M. et al. (2020). Kraft lignin solubility and its chemical modification in deep eutectic solvents. *ACS Sustainable Chemistry & Engineering* 8 (50): 18577–18589.

33 Huang, C., Cheng, J., Zhan, Y. et al. (2022). Utilization of guaiacol-based deep eutectic solvent for achieving a sustainable biorefinery. *Bioresource Technology* 362: 127771.

34 Kosyakov, D.S., Anikeenko, E.A., Ul'yanovskii, N.V. et al. (2018). Ionic liquid matrices for MALDI mass spectrometry of lignin. *Analytical and Bioanalytical Chemistry* 410 (28): 7429–7439.

35 Chambon, C.L., Fitriyanti, V., Verdía, P. et al. (2020). Fractionation by sequential antisolvent precipitation of grass, softwood, and hardwood lignins isolated using low-cost ionic liquids and water. *ACS Sustainable Chemistry & Engineering* 8 (9): 3751–3761.

36 Zhao, W., Xiao, L.P., Song, G. et al. (2017). From lignin subunits to aggregates: insights into lignin solubilization. *Green Chemistry* 19 (14): 3272–3281.

37 Deng, Y., Feng, X., Yang, D. et al. (2012). π-π Stacking of the aromatic groups in lignosulfonates. *BioResources* 7 (1): 1145–1156.

38 Myrvold, B.O. (2015). Evidence for a very slow disaggregation of lignosulfonates. *Holzforschung* 69 (1): 9–16.

39 Zhu, W. and Theliander, H. (2015). Precipitation of lignin from softwood black liquor: an investigation of the equilibrium and molecular properties of lignin. *BioResources* 10 (1): 1696–1714.

40 Sewring, T., Durruty, J., Schneider, L. et al. (2019). Acid precipitation of kraft lignin from aqueous solutions: the influence of pH, temperature, and xylan. *Journal of Wood Chemistry and Technology* 39 (1): 1–13.

41 Qiu, X., Kong, Q., Zhou, M., and Yang, D. (2010). Aggregation behavior of sodium lignosulfonate in water solution. *Journal of Physical Chemistry B* 114 (48): 15857–15861.

42 Vainio, U., Lauten, R.A., and Serimaa, R. (2008). Small-angle X-ray scattering and rheological characterization of aqueous lignosulfonate solutions. *Langmuir* 24 (15): 7735–7743.

43 Vainio, U., Lauten, R.A., Haas, S. et al. (2012). Distribution of counterions around lignosulfonate macromolecules in different polar solvent mixtures. *Langmuir* 28 (5): 2465–2475.

44 Ruwoldt, J., Planque, J., and Øye, G. (2020). Lignosulfonate salt tolerance and the effect on emulsion stability. *ACS Omega* 5 (25): 15007–15015.

45 Ouyang, X., Deng, Y., Qian, Y. et al. (2011). Adsorption characteristics of lignosulfonates in salt-free and salt-added aqueous solutions. *Biomacromolecules* 12 (9): 3313–3320.

46 Myrvold, B.O. (2013). Salting-out and salting-in experiments with lignosulfonates (LSs). *hfsg* 67 (5): 549–557.

47 Sewring, T. and Theliander, H. (2019). Acid precipitation of kraft lignin from aqueous solutions: the influence of anionic specificity and concentration level of the salt. *Holzforschung* 73 (10): 937–945.

48 Korntner, P., Sumerskii, I., Bacher, M. et al. (2015). Characterization of technical lignins by NMR spectroscopy: optimization of functional group analysis by 31P NMR spectroscopy. *Holzforschung* 69 (6): 807–814.

49 Asikkala, J., Tamminen, T., and Argyropoulos, D.S. (2012). Accurate and reproducible determination of lignin molar mass by acetobromination. *Journal of Agricultural and Food Chemistry* 60 (36): 8968–8973.

50 Andrianova, A.A., Yeudakimenka, N.A., Lilak, S.L. et al. (2017). Size exclusion chromatography of lignin: the mechanistic aspects and elimination of undesired secondary interactions. *Journal of Chromatography A* 1534: 101–110.

51 Salanti, A., Orlandi, M., and Zoia, L. (2020). Fluorescence labeling of technical lignin for the study of phenolic group distribution as a function of the molecular weight. *ACS Sustainable Chemistry & Engineering* 8 (22): 8279–8287.

52 Salanti, A., Orlandi, M., Lange, H. et al. (2022). Phenolic group distribution as a function of molecular weight in lignins and their fractions. *ACS Sustainable Chemistry & Engineering* 10 (35): 11680–11691.

53 Tomaz, C.T. and Queiroz, J.A. (2013). Hydrophobic interaction chromatography. In: *Liquid Chromatography: Fundamentals and Instrumentation* (ed. S. Fanali, P.R. Haddad, C.F. Poole, et al.), 121–142. Elsevier Inc.

54 Baumberger, S., Abaecherli, A., Fasching, M. et al. (2007). Molar mass determination of lignins by size-exclusion chromatography: towards standardisation of the method. *Holzforschung* 61 (4): 459–468.

55 Zinovyev, G., Sulaeva, I., Podzimek, S. et al. (2018). Getting closer to absolute molar masses of technical lignins. *ChemSusChem* 11 (18): 3259–3268.

56 Sulaeva, I., Zinovyev, G., Plankeele, J.M. et al. (2017). Fast track to molar-mass distributions of technical lignins. *ChemSusChem* 10 (3): 629–635.

57 Musl, O., Sulaeva, I., Sumerskii, I. et al. (2021). Mapping of the hydrophobic composition of lignosulfonates. *ACS Sustainable Chemistry & Engineering* 9 (49): 16786–16795.

58 Duval, A., Vilaplana, F., Crestini, C., and Lawoko, M. (2016). Solvent screening for the fractionation of industrial kraft lignin. *Holzforschung* 70 (1): 11–20.

59 Striegel, A.M., Yau, W.W., Kirkland, J.J., and Bly, D.D. (2009). *Modern Size-Exclusion Liquid Chromatography: Practice of Gel Permeation and Gel Filtration Chromatography*, 2e. Wiley.

60 Fredheim, G.E., Braaten, S.M., and Christensen, B.E. (2002). Molecular weight determination of lignosulfonates by size-exclusion chromatography and multi-angle laser light scattering. *Journal of Chromatography A* 942 (1–2): 191–199.

61 Podzimek, S. (2011). *Light Scattering, Size Exclusion Chromatography and Asymmetric Flow Field Flow Fractionation: Powerful Tools for the Characterization of Polymers, Proteins and Nanoparticles*. Wiley.

62 Ringena, O., Lebioda, S., Lehnen, R., and Saake, B. (2006). Size-exclusion chromatography of technical lignins in dimethyl sulfoxide/water and dimethylacetamide. *Journal of Chromatography A* 1102 (1–2): 154–163.

63 Fredheim, G.E., Braaten, S.M., and Christensen, B.E. (2003). Comparison of molecular weight and molecular weight distributions of softwood and hardwood lignosulfonates. *Journal of Wood Chemistry and Technology* 23 (2): 197–215.

64 Skrebets, T.E. and Bogolitsin, K.G. (2010). Behavior of lignin in a dimethylsulfoxide – water mixture. *Russian Journal of Bioorganic Chemistry* 36 (7): 835–837.

65 Dong, D. and Fricke, A.L. (1995). Intrinsic viscosity and the molecular weight of kraft lignin. *Polymer* 36 (10): 2075–2078.

66 Podzimek, S. (2019). Molar mass distribution by size exclusion chromatography: comparison of multi-angle light scattering and universal calibration. *Journal of Applied Polymer Science* 136 (21): 47561.

67 Ji, L., Liu, L.-Y., Cho, M. et al. (2022). Revisiting the molar mass and conformation of derivatized fractionated softwood kraft lignin. *Biomacromolecules* 23 (3): 708–719.

68 Musl, O., Galler, S., Wurzer, G. et al. (2022). High-resolution profiling of the functional heterogeneity of technical lignins. *Biomacromolecules* 23 (3): 1413–1422.

69 Potthast, A., Radosta, S., Saake, B. et al. (2015). Comparison testing of methods for gel permeation chromatography of cellulose: coming closer to a standard protocol. *Cellulose* 22 (3): 1591–1613.

70 Striegel, A.M. (2017). Specific refractive index increment (∂n/∂c) of polymers at 660 nm and 690 nm. *Chromatographia* 80 (6): 989–996.

71 Wang, L., Uraki, Y., Koda, K. et al. (2019). Determination of the absolute molar mass of acetylated eucalyptus kraft lignin by two types of size-exclusion chromatography combined with multi-angle laser light-scattering detectors. *Holzforschung* 73 (4): 363–369.

72 Pittman, Z.A., Lynn, B., Thies, M.C., and Kitchens, C.L. (2023). Improved multiangle light scattering for molar mass determination of fluoresecent lignins. *ACS Sustainable Chemistry & Engineering* .

73 Pittman, Z.A., McCarthy, M.E., Birtwistle, M.R., and Kitchens, C.L. (2022). Method for improved fluorescence corrections for molar mass characterization by multiangle light scattering. *Biomacromolecules* 23 (9): 3743–3751.

74 Balakshin, M., Capanema, E.A., Zhu, X. et al. (2020). Spruce milled wood lignin: linear, branched or cross-linked? *Green Chemistry* 22 (13): 3985–4001.

75 Lu, Y., Joosten, L., Donkers, J. et al. (2021). Characterisation of mass distributions of solvent-fractionated lignins using analytical ultracentrifugation and size exclusion chromatography methods. *Scientific Reports* 11 (1): 1–12.

76 Jacobs, A. and Dahlman, O. (2000). Absolute molar mass of lignins by size exclusion chromatography and MALDI-TOF mass spectroscopy. *Nordic Pulp & Paper Research Journal* 15 (2): 120–127.

77 Kosyakov, D.S., Ul'yanovskii, N.V., Sorokina, E.A., and Gorbova, N.S. (2014). Optimization of sample preparation conditions in the study of lignin by MALDI mass spectrometry. *Journal of Analytical Chemistry* 69 (14): 1344–1350.

78 Meng, X., Crestini, C., Ben, H. et al. (2019). Determination of hydroxyl groups in biorefinery resources via quantitative 31P NMR spectroscopy. *Nature Protocols* 14 (9): 2627–2647.

79 Balakshin, M. and Capanema, E. (2015). On the quantification of lignin hydroxyl groups with 31P and 13C NMR spectroscopy. *Journal of Wood Chemistry and Technology* 35 (3): 220–237.

80 Korntner, P., Schedl, A., Sumerskii, I. et al. (2018). Sulfonic acid group determination in lignosulfonates by headspace gas chromatography. *ACS Sustainable Chemistry & Engineering* 6 (5): 6240–6246.

81 Sumerskii, I., Zweckmair, T., Hettegger, H. et al. (2017). A fast track for the accurate determination of methoxyl and ethoxyl groups in lignin. *RSC Advances* 7 (37): 22974–22982.

82 Striegel, A.M. (2020). Method development in interaction polymer chromatography. *TrAC Trends in Analytical Chemistry* 130: 115990.

83 Pasch, H. and Trathnigg, B. (1999). *HPLC of Polymers*. Springer.

84 Zinovyev, G., Sumerskii, I., Korntner, P. et al. (2017). Molar mass-dependent profiles of functional groups and carbohydrates in kraft lignin. *Journal of Wood Chemistry and Technology* 37 (3): 171–183.

85 Costa, C.A.E., Pinto, P.C.R., and Rodrigues, A.E. (2018). Lignin fractionation from *E. Globulus* kraft liquor by ultrafiltration in a three stage membrane sequence. *Separation and Purification Technology* 192 (September 2017): 140–151.

86 Sevastyanova, O., Helander, M., Chowdhury, S. et al. (2014). Tailoring the molecular and thermo-mechanical properties of kraft lignin by ultrafiltration. *Journal of Applied Polymer Science* 131 (18).

87 Domínguez-Robles, J., Tamminen, T., Liitiä, T. et al. (2018). Aqueous acetone fractionation of kraft, organosolv and soda lignins. *International Journal of Biological Macromolecules* 106: 979–987.

88 Jääskeläinen, A.S., Liitiä, T., Mikkelson, A., and Tamminen, T. (2017). Aqueous organic solvent fractionation as means to improve lignin homogeneity and purity. *Industrial Crops and Products* 103: 51–58.

89 Kiyota, E., Mazzafera, P., and Sawaya, A.C.H.F. (2012). Analysis of soluble lignin in sugarcane by ultrahigh performance liquid chromatography and tandem mass spectrometry with a do-it-yourself oligomer database. *Analytical Chemistry* 84 (16): 7015–7020.

90 Brudin, S., Berwick, J., Duffin, M., and Schoenmakers, P. (2008). One-dimensional and two-dimensional liquid chromatography of sulphonated lignins. *Journal of Chromatography A* 1201 (2): 196–201.

91 Hutta, M. and Góra, R. (2003). Novel stepwise gradient reversed-phase liquid chromatography separations of humic substances, air particulate humic-like substances and lignins. *Journal of Chromatography A* 1012 (1): 67–79.

92 Pojana, G., Carrer, C., Cammarata, F. et al. (2003). HPLC determination of sulphonated melamines-formaldehyde condensates (SMFC) and lignosulphonates (LS) in drinking and ground waters. *International Journal of Environmental Analytical Chemistry* 83 (1): 51–63.

93 Reemtsma, T. (1996). Methods of analysis of polar aromatic sulfonates from aquatic environments. *Journal of Chromatography A* 733 (1–2): 473–489.

94 Takahashi, N., Azuma, J.I., and Koshijima, T. (1982). Fractionation of lignin-carbohydrate complexes by hydrophobic-interaction chromatography. *Carbohydrate Research* 107 (2): 161–168.

95 Ekeberg, D., Gretland, K.S., Gustafsson, J. et al. (2006). Characterisation of lignosulphonates and kraft lignin by hydrophobic interaction chromatography. *Analytica Chimica Acta* 565 (1): 121–128.

96 Musl, O., Sulaeva, I., Bacher, M. et al. (2020). Hydrophobic interaction chromatography in 2 D liquid chromatography characterization of lignosulfonates. *ChemSusChem* 13 (17): 4595–4604.

97 Winowiski, T., Lebo, S., Gretland, K., and Gustafsson, J. (2005). Characterization of sulfonated lignin dispersants by hydrophobic interactive chromatography. *Journal of ASTM International* 2 (9): 12915.

98 Pasch, H. (2013). Hyphenated separation techniques for complex polymers. *Polymer Chemistry* 4 (9): 2628.

4

Degradative Methods for Lignin Valorization

Elisabetta Borsella[1], *Pietro Colucci*[2], *Giuseppe Lembo*[3] *and Heiko Lange*[3]

[1] *Research Centre ENEA Casaccia, Via Anguillarese 301, 00123, Rome, Italy*
[2] *Research Centre ENEA Trisaia, SS 106 Jonica km 419+500, 75026, Rotondella, Italy*
[3] *Department of Earth and Environmental Sciences, University of Milano-Bicocca, Piazza della Scienza 1, 20126, Milan, Italy*

4.1 Introduction

Polyphenolic secondary metabolites represent an interesting renewable pool of a variety of structurally different entities: small molecules in form of lignans, oligomers and smaller polymeric structures in form of tannins, and the truly polymeric lignin, with each of them displaying specific intrinsic characteristics that are interesting with respect to a wide variation of applications within the transition toward a bio-based and ideally circular economy. The most abundant of these species, i.e. lignin, represents de facto the only polymeric aromatic polymer in nature. This fact renders it especially interesting with respect to the quest of substituting common aromatic polymers in the area of platform chemicals, phenolic resins, carbon fibers, etc. Generally, valorizations can be roughly distinguished into valorizations that make use of the lignin in polymeric form, and those that start effectively with a degradation of the polymer into ideally monomeric building blocks. Both ways are determined by the challenging structural inhomogeneity of lignins, that is in part a reminiscence of the biomass it comes from, and in part an artifact of the isolation process producing technical lignins.

This chapter will highlight the possibilities that exist with respect to a valorization of lignin in depolymerized form; hence it will give an overview of degradative methods in lignin valorization. The methods described in the following paragraphs span the range from rather traditional wet-chemical methods to modern catalytic and thermal degradations. The chapter will outline how the chemical nature of the lignin, i.e. the structural features like functional groups present in isolated, i.e. technical lignins influence any depolymerization approach. On the basis of the

Lignin Chemistry: Characterization, Isolation, and Valorization, First Edition.
Edited by Yuhe Liao and Bert F. Sels.

discussed possibilities, recent trends are described, including a discussion of their potential within an economically sustainable circular bioeconomy.

4.2 Basics Aspects Regarding Structural Features in (Isolated) Lignins

Before entering n the discussion regarding the depolymerization possibilities, it is important to understand and actually acknowledge important structural features present in lignin, i.e. in the context of the chapter especially in the various lignins obtained from various industrial biomass fractionation efforts [1, 2].

4.2.1 Occurrence and Isolation

The structural characterization, and partial structural control during isolation of lignin from biomasses, is still an active field of research, and refined measurements and the development of novel techniques allow regularly for structural refinements that improve the structural picture of this natural polymer and the emerging application potential. Several reviews and monographs can be consulted to get a head start at least on a theoretical level [3–8]. The rather hydrophobic phenolic, i.e. aromatic polymer lignin, summing up to 15–35% of the dry mass of lignocellulosic material and thus representing the second most abundant polymer of land-based biomass [2, 9–11], is located mainly in the plant cell walls and is chemically and physically interacting with hemicelluloses, here giving rise to the so-called lignin carbohydrate complexes [12], and cellulose, respectively [10].

Lignins distribution across different plant parts, i.e. stem, branching points, branches, and leafs and across the different plant cell walls differs [2, 13], making it important to understand what part of the biomass is refined and or directly valorized in thermal processes.

Lignin does not display a regular repetition motif, as one might expect for a natural polymer closely interacting with other regularly structured polymers. Furthermore, its structure does not only change between different plant types but also within the same plant presumably as a function of function. Lignin formation in cells has been proposed to be a postmortem process in plant cells [14, 15]. The monomers consumed in the radical polymerizations are the three monolignols *p*-coumaryl alcohol (H-type lignin), coniferyl alcohol (G-type lignin), and sinapyl alcohol (S-type lignin), which are synthesized by the plants starting from L-tyrosine or L-phenylalanine [16–19]. Being *de facto* produced only after the death of the plant cells, with the monomers delivered to the dead cell by adjacent living cells, structural changes, and abundances of lignin are not governed by enzymes, but most probably by microenvironments present, and thus indirectly by the need of the plant for, e.g. stability or growth [13, 20]. Even the biosynthetic pathway is not yet fully understood (Figure 4.1) [17, 21–24].

Depending on the natural source from which the lignin is isolated, and also depending on the method with which the lignin has been extracted from the lignocellulolytic biomass, the ratios of the different monomer types, as well as the nature and the distribution of the interunit bonding motifs *via* which they are linked differ [9, 16, 17, 19, 25]. Three different general types of lignin are distinguishable on the basis of the monomeric units employed: (i) softwood lignin, which is mainly comprised of G-type monomers and a small percentage of H-type monomers; (ii) hardwood lignin, which shows all three types of monomers in the typical ratio H/G/S = 1/1/3; and (iii) grass lignin, which contains all three monomer types with H-type monomers being the dominant one [10].

Oxidative coupling of the monomers, leading to the various binding motifs, proceeds preferentially at only a limited number of positions, namely, O-4, C-1, C-3, C-5, and C-β (IUPAC numbering), thus eventually reflecting some selectivity of the enzymes putatively involved in the biosynthesis at least for the activation steps. Based on the theoretically possible coupling modes, it is still controversial whether lignin *in planta* is a branched, and thus three-dimensional polymer, or whether it consists of linear chains that are rather oligomeric than polymeric (Figure 4.2) [26].

Although lignin contains different asymmetric carbon atoms that are formed during its biosynthesis, a preferred absolute configuration, as one might expect from the close interplay with the stereogenic polymeric sugars, does not exist – another fact that suggests that the *in vivo* lignin formation is probably not controlled in detail. It has been shown, however, that the ratio of the different aromatic units does correlate with the ratio between *threo*- and *erythro*-configured aliphatic parts in lignin [27, 28]. This point is obviously another interesting difference between lignin and carbohydrate-based polymers in plants. With respect to the use in a pharmaceutical context, this in the macroscopic average "hidden" chirality displays an interesting aspect yet to be elucidated in its molecular consequences.

Since any isolation process can affect the natural structure of lignin, causing the lignins isolated *via* each of these processes to be more nature-derived than natural, strictly speaking, it is important to acknowledge the source of any eventually exploited lignin in form of the biomass it comes from and in form of the process used for its isolation. The three different processes are established for the isolation of lignins in meaningful scales: (i) lignosulfonate [29–31]; (ii) kraft lignin [32–34]; (iii) organosolv lignin [35–37]. Figure 4.2 shows the structures of three main lignin types. Lignosulfonates are structurally different from the other two in the fact that they contain sulfonate groups, allowing them to exist in either salt of acid form. In light of the abundant use of this technical lignin, it is all the more astonishing that rather little is known regarding structural details. Kraft lignins, abundantly available technical lignins, are structurally better understood; they comprise a fraction of truly polymeric chains, slightly branched, and a fraction of small, branched, and presumably oxidatively coupled oligomers. In terms of availability of industrial quantities emerging organosolv, lignins comprise rather linear oligomeric chains.

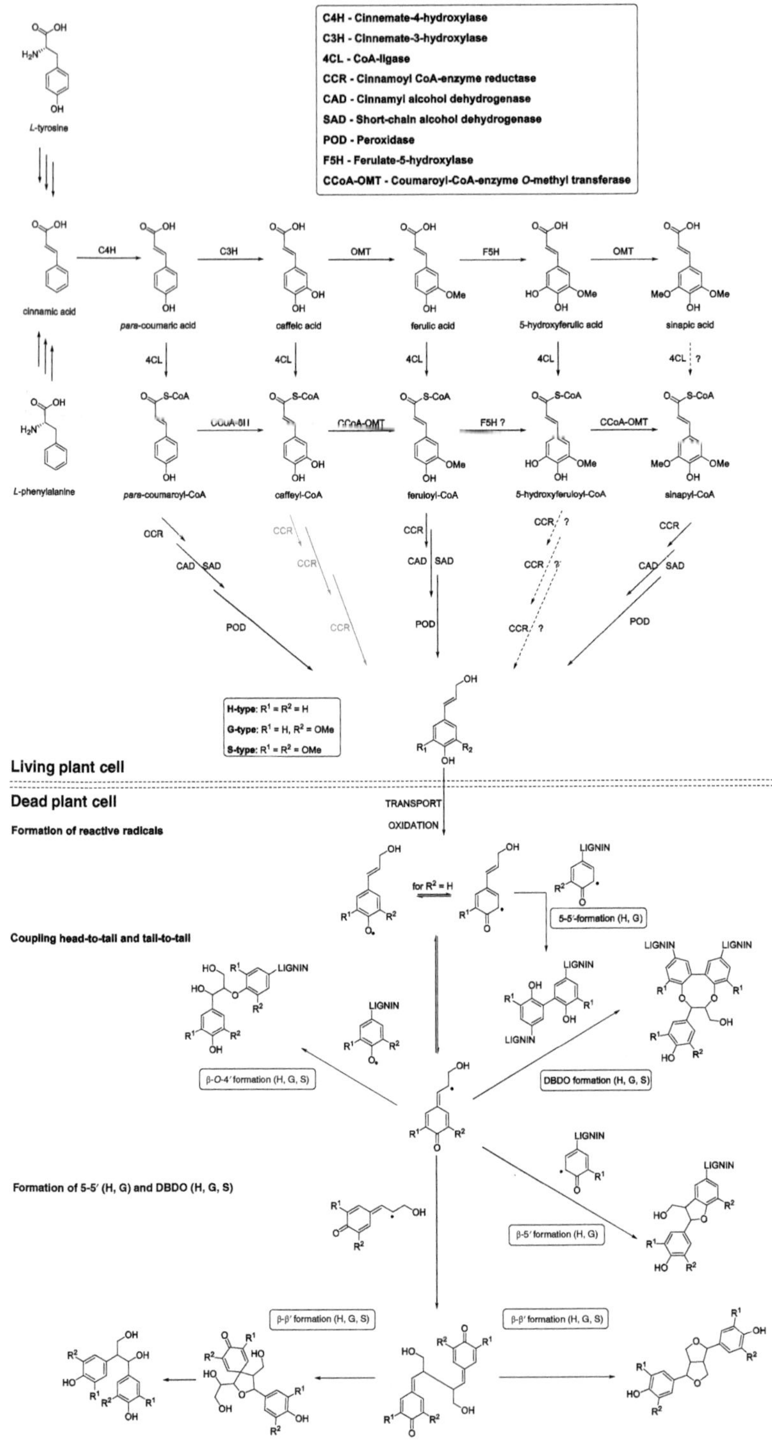

Figure 4.1 Main elements of the biosynthesis of lignin and formation of typical bonding motifs found in lignins [14, 17, 21–24].

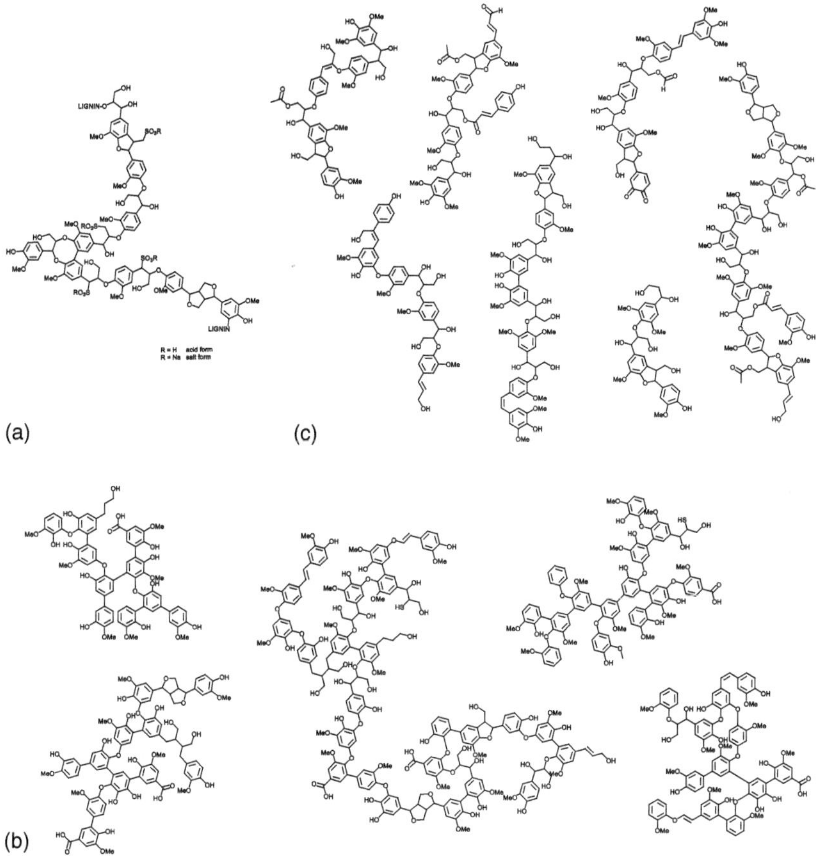

Figure 4.2 Lignin structures showing characteristic interunit bonding motifs and functional groups for different types of lignin: (a) (generic) lignosulphonate softwood; (b) pine kraft lignin; (c) wheat straw organosolv lignin.

4.2.2 Standard Means for Elucidating Basic Structural Features of (Isolated) Lignins

A minimum of structural understanding of the isolated lignin at hand is essential for designing functioning and winning valorization strategies, since it is crucial for understanding and elucidating potential and promising mechanisms for formation of derivatives for targeted applications on the basis of lignin as polymer of for meaningful degradative valorization. Classical wet-chemical and spectroscopic techniques can provide qualitative and quantitative information on functional groups and linkages of constituents in lignin as well as the degradation products. Modern nuclear magnetic resonance (NMR)-based techniques allow analysis of isolated lignins without touching the oligo- or polymeric structure, and give more detailed insights. Efforts are regularly summarized and reviewed, and the reader is referred to these more dedicated works for detailed and more complete information [3, 38–40].

Only the most important parts of the state-of-the-art toolbox available for structural elucidation are listed in the following. The named analyses allow for a basic structural insight that might be already sufficient in case valorization upon degradation is opted for.

4.2.2.1 Nuclear Magnetic Resonance (NMR) Spectroscopy-Based Analysis Methods

Both the number and the nature of free hydroxyl groups can be determined using quantitative ^{31}P NMR spectroscopy after phosphitylation of lignins in the presence of an internal standard [7, 38, 41–43]. Quantitative ^{31}P NMR can be seen as one of the standard analytical tools in lignin characterization. Advantages in the use of ^{31}P NMR lie in the low amount of sample needed, the possibility of using ionic liquids (ILs) for samples that dissolve only scarcely, reduced analysis time, ease in sample preparation, and high reproducibility. The different chemical environments around the phosphorus atoms lead to characteristic shifts, which in turn allows for differentiation and quantification.

QQ-HSQC is a potential method for the acquisition of quantitative HSQC spectra, but often the available NMR equipment dictates the method that should be employed. In case a high-field (400 MHz or more) NMR machine equipped with a cryo-probe is available, either a QQ-HSQC pulse sequence [44], or an approach consisting of a series of HSQC-measurements with incremented repetition times and mathematical backward-extrapolation, called $HSQC_0$ [45], can be adopted for the acquisition of quantitative HSQC lignin spectra. In both QQ-HSQC and $HSQC_0$, quantification is achieved based on the intrinsic standard represented by the distinct shifts of the aromatic hydrogen atoms in 2-position, which can be easily detected, summed up, and used as reference. As a third option, quantification of a high-quality standard ^{1}H–^{13}C HSQC spectrum can be achieved on the basis of a quantitative ^{13}C NMR analysis of the very same sample used for acquisition of the HSQC, given the sample exhibits the necessary structural stability in the solvent of choice in the time frames needed for the separate measurements [46].

4.2.2.2 Fourier-Transform Infrared Spectroscopy and Raman Spectroscopy

FT-IR remains an important tool for the structural characterization also in the field of lignins and its derivatives due to its facile applicability, nowadays in form of the ATR technique. Careful application of band deconvolution [47, 48] and/or band fitting [49, 50] allows for more detailed insights within a certain context but does normally not allow for insights as detailed as obtainable by NMR techniques.

In 1984, the first reports were published regarding the use of Raman spectroscopy for characterizing structural features in lignin [51, 52]. Since then, numerous efforts have been made to overcome initial low sensitivity stemming from the interfering natural fluorescence of lignin, and from the visible light absorption by lignin, which made it necessary to develop tailor-made irradiation sources [53] and to assist analysis by in silico approaches [54].

4.2.2.3 Size Exclusion and Gel Permeation Chromatographic Methods

Mean average molecular weight (Mn), weight average molecular weight (Mw), and the polydispersity (PD) of lignins can be determined using size exclusion chromatography (SEC) or gel permeation chromatography (GPC) [55–61]. Calibration is normally achieved based on commercially available polystyrene standards, working with the acetylated form of technical lignins in THF as master solvent. When DMSO or DMF are used as solvents, polystyrene sulfonates in acid form are used as standards, and minimum amounts of inorganic salts such as lithium chloride are normally added for breaking aggregates. Different detector-types are known to lead to different results: commonly used UV-based detectors; refractive index detectors have been shown to perform poorly for lignin, while being able to better highlight potential carbohydrate-stemming impurities. Molecular weight-sensitive detectors based on viscosimetry [62] or laser light scattering [63–66] represent currently the comparably costly state-of-the-art in terms of detection.

4.2.3 "Purity" of Isolated Lignins

Depending on the objective underlying a valorization of lignin based on decomposition, the presence of structures alien to a polyphenolic material might represent an important potential pitfall that renders valorization attempts less successful than they could intrinsically be. Such "unexpected" structures can eventually cause side reactions, poison (bio)catalysts, or complicate purification and funneling efforts on the level of produced product mixtures. Isolated lignins should not contain residues from the other parts of the plant cells and they should be free of residual carbohydrates, carbohydrate residues, and salts. While the first two aspects are commonly looked at during the optimization of lignin production according to the various biorefinery schemes, the second aspect is gaining importance since novel feedstocks are explored, i.e. halophytes [67, 68], with these feedstocks being plants that grow in areas not suitable for agriculture or industrial foresting, such as coastal areas and swamps. This idealized case of pure lignins is, however, often hard to achieve since the various methods of isolating lignin are more or less suitable to remove non-lignin plant residues, or since the starting materials suffer from significantly changing qualities due to their nature and origin [69]. Carbohydrate impurities most commonly pose a challenge, because lignins *in planta* are often covalently linked to hemicelluloses *via* ether and ester-linkages involving hydroxycinnamates such as ferulates and its derivatives. This observation has long been known for grass lignins [69–75], and newer studies have discovered LCC formation in woody biomass [76]. At least some of the putative LCC linkages are thought to be recalcitrant to standard biorefinery processes, and thus presumably cause carbohydrate impurities in isolated lignins [2, 77–79]. Additionally, besides being potentially recalcitrant linkages, insufficient methods for removing even cleaved carbohydrate residues during lignin isolation represent a main cause for a decrease of lignin purity and quality. Recent years have seen the description of what seems to be omnipresent carbohydrate-derived impurities in lignin: (i) pseudolignin [80]; (ii) humins [81]; (iii) covalently linked humin-lignin structures [82]. While these impurities can be detected using NMR-based analyses, also chromatography-based approaches in combination with chemical modification represent a powerful tool for the detection of all sorts of impurities in isolated lignins: GPC, when coupled with combination of refractive index and UV–Vis detectors, allows, under otherwise optimized conditions and suitable sample preparations [57, 83], to delineate whether carbohydrates are covalently linked to the lignin bulk material: depending on whether a time-resolved or a simultaneous signal detection occurs in the RI- and the UV–Vis-detector, carbohydrates are present as insufficiently removed impurities or are connected more likely covalently to the lignin bulk material. SEC used in connection with wet-chemical analysis methods such as acidolysis and thioacidolysis, which represent two of the most prominent wet-chemical analysis techniques used for the structural elucidation of lignins (*vide infra*), provides a possibility to identify carbohydrate residues also with respect to LCCs while the lignin fraction in the putative LCCs is degraded in a predictable way to mainly monomeric species, the carbohydrate residues stay largely untouched and remain thus identifiable as polymers in molecular size-based analysis techniques using appropriate detectors.

4.2.4 Fractionation of Isolated Lignins for Structurally more Homogeneous Starting Materials

An ever more important aspect of lignin valorization is the fractionation of isolated lignins. Fractionation of technical lignins is currently the only scalable procedure to approach the issue of the complex structural inhomogeneity without additionally chemically modifying the structures. Various fractionation techniques have been developed for refining for structural and/or molecular size features [84, 85]. Fractionation can be achieved using: (i) sequential precipitation out of alkaline solutions [86]; (ii) fractional precipitation of re-dissolved kraft lignin and a wheat straw lignin in a gradually changed binary solvent system [87, 88]; (iii) (sequential) extractions using different solvents [86, 89–97]; and (iv) fractionation by ultrafiltration of black liquor using ceramic membranes [98–100]. Figure 4.3 summarizes the efforts. Efforts have recently been summarized in a couple of dedicated reviews, and

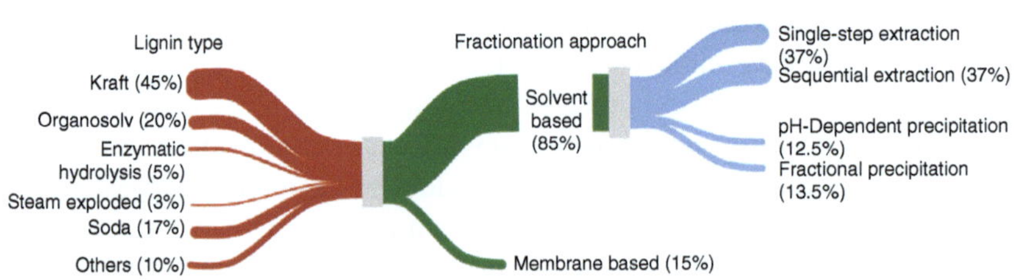

Figure 4.3 Main methods for fractionation of isolated lignins. Source: Graphic adapted with permission from [101].

readers shall be once more referred to these dedicated articles [101, 102]. While yet poorly studied in a systematic way, pyrolysis efforts on lignin might actually benefit from being done on structurally "cleaned" lignin fractions of lignin.

4.3 Classic Wet-Chemical Degradation Methods

Lignin has been degraded into monomeric building blocks, more or less chemically altered, since research began elucidating this material. While the wet-chemical methods were initially used for understanding structural features, modern methods aim at the generation of monomers suitable to interface with the chemical industries in form of solvents or aromatic buildings in the sense of platform chemicals. In this paragraph, classical methods are discussed mainly with respect to their overall and/or historical significance in lignin research. The modern methods screened highlight general possibilities in catalytic approaches and the challenges faced up till now.

An organic solvent suitable to solubilize lignin, at elevated temperature, can be used to trigger lignin depolymerization. Solvents used in such processes can be water or alcohols, i.e. protic, or dioxane, acetone, tetrahydrofuran, and tetralin, and hence aprotic. Solvent systems have also been applied. Lower solvolysis temperatures favor the formation of alkylated phenols and guaiacols, while higher temperatures lead to less substituted phenols or catechols. Playing with the solvent characteristics, and the interplay between the solvents and the lignin oligomers and polymers [103], allows for guiding eventually product formation. In order to device more sustainable processes in terms of energy consumption and solvent usage, classic solvents can be substituted by ILs [104], and eventual deep eutectic solvents (DESs) [105]. ILs have demonstrated the conversion of commercial lignin into specialty chemicals, namely, homovanillic and vanillic acid, using the combination of [EMIM][triflate] and triethylammonium methanesulfonate (TMS) with water and hydrogen peroxide [106]. Also, usage of ILs is known, however, to cause eventually a reduction in β-*O*-4′ linkages, promoting more stable C—C bond formation, which hampers depolymerization efforts.

Systematic degradation and chemical modification using solvolytic methods such as acidolysis [107, 108] thioacidolysis [109, 110], or oxidative methods like ozonolysis [111, 112] and alkaline nitrobenzene oxidation (NBO) [113] in combination with downstream GC analysis have all contributed significantly to determine the structural features and compositions of lignins in terms of the H/G/S-ratios.

Acidolysis and thioacidolysis target interunit bonding motifs and yield different C9-fragments that allow analyses and thus inference to the structural motifs originally present in a lignin sample. While the parent method acidolysis uses aqueous hydrochloric acid for degrading polymeric structures, thioacidolysis uses thioethanol in the presence of Lewis-acidic boron trifluoride etherate (Figure 4.4a). Thioacidolysis has become *the* solvolytic method used for the structural determination of lignins due to an overall superior performance in terms of yields [109, 110] and due to the possibility to combine it with a preceding alkylation to deduce additional information regarding the degree of etherification. Typical structural motifs obtained in thioacidolysis of lignins are depicted in Figure 4.4a.

Combinations of refined degradation methods such as thioacidolysis and "derivatization followed by reductive cleavage" (DFRC) [114] with non-chromatography nondestructive analysis tools are used by default nowadays in lignin analysis. In Figure 4.4b, the chemical transformations during a DFRC treatment of lignin are given (Figure 4.4b). Here, β-aryl ethers are preferentially reacted to induce defined fragmentation for analysis. Important structural findings were possible using DFRC. Particularly noteworthy is the fact that it was possible to verify the aforementioned racemic nature of lignin due to the mild chemistry used in this method. DFRC also serves in combination with nondestructive advanced NMR spectroscopy on lignins (*vide infra*), and it can be used to determine whether a given lignin represents a linear or crosslinked polymer [35, 115].

Oxidative lignin degradations are important tools for structural understanding, especially in the past. Formation of erythronic acid and threonic acid is obtained as result of ozonolytically degrading the aromatic rings and other unsaturated moieties in lignin, e.g. stilbenes, while leaving the side-chains intact [27, 28]. It has been shown that ozonolysis enables identification of interunit bonding motifs and, in addition, determination of the ratio between *threo*- and *erythro*-configured aliphatic moieties in lignins, which can give rise to stereochemical considerations.

A traditional, but still important oxidative lignin depolymerization technique is the alkaline NBO (Figure 4.4c) [113]. The technique was used to verify the aromatic nature of the lignin biopolymer, given that the phenylpropanoid structures are oxidatively cleaved by the nitrobenzene in strongly alkaline environments at elevated temperatures. Mechanistically, two modes of actions are nowadays believed to co-exist. With one of them attacking free phenolic groups, and the other one reacting at the benzylic hydroxyl in case the phenolic function is blocked [116, 117]. In case of free phenols, a two-electron transfer generating a quinone-methide intermediate is proposed. In case of blocked, i.e. internal phenolics, a one-electron transfer activated the benzylic OH-group. In the following cascades delocalizing the electron and causing the degradation, typical monomers such as benzaldehyde derivatives, vanillin, and syringaldehydes form the corresponding monomeric motifs (*vide supra*). The degradation chemistries are rather specific, so that results could eventually be seen as taxonomically specific, acknowledging certain limitations though (*vide infra*) [10], a fact that contributes to the ongoing use of this degradation method as analytical tool. The latter is additionally favored by the fact that the products can obviously conveniently be analyzed by standard high-pressure liquid chromatography (HPLC) or gas chromatographic (GC) set-ups coupled to suitable detectors. Careful analyses of the products formed and their relative abundances allow for delineating basic structural features of the lignins also in terms of how condensed the lignin is. The rather linear chemical connections of this degradation method are contrasted by the sensitivity of the method. Interlaboratory reproducibility is low; standard deviations easily arrive at 30%. The method is thus rather of use for relative comparisons across samples in one

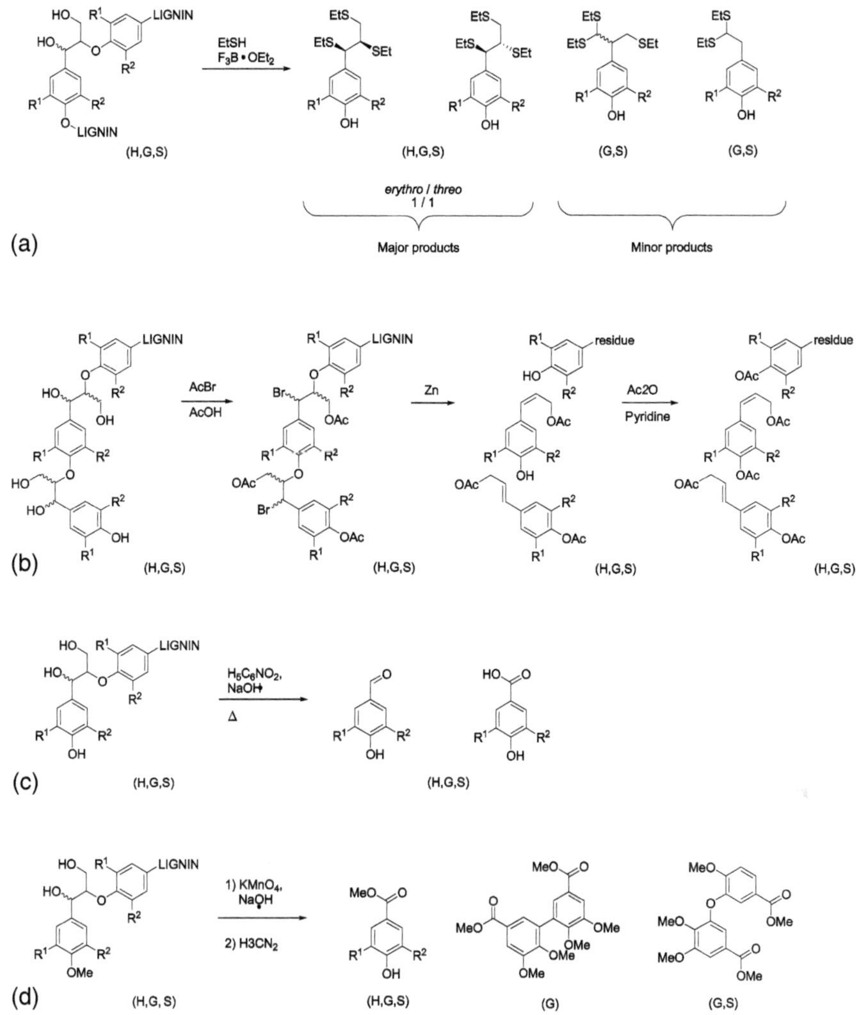

Figure 4.4 Conversion of the most common interunit bonding motif arylglycerol-β-aryl ethers (β-*O*-4′) upon: (a) thioacidolysis; (b) DFRC procedure; (c) nitrobenzene-oxidation; (d) potassium permanganate-oxidation.

laboratory, with the experiments run by a single experimentalist. Another issue related to the NBO as analysis tool is the non-specificity with respect to the nature of the phenol reacting. Problems of this kind are observed especially in grass lignins, in which, for example, also *p*-coumaric acid esters are eventually degraded to *p*-hydroxybenzaldehydes, thus eventually suggesting an artificially high content of H-type monomers in the sample [118].

This limitation of the NBO can be eventually corrected by running modified versions of the procedure [119], for example a cupric oxide-mediated oxidative degradation of the same lignin samples [120, 121]. The chemistries initiated by the CuO are less harsh, and oxidation is less destructive to the backbone structures in lignin, thus generating a larger set of oxidation products. Unifying degradation of the various cinnamic acid esters is limited, and overoxidation of formed aldehydes is avoided.

Another classic wet-chemical lignin degradation procedure employs potassium permanganate and targets the aromatic moieties in lignin (Figure 4.4d) [5]. This permanganate oxidation was initially developed nearly a century ago and subsequently improved. The rather complex procedures comprise four steps: (i) peralkylation, most often permethylation, of the phenolic OH-groups; (ii) oxidation with potassium permanganate; (iii) oxidation of the initial oxidation products with hydrogen peroxide to generate mainly carboxylic acid derivatives from the oxidative cleavages of side chains; and (iv) methylation of the so-generated benzoic acid derivatives to arrive at compounds easily analyzable in chromatographic methods like HPLC or GC. Obviously, the method works better for the non-condensed units, such as β-*O*-4′ units (*vide supra*). The degradative cleavage works in form of an *exo*-cleavage, hence only the end-standing, in the first step alkylated phenolic units are cut of the lignin chain. The method is thus suitable to get an idea of the amount of free phenolics as polymer or oligomer end-groups. The method thus allows for a step-wise structural understanding of lignins ample, but is eventually, in its original form, not suitable for analyzing the lignin in detail. Improvements introduced in the practical work flow see a preceding set of additional alkylation and initial, more severe oxidation, using, for example, the cupric acid method [113].

4.4 Catalytic Oxidative Degradation of Lignins

The aforementioned wet-chemical methods require at least stoichiometric amounts of reactants. Such chemistries are acceptable nowadays only in the analytical context. These methods are not suitable as means for the valorization of lignin based on its degradation into useful aromatic monomers. For this, degradation chemistries on the basis of chemical or biological catalysts are more suitable. The following paragraphs will highlight important examples.

4.4.1 Catalytic Chemical Oxidative Degradation

Oxidative valorization of lignin can be achieved by different means, ranging from classic metal-catalyzed approaches over biomimetic strategies to the exploitation of enzymatic activities. These various routes are regularly reviewed more specifically, and in greater detail than could be discussed in the context of a book chapter [122–129]. The following paragraphs will mention the basic notions, such as to map this interesting field.

4.4.1.1 Metal-Free Oxidations

A 2,2,6,6-tetramethylpiperidinyloxy-based catalyst system consisting of 4-acetamido-TEMPO (5 mol%; TEMPO = 2,2,6,6-tetramethylpiperidine-*N*-oxyl) in combination with nitric acid HNO_3 and hydrochloric acid, HCl (10 mol% each), proofed to be a versatile system for the chemoselective aerobic oxidation of secondary benzylic alcohols within lignin model compounds and lignins [130]. The resulting aryl ketones can then serve as substrates for further catalytic cleavages; this step could be done photolytically, as has been demonstrated recently [131].

4.4.1.2 Organometallic Catalysts

One of the simplest structures that can serve to catalytically activate the truly oxidizing species, molecular oxygen, or hydrogen peroxide is methyltrioxo rhenium (MTO) [132, 133]. Activation of hydrogen peroxide proceeds *via* the formation of two peroxorhenium intermediates, a mono-peroxo η^2-complex ([$MeRe(O_2)O_2$], and a bis-peroxo η^2-complex ([$MeRe(O_2)_2O$]. The stability and the reactivity of these species highly depend on the reaction conditions (Figure 4.5a) [134]. In order to improve, and fine-tune catalyst performance, MTO can be immobilized using polystyrene or poly(4-vinylpyridine) beads as support, or as material for encapsulation (Figure 4.5b). The transfer of oxygen from these peroxo-complexes to the substrate occurs *via* a concerted mechanism that includes a butterfly like transition state that effectively prevents the formation of intermediate radical species (Figure 4.5a). The concerted mechanism makes MTO a rather unique reagent for the oxidative valorization of lignin, since it allows a more controlled chemistry by suppressing the presence of various species of very reactive oxygen-based radicals in the reaction mixture. MTO-activated hydrogen peroxide is able to oxidize even challenging substrates such as activated phenols or methoxybenzene and is proofed suitable for oxidatively transforming lignin model compounds [135–137] and lignin samples. The polymer-supported variant oxidized both phenolic and non-phenolic lignin model compounds, and due to the lowered Lewis acidity of the MTO catalyst upon immobilization, also aliphatic groups can be oxidized while side-reactions targeting the aromatic rings and leading to over-oxidation can be suppressed.

Cobalt salen ([Co(salen)]) complexes, which had been shown to be compatible with aqueous reaction media [138], were successfully used in various studies on oxidative lignin transformations. EPR studies revealed that oxidation proceeds start with the formation of a phenoxy-radical, which reacts with molecular oxygen to ultimately form oxidized lignin model compounds (Figure 4.5b) [139–142]. More recently new Co-Schiff base catalysts bearing a bulky heterocyclic nitrogen base as a substituent have been synthesized, which was able to convert both S- and G-type lignin models in high yield, a process that remained elusive so far but would be essential for an effective use of lignin as a chemical feedstock [143]. Cobalt was supported on TiO_2, CeO_2, and ZrO_2 for the oxidative depolymerization of lignins [144]. Results showed that high bio-oil yield was obtained when Co/CeO_2 was used as the catalyst, and high selectivity of acetosyringone was achieved with Co/TiO_2 as catalyst.

Screening of the catalytic performances of $CoCl_2{\cdot}6H_2O$, $CuSO_4$, $MnN_2O_6{\cdot}4H_2O$, $CrKO_8S_2{\cdot}12H_2O$, and $NiCl_2{\cdot}xH_2O$ in the catalytic oxidative depolymerization of lignin in aqueous 1-ethyl-3-methylimidazolium acetate [C_2C_1Im][OAc] with H_2O_2 revealed that $CoCl_2$ was the most active catalyst for the production of aromatic compounds from lignin tested [145].

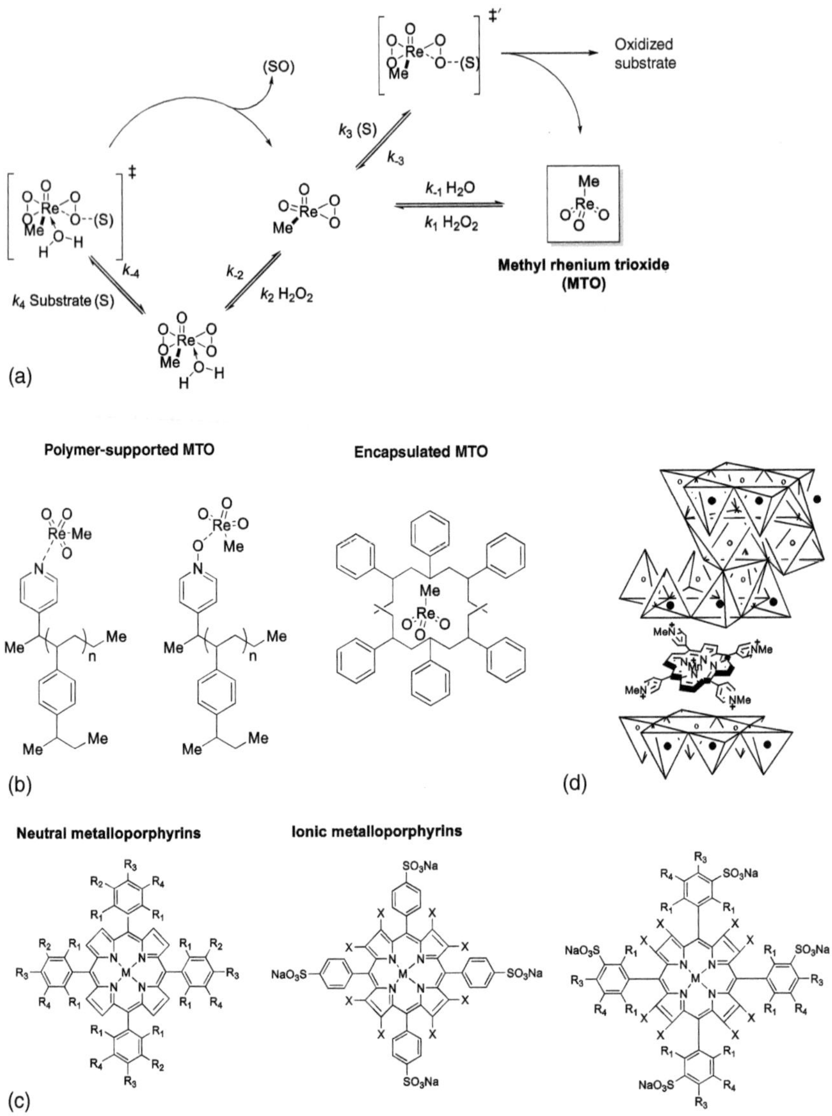

Figure 4.5 Organometallic catalysts for the oxidative valorization of lignin: (a) Methyltrioxo rhenium (MTO) and mechanistic details of radical oxidation. (b) Polymer-supported versions of MTO. (c) Neutral and ionic metalloporphyrins. (d) Schematic illustrating a polymer-supported metalloporphyrin. For details and references, see main text.

Au and Pt particles were supported on TiO_2 for the oxidative depolymerization of Kraft and ethanol organosolv lignins [146] and showed that Au–TiO_2 can further oxidize the aromatic hydrocarbons produced by the depolymerization of lignin, resulting in a lower yield of the target product, while Pt–TiO_2 can promote the conversion of lignin into high value-added aromatic hydrocarbons. $VO(acac)_2$- and

$Cu(OAc)_2$-proofed versatile homogeneous catalysts for the oxidative depolymerization of kraft lignin in the presence of oxygen [147]. A 50% bio-oil yield and increased production of aromatic monomers were achieved by the use of both catalysts, suggesting the favorable combined effect.

Soft- and hardwood lignins were degraded into monomeric aromatic compounds using biomimetic vanadate and molybdate catalyst systems. Monomers were generated with remarkable, catalyst-dependent selectivity and high depolymerization yields through oxidative *exo-* and *endo*-depolymerization processes. Using the vanadate–hydrogen peroxide system on acetosolv pine lignin, vanillin, and isovanillin were produced as main products with depolymerization yields of 31%. Using the molybdate system on acetosolv and ionosolv lignin, vanillic acid was the practically exclusive product in depolymerization yields of up to 72%. Eucalyptus acetosolv lignin showed depolymerization yields of 50% in form of vanillin, syringaldehyde, or vanillic acid employing a vanadate or molybdate catalyst, respectively [148].

Keggin-type polyoxometalates, i.e. POMs, with the general formula $[PM^1xM^2{}_{12}{-}xO_{40}]$ can activate hydrogen peroxide and molecular oxygen for the oxidative valorization of lignin and its various model compounds. Different protocols, including the catalytic use of POMs have been developed as "green" alternatives for the oxidative valorization of lignin [149–152]. Polyoxometalate IL (POM-IL) hybrid catalysts were tested in the oxidative cleavage of lignin aromatic units [153]. The high yield of diethyl maleate was attributed to the combined catalytic effect with acidic and redox active sites from the POM and IL.

Phenolic lignin model compounds were shown to be oxidized by C—C bond cleavage between the aryl ring and the adjacent hydroxyl-substituted carbon atom using vanadium catalysts as the one shown in Figure 4.6 [154]. Labeling experiments indicate key mechanistic differences to a previously reported related C—O bond cleavage reaction that was effected by a different vanadium catalyst and that leads to a cleavage of lignin bonding motifs in a reductive fashion (*vide infra*) [155].

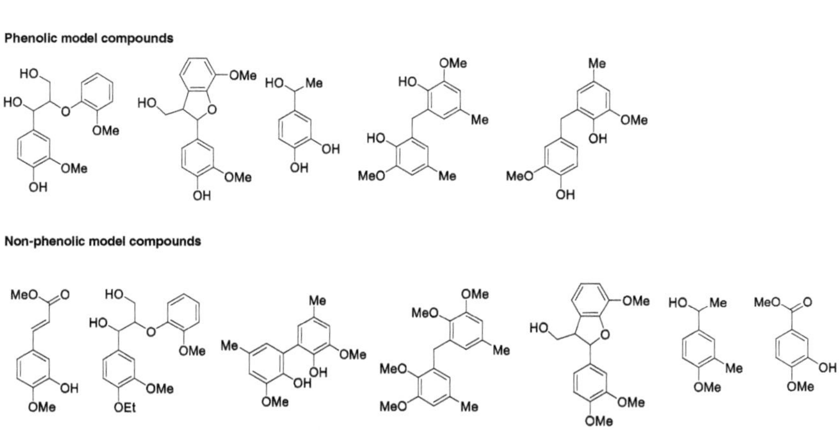

Figure 4.6 Monomeric and dimeric models used to mimic the typical interunit linkages and functional groups in lignin (*vide supra*, Figure 4.1).

It has to be clearly stated at this point that the above-mentioned reactivities of the different catalysts have been elucidated mainly using lignin model compounds (Figure 4.6). These model compounds are usually synthetically derived dimers or dimer-surrogates mimicking not only a certain binding motif but also the impact of the substitution pattern of the aromatic moiety in correspondence with the three lignin monomer units (*vide supra*). The use of models is necessary, unfortunately, since the structural variety encountered in isolated lignins will not allow for an easy discovery of mechanistic insights. On the other hand, model studies might indicate a promising selectivity of a novel catalyst that, when applied to the chemical complexity of an isolated lignin, does not necessarily verify itself. On the basis of this discussion, it should also be evident that not all of the available isolated lignins are equally suitable as starting materials for oxidative valorizations, and that not all valorization approaches in this field fit the structural features present in a technical lignin (*vide supra*).

The activity of the above-mentioned catalysts is shown in Figure 4.7 on the basis of the most common interunit bonding motif, i.e. the aryl glycerol-β-aryl ethers, i.e. β-*O*-4′ structures. This overview alone indicates already the complexity actually encountered in degradative oxidative lignin valorization. It is thus yet necessary to test the desired oxidative upgrading against the technical lignin at hand, in order to decide on the viability of the process on the basis of hard experimental facts.

4.4.1.3 Biomimetic Catalysts

In nature, the selective modification of lignin is mainly accomplished by enzymes such as laccase, lignin peroxidase (LiP), and manganese-dependent peroxidases (MnP), which use molecular oxygen [156]. Synthetic metalloporphyrins (Figure 4.5c) represent biomimetic systems for both LiP and MnP, since they can yield highly oxidized metallo-oxo species similar to those found in LiP I, and LiP II [157]. Highly functionalized porphyrins, bearing aryl substituents in the *meso*-positions of the heme motif, represent tunable catalyst systems: both the redox potential and the solubility of the complexes can be tuned by carefully varying the nature of the *meso*-substituents. These catalyst systems exhibit an advantage over the enzyme-based variants employing the same reactive metal core since the metalloporphyrins are not as sensitive toward an excess of hydrogen peroxide and are thus potentially more interesting with respect to industrial applications.

Anionic manganese and iron *meso*-tetra(2,6-dichloro-3-sulphonatophenyl porphyrin chlorides (TDCSPPMnCl and TDCSPPFeCl, respectively), anionic *meso*-tetra-4-sulphonatophenyl porphyrin chloride (TSPPMnCl), and cationic manganese *meso*-tetra(*N*-methylpyridinio)porphorin pentaacetate (TPyMePMn$(MeCOO)_5$) (Figure 4.5c) were shown to effectively oxidize residual kraft lignin and lignin model compounds yielding the usual degradation products – benzoquinones, benzyl alcohols, and phenyl alkyl ketones in acidic environments (pH 3–6) [158].

All porphyrin catalysts produced comparable mixtures of products, hinting at comparable reaction mechanisms despite the different metal centers and porphyrin moieties. The cationic manganese complex TPyMPMn$(MeCOO)_5$, however,

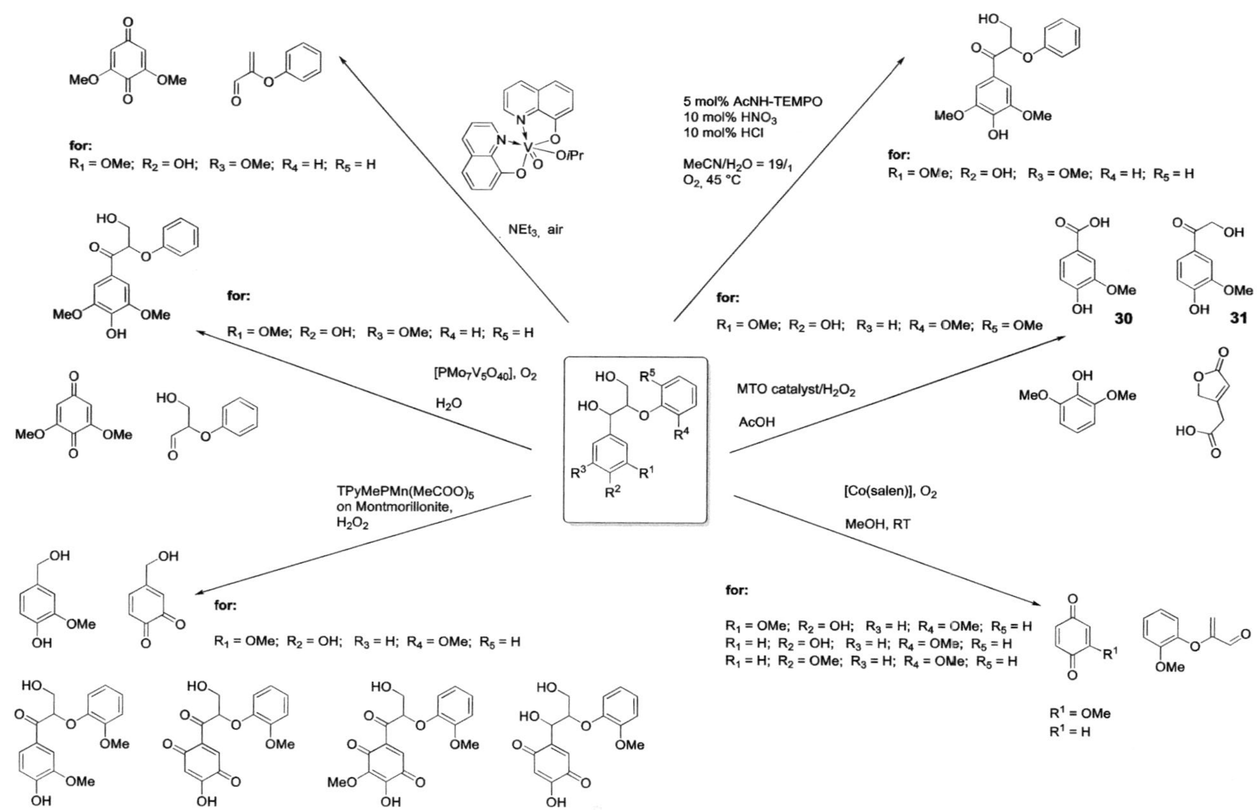

Figure 4.7 Conversion of the most common interunit bonding motif aryl glycerol-β-aryl ethers (β-*O*-4′) upon oxidative treatment with different organometallic catalysts and under metal-free conditions. Source: References are given in discussion in the main text.

performed best, and in general, manganese complexes performed better than the corresponding iron complexes. The water-soluble manganese porphyrin derivative distinguished itself through high yields of products originating from oxidations of both the aromatic ring and the side chains, as well as products indicative of demethylation reactions; only low yields of undesired products resulting from radical recombination reactions.

The major disadvantage of the porphyrin complexes lies in their liability under the reaction conditions typically used for lignin degradation processes. This problem can be solved by developing a mimic of the polypeptide envelope that protects the active center in natural enzymes. One simple way to achieve such an envelope mimic is to immobilize the porphyrin-based catalyst systems on solid supports such as silica gel [159], naturally occurring clays [158], or artificial polymers [160]. These immobilization methods, which also significantly ease recovery of the catalysts, do not cause any loss in reactivity. Clays from the smectite family proved to be very versatile due to an inherent structural feature: two-dimensional oxyanions are separated by layers of hydrated metal cations, which can be exchanged using simple ion-exchange protocols. For example, Montmorillonite-supported TPyMPMn(MeCOO)$_5$ (Figure 4.5d) activated hydrogen peroxide for the oxidative decomposition of lignin model compounds. The identical compositions of the product mixtures obtained for both the immobilized catalyst and its free analog suggest identical reaction mechanisms underlying the oxidations.

It has been reported that the porphyrin-skeleton can be attached to typical IL cations such as imidazolium derivatives, which allow solubilizing the porphyrin-based catalysts in ILs [161], which have been used before in studies on lignin degradation.

The immobilization of metalloporphyrins on the surface of porous clays, despite all the advantages mentioned above, reduces the accessibility of the active metal core for bulkier substrates. In order to maintain high turnover numbers, mediators can be used. These are ideally small enough to easily diffuse through the pores of the clay to access the activated metal core, become activated, and subsequently diffuse back into the medium where they can readily react with the bulkier substrates. The use of meta-stable mediator systems is generally beneficial with respect to the oxidative functionalization and depolymerization of lignin, since they can readily diffuse into/in between the polymeric structures of lignin. Immobilized metalloporphyrin-mediator systems were shown to oxidize phenols and lignins by side-chain oxidation/fragmentation and depolymerization, as spectroscopically (^{31}P NMR spectroscopy) validated by decreasing intensities of the signals corresponding to aliphatic hydroxyl groups, and increasing intensities of signals corresponding to carboxyl groups [162, 163]. Noteworthy, no products originating from oxidative coupling reactions were observed. More detailed mechanistic studies on the effects of the mediators on the oxidative functionalization of lignin are covered in the following paragraphs.

4.4.2 Enzymatic Oxidative Degradation

In natural environments, lignin is effectively degraded through the action of numerous enzymes secreted by fungi and bacteria. Based on the lignin degradation mechanism, fungal systems are classified as soft-rot, white-rot, and brown-rot fungi, respectively [164]. Among these three systems, the white-rot fungi system has been observed to be the most efficient. White-rot fungi were observed to catalyze the complete degradation of lignin to carbon dioxide and water. Moreover, aerobic and anaerobic bacteria with the ability to degrade lignin have been isolated from soil, gut, wastewater, and wood [165, 166]. While bacteria are less efficient in lignin degradation compared to fungi, their environmental adaptability is significantly higher [167].

The enzymatic depolymerization of lignin is considered a highly sustainable and environmentally friendly approach, primarily due to its mild conditions and absence of harsh chemicals; moreover, this method avoids the release of toxic compounds into the environment.

In both fungi and bacteria, the most active lignin-degrading enzymes are laccases [EC 1.10.3.2], manganese peroxidases [EC 1.11.1.13], and lignin peroxidases [EC 1.11.1.14] [168]. In the following, important mechanistic aspects of his natural depolymerization of lignin are highlighted. It is further discussed which possibilities exist to further enhance the industrial exploitation of enzymatic depolymerization processes (Table 4.1).

4.4.2.1 Laccase-Mediated Oxidative Depolymerization

The multicopper oxidase Laccase [EC 1.10.3.2] activates suitable substrates under concomitant reduction of oxygen to water. Within an outer-sphere electron-transfer mechanism, radical cationic species are generated initially from a phenolic substrate. This radical cation undergoes subsequent deprotonation to yield a phenoxy radical, which initiates the depolymerization of lignin following an *exo*-depolymerization mechanism [156, 180–182]. Mainly the oxidation potential of the substrate, and only to a lesser extent on the steric demand of a potential substrate dictates whether a laccase can transform it. Although different fungi produce laccases with different redox potentials, ranging from $E° = 0.43\,V$ (tree laccase from *Rhus vernicifera*) to 0.78 V (fungal laccase from *Polyporus versicolor*) [183], a laccase suitable to oxidize blocked phenolic substrates is missing.

Blocked phenolic substrates can, however, be indirectly oxidized by laccases, when these are used in combination with a radical mediator species, such as 1-hydroxybenzotriazole (HBT) [184], *N*-hydroxyacetanilide (NHA) [185], violuric acid, or 2,2′-azinobis-3-ethyl-benzthiazoline-6-sulfonate (ABTS) [186, 187] (Figure 4.8). By means of the mediators, the efficacy of laccase-based depolymerization of lignin is enhanced since these can readily diffuse into the lignin fibers after being activated by the enzyme, and since they can initiate depolymerization *via* an additional mechanistic route, i.e. an additional starting point for the depolymerizing cascade of reactions. Mechanistic studies involving the mediator HBT revealed that laccase initially converts HBT to an oxybenzotriazolyl radical, which subtracts a hydrogen atom from either the phenolic or the benzylic position. Rather simple semiempirical calculations (PM3) suggested a comparable spin density for the two radical species. Oxidation presumably does thus no longer occur through an electron transfer process in these cases, but in enzyme-mediator systems (also) *via* a hydrogen atom abstraction process [188].

Table 4.1 Lignin depolymerizing enzymes.

Enzyme	Type (native/ recombinant)	Strain name or expression host	Optimum conditions (pH, *T*[°C])	Molecular weight (kDa)	References
Laccase	Native	*Thielavia* sp.	5.0, 70	70	[169]
	Native	*Trichoderma asperellum* BPLMBT1	3.0, 80	65	[170]
	Native	*Pseudomonas* sp. S2	9.0, 80	38	[171]
	Native	*Deconica castanella*	2.5, 55	64	[172]
	Native	*Oudemansiella canarii*	5.5, 30	31	[173]
	Native	*Bacillus* sp.	2.0, 60	36	[174]
	Metagenomic origin	*E. coli* Rosetta (DE3)	5.0, 50	–	[175]
	Recombinant	*E. coli* BL-21(DE3)	5.0, 60	59	[176]
Lignin peroxidase	Native	*Acinetobacter* sp.	2.0, 60	99	[177]
	Native	*Phanerochaete chrysosporium* NK-1	3.0, 40	46	[178]
	Native	*Klebsiella pneumonia*	5.0, 50	66	[179]
Manganese peroxidase	Native	*Mycobacterium* sp.	6.0, 31	–	[178]
	Native	*Klebsiella pneumonie*	5.0, 60	48	[179]

Source: Adapted from Ref. [167].

Most noteworthy in light of a valorization of the depolymerization products, not only the mechanistic pathways are affected by the presence of mediators but also the structural motifs found across the depolymerization products. In laccase-only systems, products originating from oxidation of the alkyl side-chain oxidation, as well as oxidative coupling reactions were obtained (Figure 4.8c). These products suggest that laccases equally well oxidize side chains, observable by the decrease of aliphatic hydroxyl groups present in lignin side chains by quantitative ^{31}P NMR (*vide supra*) and aromatic rings of the lignin structure, as indicated by the increase of condensed phenolic species in the ^{31}P NMR analyses. When HBT is added, products originating from oxidation of the aromatic rings (Figure 4.8c) were found in significant yields, beside products indicative of oxidations of the side chains. The formation of the products found upon addition of HBT cannot be explained on the basis of the reaction of oxygen-centered radicals such as phenoxy radicals with molecular oxygen, since the addition of molecular oxygen to phenoxy radicals is slow [189]; the addition of a superoxide anion to a phenoxy species is faster, and thus more likely to be the reason for the formation of the observed product mixtures. The additionally postulated benzyl radical would kinetically be suitable to bind molecular oxygen, resulting in a peroxo radical species that could decompose under formation of a superoxide anion, which would then act as reaction partner for the phenoxy radical present in the mixture.

4.4.2.2 Peroxidase-Mediated Oxidative Depolymerization

Manganese peroxidases (MPs) oxidatively depolymerize lignin under concomitant reduction of molecular oxygen or hydrogen peroxide to water [190]. Activation of molecular oxygen or hydrogen peroxide is achieved *via* a two-step process, in which

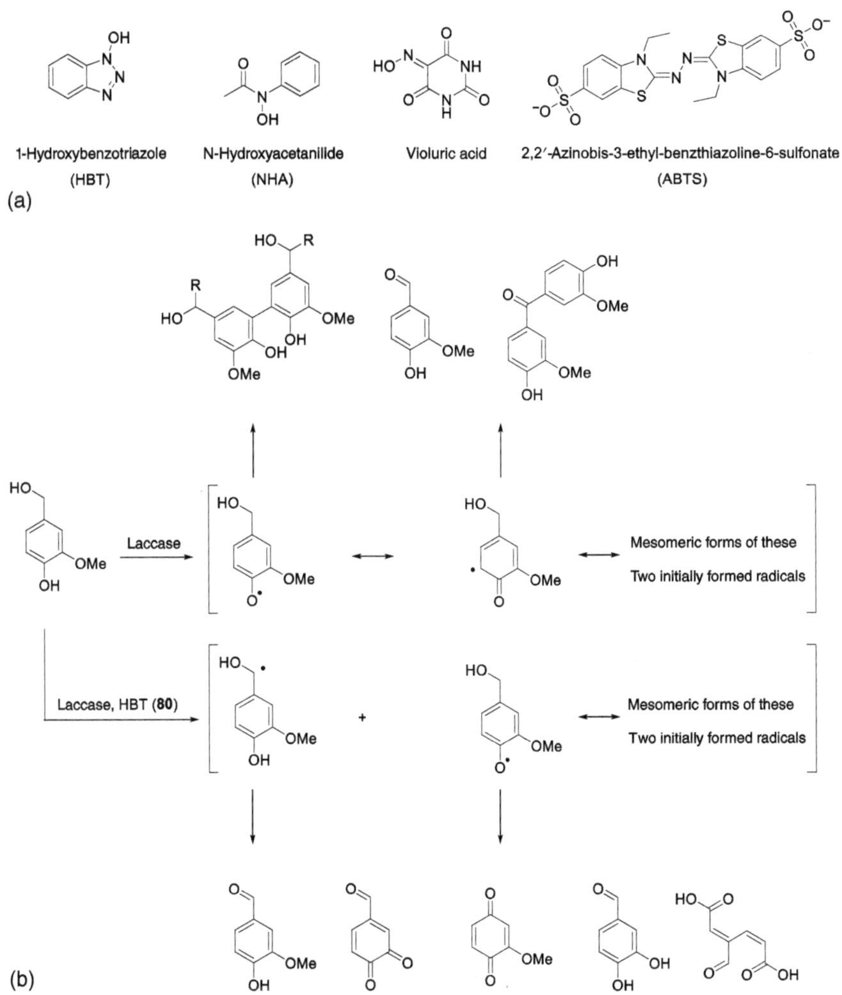

Figure 4.8 (a) Common mediators. (b) Mechanistic aspects of laccase and laccase-mediator systems. (c) Product distributions of laccase and laccase-mediator systems.

an iron protoporphyrin IX activates the oxidant first for the oxidation of the manganese co-factor from Mn(II) to Mn(III) [191]. The Mn(III)-center is subsequently chelated by carboxylic acid anions, and forms thus a small, freely diffusible species – comparable to the activated mediator in the laccase-mediator system – that ultimately oxidatively depolymerizes lignin and lignin model compounds [192, 193].

The depolymerization of lignin is initiated by the formation of a phenoxy radical species *via* hydrogen abstraction [190]. Different lignin model compounds, mimicking the recalcitrant phenolic arylglycerol β-aryl ether and diarylpropane motifs, as well as 5–5′, β-5′, and diphenylmethane subunits (Figure 4.6), could be cleaved successfully using MnP [192–194].

The combined MnP results suggest that the reactive manganese species is selectively attacking methyl and methylene groups in *para*-position to the phenolic OH-groups in lignin. As in case of laccase, an *exo*-depolymerization mechanism is assumed based on the analyses of the residual lignin.

4.4.2.3 Technological Advancements to Enable Enzymatic Lignin Degradation

Since enzymes require very specific technology and there is a growing industrial demand, these bioproducts have high market value. Consequently, a new area of research has emerged where studies have been conducted to enable enzyme immobilization [195–199].

Enzyme immobilization enhances their catalytic capabilities, improving stability and resistance to denaturation in an organic medium, facilitating the separation of the reaction medium, increasing resistance to different temperatures, pH, and organic solvents, and increasing its storage capacity and even allowing its reuse [200–205]. Enzyme immobilization can be achieved through various methods, either physical or chemical [206]. The commonly employed methods can be categorized into entrapment, where the enzyme is immobilized within an inert support, direct binding to the support through physical adsorption or binding/covalent coupling, and crosslinking, which involves the intermolecular binding of insoluble enzymes or proteins [207]. It is important to note that each method possesses its own set of advantages and disadvantages.

Moreover, it is essential to maintain the conformational structure and specific activity of the enzyme during immobilization, without compromising its functionality upon contact with the support surface [208]. Consequently, the careful selection of an appropriate support material can greatly improve the system's performance. From a morphological perspective, solid supports can be categorized as either porous or nonporous [209]. From a chemical composition standpoint, solid supports can be classified as either organic or inorganic materials [210].

The enzymes immobilization, in addition to enhancing their catalytic activity, has enabled the implementation of innovative production processes in the field of lignin degradation. The development of new technologies and chemistry for lignin valorization is a crucial aspect in realizing the potential of biomass as a chemical feedstock.

Continuous flow technology has emerged as a well-established technique for chemists, enabling them to enhance their control over reactions and explore novel reactions that may be challenging or impractical using traditional methods [211]. Flow chemistry offers several advantages, including enhanced mixing efficiency for gas/liquid reactions, improved efficiency for photochemical transformations, reduced environmental impact, increased reaction safety, and enhanced scalability. Compared with traditional batch techniques [212, 213]. Moreover, the sustainability factor associated with continuous flow processes is a key driver behind the recognition of continuous flow chemistry as one of the top 10 emerging technologies in chemistry by IUPAC [214].

A characteristic of all laccases is their strong pH dependence. Most laccases, both of fungal and bacterial origin, exhibit an optimum activity toward phenolic compounds in the pH range of 3–5.5 and become inactive at neutral and/or alkaline pH [215]. This particular characteristic greatly limits their industrial application, given the low solubility of lignin in water at this acid pH. This gap has been partially overcome through the development and utilization of extremozymes, namely, enzymes active at alkaline pH values, such as pH 10–11, and high temperatures, targeting lignin at its soluble state.

Numerous extremozymes have been directly isolated or genetically modified from extremophile microorganisms, which are known for their ability to survive in extreme environment. The development of enzymes or microorganisms that can withstand the extreme conditions of high temperature, high osmotic pressure, high acidity/alkalinity, and the toxicity of phenolic compounds is required in the lignin biorefinery system. Consequently, extremophiles and extremozymes hold significant potential in the field of lignin bioprocessing [216].

4.4.3 Catalytic Chemical Reductive Degradation

Reductive modification of lignin has been a research subject for decades [217]. The approach is, in certain chemical sense, of a more destructive nature due to the removal of versatile functional groups. Also in this area, findings are regularly updated in comprehensive review articles [218].

4.4.3.1 Metal-Free Reductions

Hydrosilanes were shown to convert aryl ethers to hydrolyzable silyl ethers in the presence of $B(C_6F_5)_3$ in a general process that was used to convert model lignin compounds to both aryl silyl ethers and alkanes. Softwood lignin itself could only be partially solubilized (~30%) but hardwood lignin was efficiently reductively decomposed by the mono-functional hydrosilane $HMe_2SiOSiMe_3$ [219].

Ethanol/water mixtures were shown to readily solubilize lignin under moderate temperatures and pressures with little residual solids. The molecular weight of the dissolved lignins was shown to be reduced by GPC and NMR methods [125].

4.4.3.2 Heterogenic Transition Metal Catalysts

Cleavage and hydrodeoxygenation (HDO) of C—O bonds in lignin model compounds was achieved using a Pd/Zn synergistic catalysis at high temperatures (150 °C) and high hydrogen pressure (20 bar) [220]. The catalyst system showed a high selectivity and was recyclable. Tests on whole lignins, however, were not reported.

A heterogeneous nickel catalyst for the selective hydrogenolysis of aryl ethers to arenes and alcohols generated in situ from soluble nickel precursor $Ni(COD)_2$ or $Ni(CH_2TMS)_2(TMEDA)$ (COD – cyclooctadiene; TMS – trimethylsilyl; TMEDA – tetramethyl ethylenediamine) in the presence of a base additive [221]. The catalyst selectively cleaves C—O bonds in aryl ether models of lignin without hydrogenation of aromatic rings, and it operates at loadings down to 0.25 mol% at atmospheric hydrogen (Figure 4.9a).

4.4.3.3 Organometallic Catalysts

The conversion of lignin into low-boiling-point arenes instead of high-boiling-point phenols was realized by depolymerizing lignin under simultaneous conversion of phenols into arenes using a tandem reaction catalyzed by a tailored nickel-centered soluble organometallic catalyst [222]. Earlier studies using homogeneous Ni-catalyzes for these purposes were not tried on lignin itself [223]. An earlier study accomplished c. 4 decades ago on lignin model compounds used a homogeneous nickel-hydride complex [224]. Selectivity in this system depended highly on the substrate specificities.

A vanadium species has also been reported to reductively degrade lignin by selectively attacking the C—O bond motifs of the aryl ethers building the lignin backbone (Figure 4.9b) [155, 225]. The studies were accompanied by a detailed structural analysis of the lignin that remains after catalysis, which confirmed the selectivity of the catalyst. It was also shown explicitly, however, that the history of the lignin has an influence on the effectiveness of the transformation.

4.5 Thermally Induced Lignin Depolymerization

Thermal treatment of lignin concerns the transformation of solid lignin into liquid or gas products by supplying thermal energy. Depending on the temperature profile and the reaction environment, the degree of lignin fragmentation can be adjusted. Pyrolysis and gasification are the two most common examples in the category. Pyrolysis of biomass is typically performed in an oxygen-free environment, and the temperature can rise from room to 400–600 °C in several seconds. The rate of heating, final temperature, and the residence time can be used to control the resulting product. Compared to chemical-based fragmentation, pyrolysis, done in the presence or absence of a catalyst (*vide infra*), is a low-cost method to achieve lignin depolymerization efficiently, but the resulting liquid products, such as bio-oil, typically exhibit multiple phase separation issues and may contain highly reactive carbonyl groups that are prone to uncontrolled polymerization. The chemical instability of the product requires additional hydrostabilization treatment before the bio-oil can be used for further upgrading or storage [226]. The recent years saw a drastic increase in research targeting the valorization of lignin using thermal degradation methods. Among these, pyrolysis is probably the best known one, since pyrolysis coupled to GC separation of products and their subsequent

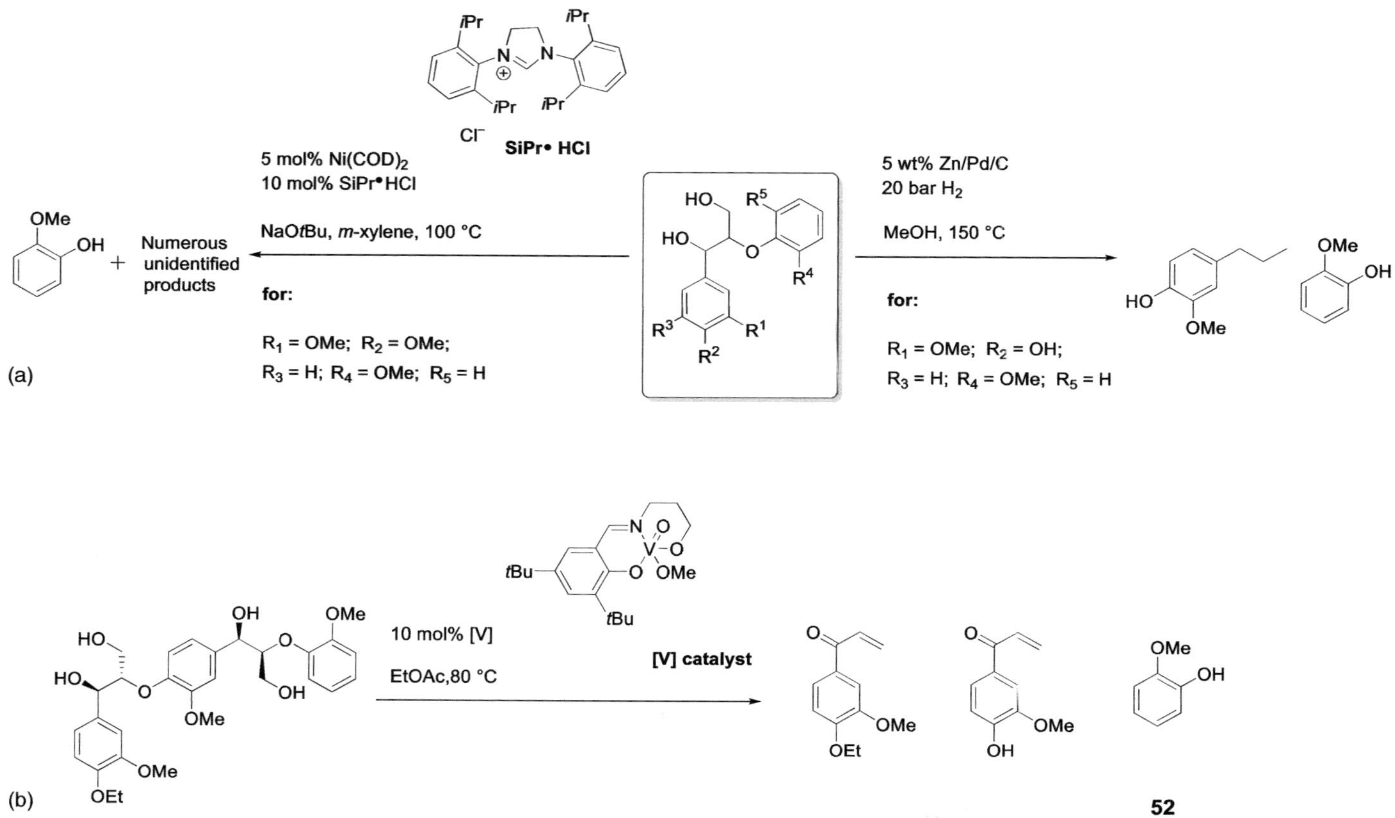

Figure 4.9 (a) Conversion of the most common interunit bonding motif aryl glycerol-β-aryl ethers (β-*O*-4′) upon reductive treatment with different organometallic catalysts (COD – cyclooctadiene). (b) Vanadium-catalyzed reductive cleavage of a trimeric lignin model. Source: References are given in discussion in the main text.

detection using mass spectrometry, i.e. pyrolysis-GC–MS, is a standard analytical approach nowadays in lignocellulolytic research. Other techniques, in part also directly applicable to non-fractionated biomass, are fast developing, not only due to the drastically changing energy landscape. While the processes discussed so far were designed for isolated lignins, the processes discussed in the paragraphs of this section can be used either for isolated lignins, or for entire, i.e. unfractionated biomass. In the latter case, the described efforts can be seen as part of the "lignin-first" biorefinery approaches.

4.5.1 Hydro Treatment-Based Lignin Depolymerization

The main types of reactions involved in the hydrotreating process of lignin include hydrogenolysis, hydrodeoxygenation, hydrogenation, and integrated hydrogen-processing. In the hydrogenolysis process, hydrogen or other hydrogen sources have been used to reduce lignin at high temperatures. Through the hydrogenation treatment, depolymerized lignin, phenols, and other chemicals with high added value are obtained, as well as low molecular weight hydrocarbons. Careful catalyst and solvent or solvent system choices are necessary to arrive at viable processes. In the hydrodeoxygenation of lignin, various pillared clays have been studied as catalysts. Some examples of pillared clays used include montmorillonite-based pillared clays characterized by a layered structure of aluminum and magnesium silicate, chemically modified to enhance their catalytic activity in the hydrodeoxygenation of lignin, bentonite-based pillared clays, that has been used as a catalyst in the conversion of lignin to produce valuable chemical compounds and, finally, zeolite-based pillared clays, that thanks of zeolites crystalline structures with highly regular pore sizes, can be conveniently pillared to form highly active catalytic materials. These zeolites used as catalysts in the hydrodeoxygenation of lignin allow obtaining valuable aromatic compounds. Table 4.2 lists processes in which monomer formations reached significant levels.

For the most part, hydrogenolysis by reductive catalytic depolymerization has received more attention than the oxidative approach because of the preservation of the linkages during the lignin-first biorefinery efforts, such as β-*O*-4′. An increasing number of research articles have also demonstrated convenient one-pot lignin isolation and depolymerization in a reductive environment; these works have been summarized elsewhere before [239].

4.5.2 Pyrolytic Lignin Degradation

4.5.2.1 Torrefaction of Lignin

Torrefaction is the thermochemical biomass conversion technology, which requires heating the biomass at temperatures between 200 and 300 °C for one hour or longer at atmospheric pressure and under an oxygen-deficient environment [240]. The process reduces effectively hydrogen and oxygen contents in biomass while increasing its carbon content, which results in the partial elimination of volatile matter, thereby improving its energy density [241]. Torrefied biomass, and hence also torrefied lignin can directly be used for combustion, gasification, and pyrolysis (*vide infra*). Torrefaction

Table 4.2 Representative pyrolytic lignin degradation results.

Lignin	Temperature	Reactor	Catalyst/additive	Results	References
Softwood kraft lignin	600 °C	Quartz pyrolysis tube reactor	HZSM-5, HY, Hbeta, Ferrierite, Mordenite	Decrease of molecular weight and oxygen-containing functional group was observed	[227]
Asian lignin, lignol lignin, sugarcane bagasse lignin, cellulose and lignin mixture	650 °C	Py-GC/MS	HZSM-5	Aromatic hydrocarbons yields of about 6 wt% lignin feedstock	[228]
Softwood kraft lignin	500 °C	Bench scale fixed bed reactor	spent FCC catalyst	Bio-oil and phenolics yields of 35.95 and 7.56 wt%, respectively	[229]
Alkali lignin	600 °C	Py-GC/MS	$CaCl_2$	Phenolics and aromatic hydrocarbons yields of about 67.18 area% and 5.02 area%, respectively	[230]
Softwood alkali lignin	650 °C	Quartz reactor-GC/MS	(Al)-MCM-41	Bio-oil and aromatic hydrocarbons with yields of about 54 wt% and 80 area%, respectively	[231]
Softwood milled wood lignin, organosolv lignin	500–700 °C	Py-GC/MS	HZSM-5 (SiO2/Al2O3 = 30–280)	Aromatic hydrocarbons with yields up to 35.7 mg/g of lignin	[232]
Alkaline lignin	650 °C	Py-GC/MS	H-USY, Amorphous silica alumina	Aromatic hydrocarbon with circa 40 wt% of carbon yield	[233]

(Continued)

Table 4.2 (Continued)

Lignin	Temperature	Reactor	Catalyst/additive	Results	References
Kraft lignin	500 °C	Py-GC/MS	Mesoporous HY (pseudomorphic synthesis)	Mono-aromatic hydrocarbons with yields of 18 area%	[234]
Kraft lignin	500 °C	Py-GC/MS	Mesoporous ZSM-5	Mono-aromatic hydrocarbon yield of 24 area%	[235]
Kraft lignin	500 °C	Py-GC/MS	Alkaline treated hierarchical HZSM-5	Selectivity of oxygenates was reduced from 27.2% to 3.3%	[236]
Alkaline lignin	650 °C	Py-GC/MS	Zeolite (HZSM-5, HUSY) supported transition metal (Ni, Co)	Selectivity of aromatic hydrocarbons yields of about 70%	[237]

Source: Adapted from Ref. [238].

weakens the inherent structural chemistry of biomass, thus making it easier to thermally degrade through other thermochemical conversion technologies.

As a pretreatment, torrefaction lowers the overall power and time consumption as well as process expenditures otherwise required for biomass drying and grinding [242]. Torrefaction of biomass also eliminates certain volatile compounds such as carboxylic acids, monomers, and low molecular weight species responsible for rendering high acidity to pyrolysis oil [243]. The non-condensable gases from torrefaction usually include hydrogen, carbon monoxide, carbon dioxide, and methane, hence syngas equivalents and reduced and highly oxidized single carbon compounds, while the condensed liquid products contain water, acetic acid, alcohols, aldehydes, ketones, furfural, and other mono-aromatic compounds [244, 245]. Torrefaction of lignin is not a mainstream process, normally lignin is directly used for pyrolysis, not only for cost-related reasons.

In studies targeting the effect of torrefaction temperature on the subsequent pyrolysis behavior of lignin, it was found that the oxygen-containing functional groups in torrefied lignin were sharply decreased, resulting in the decrease of the total oxygen-containing phenolic content in pyrolytic bio-oil [246, 247]. Following the quantitative evolution of key functional groups in lignin, i.e. interunit bonding motifs, aromatic structures, and hydroxyl groups, during torrefaction processes by ^{1}H–^{13}C-HSQC and ^{31}P NMR analysis, most fragile structures could be delineated. Another recent study highlighted the connections between structural features of lignins and the torrefaction products: the contents of oxygen-rich structures, i.e. β-*O*-4′ linkages, aliphatic OH-groups and aliphatic carboxylic acids, and methoxy groups, decreased with the increase of the torrefaction temperature from 210 to 300 °C. The oxygen removal efficiency of lignin torrefaction reached the maximum value of 25.53% at 300 °C. The removed oxygen was transferred into the gaseous fraction in form of water, carbon monoxide, and carbon dioxide, and the liquid products in form of H- and G-type phenols [247].

4.5.2.2 Catalyst-Free Pyrolysis of Lignins

Pyrolysis can thermally decompose polymeric materials at medium to high temperatures, i.e. between 400 and 600–700 °C, considering active and passive pyrolysis, and under an inert, i.e. non-oxidative atmosphere. Typical products are pyrolysis gas, pyrolysis oil, and char. Pyrolysis of biomass is known to produce high oil yields and is thus interesting for the energy as well as the platform chemical sectors of lignin valorization [248]. Efficiency of lignin pyrolysis has been increased by applying different lignin sources, extraction methods, reaction systems, pyrolysis oil upgrading technologies, and co-feeding hydrogen-rich feedstock to lignin pyrolysis [249].

The pyrolysis properties of lignin obtained from different biomass and extracted by different methods have been studied widely using microscale pyrolysis instruments, such as TGA, micro-scale Py-GC/MS, tandem-micro reactor-GC/MS, and TG- FTIR [250–253]. It is found generally that hardwood lignins decompose at lower temperatures and produce larger amounts of char than softwood lignins. This can be understood in light of the higher contents of condensed-type linkages and methoxy groups present in hardwood lignin (*vide supra*).

Interestingly, the mildness of the biorefinery method used for lignin isolation seems to play a role. In a comparative study, the lignin isolated by Björkman's procedure, i.e. gentle milling of biomass, lower temperatures for cooking, etc. [254], showed the lowest biochar and highest phenolic yield [255]. The pyrolysis of pine wood lignins extracted using different isolation methods revealed that alkali and milled wood lignins decompose at lower temperatures than klason and organosolv lignins, most probably due to differences in intrinsic thermal stabilities that can be ultimately traced back to the different abundancies of key functional motifs, i.e. interunit bondings (*vide supra*). Also in this field, studies on lignin models helped to delineate causal relationships [256, 257].

4.5.2.3 Catalytic Pyrolysis of Lignin

Lignin pyrolysis oil is a promising source for the production of valuable chemicals because of its high content of phenolics. Yet, this high content in phenolics means also a high oxygen content, notable viscosity, and chemical instability [258]. In order to tackle these issues, but especially the instability, catalytic pyrolysis was developed using as catalysts zeolites [227], metal oxide [228], alumina [229], metal chlorides [230], and mesoporous materials [231]. Proton Zeolite Socony Mobil-5, i.e. HZSM-5, is the most effective catalyst for aromatic hydrocarbon production *via* catalytic pyrolysis of lignin [234, 236, 237, 259]. Higher temperatures and catalyst-to-lignin ratios are essential to increase the yield of aromatic hydrocarbons, exploiting the fact that the catalyst pores can be expanded at higher temperatures and that the overall number of acid sites, essential for beneficial catalytic activity [259], can be maximized by simple increasing the amount of catalyst.

Mechanistically, two scenarios are currently discussed as most probable: (i) hydrocarbon pool [233]; and (ii) phenolic pool [260]. The hydrocarbon pool mechanism is related to the formation of light olefins by the de-alkylation of lignin pyrolyzates. These olefins are finally converted to aromatic hydrocarbons by their oligomerization inside the catalyst pore [261]. The phenolic pool mechanism was suggested by a study for the conversion of lignin model compounds. In the case of production of aromatics *via* the phenolic pool mechanism, the phenolic pyrolyzates of lignin undergo isomerization, transalkylation, and condensation over the catalysts. The phenolic pool mechanism was explained by the formation of a phenolic pool consisting of multi-phenolic species on the catalyst surface. These phenolic pools produce small aromatic hydrocarbons by additional cracking and hydrogen atom transfer. Detailed studies using the simplest possible monomeric lignin model (*vide supra*), i.e. guaiacol, indicated that additional mechanism involve demethylation and dehydration steps as well [262].

Transition metals such as cobalt, molybdenum, nickel, iron, manganese, or copper, but equally well magnesium as alkali earth metal were impregnated on various supports, such as zeolites [237] and Al-Mobil Composition Matter-41 (Al-MCM-41) [263] were tested for production of "stable" pyrolysis oils [237].

Pyrolysis oil manipulation in lignin pyrolysis was also investigated using torrefied biomass, chemical treatment, and co-feeding of hydrogen-rich feedstocks to the catalytic pyrolysis of lignin. The use of pretreated lignins, which had been torrefied and pretreated with hydrochloric acid, sodium hydroxide, and hydrogen peroxide, caused the generation of aromatic hydrocarbons with good product selectivity [264]. The catalytic co-pyrolysis of lignin and hydrogen-rich plastics, such as polyethylene, polypropylene, polystyrene, or polyethylene terephthalate yielded stable oils [263].

Despite the promising results obtained on a laboratory, i.e. research level scale, scaling-up of lignin pyrolysis to industrial levels is difficult yet. Continuous lignin pyrolysis is difficult due to lignin melting and char foaming/agglomeration. More research is needed in this area in order to optimize the various steps separately and to arrive at in-situ, or in-line solutions for the mentioned issues. Another issue is the fact that even the stable pyrolysis oils obtained from lignin pyrolysis need most often an additional treatment. Hydrodeoxygenation (HDO) is considered to be the most effective method to upgrade lignin pyrolysis oil to deoxygenated hydrocarbon fuels or high-value chemicals such as cyclohexanol or cyclohexanone [265]. In the presence of supported metal catalyst and molecular hydrogen, oxygen functionalities in the aromatic rings of lignin pyrolysis, oils such as methoxy and hydroxyl groups, are selectively removed in the form of water and methanol [266]. The design of highly active and selectively supported metal catalysts for the HDO of lignin pyrolysis oil is actually one of the main research subjects in the area of lignin valorization. Recently efforts were summarized, and the interested reader is referred to these review articles [238, 267]. Yet, most studies on the HDO of lignin pyrolysis are realized using once more simplest model compounds, such as guaiacol or phenol. HDO studies using real lignin pyrolysis oil are relatively limited. Due to its complex composition, i.e. a mixture of phenolic monomers and oligomers and mineral impurities such as sulfur, the activity and stability of the catalysts for the HDO of pyrolysis oil are significantly different from those for the model compounds. Available studies [268–270] suggest that catalyst stability is the main issue rather than catalyst activity for the HDO of real lignin oils.

4.5.2.4 Advanced Pyrolysis Technologies

In light of the shortcomings connected to current pyrolysis technologies, novel variations were developed recently. These pyrolysis technologies focus on enhancing the efficiency, selectivity, and versatility of lignin conversion, aiming thus at maximizing the production of valuable products while minimizing waste and environmental impact. They often involve novel technological approaches, advanced catalysts, and catalyst systems, and tailored process conditions to optimize lignin depolymerization as well as product formation and collection. Five important novel developments are briefly discussed in the following.

Liquid-phase pyrolysis involves converting lignin in form of a solution; suitable solvents include water, alcohols, or oils. Lignin is dissolved and heated to induce thermal decomposition. This technology can enhance selectivity in producing desired compounds such as phenols, furfurals, and carboxylic acids [271–273].

Supercritical-phase pyrolysis takes place under conditions of temperature and pressure above the critical point of the solvent used for lignin solubilization. This increases the solubility of reactants and products, improving lignin conversion and selectivity for desired products such as hydrocarbons and aromatic compounds [274].

In modified atmosphere pyrolysis, the pyrolysis environment is controlled by using dedicated atmospheres, ranging from inert to reactive conditions. The controlled atmosphere influences lignin decomposition and product formation [275–277].

Pulsed electric field pyrolysis utilizes high-voltage, short-duration electric pulses to rapidly heat lignin and induce pyrolysis. The electric pulses generate rapid heating and can promote the breaking of lignin polymer chains, producing desired chemical products such as hydrocarbons, phenols, and aldehydes [278, 279].

4.6 Electrochemical Lignin Degradation

Electrochemical depolymerization of lignin is an intrinsically interesting approach, since potentials for the reduction–oxidation reactions can be technically tuned [280]. Working in alkaline electrolytes, suitable for dissolving the lignin easily, depolymerization often occurs in oxidative environments, while electrochemical hydrogenation is usually applied toward upgrading the degraded aromatic units [226, 280]. Monoaromatic specialty chemicals, i.e. vanillin, syringaldehyde, guaiacol, ferulic acid, syringaldehyde, and acetosyringone could be generated, but the overall yields remained low [281]. The mechanisms of electro-oxidative depolymerization of lignin are still unclear in detail. It is known that in the electro-oxidation system, lignin in alkaline solution is electro-oxidized directly on the anode surface, resulting in intermediate radical cations [282, 283]. The destiny of lignin radical cations depends on the selected anode [284], initial lignin concentration, alkali concentration, and reaction conditions, such as current density, reaction time, and temperature [285]. Reaction rates are a function of current flow, and this needs to be low since side reactions such as water oxidation need to be suppressed/circumvented. Here, the developments in adjacent fields such as electrode design will probably be crucial: Nickel-based anodes are promising because of their enhanced stability against corrosion and relatively low cost, whereas Co-based anodes show some corrosion even if they provide higher yields of phenolic aldehydes [286]. The electrode structure is important to facilitate the accessibility of lignin macromolecules and obtain a high surface area for electro-oxidation.

Over-oxidation of the products, occurring normally when the chosen electrocatalysts are nonselective, was shown to be circumvented using a membrane reactor [287]. However, oxidatively degrading the lignin by a reactive oxygen species generated in-situ during electro-oxidation as well as direct anodic oxidation was achieved using a non-diaphragm cylindrical electrolytic cell [283]. These two types of oxidation routes produced higher yields of low molecular weight products in short time frames. The latter is beneficial for obtaining meaningful yields of aldehydes and ketones. Phenols and acids, on the other hand, are generally obtained in higher amounts only by increasing electrolysis time. To arrive at high concentrations of both times and circumventing undesired over-oxidations, electrochemical depolymerization was coupled to product separation by means of a suitable ultrafiltration [288].

Also in the area of electrochemical lignin depolymerization, the unique properties of ILs and DESs were tested with respect to synergistically enhance depolymerization performance: DES was suggested to dissolve the lignin [231], in combination with an in situ product extraction concept for the electrochemical depolymerization of kraft lignin [289].

A common, rather general drawback is yet the efforts necessary for the recovery of the electrolytes. More interested readers are referred to more recent reviews that go into additional details for the various approaches [218, 238, 290].

4.7 Concluding Remarks and Outlook

The summarizing chapter on lignin valorization upon degradation demonstrates the broad variety of possibilities that have been developed within the scientific community targeting lignin valorization. It is evident that some well-known, but nevertheless old-fashioned and in terms of chemical processing standards in 2023 outdated methods that use stoichiometric reagents are important only for the analytical aspects connected to them with respect to lignin structure elucidation. Depolymerization-based valorizations of lignin employ nowadays catalysts, be it for classical wet-chemical procedures, be it for biotechnological approaches. Using appropriate (bio)catalysts, oxidative and reductive degradations of lignin can be run. Yet, the major drawbacks continue persisting in form of low depolymerization yields and low product selectivity when "real," i.e. commercial lignins are used. While mechanistic understanding made and still makes rapid progress in the area, conversion of this acquired mechanistic knowledge into more selective and higher-yielding processes struggles when it comes to conversions of technical lignins with biomass-typical batch-to-batch variations.

Giving seemingly credits to the ill-famous saying that "lignin is not good for anything but burning it," only thermal valorizations in form of pyrolysis and connected processes like HDO seem to hold a realistic future potential for winning lignin degradation into monomeric species compatible with the needs of the energy sector and the chemical industry.

In light of the pressing socio-economic and environmental boundary conditions that characterize our time, efforts should be focused to further optimize and mature the most promising approaches, e.g. pyrolysis, such as to have sustainable options for monomeric aromatics and biofuels in the nearer future. In longer time-frames, chemical methods based on well-performing catalysts might become universally applicable in terms of "preferred" "catalyst-biopolymer" combinations, as part of a thinned out and cleaned up, i.e. more targeted and overall more specific, realistic biorefinery landscape.

References

1 Beckham, G.T. (ed.) (2018). *Lignin Valorization: Emerging Approaches*. Cambridge: Royal Society of Chemistry https://doi.org/10.1039/9781788010351.

2 Aresta, M., Dibenedetto, A., and Dumeignil, F. (2015). *Biorefineries, An Introduction*. Berlin, Boston: De Gruyter.

3 Lange, H., Giannì, P., and Crestini, C. (2018). Lignin analytics, Chap. 15. In: *Energy and Environment* (ed. G.T. Beckham), 413–476. Cambridge: Royal Society of Chemistry https://doi.org/10.1039/9781788010351-00413.

4 Lupoi, J.S., Singh, S., Parthasarathi, R. et al. (2015). Recent innovations in analytical methods for the qualitative and quantitative assessment of lignin. *Renewable and Sustainable Energy Reviews* 49: 871–906. https://doi.org/10.1016/j.rser.2015.04.091.

5 Lapierre, C. (2010). Determining lignin structure by chemical degradations. In: *Lignin Lignans* (ed. C. Heitner, D. Dimmel, and J. Schmidt), 11–48. CRC Press https://doi.org/10.1201/EBK1574444865-c2.

6 Schmidt, J. (2010). Electronic spectroscopy of Lignins. In: *Lignin Lignans* (ed. C. Heitner, D. Dimmel, and J. Schmidt), 49–102. CRC Press https://doi.org/10.1201/EBK1574444865-c3.

7 Argyropoulos, D. and Heteronuclear, N.M.R. (2010). Spectroscopy of Lignins. In: *Lignin Lignans* (ed. C. Heitner, D. Dimmel, and J. Schmidt), 245–265. CRC Press https://doi.org/10.1201/EBK1574444865-c6.

8 Hatakeyama, H. and Hatakeyama, T. (2010). Thermal properties of isolated and in situ lignin. In: *Lignin Lignans* (ed. C. Heitner, D. Dimmel, and J. Schmidt), 301–319. CRC Press https://doi.org/10.1201/EBK1574444865-c8.

9 Lewis, N.G. and Yamamoto, E. (1990). Lignin: occurrence, biogenesis and biodegradation. *Annual Review of Plant Physiology and Plant Molecular Biology* 41: 455–496. https://doi.org/10.1146/annurev.pp.41.060190.002323.

10 Sarkanen, K.V. and Ludwig, C.H. (1971). *Lignins: Occurrence, Formation, Structure and Reactions*. Wiley-Interscience.

11 Adler, E. (1977). Lignin chemistry—past, present and future. *Wood Science and Technology* 11: 169–218. https://doi.org/10.1007/BF00365615.

12 Zhao, Y., Shakeel, U., Rehman, M.S.U. et al. (2020). Lignin-carbohydrate complexes (LCCs) and its role in biorefinery. *Journal of Cleaner Production* 253: 120076. https://doi.org/10.1016/j.jclepro.2020.120076.

13 Glasser, W.G. and Sarkanen, S. (1989). *Lignin: Properties and Materials*. Washington, DC: American Chemical Society.

14 Pesquet, E., Zhang, B., Gorzsás, A. et al. (2013). Non-cell-autonomous Postmortem lignification of Tracheary elements in *Zinnia elegans*. *Plant Cell* 25: 1314–1328. https://doi.org/10.1105/tpc.113.110593.

15 Decou, R., Serk, H., Ménard, D., and Pesquet, E. (2017). Analysis of lignin composition and distribution using fluorescence laser confocal microspectroscopy. In: *Xylem* (ed. M. de Lucas, J.P. Etchhells, and M. de Lucas), 233–247. New York: Springer https://doi.org/10.1007/978-1-4939-6722-3_17.

16 Vanholme, R., Demedts, B., Morreel, K. et al. (2010). Lignin biosynthesis and structure. *Plant Physiology* 153: 895–905. https://doi.org/10.1104/pp.110.155119.

17 Boerjan, W., Ralph, J., and Baucher, M. (2003). Lignin biosynthesis. *Annual Review of Plant Biology* 54: 519–546. https://doi.org/10.1146/annurev.arplant.54.031902.134938.

18 Donaldson, L.A. (2001). Lignification and lignin topochemistry—an ultrastructural view. *Phytochemistry* 57: 859–873. https://doi.org/10.1016/S0031-9422(01)00049-8.

19 Whetten, R. and Sederoff, R. (1995). Lignin biosynthesis. *Plant Cell* 7: 1001–1013. https://doi.org/10.1105/tpc.7.7.1001.

20 Heitner, C., Dimmel, D., and Schmidt, J. (2010). *Lignin and Lignans: Advances in Chemistry*. CRC Press.

21 Martarello, D.C.I., Almeida, A.M., Sinzker, R.C. et al. (2023). The known unknowns in lignin biosynthesis and its engineering to improve lignocellulosic saccharification efficiency. *Biomass Convers Biorefinery* 13: 2497–2515. https://doi.org/10.1007/s13399-021-01291-6.

22 Weng, J.-K. and Chapple, C. (2010). The origin and evolution of lignin biosynthesis. *The New Phytologist* 187: 273–285. https://doi.org/10.1111/j.1469-8137.2010.03327.x.

23 Hussain, K., Jaweed, T.H., and Kamble, A.C. (2023). Modulation of phenylpropanoid and lignin biosynthetic pathway is crucial for conferring resistance in pigeon pea against fusarium wilt. *Gene* 851: 146994. https://doi.org/10.1016/j.gene.2022.146994.

24 Vanholme, R., De Meester, B., Ralph, J., and Boerjan, W. (2019). Lignin biosynthesis and its integration into metabolism. *Current Opinion in Biotechnology* 56: 230–239. https://doi.org/10.1016/j.copbio.2019.02.018.

25 Humphreys, J.M. and Chapple, C. (2002). Rewriting the lignin roadmap. *Current Opinion in Plant Biology* 5: 224–229. https://doi.org/10.1016/S1369-5266(02)00257-1.

26 Hatfield, R. and Fukushima, R.S. (2005). Can lignin be accurately measured? *Crop Science* 45: 832. https://doi.org/10.2135/cropsci2004.0238.

27 Akiyama, T., Goto, H., Nawawi, D.S. et al. (2005). Erythro/threo ratio of β-O-4-5 structures as an important structural characteristic of lignin. Part 4: variation in the erythro/threo ratio in softwood and hardwood lignins and its relation to syringyl/guaiacyl ratio. *Holzforschung* 59: 276–281. https://doi.org/10.1515/HF.2005.045.

28 Akiyama, T., Sugimoto, T., Matsumoto, Y., and Meshitsuka, G. (2002). Erythro/threo ratio of β-O-4 structures as an important structural characteristic of lignin. I: improvement of ozonation method for the quantitative analysis of lignin side-chain structure. *Journal of Wood Science* 48: 210. https://doi.org/10.1007/BF00771369.

29 Howard, G.C. (1934). Utilization of sulfite liquor. *Industrial and Engineering Chemistry* 26: 614–617. https://doi.org/10.1021/ie50294a007.

30 Ruwoldt, J. (2020). A critical review of the physicochemical properties of lignosulfonates: chemical structure and behavior in aqueous solution, at surfaces and interfaces. *Surfaces* 3: 622–648. https://doi.org/10.3390/surfaces3040042.

31 Kienberger, M., Maitz, S., Pichler, T., and Demmelmayer, P. (2021). Systematic review on isolation processes for technical lignin. *Processes* 9: 804. https://doi.org/10.3390/pr9050804.

32 Chakar, F.S. and Ragauskas, A.J. (2004). Review of current and future softwood Kraft lignin process chemistry. *Industrial Crops and Products* 20: 131–141. https://doi.org/10.1016/j.indcrop.2004.04.016.

33 Gierer, J. (1980). Chemical aspects of Kraft pulping. *Wood Science and Technology* 14: 241–266. https://doi.org/10.1007/BF00383453.

34 Tomani, P. (2010). The lignoboost process. *Cellulose Chemistry and Technology* 44: 53–58.

35 Guerra, A., Filpponen, I., Lucia, L.A., and Argyropoulos, D.S. (2006). Comparative evaluation of three lignin isolation protocols for various wood species. *Journal of Agricultural and Food Chemistry* 54: 9696–9705. https://doi.org/10.1021/jf062433c.

36 McDonough, T.J. (1993). The chemistry of Organosolv delignification. *Tappi Journal* 76: 186–193.

37 Pye, E.K. and Lora, J.H. (1991). The Alcell process. A proven alternative to Kraft pulping. *Tappi Journal* 74: 113–118.

38 Meng, X., Crestini, C., Ben, H. et al. (2019). Determination of hydroxyl groups in biorefinery resources via quantitative 31P NMR spectroscopy. *Nature Protocols* 14: 2627–2647. https://doi.org/10.1038/s41596-019-0191-1.

39 Rönnols, J., Danieli, E., Freichels, H., and Aldaeus, F. (2020). Lignin analysis with benchtop NMR spectroscopy. *Holzforschung* 74: 226–231. https://doi.org/10.1515/hf-2018-0282.

40 Argyropoulos, D.S., Pajer, N., and Crestini, C. (2021). Quantitative ^{31}P NMR analysis of Lignins and tannins. *Journal of Visualized Experiments* 62696: https://doi.org/10.3791/62696.

41 Saake, B., Argyropoulos, D.S., Beinhoff, O., and Faix, O. (1996). A comparison of lignin polymer models (DHPs) and lignins by 31P NMR spectroscopy. *Phytochemistry* 43: 499–507. https://doi.org/10.1016/0031-9422(96)00283-X.

42 Granata, A. and Argyropoulos, D.S. (1995). 2-Chloro-4,4,5,5-tetramethyl-1,3,2-dioxaphospholane, a reagent for the accurate determination of the uncondensed and condensed phenolic moieties in Lignins. *Journal of Agricultural and Food Chemistry* 43: 1538–1544. https://doi.org/10.1021/jf00054a023.

43 Wroblewski, A.E., Lensink, C., Markuszewski, R., and Verkade, J.G. (1988). Phosphorus-31 NMR spectroscopic analysis of coal pyrolysis condensates and extracts for heteroatom functionalities possessing labile hydrogen. *Energy & Fuels* 2: 765–774. https://doi.org/10.1021/ef00012a008.

44 Peterson, D.J. and Loening, N.M. (2007). QQ-HSQC: a quick, quantitative heteronuclear correlation experiment for NMR spectroscopy. *Magnetic Resonance in Chemistry* 45: 937–941. https://doi.org/10.1002/mrc.2073.

45 Hu, K., Westler, W.M., and Markley, J.L. (2011). Simultaneous quantification and identification of individual chemicals in metabolite mixtures by two-dimensional extrapolated time-zero 1H–13C HSQC (HSQC0). *Journal of the American Chemical Society* 133: 1662–1665. https://doi.org/10.1021/ja1095304.

46 Zhang, L. and Gellerstedt, G. (2007). Quantitative 2D HSQC NMR determination of polymer structures by selecting suitable internal standard references. *Magnetic Resonance in Chemistry* 45: 37–45. https://doi.org/10.1002/mrc.1914.

47 Yang, W.-J., Griffiths, P.R., Byler, D.M., and Susi, H. (1985). Protein conformation by infrared spectroscopy: resolution enhancement by Fourier self-deconvolution. *Applied Spectroscopy* 39: 282–287.

48 Kauppinen, J.K., Moffatt, D.J., Mantsch, H.H., and Cameron, D.G. (1981). Fourier self-deconvolution: a method for resolving intrinsically overlapped bands. *Applied Spectroscopy* 35: 271–276. https://doi.org/10.1366/0003702814732634.

49 Maddams, W.F. (1980). The scope and limitations of curve fitting. *Applied Spectroscopy* 34: 245–267.

50 Pierce, J.A., Jackson, R.S., Van Every, K.W. et al. (1990). Combined deconvolution and curve fitting for quantitative analysis of unresolved spectral bands. *Analytical Chemistry* 62: 477–484. https://doi.org/10.1021/ac00204a011.

51 Hergert, H.L. (1984). *An Investigation of the Vibrational Spectra of Lignin Model Compounds*. Georgia Institute of Technology.

52 Agarwal, U. and Atalla, R. (2010). Vibrational spectroscopy. In: *Lignin Lignans* (ed. C. Heitner, D. Dimmel, and J. Schmidt), 103–136. CRC Press https://doi.org/10.1201/EBK1574444865-c4.

53 Meyer, M.W., Lupoi, J.S., and Smith, E.A. (2011). 1064 nm dispersive multichannel Raman spectroscopy for the analysis of plant lignin. *Analytica Chimica Acta* 706: 164–170. https://doi.org/10.1016/j.aca.2011.08.031.

54 Larsen, K.L. and Barsberg, S. (2010). Theoretical and Raman spectroscopic studies of phenolic lignin model monomers. *The Journal of Physical Chemistry. B* 114: 8009–8021. https://doi.org/10.1021/jp1028239.

55 Obiaga, T.I. and Wayman, M. (1974). Improved calibration procedure for gel permeation chromatography of lignins. *Journal of Applied Polymer Science* 18: 1943–1952. https://doi.org/10.1002/app.1974.070180704.

56 Chen, F. and Li, J. (2000). Aqueous gel permeation chromatographic methods for technical Lignins. *Journal of Wood Chemistry and Technology* 20: 265–276. https://doi.org/10.1080/02773810009349636.

57 Salanti, A., Zoia, L., Tolppa, E.-L., and Orlandi, M. (2012). Chromatographic detection of lignin–carbohydrate complexes in annual plants by derivatization in ionic liquid. *Biomacromolecules* 13: 445–454. https://doi.org/10.1021/bm2014763.

58 Wong, K.K.Y. and de Jong, E. (1996). Size-exclusion chromatography of lignin- and carbohydrate-containing samples using alkaline eluents. *Journal of Chromatography A* 737: 193–203. https://doi.org/10.1016/0021-9673(95)01382-2.

59 Pellinen, J. and Salkinoja-Salonen, M. (1985). Aqueous size exclusion chomatography of industrial lignins. *Journal of Chromatography A* 322: 129–138. https://doi.org/10.1016/S0021-9673(01)97665-4.

60 Chum, H.L., Johnson, D.K., Tucker, M.P., and Himmel, M.E. (1987). Some aspects of lignin characterization by high performance size exclusion chromatography using styrene Divinylbenzene copolymer gels. *Holzforschung* 41: 97–108. https://doi.org/10.1515/hfsg.1987.41.2.97.

61 Bikova, T., Treimanis, A., Rossinska, G., and Telysheva, G. (2004). On-line study of lignin behaviour in dilute alkaline solution by the SEC-UV method. *Holzforschung* 58: https://doi.org/10.1515/HF.2004.074.

62 Ringena, O., Lebioda, S., Lehnen, R., and Saake, B. (2006). Size-exclusion chromatography of technical lignins in dimethyl sulfoxide/water and dimethylacetamide. *Journal of Chromatography A* 1102: 154–163. https://doi.org/10.1016/j.chroma.2005.10.037.

63 Gidh, A.V., Decker, S.R., Vinzant, T.B. et al. (2006). Determination of lignin by size exclusion chromatography using multi angle laser light scattering. *Journal of Chromatography A* 1114: 102–110. https://doi.org/10.1016/j.chroma.2006.02.044.

64 Braaten, S.M., Christensen, P.B.E., and Fredheim, G.E. (2003). Comparison of molecular weight and molecular weight distributions of softwood and hardwood lignosulfonates. *Journal of Wood Chemistry and Technology* 23: 197–215. https://doi.org/10.1081/WCT-120021925.

65 Cathala, B., Saake, B., Faix, O., and Monties, B. (2003). Association behaviour of lignins and lignin model compounds studied by multidetector size-exclusion chromatography. *Journal of Chromatography A* 1020: 229–239. https://doi.org/10.1016/j.chroma.2003.08.046.

66 Fredheim, G.E., Braaten, S.M., and Christensen, B.E. (2002). Molecular weight determination of lignosulfonates by size-exclusion chromatography and multi-angle laser light scattering. *Journal of Chromatography A* 942: 191–199. https://doi.org/10.1016/S0021-9673(01)01377-2.

67 Cheeseman, J.M. (2015). The evolution of halophytes, glycophytes and crops, and its implications for food security under saline conditions. *The New Phytologist* 206: 557–570. https://doi.org/10.1111/nph.13217.

68 Glenn, E.P., Brown, J.J., and Blumwald, E. (1999). Salt tolerance and crop potential of halophytes. *Critical Reviews in Plant Sciences* 18: 227–255. https://doi.org/10.1080/07352689991309207.

69 Helm, R.F. (1999). Lignin—polysaccharide interactions in Woody plants. In: *Lignin: Historical, Biological, and Materials Perspectives*, vol. 742 (ed. W.G. Glasser, R.A. Northey, and T.P. Schultz), 161–171. Washington, DC: American Chemical Society https://doi.org/10.1021/bk-2000-0742.ch005.

70 Zikeli, F., Ters, T., Fackler, K. et al. (2016). Wheat straw lignin fractionation and characterization as lignin-carbohydrate complexes. *Industrial Crops and Products* 85: 309–317. https://doi.org/10.1016/j.indcrop.2016.03.012.

71 Hatfield, R.D., Rancour, D.M., and Marita, J.M. (2017). Grass cell walls: a story of cross-linking. *Frontiers in Plant Science* 7: https://doi.org/10.3389/fpls.2016.02056.

72 Lam, T.B.T., Kadoya, K., and Iiyama, K. (2001). Bonding of hydroxycinnamic acids to lignin: ferulic and p-coumaric acids are predominantly linked at the benzyl position of lignin, not the β-position, in grass cell walls. *Phytochemistry* 57: 987–992. https://doi.org/10.1016/S0031-9422(01)00052-8.

73 Ghaffar, S.H. and Fan, M. (2013). Structural analysis for lignin characteristics in biomass straw. *Biomass and Bioenergy* 57: 264–279. https://doi.org/10.1016/j.biombioe.2013.07.015.

74 Buranov, A.U. and Mazza, G. (2008). Lignin in straw of herbaceous crops. *Industrial Crops and Products* 28: 237–259. https://doi.org/10.1016/j.indcrop.2008.03.008.

75 Ralph, J., Hatfield, R.D., Grabber, J.H. et al. (1998). Cell Wall cross-linking in grasses by Ferulates and Diferulates. In: *Lignin Lignan Biosynth*, American Chemical Society, vol. 697, 209–236. https://doi.org/10.1021/bk-1998-0697.ch016.

76 Oinonen, P., Zhang, L., Lawoko, M., and Henriksson, G. (2015). On the formation of lignin polysaccharide networks in Norway spruce. *Phytochemistry* 111: 177–184. https://doi.org/10.1016/j.phytochem.2014.10.027.

77 Jeffries, T.W. (1991). Biodegradation of lignin-carbohydrate complexes. In: *Physiology of Biodegradative Microorganisms* (ed. C. Ratledge), 163–176. Dordrecht: Springer https://doi.org/10.1007/978-94-011-3452-1_7.

78 Dimmel, D. and Gellerstedt, G. (2010). Chemistry of alkaline pulping. In: *Lignin Lignans* (ed. C. Heitner, D. Dimmel, and J. Schmidt), 349–391. CRC Press https://doi.org/10.1201/EBK1574444865-c10.

79 Eriksson, K.-E. (2010). Lignin biodegradation. In: *Lignin Lignans* (ed. C. Heitner, D. Dimmel, and J. Schmidt), 495–520. CRC Press https://doi.org/10.1201/EBK1574444865-c14.

80 Shinde, S.D., Meng, X., Kumar, R., and Ragauskas, A.J. (2018). Recent advances in understanding the pseudo-lignin formation in a lignocellulosic biorefinery. *Green Chemistry* 20: 2192–2205. https://doi.org/10.1039/C8GC00353J.

81 Liu, S., Zhu, Y., Liao, Y. et al. (2022). Advances in understanding the humins: formation, prevention and application. *Applications in Energy and Combustion Science* 10: 100062. https://doi.org/10.1016/j.jaecs.2022.100062.

82 Thoresen, P.P., Lange, H., Rova, U. et al. (2023). Covalently bound humin-lignin hybrids as important novel substructures in organosolv spruce lignins. *International Journal of Biological Macromolecules* 233: 123471. https://doi.org/10.1016/j.ijbiomac.2023.123471.

83 Lange, H., Rulli, F., and Crestini, C. (2016). Gel permeation chromatography in determining molecular weights of Lignins: critical aspects revisited for improved utility in the development of novel materials. *ACS Sustainable Chemistry & Engineering* 4: 5167–5180. https://doi.org/10.1021/acssuschemeng.6b00929.

84 Lundquist, K. and Kirk, T.K. (1980). Fractionation-purification of an industrial Kraft lignin. *Tappi Journal* 63: 1.

85 Schuerch, C. (1952). The solvent properties of liquids and their relation to the solubility, swelling, isolation and fractionation of lignin. *Journal of the American Chemical Society* 74: 5061–5067. https://doi.org/10.1021/ja01140a020.

86 Mousavioun, P. and Doherty, W.O.S. (2010). Chemical and thermal properties of fractionated bagasse soda lignin. *Industrial Crops and Products* 31: 52–58. https://doi.org/10.1016/j.indcrop.2009.09.001.

87 Cui, C., Sun, R., and Argyropoulos, D.S. (2014). Fractional precipitation of softwood Kraft lignin: isolation of narrow fractions common to a variety of Lignins. *ACS Sustainable Chemistry & Engineering* 2: 959–968. https://doi.org/10.1021/sc400545d.

88 Lange, H., Schiffels, P., Sette, M. et al. (2016). Fractional precipitation of wheat straw Organosolv lignin: macroscopic properties and structural insights. *ACS Sustainable Chemistry & Engineering* 4: 5136–5151. https://doi.org/10.1021/acssuschemeng.6b01475.

89 Thring, R.W., Vanderlaan, M.N., and Griffin, S.L. (1996). Fractionation of Alcell® lignin by sequential solvent extraction. *Journal of Wood Chemistry and Technology* 16: 139–154. https://doi.org/10.1080/02773819608545815.

90 Teng, N.-Y., Dallmeyer, I., and Kadla, J.F. (2013). Effect of softwood Kraft lignin fractionation on the dispersion of multiwalled carbon nanotubes. *Industrial and Engineering Chemistry Research* 52: 6311–6317. https://doi.org/10.1021/ie303261z.

91 Gosselink, R.J.A., van Dam, J.E.G., de Jong, E. et al. (2010). Fractionation, analysis, and PCA modeling of properties of four technical lignins for prediction of their application potential in binders. *Holzforschung* 64: 193–200. https://doi.org/10.1515/hf.2010.023.

92 Zikeli, F., Ters, T., Fackler, K. et al. (2014). Successive and quantitative fractionation and extensive structural characterization of lignin from wheat straw. *Industrial Crops and Products* 61: 249–257. https://doi.org/10.1016/j.indcrop.2014.07.013.

93 Duval, A., Vilaplana, F., Crestini, C., and Lawoko, M. (2015). Solvent screening for the fractionation of industrial Kraft lignin. *Holzforschung* 70: 11–20. https://doi.org/10.1515/hf-2014-0346.

94 Griffini, G., Passoni, V., Suriano, R. et al. (2015). Polyurethane coatings based on chemically unmodified fractionated lignin. *ACS Sustainable Chemistry & Engineering* 3: 1145–1154. https://doi.org/10.1021/acssuschemeng.5b00073.

95 Passoni, V., Scarica, C., Levi, M. et al. (2016). Fractionation of industrial softwood Kraft lignin: solvent selection as a tool for tailored material properties. *ACS Sustainable Chemistry & Engineering* 4: 2232–2242. https://doi.org/10.1021/acssuschemeng.5b01722.

96 Majdar, R.E., Crestini, C., and Lange, H. (2020). Lignin fractionation in segmented continuous flow. *ChemSusChem* 13: 4735–4742. https://doi.org/10.1002/cssc.202001138.

97 Gosselink, R.J.A., Van, D.P., and Van, E. (2015). Fractionation of Technical Lignin. WO2015178771 (A1).

98 Toledano, A., García, A., Mondragon, I., and Labidi, J. (2010). Lignin separation and fractionation by ultrafiltration. *Separation and Purification Technology* 71: 38–43. https://doi.org/10.1016/j.seppur.2009.10.024.

99 Sevastyanova, O., Helander, M., Chowdhury, S. et al. (2014). Tailoring the molecular and thermo-mechanical properties of Kraft lignin by ultrafiltration. *Journal of Applied Polymer Science* 131: n/a–n/a. https://doi.org/10.1002/app.40799.

100 Duval, A., Molina-Boisseau, S., and Chirat, C. (2014). Fractionation of lignosulfonates: comparison of ultrafiltration and ethanol solubility to obtain a set of fractions with distinct properties. *Holzforschung* 69: 127–134. https://doi.org/10.1515/hf-2014-0082.

101 Gigli, M. and Crestini, C. (2020). Fractionation of industrial lignins: opportunities and challenges. *Green Chemistry* 22: 4722–4746. https://doi.org/10.1039/D0GC01606C.

102 Sadeghifar, H. and Ragauskas, A. (2020). Perspective on technical lignin fractionation. *ACS Sustainable Chemistry & Engineering* 8: 8086–8101. https://doi.org/10.1021/acssuschemeng.0c01348.

103 Paulsen Thoresen, P., Lange, H., Rova, U. et al. (2023). Role and importance of solvents for the fractionation of lignocellulosic biomass. *Bioresource Technology* 369: 128447. https://doi.org/10.1016/j.biortech.2022.128447.

104 Singh, S.K. (2022). Ionic liquids and lignin interaction: an overview. *Bioresource Technology Reports* 17: 100958. https://doi.org/10.1016/j.biteb.2022.100958.

105 Wang, Z.-K., Hong, S., Wen, J. et al. (2020). Lewis acid-facilitated deep eutectic solvent (DES) pretreatment for producing high-purity and antioxidative lignin. *ACS Sustainable Chemistry & Engineering* 8: 1050–1057. https://doi.org/10.1021/acssuschemeng.9b05846.

106 Dier, T.K.F., Rauber, D., Durneata, D. et al. (2017). Sustainable electrochemical depolymerization of lignin in reusable ionic liquids. *Scientific Reports* 7: 5041. https://doi.org/10.1038/s41598-017-05316-x.

107 Pometto, A.L. and Crawford, D.L. (1985). Simplified procedure for recovery of lignin acidolysis products for determining the lignin-degrading abilities of microorganisms. *Applied and Environmental Microbiology* 49: 879–881.

108 Lundquist, K. (1992). Acidolysis. In: *Methods in Lignin Chemistry* (ed. L. DSY and P.E.D.C.W. Dence), 289–300. Berlin Heidelberg: Springer https://doi.org/10.1007/978-3-642-74065-7_20.

109 Lapierre, C., Monties, B., and Rolando, C. (2009). Preparative thioacidolysis of spruce lignin: isolation and identification of main monomeric products. *Holzforschung - International Journal of the Biology, Chemistry, Physics and Technology of Wood* 40: 47–50. https://doi.org/10.1515/hfsg.1986.40.1.47.

110 Rolando, C., Monties, B., and Lapierre, C. (1992). Thioacidolysis. In: *Methods in Lignin Chemistry* (ed. L. DSY and D. PEDCW), 334–349. Berlin, Heidelberg: Springer https://doi.org/10.1007/978-3-642-74065-7_23.

111 Khudoshin, A.G., Mitrofanova, A.N., and Lunin, V.V. (2012). Lignin transformations and reactivity upon ozonation in aqueous media. *Russian Journal of Physical Chemistry A* 86: 360–365. https://doi.org/10.1134/S0036024412030156.

112 Sarkanen, K.V., Islam, A., and Anderson, C.D. (1992). Ozonation. In: *Methods in Lignin Chemistry* (ed. L. DSY and D. PEDCW), 387–406. Berlin, Heidelberg: Springer https://doi.org/10.1007/978-3-642-74065-7_26.

113 Chen, C.-L. (1992). Nitrobenzene and cupric oxide oxidations. In: *Methods in Lignin Chemistry* (ed. L. DSY and P.E.D.C.W. Dence), 301–321. Berlin, Heidelberg: Springer https://doi.org/10.1007/978-3-642-74065-7_21.

114 Lu, F. and Ralph, J. (1997). DFRC method for lignin analysis. 1. New method for β-aryl ether cleavage: lignin model studies. *Journal of Agricultural and Food Chemistry* 45: 4655–4660. https://doi.org/10.1021/jf970539p.

115 Crestini, C., Melone, F., Sette, M., and Saladino, R. (2011). Milled wood lignin: a linear oligomer. *Biomacromolecules* 12: 3928–3935. https://doi.org/10.1021/bm200948r.

116 Schultz, T.P. and Templeton, M.C. (1986). Proposed mechanism for the nitrobenzene oxidation of lignin. *Holzforschung* 40: 93–97. https://doi.org/10.1515/hfsg.1986.40.2.93.

117 Chan, F.D., Nguyen, K.L., and Wallis, A.F.A. (1995). Contribution of lignin sub-structures to nitrobenzene oxidation products. *Journal of Wood Chemistry and Technology* 15: 329–347. https://doi.org/10.1080/02773819508009514.

118 Higuchi, T., Ito, Y., and Kawamura, I. (1967). p-Hydroxyphenylpropane component of grass lignin and role of tyrosine-ammonia lyase in its formation. *Phytochemistry* 6: 875–881. https://doi.org/10.1016/S0031-9422(00)86035-5.

119 Tamai, A., Goto, H., Akiyama, T., and Matsumoto, Y. (2015). Revisiting alkaline nitrobenzene oxidation: quantitative evaluation of biphenyl structures in cedar wood lignin (*Cryptomeria japonica*) by a modified nitrobenzene oxidation method. *Holzforschung* 69: 951–958. https://doi.org/10.1515/hf-2014-0153.

120 Karhunen, P., Rummakko, P., Sipilä, J. et al. (1995). The formation of dibenzodioxocin structures by oxidative coupling. A model reaction for lignin biosynthesis. *Tetrahedron Letters* 36: 4501–4504. https://doi.org/10.1016/0040-4039(95)00769-9.

121 Hedges, J.I. and Ertel, J.R. (1982). Characterization of lignin by gas capillary chromatography of cupric oxide oxidation products. *Analytical Chemistry* 54: 174–178. https://doi.org/10.1021/ac00239a007.

122 Crestini, C., Crucianelli, M., Orlandi, M., and Saladino, R. (2010). Oxidative strategies in lignin chemistry: a new environmental friendly approach for the functionalisation of lignin and lignocellulosic fibers. *Catalysis Today* 156: 8–22. https://doi.org/10.1016/j.cattod.2010.03.057.

123 Lange, H., Decina, S., and Crestini, C. (2013). Oxidative upgrade of lignin – recent routes reviewed. *European Polymer Journal* 49: 1151–1173. https://doi.org/10.1016/j.eurpolymj.2013.03.002.

124 Zakzeski, J., Bruijnincx, P.C.A., Jongerius, A.L., and Weckhuysen, B.M. (2010). The catalytic valorization of lignin for the production of renewable chemicals. *Chemical Reviews* 110: 3552–3599. https://doi.org/10.1021/cr900354u.

125 Zakzeski, J., Jongerius, A.L., Bruijnincx, P.C.A., and Weckhuysen, B.M. (2012). Catalytic lignin valorization process for the production of aromatic chemicals and hydrogen. *ChemSusChem* 5: 1602–1609. https://doi.org/10.1002/cssc.201100699.

126 Liu, C., Wu, S., Zhang, H., and Xiao, R. (2019). Catalytic oxidation of lignin to valuable biomass-based platform chemicals: a review. *Fuel Processing Technology* 191: 181–201. https://doi.org/10.1016/j.fuproc.2019.04.007.

127 Behling, R., Valange, S., and Chatel, G. (2016). Heterogeneous catalytic oxidation for lignin valorization into valuable chemicals: what results? What limitations?

What trends? *Green Chemistry* 18: 1839–1854. https://doi.org/10.1039/C5GC03061G.

128 Dai, J., Patti, A.F., and Saito, K. (2016). Recent developments in chemical degradation of lignin: catalytic oxidation and ionic liquids. *Tetrahedron Letters* 57: 4945–4951. https://doi.org/10.1016/j.tetlet.2016.09.084.

129 Ma, R., Guo, M., and Zhang, X. (2018). Recent advances in oxidative valorization of lignin. *Catalysis Today* 302: 50–60. https://doi.org/10.1016/j.cattod.2017.05.101.

130 Rahimi, A., Azarpira, A., Kim, H. et al. (2013). Chemoselective metal-free aerobic alcohol oxidation in lignin. *Journal of the American Chemical Society* 135: 6415–6418. https://doi.org/10.1021/ja401793n.

131 Nguyen, J.D., Matsuura, B.S., and Stephenson, C.R.J. (2014). A photochemical strategy for lignin degradation at room temperature. *Journal of the American Chemical Society* 136: 1218–1221. https://doi.org/10.1021/ja4113462.

132 Hermann, W.A. and Fischer, R.W. (1995). Multiple bonds between main-group elements and transition metals. 136. "Polymerization" of an organometal oxide: the unusual behavior of methyltrioxorhenium(VII) in water. *Journal of the American Chemical Society* 117: 3223–3230. https://doi.org/10.1021/ja00116a026.

133 Genin, H.S., Lawler, K.A., Hoffmann, R. et al. (1995). Multiple bonds between main-group elements and transition metals. 138. Polymeric methyltrioxorhenium: some models for its electronic structure. *Journal of the American Chemical Society* 117: 3244–3252. https://doi.org/10.1021/ja00116a028.

134 Herrmann, W.A., Fischer, R.W., Scherer, W., and Rauch, M.U. (1993). Methyltrioxorhenium(VII) as catalyst for epoxidations: structure of the active species and mechanism of catalysis. *Angewandte Chemie (International Ed. in English)* 32: 1157–1160. https://doi.org/10.1002/anie.199311571.

135 Bianchini, G., Crucianelli, M., Crestini, C., and Saladino, R. (2006). Catalytic MTO-based C–H insertion reactions of hydrogen peroxide: an investigation on the polymeric support role in heterogeneous conditions. *Topics in Catalysis* 40: 221–227. https://doi.org/10.1007/s11244-006-0123-5.

136 Crestini, C., Pro, P., Neri, V., and Saladino, R. (2005). Methyltrioxorhenium: a new catalyst for the activation of hydrogen peroxide to the oxidation of lignin and lignin model compounds. *Bioorganic & Medicinal Chemistry* 13: 2569–2578. https://doi.org/10.1016/j.bmc.2005.01.049.

137 Argyropoulos, D.S., Sun, Y., Berry, R.M., and Bouchard, J. (n.d.). Reaction of dimethyldioxirane with lignin model compounds. *Journal of Pulp and Paper Science* 22: 84–90.

138 Haikarainen, A., Sipilä, J., Pietikäinen, P. et al. (2001). Synthesis and characterization of bulky salen-type complexes of Co, Cu, Fe, Mn and Ni with amphiphilic solubility properties. *Journal of the Chemical Society Dalton Transactions* 991–995. https://doi.org/10.1039/B008167L.

139 Canevali, C., Orlandi, M., Pardi, L. et al. (2002). Oxidative degradation of monomeric and dimeric phenylpropanoids: reactivity and mechanistic investigation. *Journal of the Chemical Society Dalton Transactions* 3007–3014. https://doi.org/10.1039/B203386K.

140 Zanini, S., Canevali, C., Orlandi, M. et al. (2008). Radical formation on CTMP FIBERS by argon plasma treatments and related lignin chemical changes. *BioResources* 3: 995–1006. https://doi.org/10.15376/biores.3.4.995-1006.

141 Bolzacchini, E., Chiavetto, L.B., Canevali, C. et al. (1996). Oxidation of propenoidic phenols catalysed by *N,N′*-ethylenebis(salicylideneiminato)cobalt(II) [Cosalen]: reactivity and spectroscopic studies. *Journal of Molecular Catalysis A: Chemical* 112: 347–351. https://doi.org/10.1016/1381-1169(96)00110-0.

142 Bolzacchini, E., Canevali, C., Morazzoni, F. et al. (1997). Spectromagnetic investigation of the active species in the oxidation of propenoidic phenols catalysed by [*N,N′*-bis(salicylidene)ethane-1,2-diaminato]cobalt(II). *Journal of the Chemical Society Dalton Transactions* 4695–4700. https://doi.org/10.1039/a705188c.

143 Biannic, B. and Bozell, J.J. (2013). Efficient cobalt-catalyzed oxidative conversion of lignin models to benzoquinones. *Organic Letters* 15: 2730–2733. https://doi.org/10.1021/ol401065r.

144 Kumar, A., Biswas, B., and Bhaskar, T. (2020). Effect of cobalt on titania, ceria and zirconia oxide supported catalysts on the oxidative depolymerization of prot and alkali lignin. *Bioresource Technology* 299: 122589. https://doi.org/10.1016/j.biortech.2019.122589.

145 Das, L., Xu, S., and Shi, J. (2017). Catalytic oxidation and depolymerization of lignin in aqueous ionic liquid. *Frontiers in Energy Research* 5.

146 Cabral Almada, C., Kazachenko, A., Fongarland, P. et al. (2021). Supported-metal catalysts in upgrading lignin to aromatics by oxidative depolymerization. *Catalysts* 11: 467. https://doi.org/10.3390/catal11040467.

147 Walch, F.Y., Abdelaziz, O., Meier, S. et al. (2021). Oxidative depolymerization of Kraft lignin to high-value aromatics using a homogeneous vanadium–copper catalyst. *Catalysis Science & Technology* 11: 1843–1853. https://doi.org/10.1039/D0CY02158J.

148 Penín, L., Gigli, M., Sabuzi, F. et al. (2020). Biomimetic vanadate and molybdate systems for oxidative upgrading of iono- and organosolv hard- and softwood lignins. *Processes* 8: 1161. https://doi.org/10.3390/pr8091161.

149 Lv, H., Geletii, Y.V., Zhao, C. et al. (2012). Polyoxometalate water oxidation catalysts and the production of green fuel. *Chemical Society Reviews* 41: 7572–7589. https://doi.org/10.1039/C2CS35292C.

150 Bujanovic, B., Reiner, R.S., Ralph, S.A., and Atalla, R.H. (2011). Polyoxometalate delignification of birch Kraft pulp and effect on residual lignin. *Journal of Wood Chemistry and Technology* 31: 121–141. https://doi.org/10.1080/02773813.2010.503980.

151 Bujanovic, B., Ralph, S., Reiner, R. et al. (2010). Polyoxometalates in oxidative delignification of chemical pulps: effect on lignin. *Materials* 3: 1888–1903. https://doi.org/10.3390/ma3031888.

152 Tonucci, L., Coccia, F., Bressan, M., and d'Alessandro, N. (2012). Mild Photocatalysed and catalysed green oxidation of lignin: a useful pathway to low-molecular-weight derivatives. *Waste and Biomass Valorization* 3: 165–174. https://doi.org/10.1007/s12649-011-9102-6.

153 Cai, Z., Long, J., Li, Y. et al. (2019). Selective production of diethyl maleate via oxidative cleavage of lignin aromatic unit. *Chemistry* 5: 2365–2377. https://doi.org/10.1016/j.chempr.2019.05.021.

154 Hanson, S.K., Wu, R., and Silks, L.A.P. (2012). C—C or C—O bond cleavage in a phenolic lignin model compound: selectivity depends on vanadium catalyst. *Angewandte Chemie, International Edition* 51: 3410–3413. https://doi.org/10.1002/anie.201107020.

155 Son, S. and Toste, F.D. (2010). Non-oxidative vanadium-catalyzed C—O bond cleavage: application to degradation of lignin model compounds. *Angewandte Chemie, International Edition* 49: 3791–3794. https://doi.org/10.1002/anie.201001293.

156 Ferm, R., Kringstad, K.P., and Cowling, E.B. (1972). Formation of free radicals in milled wood lignin and syringaldehyde by phenol-oxidizing enzymes. *Sven Papperstidn* 75: 859–865.

157 Tien, M. and Kirk, T.K. (1983). Lignin-degrading enzyme from the Hymenomycete *Phanerochaete chrysosporium* Burds. *Science* 221: 661–663. https://doi.org/10.1126/science.221.4611.661.

158 Shimada, M., Habe, T., Higuchi, T. et al. (1977). Biomimetic approach to lignin degradation. *Holzforschung* 41: 277–285.

159 Ghiaci, M., Molaie, F., Sedaghat, M.E., and Dorostkar, N. (2010). Metalloporphyrin covalently bound to silica. Preparation, characterization and catalytic activity in oxidation of ethyl benzene. *Catalysis Communications* 11: 694–699. https://doi.org/10.1016/j.catcom.2010.01.023.

160 Pattou, D., Labat, G., Defrance, S. et al. (1994). Preparation, catalytic activity of supported metalloporphyrin catalysts on poly(vinylpyridine) polymers and related molecular modeling studies (case of a proximal ligand provided by the polymer itself). *Bulletin de la Société Chimique de France* 131: 78–88.

161 Baker, J.H.F. (2012). *Synthesis of Novel Ionic Liquid Tagged Porphryin Media Designed for the Oxidation of Lignin*. Eastern Kentucky University.

162 Crestini, C., Pastorini, A., and Tagliatesta, P. (2004). The immobilized porphyrin-mediator system Mn(TMePyP)/clay/HBT (clay-PMS): a lignin peroxidase biomimetic catalyst in the oxidation of lignin and lignin model compounds. *European Journal of Inorganic Chemistry* 2004: 4477–4483. https://doi.org/10.1002/ejic.200400351.

163 Crestini, C., Pastorini, A., and Tagliatesta, P. (2004). Metalloporphyrins immobilized on motmorillonite as biomimetic catalysts in the oxidation of lignin model compounds. *Journal of Molecular Catalysis A: Chemical* 208: 195–202. https://doi.org/10.1016/j.molcata.2003.07.015.

164 Andlar, M., Rezić, T., Marđetko, N. et al. (2018). Lignocellulose degradation: an overview of fungi and fungal enzymes involved in lignocellulose degradation. *Engineering in Life Sciences* 18: 768–778. https://doi.org/10.1002/elsc.201800039.

165 Xu, Z., Qin, L., Cai, M. et al. (2018). Biodegradation of Kraft lignin by newly isolated *Klebsiella pneumoniae, pseudomonas putida*, and *Ochrobactrum tritici* strains. *Environmental Science and Pollution Research* 25: 14171–14181. https://doi.org/10.1007/s11356-018-1633-y.

166 Xu, R., Zhang, K., Liu, P. et al. (2018). Lignin depolymerization and utilization by bacteria. *Bioresource Technology* 269: 557–566. https://doi.org/10.1016/j.biortech.2018.08.118.

167 Reshmy, R., Athiyaman Balakumaran, P., Divakar, K. et al. (2022). Microbial valorization of lignin: prospects and challenges. *Bioresource Technology* 344: 126240. https://doi.org/10.1016/j.biortech.2021.126240.

168 Chang, A., Jeske, L., Ulbrich, S. et al. (2021). BRENDA, the ELIXIR core data resource in 2021: new developments and updates. *Nucleic Acids Research* 49: D498–D508. https://doi.org/10.1093/nar/gkaa1025.

169 Mtibaà, R., Barriuso, J., de Eugenio, L. et al. (2018). Purification and characterization of a fungal laccase from the ascomycete *Thielavia* sp. and its role in the decolorization of a recalcitrant dye. *International Journal of Biological Macromolecules* 120: 1744–1751. https://doi.org/10.1016/j.ijbiomac.2018.09.175.

170 Shanmugam, S., Hari, A., Ulaganathan, P. et al. (2018). Potential of biohydrogen generation using the delignified lignocellulosic biomass by a newly identified thermostable laccase from *Trichoderma asperellum* strain BPLMBT1. *International Journal of Hydrogen Energy* 43: 3618–3628. https://doi.org/10.1016/j.ijhydene.2018.01.016.

171 Chauhan, P.S. and Jha, B. (2018). Pilot scale production of extracellular thermo-alkali stable laccase from pseudomonas sp. S2 using agro waste and its application in organophosphorous pesticides degradation. *Journal of Chemical Technology and Biotechnology* 93: 1022–1030. https://doi.org/10.1002/jctb.5454.

172 Coelho, G.D., Ballaminut, N., Thomaz, D.V., and Gomes Machado, K.M. (2019). Characterization of a thermostable *Deconica castanella* laccase and application toward pentachlorophenol degradation. *Preparative Biochemistry & Biotechnology* 49: 908–915. https://doi.org/10.1080/10826068.2019.1636280.

173 Iark, D., dos Reis Buzzo, A.J., JAA, G. et al. (2019). Enzymatic degradation and detoxification of azo dye Congo red by a new laccase from *Oudemansiella canarii*. *Bioresource Technology* 289: 121655. https://doi.org/10.1016/j.biortech.2019.121655.

174 Sharma, V., Ayothiraman, S., and Dhakshinamoorthy, V. (2019). Production of highly thermo-tolerant laccase from novel thermophilic bacterium *Bacillus* sp. PC-3 and its application in functionalization of chitosan film. *Journal of Bioscience and Bioengineering* 127: 672–678. https://doi.org/10.1016/j.jbiosc.2018.11.008.

175 Motamedi, E., Kavousi, K., Sadeghian Motahar, S.F. et al. (2021). Efficient removal of various textile dyes from wastewater by novel thermo-halotolerant laccase. *Bioresource Technology* 337: 125468. https://doi.org/10.1016/j.biortech.2021.125468.

176 Cheng, C.-M., Patel, A.K., Singhania, R.R. et al. (2021). Heterologous expression of bacterial CotA-laccase, characterization and its application for biodegradation of malachite green. *Bioresource Technology* 340: 125708. https://doi.org/10.1016/j.biortech.2021.125708.

177 Singh, R., Shedbalkar, U.U., Nadhe, S.B. et al. (2017). Lignin peroxidase mediated silver nanoparticle synthesis in *Acinetobacter* sp. *AMB Express* 7: 226. https://doi.org/10.1186/s13568-017-0528-5.

178 Sadaqat, B., Khatoon, N., Malik, A.Y. et al. (2020). Enzymatic decolorization of melanin by lignin peroxidase from *Phanerochaete chrysosporium*. *Scientific Reports* 10: https://doi.org/10.1038/s41598-020-76376-9.

179 Gaur, N., Narasimhulu, K., and Pydisetty, Y. (2018). Biochemical and kinetic characterization of laccase and manganese peroxidase from novel *Klebsiella pneumoniae* strains and their application in bioethanol production. *RSC Advances* 8: 15044–15055. https://doi.org/10.1039/C8RA01204K.

180 Higuchi, P.T. (1990). Lignin biochemistry: biosynthesis and biodegradation. *Wood Science and Technology* 24: 23–63. https://doi.org/10.1007/BF00225306.

181 Caldwell, E.S. and Steelink, C. (1969). Phenoxy radical intermediate in the enzymatic degradation of lignin model compounds. *Biochimica et Biophysica Acta (BBA) - General Subjects* 184: 420–431. https://doi.org/10.1016/0304-4165(69)90046-4.

182 Lundquist, K. and Kristersson, P. (1985). Exhaustive laccase-catalysed oxidation of a lignin model compound (vanillyl glycol) produces methanol and polymeric quinoid products. *Biochemical Journal* 229 (1): 277. http://www.biochemj.org/bj/229/0277/bj2290277_browse.htm (accessed 24 September 2014).

183 Shleev, S., Christenson, A., Serezhenkov, V. et al. (2005). Electrochemical redox transformations of T1 and T2 copper sites in native *Trametes hirsuta* laccase at gold electrode. *The Biochemical Journal* 385: 745. https://doi.org/10.1042/BJ20041015.

184 Crestini, C., Jurasek, L., and Argyropoulos, D.S. (2003). On the mechanism of the laccase–mediator system in the oxidation of lignin. *Chemistry - A European Journal* 9: 5371–5378. https://doi.org/10.1002/chem.200304818.

185 Poppius-Levlin, K., Tamminen, T., Kalliola, A., and Ohra-aho, T. (2001). Characterization of residual lignins in pulps delignified by laccase/*N*-hydroxyacetanilide. In: *Oxidative Delignification Chemistry*, vol. 785, 358–372. American Chemical Society.

186 Johannes, C. and Majcherczyk, A. (2000). Natural mediators in the oxidation of polycyclic aromatic hydrocarbons by laccase mediator systems. *Applied and Environmental Microbiology* 66: 524–528. https://doi.org/10.1128/AEM.66.2.524-528.2000.

187 Messerschmidt, A. and Huber, R. (1990). The blue oxidases, ascorbate oxidase, laccase and ceruloplasmin modelling and structural relationships. *European Journal of Biochemistry* 187: 341–352. https://doi.org/10.1111/j.1432-1033.1990.tb15311.x.

188 Fabbrini, M., Galli, C., and Gentili, P. (2002). Comparing the catalytic efficiency of some mediators of laccase. *Journal of Molecular Catalysis B: Enzymatic* 16: 231–240. https://doi.org/10.1016/S1381-1177(01)00067-4.

189 d'Acunzo, F., Galli, C., Gentili, P., and Sergi, F. (2006). Mechanistic and steric issues in the oxidation of phenolic and non-phenolic compounds by laccase or laccase-mediator systems. The case of bifunctional substrates. *New Journal of Chemistry* 30: 583–591. https://doi.org/10.1039/B516719A.

190 Paice, M.G., Bourbonnaise, R., Reid, I.D. et al. (1995). Oxidative bleaching enzymes: a review. *Journal of Pulp and Paper Science* 21: J280–J284.

191 Wariishi, H., Akileswaran, L., and Gold, M.H. (1988). Manganese peroxidase from the basidiomycete *Phanerochaete chrysosporium*: spectral characterization of the oxidized states and the catalytic cycle. *Biochemistry* 27: 5365–5370. https://doi.org/10.1021/bi00414a061.

192 Tuor, U., Wariishi, H., Schoemaker, H.E., and Gold, M.H. (1992). Oxidation of phenolic arylglycerol .beta.-aryl ether lignin model compounds by manganese peroxidase from *Phanerochaete chrysosporium*: oxidative cleavage of an .alpha.-carbonyl model compound. *Biochemistry* 31: 4986–4995. https://doi.org/10.1021/bi00136a011.

193 Wariishi, H., Valli, K., and Gold, M.H. (1989). Oxidative cleavage of a phenolic diarylpropane lignin model dimer by manganese peroxidase from *Phanerochaete chrysosporium*. *Biochemistry* 28: 6017–6023. https://doi.org/10.1021/bi00440a044.

194 Crestini, C., D'Annibale, A., Sermanni, G.G., and Saladino, R. (2000). The reactivity of phenolic and non-phenolic residual Kraft lignin model compounds with Mn(II)-peroxidase from *Lentinula edodes*. *Bioorganic & Medicinal Chemistry* 8: 433–438. https://doi.org/10.1016/s0968-0896(99)00303-x.

195 Rodrigues, A.F.S., da Silva, A.F., da Silva, F.L.B. et al. (2023). A scientometric analysis of research progress and trends in the design of laccase biocatalysts for the decolorization of synthetic dyes. *Process Biochemistry* 126: 272–291. https://doi.org/10.1016/j.procbio.2023.01.014.

196 Cavalcante, A.L.G., Chaves, A.V., Fechine, P.B.A. et al. (2022). Chemical modification of clay nanocomposites for the improvement of the catalytic properties of lipase a from *Candida antarctica*. *Process Biochemistry* 120: 1–14. https://doi.org/10.1016/j.procbio.2022.05.020.

197 Bilal, M., Rashid, E.U., Zdarta, J. et al. (2022). Engineering magnetic nanobiocatalytic systems with multipurpose functionalities for biocatalysis, biotechnology and bioprocess applications. *Sustainable Chemistry and Pharmacy* 30: 100866. https://doi.org/10.1016/j.scp.2022.100866.

198 Capecchi, E., Piccinino, D., Delfino, I. et al. (2018). Functionalized tyrosinase-lignin nanoparticles as sustainable catalysts for the oxidation of phenols. *Nanomaterials* 8: 438. https://doi.org/10.3390/nano8060438.

199 Crestini, C., Perazzini, R., and Saladino, R. (2010). Oxidative functionalisation of lignin by layer-by-layer immobilised laccases and laccase microcapsules. *Applied Catalysis A: General* 372: 115–123. https://doi.org/10.1016/j.apcata.2009.10.012.

200 Monteiro, R.R.C., de Oliveira, A.L.B., de Menezes, F.L. et al. (2022). Improvement of enzymatic activity and stability of lipase a from *Candida antartica* onto halloysite nanotubes with Taguchi method for optimized immobilization. *Applied Clay Science* 228: 106634. https://doi.org/10.1016/j.clay.2022.106634.

201 Monteiro, R.R.C., dos Santos, J.C.S., Alcántara, A.R., and Fernandez-Lafuente, R. (2020). Enzyme-coated micro-crystals: an almost forgotten but very simple and elegant immobilization strategy. *Catalysts* 10: 891. https://doi.org/10.3390/catal10080891.

202 Bezerra, R.M., Monteiro, R.R.C., Neto, D.M.A. et al. (2020). A new heterofunctional support for enzyme immobilization: PEI functionalized Fe_3O_4

MNPs activated with divinyl sulfone. Application in the immobilization of lipase from Thermomyces lanuginosus. *Enzyme and Microbial Technology* 138: 109560. https://doi.org/10.1016/j.enzmictec.2020.109560.

203 Monteiro, R.R.C., Virgen-Ortiz, J.J., Berenguer-Murcia, Á. et al. (2021). Biotechnological relevance of the lipase a from Candida antarctica. *Catalysis Today* 362: 141–154. https://doi.org/10.1016/j.cattod.2020.03.026.

204 Silva, A.R.M., Alexandre, J.Y.N.H., Souza, J.E.S. et al. (2022). The chemistry and applications of metal–organic frameworks (MOFs) as industrial enzyme immobilization systems. *Molecules* 27: 4529. https://doi.org/10.3390/molecules27144529.

205 da Fonseca, A.M., Cleiton Sousa dos Santos, J., de MCM, S. et al. (2020). *Industrial Crops and Products* 145: 111890. https://doi.org/10.1016/j.indcrop.2019.111890.

206 Homaei, A.A., Sariri, R., Vianello, F., and Stevanato, R. (2013). Enzyme immobilization: an update. *Journal of Chemical Biology* 6: 185–205. https://doi.org/10.1007/s12154-013-0102-9.

207 Farhan, L.O., Mehdi, W.A., Taha, E.M. et al. (2021). Various type immobilizations of Isocitrate dehydrogenases enzyme on hyaluronic acid modified magnetic nanoparticles as stable biocatalysts. *International Journal of Biological Macromolecules* 182: 217–227. https://doi.org/10.1016/j.ijbiomac.2021.04.026.

208 Rodrigues, R.C., Virgen-Ortíz, J.J., dos Santos, J.C.S. et al. (2019). Immobilization of lipases on hydrophobic supports: immobilization mechanism, advantages, problems, and solutions. *Biotechnology Advances* 37: 746–770. https://doi.org/10.1016/j.biotechadv.2019.04.003.

209 Chen, M., Mou, X., Wang, S. et al. (2020). Porous organic polymer-supported palladium catalyst for hydroesterification of olefins. *Molecular Catalysis* 498: 111239. https://doi.org/10.1016/j.mcat.2020.111239.

210 Lou, W.-Y., Fernández-Lucas, J., Ge, J., and Wu, C. (2021). Editorial: enzyme or whole cell immobilization for efficient biocatalysis: focusing on novel supporting platforms and immobilization techniques. *Frontiers in Bioengineering and Biotechnology* 9: 620292. https://doi.org/10.3389/fbioe.2021.620292.

211 Brown, E.E. (2022). Minireview: recent efforts toward upgrading lignin-derived phenols in continuous flow. *Journal of Flow Chemistry* https://doi.org/10.1007/s41981-022-00248-8.

212 Gérardy, R., Morodo, R., Estager, J. et al. (2018). Sustaining the transition from a Petrobased to a biobased chemical industry with flow chemistry. *Topics in Current Chemistry* 377: 1. https://doi.org/10.1007/s41061-018-0222-3.

213 Gérardy, R., Debecker, D.P., Estager, J. et al. (2020). Continuous flow upgrading of selected C_2–C_6 platform chemicals derived from biomass. *Chemical Reviews* 120: 7219–7347. https://doi.org/10.1021/acs.chemrev.9b00846.

214 Gomollón-Bel, F. (2019). Ten chemical innovations that will change our world: IUPAC identifies emerging technologies in chemistry with potential to make our planet more sustainable. *Chemistry International* 41: 12–17. https://doi.org/10.1515/ci-2019-0203.

215 Hämäläinen, V., Grönroos, T., Suonpää, A. et al. (2018). Enzymatic processes to unlock the lignin value. *Frontiers in Bioengineering and Biotechnology* 6: 20. https://doi.org/10.3389/fbioe.2018.00020.

216 Zhu, D., Qaria, M.A., Zhu, B. et al. (2022). Extremophiles and extremozymes in lignin bioprocessing. *Renewable and Sustainable Energy Reviews* 157: 112069. https://doi.org/10.1016/j.rser.2021.112069.

217 Du, X., Wu, S., and Li, P. (2023). Catalytic hydrogenolysis lignin to obtain phenols: A review of selective cleavage of ether bonds. *BioResources* 18 (2): 4231–4261. https://doi.org/10.15376/biores.18.2.Du.

218 Tian, Q., Xu, P., Huang, D. et al. (2023). The driving force of biomass value-addition: selective catalytic depolymerization of lignin to high-value chemicals. *Journal of Environmental Chemical Engineering* 11: 109719. https://doi.org/10.1016/j.jece.2023.109719.

219 Zhang, J., Chen, Y., and Brook, M.A. (2014). Reductive degradation of lignin and model compounds by Hydrosilanes. *ACS Sustainable Chemistry & Engineering* 2: 1983–1991. https://doi.org/10.1021/sc500302j.

220 Parsell, T.H., Owen, B.C., Klein, I. et al. (2013). Cleavage and hydrodeoxygenation (HDO) of C–O bonds relevant to lignin conversion using Pd/Zn synergistic catalysis. *Chemical Science* 4: 806–813. https://doi.org/10.1039/C2SC21657D.

221 Sergeev, A.G., Webb, J.D., and Hartwig, J.F. (2012). A heterogeneous nickel catalyst for the Hydrogenolysis of aryl ethers without Arene hydrogenation. *Journal of the American Chemical Society* 134: 20226–20229. https://doi.org/10.1021/ja3085912.

222 Wang, X. and Rinaldi, R. (2013). A route for lignin and bio-oil conversion: dehydroxylation of phenols into Arenes by catalytic tandem reactions. *Angewandte Chemie, International Edition* 52: 11499–11503. https://doi.org/10.1002/anie.201304776.

223 Sergeev, A.G. and Hartwig, J.F. (2011). Selective, nickel-catalyzed hydrogenolysis of aryl ethers. *Science* 332: 439–443. https://doi.org/10.1126/science.1200437.

224 Eachus, S.W. and Dence, C.W. (2009). Hydrogenation of lignin model compounds in the presence of a homogeneous catalyst. *Holzforschung - International Journal of the Biology, Chemistry, Physics and Technology of Wood* 29: 41–48. https://doi.org/10.1515/hfsg.1975.29.2.41.

225 Chan, J.M.W., Bauer, S., Sorek, H. et al. (2013). Studies on the vanadium-catalyzed nonoxidative depolymerization of *Miscanthus giganteus*-derived lignin. *ACS Catalysis* 3: 1369–1377. https://doi.org/10.1021/cs400333q.

226 Lam, C.H., Deng, W., Lang, L. et al. (2020). Minireview on bio-oil upgrading via electrocatalytic hydrogenation: connecting biofuel production with renewable power. *Energy & Fuels* 34: 7915–7928. https://doi.org/10.1021/acs.energyfuels.0c01380.

227 Ben, H. and Ragauskas, A.J. (2012). One step thermal conversion of lignin to the gasoline range liquid products by using zeolites as additives. *RSC Advances* 2: 12892–12898. https://doi.org/10.1039/C2RA22616B.

228 Mullen, C.A. and Boateng, A.A. (2010). Catalytic pyrolysis-GC/MS of lignin from several sources. *Fuel Processing Technology* 91: 1446–1458. https://doi.org/10.1016/j.fuproc.2010.05.022.

229 Shafaghat, H., Rezaei, P.S., Ro, D. et al. (2017). In-situ catalytic pyrolysis of lignin in a bench-scale fixed bed pyrolyzer. *Journal of Industrial and Engineering Chemistry* 54: 447–453. https://doi.org/10.1016/j.jiec.2017.06.026.

230 Wang, W.-L., Ren, X.-Y., Li, L.-F. et al. (2015). Catalytic effect of metal chlorides on analytical pyrolysis of alkali lignin. *Fuel Processing Technology* 134: 345–351. https://doi.org/10.1016/j.fuproc.2015.02.015.

231 Custodis, V.B.F., Karakoulia, S.A., Triantafyllidis, K.S., and van Bokhoven, J.A. (2016). Catalytic fast pyrolysis of lignin over high-surface-area mesoporous aluminosilicates: effect of porosity and acidity. *ChemSusChem* 9: 1134–1145. https://doi.org/10.1002/cssc.201600105.

232 Kim, J.-Y., Lee, J.H., Park, J. et al. (2015). Catalytic pyrolysis of lignin over HZSM-5 catalysts: effect of various parameters on the production of aromatic hydrocarbon. *Journal of Analytical and Applied Pyrolysis* 114: 273–280. https://doi.org/10.1016/j.jaap.2015.06.007.

233 Ma, Z., Troussard, E., and van Bokhoven, J.A. (2012). Controlling the selectivity to chemicals from lignin via catalytic fast pyrolysis. *Applied Catalysis A: General* 423–424: 130–136. https://doi.org/10.1016/j.apcata.2012.02.027.

234 Lee, H.W., Kim, T.H., Park, S.H. et al. (2013). Catalytic fast pyrolysis of lignin over mesoporous Y zeolite using Py-GC/MS. *Journal of Nanoscience and Nanotechnology* 13: 2640–2646. https://doi.org/10.1166/jnn.2013.7421.

235 Kim, S.-S., Lee, H.W., Ryoo, R. et al. (2014). Conversion of Kraft lignin over hierarchical MFI zeolite. *Journal of Nanoscience and Nanotechnology* 14: 2414–2418. https://doi.org/10.1166/jnn.2014.8545.

236 Bi, Y., Lei, X., Xu, G. et al. (2018). Catalytic fast pyrolysis of Kraft lignin over hierarchical HZSM-5 and Hβ zeolites. *Catalysts* 8: 82. https://doi.org/10.3390/catal8020082.

237 Ma, Z., Custodis, V., and van Bokhoven, J.A. (2014). Selective deoxygenation of lignin during catalytic fast pyrolysis. *Catalysis Science & Technology* 4: 766–772. https://doi.org/10.1039/C3CY00704A.

238 Ha, J.-M., Hwang, K.-R., Kim, Y.-M. et al. (2019). Recent progress in the thermal and catalytic conversion of lignin. *Renewable and Sustainable Energy Reviews* 111: 422–441. https://doi.org/10.1016/j.rser.2019.05.034.

239 Renders, T., den Bosch, S.V., Koelewijn, S.-F. et al. (2017). Lignin-first biomass fractionation: the advent of active stabilisation strategies. *Energy & Environmental Science* 10: 1551–1557. https://doi.org/10.1039/C7EE01298E.

240 Sarker, T.R., Nanda, S., Dalai, A.K., and Meda, V. (2021). A review of torrefaction technology for upgrading lignocellulosic biomass to solid biofuels. *Bioenergy Research* 14: 645–669. https://doi.org/10.1007/s12155-020-10236-2.

241 Acharya, B. and Dutta, A. (2016). Fuel property enhancement of lignocellulosic and nonlignocellulosic biomass through torrefaction. *Biomass Convers Biorefinery* 6: 139–149. https://doi.org/10.1007/s13399-015-0170-x.

242 Ribeiro, J.M.C., Godina, R., JCDO, M., and LJR, N. (2018). Future perspectives of biomass torrefaction: review of the current state-of-the-art and research development. *Sustainability* 10: 2323. https://doi.org/10.3390/su10072323.

243 Wen, J.-L., Sun, S.-L., Yuan, T.-Q. et al. (2014). Understanding the chemical and structural transformations of lignin macromolecule during torrefaction. *Applied Energy* 121: 1–9. https://doi.org/10.1016/j.apenergy.2014.02.001.

244 Chen, D., Mei, J., Li, H. et al. (2017). Combined pretreatment with torrefaction and washing using torrefaction liquid products to yield upgraded biomass and pyrolysis products. *Bioresource Technology* 228: 62–68. https://doi.org/10.1016/j.biortech.2016.12.088.

245 Ren, S., Lei, H., Zhang, Y. et al. (2019). Furfural production from microwave catalytic torrefaction of Douglas fir sawdust. *Journal of Analytical and Applied Pyrolysis* 138: 188–195. https://doi.org/10.1016/j.jaap.2018.12.023.

246 Mahadevan, R., Adhikari, S., Shakya, R. et al. (2016). Effect of torrefaction temperature on lignin macromolecule and product distribution from HZSM-5 catalytic pyrolysis. *Journal of Analytical and Applied Pyrolysis* 122: 95–105. https://doi.org/10.1016/j.jaap.2016.10.011.

247 Ma, Z., Wang, J., Li, C. et al. (2019). New sight on the lignin torrefaction pretreatment: relevance between the evolution of chemical structure and the properties of torrefied gaseous, liquid, and solid products. *Bioresource Technology* 288: 121528. https://doi.org/10.1016/j.biortech.2019.121528.

248 Kabir, G. and Hameed, B.H. (2017). Recent progress on catalytic pyrolysis of lignocellulosic biomass to high-grade bio-oil and bio-chemicals. *Renewable and Sustainable Energy Reviews* 70: 945–967. https://doi.org/10.1016/j.rser.2016.12.001.

249 Fan, L., Zhang, Y., Liu, S. et al. (2017). Bio-oil from fast pyrolysis of lignin: effects of process and upgrading parameters. *Bioresource Technology* 241: 1118–1126. https://doi.org/10.1016/j.biortech.2017.05.129.

250 Shen, D., Liu, G., Zhao, J. et al. (2015). Thermo-chemical conversion of lignin to aromatic compounds: effect of lignin source and reaction temperature. *Journal of Analytical and Applied Pyrolysis* 112: 56–65. https://doi.org/10.1016/j.jaap.2015.02.022.

251 Wang, S., Wang, K., Liu, Q. et al. (2009). Comparison of the pyrolysis behavior of lignins from different tree species. *Biotechnology Advances* 27: 562–567. https://doi.org/10.1016/j.biotechadv.2009.04.010.

252 Lin, X., Sui, S., Tan, S. et al. (2015). Fast pyrolysis of four Lignins from different isolation processes using Py-GC/MS. *Energies* 8: 5107–5121. https://doi.org/10.3390/en8065107.

253 Wang, S., Lin, H., Ru, B. et al. (2014). Comparison of the pyrolysis behavior of pyrolytic lignin and milled wood lignin by using TG–FTIR analysis. *Journal of Analytical and Applied Pyrolysis* 108: 78–85. https://doi.org/10.1016/j.jaap.2014.05.014.

254 Björkman, A. (1956). Studies on finely divided wood. Part 1. Extraction of lignin with neutral solvents. *Sven Papperstidn* 59: 477–485.

255 Zhang, J., Kim, K.H., Choi, Y.S. et al. (2017). Comparison of fast pyrolysis behavior of cornstover lignins isolated by different methods. *ACS Sustainable Chemistry & Engineering* 5: 5657–5661. https://doi.org/10.1021/acssuschemeng.7b01393.

256 He, T., Zhang, Y., Zhu, Y. et al. (2016). Pyrolysis mechanism study of lignin model compounds by synchrotron vacuum ultraviolet photoionization mass spectrometry. *Energy & Fuels* 30: 2204–2208. https://doi.org/10.1021/acs.energyfuels.5b02635.

257 Custodis, V.B.F., Hemberger, P., Ma, Z., and van Bokhoven, J.A. (2014). Mechanism of fast pyrolysis of lignin: studying model compounds. *The Journal of Physical Chemistry. B* 118: 8524–8531. https://doi.org/10.1021/jp5036579.

258 Milovanović, J., Luque, R., Tschentscher, R. et al. (2017). Study on the pyrolysis products of two different hardwood lignins in the presence of NiO contained-zeolites. *Biomass and Bioenergy* 103: 29–34. https://doi.org/10.1016/j.biombioe.2017.05.007.

259 Xu, L., Zhang, Y., and Fu, Y. (2017). Advances in upgrading lignin pyrolysis Vapors by ex situ catalytic fast pyrolysis. *Energy Technology* 5: 30–51. https://doi.org/10.1002/ente.201600107.

260 To AT and Resasco, D.E. (2014). Role of a phenolic pool in the conversion of m-cresol to aromatics over HY and HZSM-5 zeolites. *Applied Catalysis A: General* 487: 62–71. https://doi.org/10.1016/j.apcata.2014.09.006.

261 Ma, Z. and van Bokhoven, J.A. (2012). Deactivation and regeneration of H-USY zeolite during lignin catalytic fast pyrolysis. *ChemCatChem* 4: 2036–2044. https://doi.org/10.1002/cctc.201200401.

262 Hemberger, P., Custodis, V.B.F., Bodi, A. et al. (2017). Understanding the mechanism of catalytic fast pyrolysis by unveiling reactive intermediates in heterogeneous catalysis. *Nature Communications* 8: 15946. https://doi.org/10.1038/ncomms15946.

263 Dorado, C., Mullen, C.A., and Boateng, A.A. (2014). H-ZSM5 catalyzed co-pyrolysis of biomass and plastics. *ACS Sustainable Chemistry & Engineering* 2: 301–311. https://doi.org/10.1021/sc400354g.

264 Ma, Z., Ghosh, A., Asthana, N., and van Bokhoven, J. (2017). Optimization of the reaction conditions for catalytic fast pyrolysis of pretreated lignin over zeolite for the production of phenol. *ChemCatChem* 9: 954–961. https://doi.org/10.1002/cctc.201601674.

265 Saidi, M., Samimi, F., Karimipourfard, D. et al. (2013). Upgrading of lignin-derived bio-oils by catalytic hydrodeoxygenation. *Energy & Environmental Science* 7: 103–129. https://doi.org/10.1039/C3EE43081B.

266 Li, X., Chen, G., Liu, C. et al. (2017). Hydrodeoxygenation of lignin-derived bio-oil using molecular sieves supported metal catalysts: a critical review. *Renewable and Sustainable Energy Reviews* 71: 296–308. https://doi.org/10.1016/j.rser.2016.12.057.

267 Khan, R.J., Lau, C.Y., Guan, J. et al. (2022). Recent advances of lignin valorization techniques toward sustainable aromatics and potential benchmarks to fossil refinery products. *Bioresource Technology* 346: 126419. https://doi.org/10.1016/j.biortech.2021.126419.

268 Elliott, D.C., Wang, H., Rover, M. et al. (2015). Hydrocarbon liquid production via catalytic hydroprocessing of phenolic oils fractionated from fast pyrolysis of red oak and corn Stover. *ACS Sustainable Chemistry & Engineering* 3: 892–902. https://doi.org/10.1021/acssuschemeng.5b00015.

269 Hidajat, M.J., Riaz, A., and Kim, J. (2018). A two-step approach for producing oxygen-free aromatics from lignin using formic acid as a hydrogen source. *Chemical Engineering Journal* 348: 799–810. https://doi.org/10.1016/j.cej.2018.05.036.

270 Mu, W., Ben, H., Du, X. et al. (2014). Noble metal catalyzed aqueous phase hydrogenation and hydrodeoxygenation of lignin-derived pyrolysis oil and related model compounds. *Bioresource Technology* 173: 6–10. https://doi.org/10.1016/j.biortech.2014.09.067.

271 Schwaiger, N., Witek, V., Feiner, R. et al. (2012). Formation of liquid and solid products from liquid phase pyrolysis. *Bioresource Technology* 124: 90–94. https://doi.org/10.1016/j.biortech.2012.07.115.

272 Schwaiger, N., Feiner, R., Zahel, K. et al. (2011). Liquid and solid products from liquid-phase pyrolysis of softwood. *Bioenergy Research* 4: 294–302. https://doi.org/10.1007/s12155-011-9132-8.

273 Pucher, H., Schwaiger, N., Feiner, R. et al. (2015). Biofuels from liquid phase pyrolysis oil: a two-step hydrodeoxygenation (HDO) process. *Green Chemistry* 17: 1291–1298. https://doi.org/10.1039/C4GC01741B.

274 Yamaguchi, A., Watanabe, T., Saito, K. et al. (2019). Direct conversion of lignocellulosic biomass into aromatic monomers over supported metal catalysts in supercritical water. *Molecular Catalysis* 477: 110557. https://doi.org/10.1016/j.mcat.2019.110557.

275 Pakdel, H., De Caumia, B., and Roy, C. (1992). Vacuum pyrolysis of lignin derived from steam-exploded wood. *Biomass and Bioenergy* 3: 31–40. https://doi.org/10.1016/0961-9534(92)90017-K.

276 Figueirêdo, M.B., Hita, I., Deuss, P.J. et al. (2022). Pyrolytic lignin: a promising biorefinery feedstock for the production of fuels and valuable chemicals. *Green Chemistry* 24: 4680–4702. https://doi.org/10.1039/D2GC00302C.

277 Terrell, E. and Garcia-Perez, M. (2020). Vacuum pyrolysis of hybrid poplar milled wood lignin with Fourier transform-ion cyclotron resonance mass spectrometry analysis of feedstock and products for the elucidation of reaction mechanisms. *Energy & Fuels* 34: 14249–14263. https://doi.org/10.1021/acs.energyfuels.0c02928.

278 Kiełbasa, P., Dróżdż, T., and Popardowski, E. (2021). Influence of coniferous wood conditioning by pulsed electric field on its combustion heat characteristics. *Applied Sciences* 11: 983. https://doi.org/10.3390/app11030983.

279 Yu, X., Bals, O., Grimi, N., and Vorobiev, E. (2015). A new way for the oil plant biomass valorization: polyphenols and proteins extraction from rapeseed stems and leaves assisted by pulsed electric fields. *Industrial Crops and Products* 74: 309–318. https://doi.org/10.1016/j.indcrop.2015.03.045.

280 Garedew, M., Young-Farhat, D., Jackson, J.E., and Saffron, C.M. (2019). Electrocatalytic upgrading of phenolic compounds observed after lignin pyrolysis. *ACS Sustainable Chemistry & Engineering* 7: 8375–8386. https://doi.org/10.1021/acssuschemeng.9b00019.

281 Cai, P., Fan, H., Cao, S. et al. (2018). Electrochemical conversion of corn Stover lignin to biomass-based chemicals between cu/NiMoCo cathode and Pb/PbO_2 anode in alkali solution. *Electrochimica Acta* 264: 128–139. https://doi.org/10.1016/j.electacta.2018.01.111.

282 Pardini, V.L., Smith, C.Z., Utley, J.H.P. et al. (1991). Electroorganic reactions. 38. Mechanism of electrooxidative cleavage of lignin model dimers. *The Journal of Organic Chemistry* 56: 7305–7313. https://doi.org/10.1021/jo00026a022.

283 Zhu, H., Wang, L., Chen, Y. et al. (2014). Electrochemical depolymerization of lignin into renewable aromatic compounds in a non-diaphragm electrolytic cell. *RSC Advances* 4: 29917–29924. https://doi.org/10.1039/C4RA03793F.

284 Tolba, R., Tian, M., Wen, J. et al. (2010). Electrochemical oxidation of lignin at IrO_2-based oxide electrodes. *Journal of Electroanalytical Chemistry* 649: 9–15. https://doi.org/10.1016/j.jelechem.2009.12.013.

285 Schmitt, D., Regenbrecht, C., Hartmer, M. et al. (2015). Highly selective generation of vanillin by anodic degradation of lignin: a combined approach of electrochemistry and product isolation by adsorption. *Beilstein Journal of Organic Chemistry* 11: 473–480. https://doi.org/10.3762/bjoc.11.53.

286 Smith, C.Z., Utley, J.H.P., and Hammond, J.K. (2011). Electro-organic reactions. Part 60[1]. The electro-oxidative conversion at laboratory scale of a lignosulfonate into vanillin in an FM01 filter press flow reactor: preparative and mechanistic aspects. *Journal of Applied Electrochemistry* 41: 363–375. https://doi.org/10.1007/s10800-010-0245-0.

287 Bawareth, B., Di Marino, D., Nijhuis, T.A. et al. (2019). Electrochemical membrane reactor modeling for lignin depolymerization. *ACS Sustainable Chemistry & Engineering* 7: 2091–2099. https://doi.org/10.1021/acssuschemeng.8b04670.

288 Stiefel, S., Schmitz, A., Peters, J. et al. (2016). An integrated electrochemical process to convert lignin to value-added products under mild conditions. *Green Chemistry* 18: 4999–5007. https://doi.org/10.1039/C6GC00878J.

289 Marino, D.D., Aniko, V., Stocco, A. et al. (2017). Emulsion electro-oxidation of Kraft lignin. *Green Chemistry* 19: 4778–4784. https://doi.org/10.1039/C7GC02115A.

290 Hu, R., Zhao, Y., Tang, C. et al. (2022). Electrochemical biorefinery towards chemicals synthesis and bio-oil upgrading from lignin. *Engineering* https://doi.org/10.1016/j.eng.2022.10.013.

5

Isolation of Native-Like Lignin

Zhiwen Wang, Yiwei Fan and Peter J. Deuss

Department of Chemical Engineering (ENTEG), University of Groningen, Nijenborgh 4, Groningen, 9747 AG, The Netherlands

Lignin, as the most abundant aromatic biopolymer, has boundless potential for a variety of applications such as chemical modification for novel material production and selective depolymerization oriented toward aromatic chemicals. However, the heterogeneous structure and complicated interactions with polysaccharide components in the lignocellulosic matrix, render valorization of lignin challenging. A prerequisite for value-added utilization of lignin, detailed structure elucidation, and exploration of its complex reactivity are necessary. In these cases, a detailed understanding of lignin native structure is significant, in particular in the context of understanding the effect of processing conditions on this structure. For unraveling the complex puzzle that is lignin, it is often desired to separate lignin in its native form from the lignocellulosic matrix. Furthermore, recent developments on selective depolymerization of lignin to aromatic chemicals also directly benefit from the isolation of lignin with high aryl ether content and thus preservation of its original structure [1–6]. However, in reality, it is impossible to isolate native lignin as any method used, it being mechanical or chemical will alter its structures to some extent. Thus, researchers have to settle for isolating native-like lignin with an as close as possible structural similarity to native lignin. Different separation methods and types of native-like lignin products have been reported each with its inherent uses and limitations. However, all can contribute to help us to understand native structure of lignin and explore and explain new lignin chemistry and lignin-related reactivity phenomena.

It is difficult to define a native-like lignin. First of all, a precise and reliable benchmark or standard of native-like lignin is necessary to evaluate the success of native-like lignin isolation. One important indicator for resemblance to native lignin is the retention of labile linkages such as aryl ethers and other C—O bonds. Furthermore, one should consider the retention of various rarer units such as hydroxybenzoic acid (*p*-coumaric acid, ferulic acid, sinapic acid) and acetyl groups incorporated into lignin polymer through ester linkages (Scheme 5.1a). Ester bonds are more sensitive

Lignin Chemistry: Characterization, Isolation, and Valorization, First Edition.
Edited by Yuhe Liao and Bert F. Sels.

to alkaline or acidic treatments than the most common aryl ether linkages and are thus readily lost upon biomass processing. The structural integrity of lignin can be monitored by the ester linkages or the occurrence of structural motifs containing an ester linkage. For example, the ester linkage between lignin and carbohydrate constituents (Scheme 5.1b) can also serve as marker constituents. Constitutional units such as *p*-coumarate, ferulate, and some lignin-carbohydrate-complex (LCC) units linking with lignin via easter linkage can serve as monitor units to define the structure integrations of the lignin. However, the lower abundance of such linkages and motifs limit their application on marker units for native-like lignin evaluation.

Lignin is more known for its aryl ether linking motifs such as the most common β-O-4; however, there are many more structures that are more labile and can be a target for retention during native-like lignin isolation. Special labile ether bonds in rare units such as spirodienone and dibenzodioxocin are unstable under mild physiochemical treatment as determined by experimental works and supported by theoretical calculations. These are thus often only indirectly observed. For instance, β-1 structure products derived from ester cleavage of spirodienone were observed and identified via treating spruce with 4% sodium sulfite at 140 °C (Scheme 5.1c) [7]. The rare occurrence of these structures thus also makes it hard for usage as benchmarking. Also, the α-O-4 linkage has a much lower bond dissociation enthalpies than that of β-O-4 linkage, which has been calculated by different computational methods (Scheme 5.1d) [8, 9]. That means that in theory, α-O-4 linkage is better than β-O-4 linkage to indicate the structure integration of lignin. However, current studies indicate that no individual α-O-4 linkage without cyclization is likely to present in native lignin, and it only appears as a part linkage of phenylcoumaran and dibenzodioxocin. Nevertheless, the presence of α-O-4 linkage in particular in phenylcoumaran is a more universal and sensitive indicator for lignin native structure. In general, the amount of total β-O-4 linkage is often taken as a benchmark to evaluate the native feature of lignin, as this linkage has proved to be less stable than carbon–carbon linkages and its loss in abundance allows for better quantification compared to rare linking motifs. Thus, the total aryl ether linkage in the form of β-O-4 structure in isolated lignin is widely used as a clear indicator for retention of the native state of lignin.

According to the state in which lignin is isolated, native lignin isolation is divided into two main strategies. Dissolving lignin in (organic) solvents followed by recovery from the extract is the most common method used for lignin isolation (Scheme 5.2a). Nonetheless, organic solvents not only dissolve lignin but also some other impurities. Moreover, the intrinsic large molecular weight, aromatic features, and the connections with other biomass constituents make lignin less soluble in its native state, which significantly limits the efficiency of these methods. Consequently, pretreatments and proper enhancement of the extraction intensity are essential. However, the balance between keeping the native state of lignin and proper isolation conditions is hard to establish, which needs more optimization for specific organic solvent based isolation methods and type of biomass. The other way to isolate lignin is to remove polysaccharide constituents and leave lignin in residue (Scheme 5.2b). However, options for polysaccharide removal are limited. A combination of intensive physiochemical pretreatments breaking the cell wall structure followed by mild enzyme

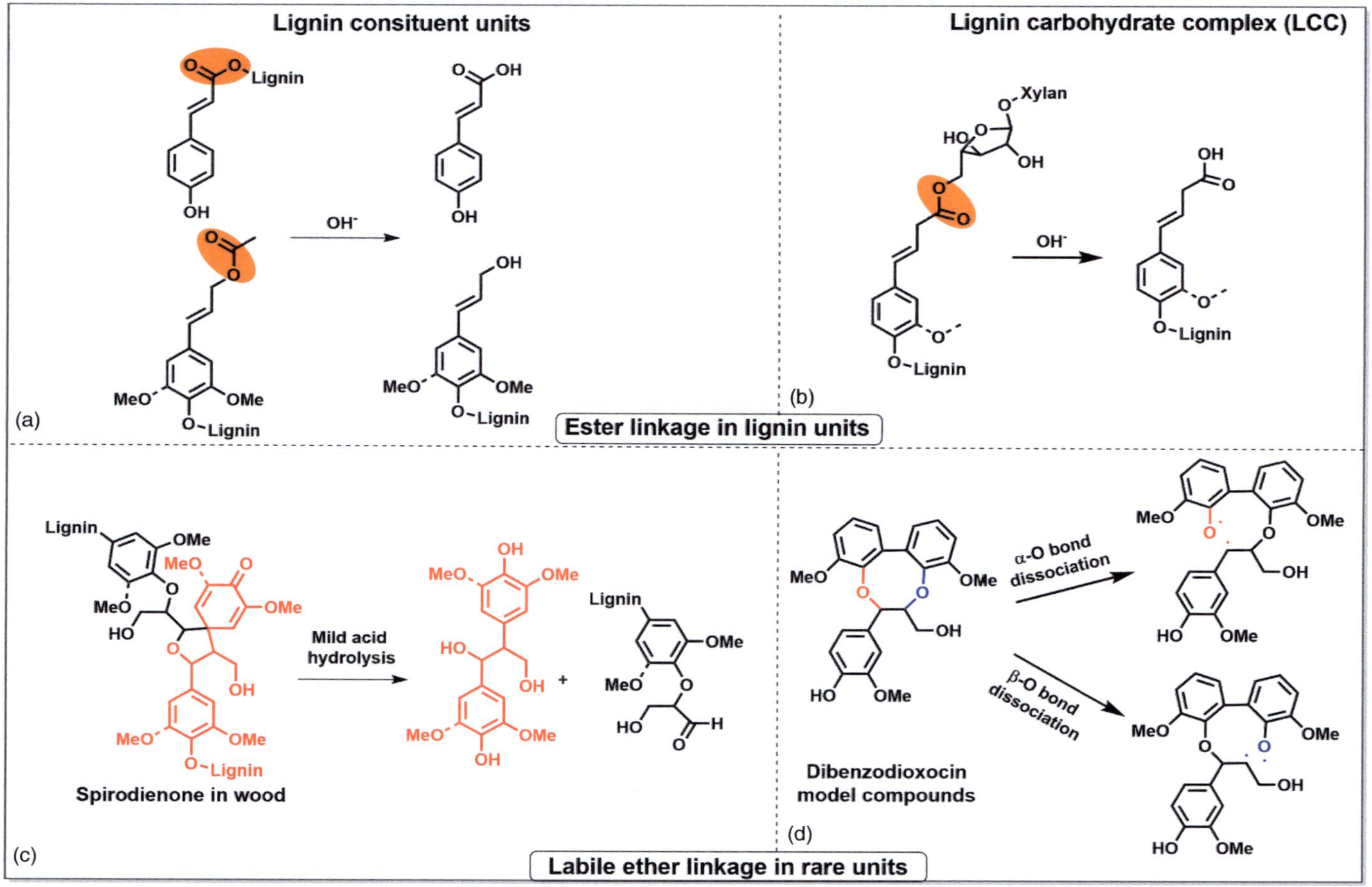

Scheme 5.1 Labile linkages used for native-like lignin evaluation.

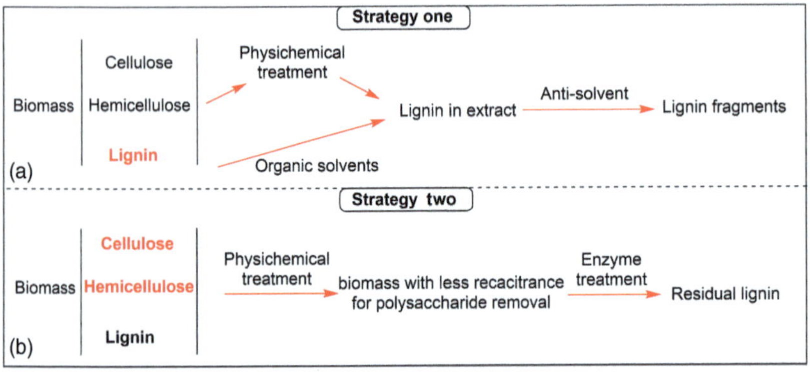

Scheme 5.2 Strategies used for lignin isolation (red color represents the component that is removed for lignin extraction).

Table 5.1 The dis/advantage of different lignin isolation strategies.

Strategies	Advantages	Disadvantages
Strategy one	Easily scale up, fast, low yield, cost, high solubility in common organic solvents	High structural alteration and low MW and retention of labile linkages
Strategy two	High yield and MW, revealing whole lignin structure, and retention of rare labile linkages	Tedious steps, high impurity and cost, low solubility in common organic solvents

hydrolysis of polysaccharide is the most common strategy used. It is necessary to point out that the polysaccharide impurities of this residual lignin might limit its application on elucidating lignin structure and exploring lignin chemistries. There are advantages and disadvantages for both isolation strategies (Table 5.1), and the choice of a particular isolation strategy depends basically on the objective of the study. Different approaches that fall under these strategies are discussed in more detail providing step-by-step procedures for the most commonly used ones. The main introduction and discussion of the following methods used for native-like lignin isolation will include first several classical methods and are then followed by upcoming methods based on the two individual strategies.

5.1 Classical Methods Used for Native-Like Lignin Isolation

5.1.1 Mild Alcohol Extracted Lignin

After the discovery of lignin, advances towards determining the native lignin chemical structure study were made by many pioneering researchers. As discussed previously, isolation of the total lignin from biomass with less structural alteration is one of the most challenging parts in studying the lignin structure. Lignin isolation by

Scheme and steps of Brauns's lignin isolation

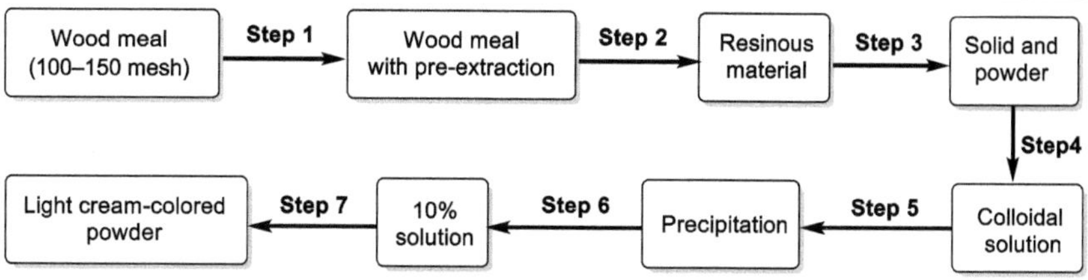

Step 1	Sequential pre-extraction with cold water and ether
Step 2	Extraction with 96% ethanol untill the solvent drains off colorless Ethanol phase combination and solvent removal
Step 3	Washed with water and ether
Step 4	Dissolved in dioxane, filtered and precipitated in water
Step 5	Addition of sodium sulfate
Step 6	Filtered, dried and redissolved in dioxane
Step 7	Filtered and added into ether

Scheme 5.3 The detailed steps for Braun's lignin isolation.[12]

alcohol extraction at elevated temperature was first developed by Klason in 1911 [10], but no detailed characterization was provided for verifying the extract to be lignin. Freudenberg also discussed that extract from alcohol-benzene contained some low molecular fragments of lignin [11]. About 80 years ago, Brauns (1939) first declared that the product extracted by neutral alcoholic solvent was native-like lignin [12]. At that time, he found some phenolic oligomers from neutral alcoholic solvent extraction of wood meal and claimed it was the native state via the presence of methoxy groups. Lignin obtained from this method was given the name Brauns' lignin. In this method, wood meal was first mixed with 96% ethanol at room temperature for three to four days (detailed steps are provided in Scheme 5.3). The obtained resinous material from the extract was purified by washing several times with water and diethyl ether followed by dissolution in 1,4-dioxane and further water precipitation and washing. The final product was light cream-colored powder. Typical tests for lignin at that time, such as lignin color reactions, methoxyl content measurement, and concentrated acid treatment, were done to confirm that the obtained product was lignin. Nonetheless, lignin obtained from this method had low yield (2–3%, based on lignin content) and still contained impurities such as oligosaccharides and other ethanol-soluble extractives, which limits application of lignin obtained by this method. Recently, ultrasound was employed to enhance the extraction efficiency. The yield almost doubled, but the actual total yield was still low [13]. Definitely, this method provides plenty of initial information about lignin, and a large amount of work about lignin structure elucidation and chemistry was developed by following this method, in particular driven by the demand of accessing native-like lignin. For instance, method used for acid-soluble lignin measurements was developed by using Brauns' lignin [14].

5.1.2 Milled Wood Lignin (MWL)

With high demand of attaining higher purity native-like lignin in a higher yield, a significant step forward was achieved in 1956 when Björkman extracted lignin coined intensive ball-milled wood lignin (MWL) [15]. In this method, intensive ball milling disrupts the compact cell wall structure, decreasing the crystallinity of cellulose, and thus enhancing the penetration of solvent in the cell wall to increase the extraction efficiency of lignin (Detailed steps about MWL isolation are indicated in Scheme 5.4). Therefore, compared with Brauns' lignin, MWL has a relatively higher yield. The yield of MWL varies between 5% and 50% (based on the lignin content), which closely connects with the intensity of ball milling and type of biomass. However, physical treatments such as intensive long-time ball milling also induce structural alteration such as oxidation and some cleavage of aryl ether linkages [16–18]. Moreover, native lignin with high molecular weight has typically poor solubility in organic solvents. From this point of view, MWL lignin can also be seen as a specific fraction of soluble lignin fragments that lack some representativeness of the whole structure of lignin. It still provides a large amount of valuable information for us to study native structure of lignin and is often used as a reference material due to the fact that it will yield lignin with high β-O-4 content and some rare units. In addition, attention should be paid to the structural alterations induced by intensive and long-time ball milling. Plenty of seminal works

Scheme and steps of milled wood lignin isolation

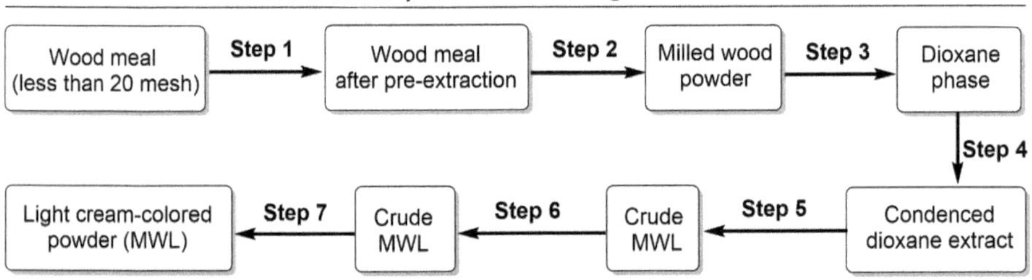

Step 1	Pre-extraction or dewax with toluene/ethanol (2/1, v/v)
Step 2	12 g material milled by lampen milling for 48 h 6 g material further milled by vibrational ball milling in toluene for 48 h
Step 3	96% (v/v) dioxane extraction (1 g : 10 mL) for 24 h and repeated for two times Separated by centrifugation to attain dioxane phase
Step 4	Concentrated under vaccum at 40 °C (20 mL) in water bath
Step 5	Dropwise precipitated in deionized water (200 mL) Separation and freeze-drying of the precipitation
Step 6	Dissolved in10 mL 90% acetic acid (9 : 1, v/v) and precipited in deionized water (200 mL) Separation and freeze-drying of the precipitation
Step 7	Dissolved in 5 mL DCM/ethanol (2 : 1, v/v) and precipitated (100 mL) and washed in ether

Scheme 5.4 The detailed steps for milled wood lignin (MWL) isolation from work of Björkman [15].

about lignin structure elucidation and general chemistry exploration are developed by using MWL. For instance, novel lignin structure elucidations such as LCC linking units and rare units of lignin were recognized and identified by structural explorations of MWL [19], as well as reaction behavior in novel reaction systems such as ionic liquids (ILs) and deep eutectic solvents [20].

5.1.3 Cellulolytic Enzyme Lignin (CEL)

In 1975, Chang et al., inspired by the isolation of MWL lignin, developed an updated method to increase the yield of lignin from neutral organic solvent extraction [21]. Detailed schematic illustration about steps for CEL isolation is profiled in Scheme 5.5. In his method, one initial step is accumulating the content of lignin in biomass by mild enzymatic hydrolysis of the polysaccharide components. In this way, the yield of lignin isolated from MWL was shown to be significantly increased from around 17% to 57% and 68% for spruce and sweetgum, respectively. In the meantime, detailed structure characterization such as molecular weight, phenolic hydroxyl content, and oxidic degradation indicated that the structure of CEL was similar to that of MWL and had a better representativeness of total lignin structure than that of MWL in terms of MW, aryl ether linkage and rare units. Later on, Zhang et al. further developed this method, and before the enzyme treatment, a step of swelling of wood meal in sulfoxide/*N*-methylimidazole was added. Through this pretreatment, the polysaccharide digestibility was enhanced to result a better yield

Scheme and steps of cellulolytic enzyme lignin isolation

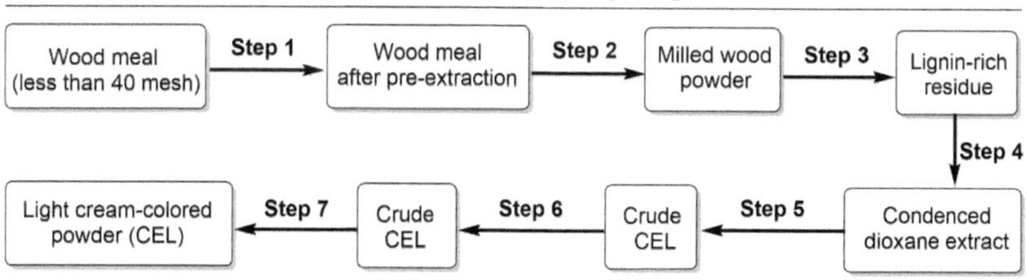

Step 1	Pre-extraction or dewax with toluene/ethanol (2 : 1, v/v)
Step 2	20 g material milled by vibrarional ball milling in toluene for 48 h
Step 3	100 g milled material was treated with 80 g Onozuka SS 1500 in 800 mL sodium acetate buffer (25 mM, pH = 4.5) and incubated at 50 °C for 3 days Residue was separated by centrifugation
Step 4	Extracted by first 96% and then 50% dioxane for 24 h and each repeated two times The dioxane phase was separately combined and condenced under vaccum
Step 5	Dropwise precipitated in deionized water Separation and freeze-drying of the precipitation
Step 6	Dissolved in 90% acetic acid (9 : 1, v/v) and precipitated in deionized water Separation and freeze-drying of the precipitation
Step 7	Dissolved in DCM/ethanol (2 : 1, v/v) and precipitated and washed in ether (only for 96% dioxane extraction)

Scheme 5.5 The detailed steps for cellulolytic enzyme lignin (CEL) isolation from work of Chang et al. [21].

of CEL [22]. The high yield and structure integrity of total lignin gives CEL more chance on lignin-related study. Study about lignin biosynthesis, novel catalytically oxidative transformation, and reductive fractionation of biomass have often used CEL as a model lignin for concept-proven test [23–27].

5.1.4 Enzymatic Mild Acidolysis Lignin (EMAL)

The pursuit for higher-yielding methods for native-like lignin still continued. Integrating a method consisting of enzymatic hydrolysis of polysaccharides and a mildly acidic organic extraction of the lignin-rich residue was applied [28]. A combined name of enzymatic mild acidolysis lignin (EMAL) was assigned for this method. Detailed steps and procedures of EMAL isolation are provided in Scheme 5.6. In this method, by taking use of lignin-rich residue and slightly acidifying the extracting media, the solubilization of larger fragments is significantly enhanced via cleaving a small number of labile linkages. Through this combined synergistic effect, the yield is further increased from around 10% for CEL to 50% for EMAL. This method was also demonstrated to isolate lignin from different types of biomasses such as White fir, *Eucalyptus globulus*, and Southern pine, with less saccharide impurities [29]. Our own lab (unpublished data) isolated EMAL from birch and carefully characterized it by linkage quantification from 2D HSQC NMR and molecular weight distribution from GPC. The results showed that in contrast to corresponding MWL, this lignin had similar linkage content and distribution with MWL, but a high yield and much larger average molecular weight.

Scheme and steps of enzymatic mild acidolysis lignin isolation

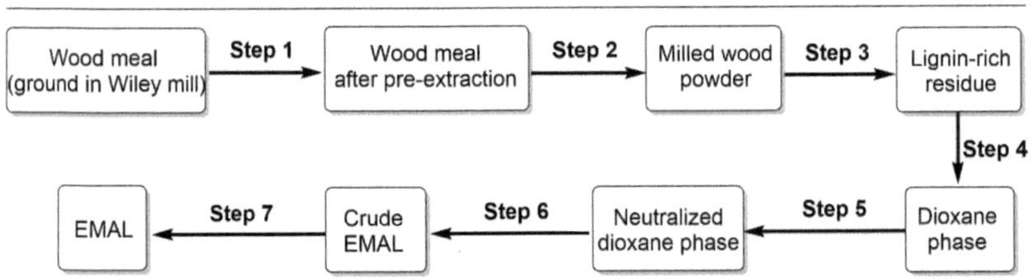

Step 1	Pre-extraction with acetone for 48 h
Step 2	80 g material milled by ball milling (ball and material accounting 18% of the taltal jar volume) at 36 rpm for 14 days
Step 3	40 g milled material was treated with 12.3 ml Iogen Industrial Cellulase (Iogen Corp., Ottawa, ON, Canada) in phosphate buffer (pH = 4.5) at 40 °C for 48 h Residue was separated by centrifugation and freeze-dried
Step 4	5 g lignin-rich residue was extracted by 100 ml dioxane-water (85 : 15, v/v) with 0.01 mol/l HCl (98.6 mg of 37% HCl) at 86 °C for 2 h
Step 5	The dioxane phase was collected by filtration and neutrolized by sodium bicarnonate The nutrolized dioxane phase was condenced at 30 °C until a thick solution was obtained
Step 6	Dropwise precipitated in deionized water (1 l, pH = 2) The precipitated lignin was isolated by centrifugation, washed and and freeze dried
Step 7	The isolated lignin was washed with HPLC grade hexane to remove extractives and vaccum dried at room temerature

Scheme 5.6 The detailed steps for enzymatic mild acidolysis lignin (EMAL) isolation from work of Wu et al. [28].

5.1.5 Residual Enzyme Lignin (REL)

Generally, native-like lignin isolation is achieved by dissolving lignin in the extracting media, and then lignin is recovered by anti-solvent precipitation. Apart from this strategy, residual lignin can be isolated by near complete removal of polysaccharides and leaving lignin as part of a solid residue. Less methods are available to apply this strategy to lignin isolation. Multiple times of mild enzymatic hydrolysis of polysaccharide with assistance of intensive physical pretreatments are typically applied [30]. A typical schematic illustration of REL isolation is indicated in Scheme 5.7. The milder extractions condition well preserves the pristine state of the lignin, but the low solubility in common organic solvents limits its analysis and application [21]. Thus, from early work, less structural information could be obtained for this type of lignin. Later on, with the development of art state of technology, more structural information was provided and indicated that native-like residual enzyme lignin is in several ways better representative of native lignin compared to lignin obtained from extractive methods [16, 31]. Unique advantages include isolation of native-like lignin in near theoretical yield, the better representativeness of the total lignin structure, high molecular weight, and retention of minor labile lignin structural motifs. However, the lignin also contains a high proportion of sugar impurities resulted from the incomplete removal of saccharides partly due to the low selectivity of the enzyme cocktail for hemicellulose components. Intensive ball milling causes some decrease in molecular weight and β-O-4 content and induces some oxidation [18, 32]. Additionally, residual protein

Scheme and steps of native-like residual enzyme lignin isolation

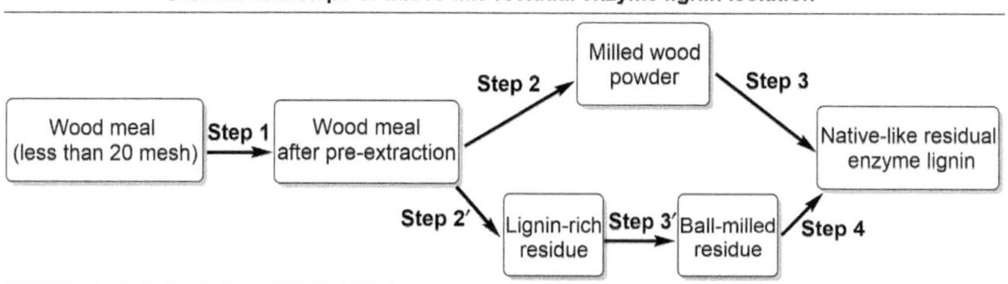

Step 1	Pre-extraction with toluene/ethanol (2 : 1, v/v) for 8 h
Step 2	30 g material milled by planetary ball milling with a 250 ml ZrO_2 jar (5 × 15 mm, 10 × 5 mm) at 450 rpm for 24 h (actual time, with 10 min milling, 15 min pause, 144 rotations)
Step 3	10 g milled material was treated with 5 ml Ctec2 (Novoenzyme) in acetate buffer (50 mM, pH = 5.5) at 50 °C for 72 h Residue was separated by centrifugation and freeze-dried
Step 2'	30 g material milled by planetary ball milling with a 250 ml ZrO_2 jar (5 × 15 mm, 10 × 5 mm) at 450 rpm for 6 h (actual time, with 10 min milling, 15 min pause, 36 rotations)
Step 3'	20 g milled material was treated with 10 ml Ctec2 (Novoenzyme) in acetate buffer (50 mM, pH = 5.5) at 50 °C for 72 h Residue was separated by centrifugation and freeze-dried
Step 4	Repeat step 2′ to step 3′ to several times according to different type of biomasses

Scheme 5.7 Detailed procedures of residual enzyme lignin (REL) isolation from our work [16].

contamination from the enzyme treatment further increases the purities and influences the analysis of *p*-hydroxylphenyl units by methods such as 2D HSQC NMR [33, 34]. Moreover, the tedious steps and high energy consumption of ball milling hinder the large-scale production and application of this type of lignin. Inexpensive DES/IL pretreatments of the starting biomass might be introduced to tackle these problems, since relevant studies indicated that milder DES or IL pretreatments only had minor influence on lignin structure but could disrupt the entangled structure of lignocellulose [35, 36]. Employment of protection strategies such as lignin benzylic alkoxylation, boric ester cyclization, and relatively harsh fractionation conditions are possible for native-like lignin isolation, which significantly enhance the efficiency of the extraction [37, 38].

5.2 Isolation of Lignin with High β-O-4 Content

5.2.1 Isolation Under Acidic Conditions

Lignin isolated under acidic condition is a process in which acidic reagents are introduced to boost lignin release from the lignocellulosic biomass. Part of the above-discussed methods such as isolation of EMAL also belong to this strategy. In this process, apart from the physical state of the starting material, the acidity, retention time, and temperature play predominant roles, since under strong acidity and high extracting temperature, not only the labile rare units but also the predominant β-O-4 linkage are cleaved. For instance, the influence of bond cleavage by acidic conditions can be readily seen upon acidic pretreatment of lignocellulosic biomass. Under a dilute acidic pretreatment (0.05 g H_2SO_4/g biomass) of switchgrass at 190 °C, a 36% cleavage of aryl ether linkage was observed as evidenced by 2D HSQC NMR and ^{31}P NMR [39]. Even a hot water treatment of wheat straw at 160 °C led to a decrease of β-O-4 linkage from 39% to 14% by the analysis REL before and after treatment [40]. Moreover, a slight increase loading of H_2SO_4 from 0.25% to 1% to treat the wheat straw at 160 °C resulted in a further decrease of the β-O-4 linkage from 9% to 2%, indicating the dominant effect of acidity on the structural integrity of lignin.

Through analyzing the extracting steps of EMAL, it is found that by combination of suitable pretreatment and carefully control of the acidity of the extraction media, directly acidic organic extraction of native-like lignin is feasible. The challenge in this process is the balance between yield and structure integrity of the lignin (see the extraction strategy in Scheme 5.8a). For instance, under a moderate temperature range of 90–94 °C and a slightly longer extraction time of 48 hours, a 25% (based on the starting material) yield of lignin was obtained from aspen, since the lack of advanced analytic techniques, linkage information was not provided for this lignin [41]. Later on, detailed structure information in terms of linkage content and distribution as well as molecular weight was provided by others who isolated the lignin by following this method [6]. The yield (1%, based on starting material) was much lower than that previously reported, which might be due to different biomass and solvent volumes used for the lignin extraction. However, lignin obtained in this case had high β-O-4 linkage and less condensation, further indicating the potential of mild acidic organic extraction on native-like lignin isolation. The lower yield of lignin is the limiting parameter for wide application of mild acidic organic extraction on isolating native-like lignin. This is more or less enhanced by intensively physical pretreatments such as ball milling as widely used in previous methods. For instance, even under a less intensive ball milling, the yield of lignin with high β-O-4 was significantly increased [42].

Organic acids serve a double role in lignin extraction of solvent and catalyst. Therefore, organic acid media have advantages for native-like lignin isolation. This is in particular the case for small molecular weight organic acids as they have low boiling points and high volatility, increasing the ease of the extraction and isolation under relatively low temperature. Many organic acids have been applied in lignin isolation. Among them, formic acid (FA) and acetic acid (EA) or their mixtures are the most commonly used. Lignin collected from stirring bamboo with boiling FA has a yield of 19.4% (based on the starting bamboo) and only exhibited a slightly less β-O-4 linkage (64%) than that of MWL (72%) [43]. After shortening the reaction time from two to one hour for boiling the biomass in FA, and then adding H_2O_2 (3%) and continuing to boil the mixture for another one hour, the obtained lignin had a similar yield (18.8%) with that of pure FA extraction but possessed almost same β-O-4 linkage content (76%) compared to MWL. This case indicated the outstanding potential of FA on isolating native-like lignin in high yield, purity, and quality.

para-Toluenesulfonic acid (PTSA) and its derivatives have also been employed in native-like lignin isolation [44]. Different from the small molecular acid, PTSA appears in a solid state under room temperature, thus concentrated aqueous PTSA (>50%) is actually used for wood dissolution. An distinguish advantage of this aqueous is the formation of a hydrotrope, which significantly facilitates the separation of lignin and polysaccharide via tuning the concentration of the final acid. Around 70% of lignin can be removed from dissolving poplar NE222 in 70% PTSA at 50 °C for a short of 20 minutes. The recovered lignin had 94% (analysis of whole cell

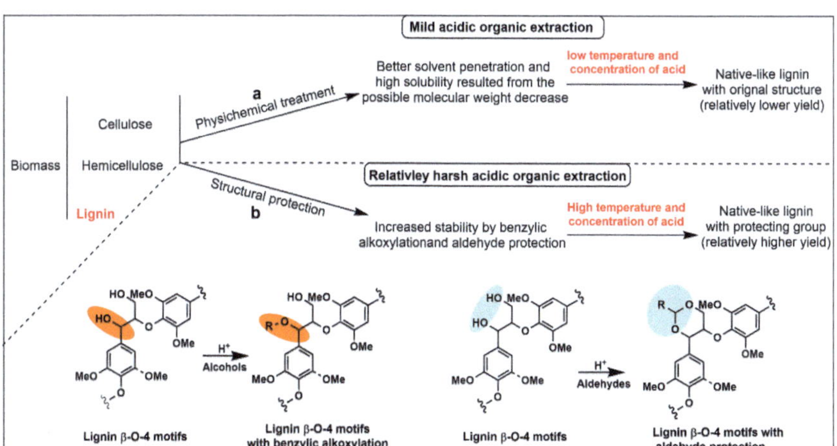

Scheme 5.8 Schematic illustration of acidic organic extraction of native-like lignin.

material) of β-O-4 retention. Excellent performance was observed when PTSA served as acid additive for alcohol extraction. For instance, both methanol and ethanol acidified by PTSA gave a more than 80% lignin removal under milder conditions [45, 46]. Among them, lignin fractionated from 3 M acid ethanol at 85 °C for five hours had a high β-O-4 retention, which could be confirmed by the near theoretical aromatic monomer yield from lignin hydrogenolysis.

As stated above, the cleavage of labile linkages and units limits the high efficacy isolation of native lignin. If one could sacrifice the loss of some rare units and linkages and only focused on conserving the β-O-4 linkage, some other methods are viable. Another strategy through which the β-O-4 linking motif is simultaneously modified during extraction allows for better stability of this new unit to resist harsher extraction conditions. This leads to a significant increase in the efficacy of isolating lignin with high β-O-4 content albeit in a modified form (see the extraction strategy in Scheme 5.8b). For example, lignin after benzylic alkoxylation has a better stability and recalcitrance toward harsh chemical conditions, which provides the possibility to isolate high β-O-4 lignin under high temperature and strong acidity. This can be achieved by alcohol-mediated acidic organic extraction and further verified by lignin extraction from different alcohols acidified by HCl [47]. A representative isolation protocol is shown in Figure 5.1. Very recently, methanol acidified by HCl (1.37 mol/l) extraction gave a 24.2% yield of lignin in contrast to 2% of the corresponding MWL, and at the same time, almost all the β-O-4 linkage (42.6% including 20.3% alkoxylated by methanol) was preserved [48]. Alcohols with different sizes are used for lignin isolation, among them, butanol showed the most promising efficacy [47]. However, recovery of lignin from the concentrated butanol phase is the limiting step for the whole process. Normally, water precipitation is used to recover the lignin from the extract, and this is also suitable to separate lignin extracted by alcohols with small size. For alcohols with large alcohols, simple water precipitation cannot efficiently disperse alcohol phase and separate lignin out of the phase. In these cases, a slight addition of salt will have a better flocculation of lignin [49]. In addition, this system might also bring butanol-alkoxylated saccharide impurities that are harder to dissolve in water than that alkoxylated from ethanol

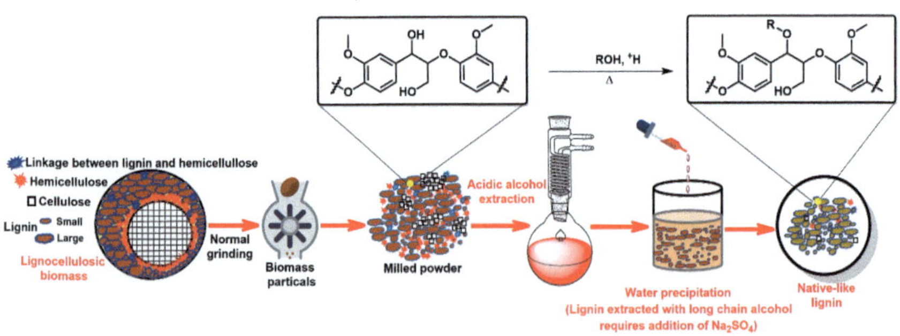

Figure 5.1 Alcohol mediated extraction of native-like lignin.

and propanol. Our group developed a simple method to remove the (poly)saccharide impurities by washing the lignin with mild alkaline solution, with almost no decrease of total aryl ether linkage [50].

Also, the acidic organic extraction with addition of aldehydes was shown to yield stabilized modified lignin via formation of a 1,3-dioxolane-structure between the aldehyde and α,γ diols of the β-O-4 linking motif [5, 51, 52]. Through this way, 5.0 g birch as starting material and propionaldehyde as protection reagents gave a 1.2 g of finally precipitated product in high structural integrity, as revealed by the intact aromatic region on the 2D HSQC NMR [52]. Similar to the protection of aldehyde, other potential reagents such as boronic acid and its derivatizes might also react with the α,γ diols of lignin to produce a six-member boronate ester rings, and thus protect the labile β-O-4 linkage [51, 53].

5.2.2 Isolation Under Alkaline Conditions

In the chemical industry, the higher aqueous solubility of phenolates compared to phenolics has always been used for their isolation and purification. This principle can also be extended lignin isolation, since once some aryl ether linkages are cleaved, phenolic end groups are released. A schematic representation of the alkaline extraction principle is illustrated in Figure 5.2. Extensive cleavage of β-O-4 linkages is of course not desired for native-like lignin isolation. However, if only limited sites of the linkage are cleaved to increase the possibility to release lignin out of the lignocellulosic matrix, and still a high proportion of β-O-4 linkage are conserved, this is a viable strategy. Pioneering researchers already made use of this method to isolate lignin. Sun et al. used water basified by potassium and lithium hydroxide to isolate lignin under milder conditions, and they found that around 60% (based on lignin content) of lignin from the wheat straw could be extracted [54]. Later, researchers provided more detailed structural insights about lignin isolated following this protocol and confirmed that the lignin had a similar content of β-O-4 linkage to that of the corresponding CEL [34]. Alkalized organic solvent mixtures such as 1,4-dioxane/water (80/20, v/v, with 0.1 M NaOH) were also applied on lignin isolation. Lignin obtained from this process contained a large amount of non-condensed ether

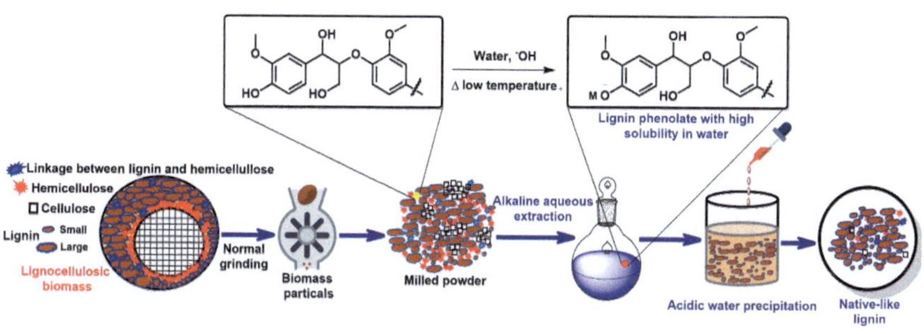

Figure 5.2 Alkaline aqueous extraction of native-like lignin.

structures [55]. Detailed quantification of the linkage content was provided by the researcher who followed this protocol with less loading of NaOH (0.05 M), and lignin was isolated in a maximum yield of 30% (based on lignin content) with higher β-O-4 linkage content compared to corresponding MWL [56].

5.2.3 Isolation Using Flow-through Extraction

Most methods used for native-like lignin isolation stated above involve extractions in batch reactors (Scheme 5.9a). In this lignin isolation strategy, lignin released from the initial stages of the extraction can have serious structural alteration under extraction conditions. Therefore, if the extracted lignin fragments are continuously removed from extraction medium, unnecessary structural alterations will be avoided. In this way, somewhat harsher extracting conditions can be applied for native-like lignin isolation. For this purpose, fast flow-through lignin extraction through a fixed bed extractor has been employed (Scheme 5.9b). As expected, the efficacy (yield) and quality (β-O-4 linkage) of lignin fractionation of all the studied cases in flow-through extraction were higher than the corresponding batch extraction, when the system applied on various biomasses via alcohols/water/H_2SO_4 system at 120 °C [57]. Even under a relatively high temperature of 225 °C, the extraction of lignin from poplar with methanol in a flow-through set up still yielded native-like lignin by analysis of the aromatic monomer yield of the lignin and starting biomass from reductive hydrogenolysis [58]. A flow-through extractor was also applied to organic acid lignin extraction. Under the same retention time, lignin yields from batch and flow-through extraction were similar with each other, but lignin obtained from the later method had a higher retention of β-O-4 linkage [59]. In addition, the flow-through extraction achieved a maximum of 83.3% yield of lignin from poplar by using 1.5 mol/l *p*-toluenesulfonic at 70 °C for a 30 minutes [60].

5.2.4 Isolation With Alternative Solvents

Due to the neglectable vapor pressure, high thermal stability, tunable acidity/basicity, viscosity, and density, ILs have been widely applied for processing complex polymers that are hard to handle in more common solvent systems [61, 62]. Their unique high capability for lignin dissolution has meant more focus on the application of ILs in lignin isolation [63, 64]. Typical ionized constituents of ILs used for native-like lignin isolation are listed in Scheme 5.10. Many factors of ILs might affect lignin isolation, mainly including sorts of anions/cations, π–π, and n-π interaction between the IL and lignin [65]. Different anion and cation components of IL have various performances on lignin solubility and dissociation from lignocellulosic biomass. It is reported that anions played the dominant role in lignin dissolution, while, cations in particular the planar ones only slightly influenced the process [64]. However, cations of ILs influence the cleavage of β-O-4 linkage when these contain acidic protons. A comparison of cations of protic ILs with different carbon chain length showed that the ones with shorter length isolated lignin in the higher β-O-4 content.

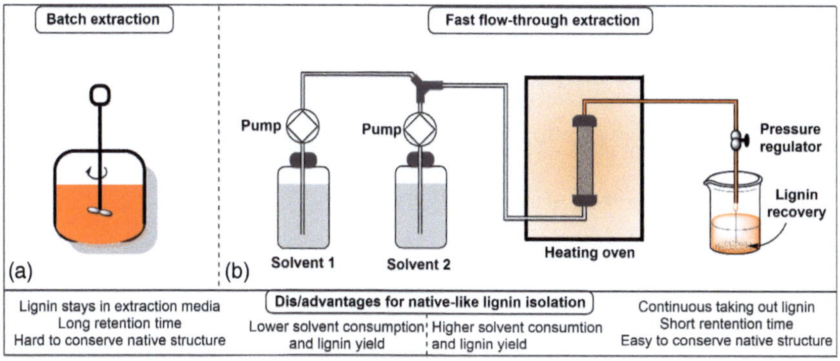

Scheme 5.9 Schematic illustration of the difference between batch and fast flow-through extraction.

Those cases indicate that a careful selection of cation/anion match is necessary for native-like lignin isolation. Various types of IL have been applied in biomass fractionation, lignin isolation and modification [66, 67], but only in several cases, lignin that can be considered native-like lignin was obtained. The acidity and usage temperature of IL are closely connected with the structure integrity of lignin. ILs with a lower acidity such as 1-ethyl-3-methyllimidazolium acetate ([C2mim]-[OAC]) showed excellent performance on isolation lignin in high content of β-O-4 linkage under the milder fractionation conditions [35, 68, 69]. Triethylammonium hydrogen sulfate ([N_{2220}][HSO_4]) is an inexpensive alternative of [C2mim]-[OAC] [70]. Through carefully changing the ratio of acid and base (a/b), the acidity was tuned to isolate lignin in different yields and quality. Among the studied cases, a relatively high temperature of 120 °C and a short treatment time of 0.5 hours under both a/b ratio of 1.02 and 0.98 of IL gave lignin with the highest β-O-4 linkage content (higher than 65%) but relatively lower yield (less than 5%). Under a proper control of the temperature under 50 °C and simultaneous protection by benzylic alkoxylation, acidified ILs formed by triflic acid (HOTf) and triethylamine, could isolate a 7.4% yield (based on starting biomass) of EG benzylic-alkoxylated lignin with a total of 61.2% aryl ether linkage [37]. A further step of enzymatic removal of polysaccharide in the residue produced a 21.3% (based on starting biomass) yield of residual enzyme lignin. This lignin also had a high degree of benzylic alkoxylation and a total of 76.2% aryl ether linkage. The combination of direct extraction and followed with the isolation of residual enzyme lignin provide an integrated and efficient strategy for native-like lignin isolation. In total, either increasing the temperature or prolongation of treatment time boosted the lignin yield but induced a decline in the β-O-4 linkage content of the recovered lignin, indicating the limitation of this IL for native lignin isolation.

As an analogous group of solvents to ILs, DES have similar applications to IL, but the lower cost and toxicity, as well as easy preparation give DES more potential on decrease of the operational cost, which is particular important for lignin as the

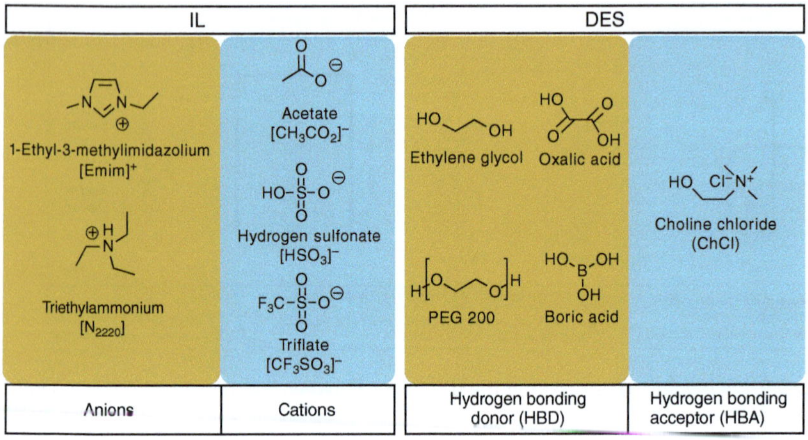

Scheme 5.10 Constitution of Ionic liquid and deep eutectic solvent used for native-like lignin isolation.

targets are usually basic chemicals [71, 72]. The performance of DES on lignin and biomass fractionation depends on its constitutions that decide the acidity/alkalinity, density, polarity, etc. Various combinations of hydrogen donor and hydrogen acceptor for DES production have been applied on biomass fractionation and while successful for attaining lignin in high yield, recovery of native-like lignin is not often reported [73–80]. Some representative DES components for native-like lignin isolation are listed in Scheme 5. Used at low temperature and a short extraction time or by introducing lignin protecting groups, DES have been shown to be able to isolate native-like lignin. For instance, boric acid-based DES had an excellent capability upon lignin isolation and achieved more than 40% lignin isolation. The lignin obtained under milder conditions from this DES showed only weak condensation signals on the spectra of 2D HSQC NMR, indicating the high structural integrity of the obtained lignin [38]. Acidified alcohol-based DES enhanced native-like lignin isolation by introducing protecting groups via benzylic alkoxylation. In this process, alcohols such as diols and glycerol not only serve as hydrogen-donor as part of the DES but also simultaneously modify lignin via benzylic alkoxylation. Around 60% lignin from Switchgrass was isolated from DES prepared by ChCl and ethylene glycol (1/2 molar ratio and 10 wt% water) and acidified by 1 wt% H_2SO_4 isolated at 130 °C for 30 minutes. The obtained lignin had intact signals on the aromatic region of the 2D HSQC NMR and an obvious signal of α-benzylic position but no detailed analysis and highlight were provided [81]. Later on, under similar protection strategy, a tunable ternary DES of ChCl/ethylene glycol/oxalic acid could directly fractionate more than 40% yield of lignin with a total 80% retention of β-O-4 content (based on MWL) at even milder conditions (100 °C, 24 hours). Detailed study via utilization of representative model compounds confirmed that benzylic alkoxylation of β-O-4 linking motif under a condition of high concentration of alcohols significantly retarded the degradation and condensation [80].

5.2.5 Alternative Resources and Special Lignins

As a byproduct, technical lignin obtained from purple industry always has undesired structural alteration such as condensation and loss of β-O-4 content. However, with the progress of new concept of biorefinery, even as a byproduct, native structure might be conserved from the process. For instance, for the bio-ethanol production via mild pretreatments such as electron beam treatment, residual lignin still had a high content non-condensed unit [82]. The stability of the lignin under the extracting condition is heavily relied on the structure of lignin. For some lignin with high stability under acidic conditions, even conditions or methods that are not suitable for native-like lignin isolation might be applied. For instance, the lignin (C type lignin) sole consisting of benzodioxane units tolerates relatively strong acidic condition [83, 84]. Thus, acidic DES such as ChCl/lactic acid, which will cause large structural alteration during fractionation, still could be used to isolate high-quality C type lignin in relatively high yield (almost close to 90% based on Klason lignin), and monomer released from the hydrogenolysis of the lignin had a lower yield but close to that of EMAL [85].

5.3 Concluding Comments

Information about lignin structure, chemistry, and interaction with other constituents is highly beneficial for the development of value-added conversion strategies for lignin. Accessing lignin in a near-pristine state plays an important role in this process. Different methods used to obtain the lignin have been discussed in this chapter. The main difference between them is found in terms of approaches and the yields, purities, and structural preservation contained within the lignin products. These aspects significantly influence their application in lignin studies. MWL is still the most widely used method, as it contains almost all the typical characteristics of native lignin in terms of linking motifs and has the advantage of being high solubility in normal organic solvents. Attention should be paid to MWL when the study requests lignin has a better representativeness of the total lignin structural diversity as only specific short fragments are extracted as well as when there are requirements of high yield. The limitation of MWL might be overcome by enhanced methods such as CEL and EMAL. However, depending on the final purpose of the native-like lignin isolation, almost all the traditional methods require carefully tuning the actual fractionation conditions to specific biomass samples as to avoid degradation. When a better representation of the overall diverse native structure of lignin in terms of combinations of larger and smaller fragments is desired, REL is the best option. For instance, this type of lignin has been found useful during the optimization and tracking of specific lignin fragments released upon depolymerization under reductive catalytic fractionation. Also, this type of lignin isolation method has proven useful in studies on genetic engineering of lignin structure to decrease its recalcitrance toward chemical depolymerization. An elucidation of the total lignin structure is necessary to track trace structure changes before and after the genetic engineering. Residual enzyme lignin also has intrinsic advantages on revealing lignin structure alteration and transformation from different pretreatments. The options for different native-like lignin isolation methods thus relate closely to the aims of the study. If rare units and linkage are the targets of the study, neutral organic extraction such as MWL is recommended. Regarding research targeting at β-O-4 linking motif and requiring high amount of quantification, acidic/basic and DES/IL extraction under the assistance of flow-through extractor are more suitable. If high yields are required and chemical modification of the β-O-4 is less of an issue, alcohol or aldehyde types extraction can give excellent lignin yields that are very suitable for further processing and actually are viable for upscaling. On the other spectrum, native lignin isolation methods will continually develop to allow for new and rare structural identification and provide insight into the diverse chemistry of lignin.

References

1 Schutyser, W., Renders, A.T., Van den Bosch, S. et al. (2018). Chemicals from lignin: an interplay of lignocellulose fractionation, depolymerisation, and upgrading. *Chemical Society Reviews* 47: 852–908.

2 Zakzeski, J., Bruijnincx, P.C.A., Jongerius, A.L., and Weckhuysen, B.M. (2010). The catalytic valorization of ligning for the production of renewable chemicals. *Chemical Reviews* 110: 3552–3599.

3 Rinaldi, R., Jastrzebski, R., Clough, M.T. et al. (2016). Paving the way for lignin valorisation: recent advances in bioengineering, biorefining and catalysis. *Angewandte Chemie, International Edition* 55: 8164–8215.

4 Sun, Z., Fridrich, B., De Santi, A. et al. (2018). Bright side of lignin depolymerization: toward new platform chemicals. *Chemical Reviews* 118: 614–678.

5 Shuai, L., Amiri, M.T., Questell-Santiago, Y.M. et al. (2016). Formaldehyde stabilization facilitates lignin monomer production during biomass depolymerization. *Science* 354: 329–333.

6 Deuss, P.J., Scott, M., Tran, F. et al. (2015). Aromatic monomers by in situ conversion of reactive intermediates in the acid-catalyzed depolymerization of lignin. *Journal of the American Chemical Society* 137: 7456–7467.

7 Gellerstedt, G. and Zhang, L. (1991). Reactive lignin structures in high yield pulping. *Nordic Pulp & Paper Research Journal* 6: 136–139.

8 Castellan, A. (2013). Christian Gardrat, Reinaldo Ruggiero, Marie-Thérèse Rayez, Jean-Claude Rayez. *Wood Science and Technology* 47: 27–41.

9 Elder, T. (2013). Bond dissociation enthalpies of a dibenzodioxocin lignin model compound. *Energy & Fuels* 27: 4785–4790.

10 Klason, P. (1911). *Beitrage zur Kenntnis der chemischen Zusammensetzung des Fichtenholzes*, 34–36. Berlin: Gebrüder Borntraeger.

11 Freudenberg, K., Janson, A., Knopf, E., and Haag, A. (1936). *Zur Kennmis des lignins Title Zur*, 1415–1425. Berlin.

12 Brauns, F.E. (1939). Native lignin 1 I. Its isolation and methylation. *Journal of the American Chemical Society* 61: 2120–2127.

13 Su, X., Fu, Y., Shao, Z. et al. (2022). Light-colored lignin isolated from poplar by ultrasound-assisted ethanol extraction: structural features and anti-ultraviolet and anti-oxidation activities. *Industrial Crops and Products* 176: 114359.

14 Pearl, I.A. and Busche, L.R. (1960). Studies on the chemistry of Aspenwood. XI. The Klason lignin determination as applied to Aspenwood with special reference to acid-soluble lignin. *Tappi* 43: 961–970.

15 Björkman, A. (1954). Isolation of lignin from finely divided wood with neutral solvents. *Nature* 174: 1057–1058.

16 Wang, Z., Zhu, X., and Deuss, P.J. (2021). The effect of ball milling on birch, pine, reed, walnut shell enzymatic hydrolysis recalcitrance and the structure of the isolated residual enzyme lignin. *Industrial Crops and Products* 167: 113493.

17 Jääskeläinen, A.S., Sun, Y., Argyropoulos, D.S. et al. (2003). The effect of isolation method on the chemical structure of residual lignin. *Wood Science and Technology* 37: 91–102.

18 Ikeda, T., Holtman, K., Kadla, J.F. et al. (2002). Studies on the effect of ball milling on lignin structure using a modified DFRC method. *Journal of Agricultural and Food Chemistry* 50: 129–135.

19 Ando, D., Nakatsubo, F., Takano, T., and Yano, H. (2016). Elucidation of LCC bonding sites via γ-TTSA lignin degradation: crude milled wood lignin (MWL) from *Eucalyptus globulus* for enrichment of lignin xylan linkages and their HSQC-NMR characterization. *Holzforschung* 70: 489–494.

20 Ogawa, S. and Miyafuji, H. (2015). Reaction behavior of milled wood lignin in an ionic liquid, 1-ethyl-3-methylimidazolium chloride. *Journal of Wood Science* 61: 285–291.

21 Chang, H., Cowling, E.B., and Brown, W. (1975). Comparative studies on cellulolytic enzyme lignin and milled wood lignin of sweetgum and spruce. *Holzforschung : International Journal of the Biology, Chemistry, Physics, and Technology of Wood* 29: 153–159.

22 Zhang, A., Lu, F., Sun, R.C., and Ralph, J. (2010). Isolation of cellulolytic enzyme lignin from wood preswollen/dissolved in dimethyl sulfoxide/N-methylimidazole. *Journal of Agricultural and Food Chemistry* 58: 3446–3450.

23 Weng, J.-K., Akiyama, T., Ralph, J., and Chapple, C. (2011). Independent recruitment of an O-methyltransferase for syringyl lignin biosynthesis in *Selaginella moellendorffii*. *Plant Cell* 23: 2708–2724.

24 Chen, C., Cai, N., Wan, C. et al. (2022). Cell wall modification and lignin biosynthesis involved in disease resistance against *Diaporthe citri* in harvested pummelo fruit elicited by carvacrol. *Journal of the Science of Food and Agriculture* 102: 3140–3149.

25 Wang, S., Li, W.X., Yang, Y.Q. et al. (2020). Unlocking structure–reactivity relationships for catalytic hydrogenolysis of lignin into phenolic monomers. *ChemSusChem* 13: 4548–4556.

26 Rahimi, A., Azarpira, A., Kim, H. et al. (2013). Chemoselective metal-free aerobic alcohol oxidation in lignin. *Journal of the American Chemical Society* 135: 6415–6418.

27 Rahimi, A., Ulbrich, A., Coon, J.J., and Stahl, S.S. (2014). Formic-acid-induced depolymerization of oxidized lignin to aromatics. *Nature* 515: 249–252.

28 Vvu, S. and Argyropoulos, D. (2003). An improved method for lsolating lignin in high yield and purity. *Journal of Pulp and Paper Science* 29 (7): 235–240.Argyropoulos, D. (2003). An improved method for lsolating lignin in high yield and purity. *Journal of Pulp and Paper Science* 29: 235–240.

29 Guerra, A., Filpponen, I., Lucia, L.A., and Argyropoulos, D.S. (2006). Comparative evaluation of three lignin isolation protocols for various wood species. *Journal of Agricultural and Food Chemistry* 54: 9696–9705.

30 Pew, J.C. (1957). Properties of powdered wood and isolation of lignin by cellulytic enzymes. *Tappi* 40: 553–558.

31 Deuss, P.J. and Wang, Z. (2021). Catalytic hydrogenolysis of lignin: the influence of minor units and saccharides. *ChemSusChem* 14: 5186–5198.

32 Hu, Z., Yeh, T.-F., Chang, H. et al. (2006). Elucidation of the structure of cellulolytic enzyme lignin. *Holzforschung* 60: 389–397.

33 Kim, H. et al. (2017). Characterization and elimination of undesirable protein residues in plant cell wall materials for enhancing lignin analysis by solution-state nuclear magnetic resonance spectroscopy. *Biomacromolecules* 18: 4184–4195.

34 Wen, J.-L., Sun, S.-L., Yuan, T.-Q., and Sun, R.-C. (2015). Structural elucidation of whole lignin from eucalyptus based on preswelling and enzymatic hydrolysis. *Green Chemistry* 17: 1589–1596.

35 Sathitsuksanoh, N. et al. (2014). Lignin fate and characterization during ionic liquid biomass pretreatment for renewable chemicals and fuels production. *Green Chemistry* 16: 1236–1247.

36 Kim, K.H., Dutta, T., Sun, J. et al. (2018). Biomass pretreatment using deep eutectic solvents from lignin derived phenols. *Green Chemistry* 20: 809–815.

37 Wang, Z., Zhang, Z., Wang, H., and Deuss, P.J. (2022). Tuning lignin properties by mild ionic-liquid-mediated selective alcohol incorporation. *Chem Catalysis* 2: 1407–1427.

38 Ci, Y., Zhang, Z., Wang, H. et al. (2020). New ternary deep eutectic solvents for effective wheat straw deconstruction into its high-value utilization under near-neutral conditions. *Green Chemistry* 22: 8713–8720.

39 Samuel, R., Pu, Y., Raman, B., and Ragauskas, A.J. (2010). Structural characterization and comparison of switchgrass ball-milled lignin before and after dilute acid pretreatment. *Applied Biochemistry and Biotechnology* 162: 62–74.

40 Jensen, A. et al. (2017). 2D NMR characterization of wheat straw residual lignin after dilute acid pretreatment with different severities. *Holzforschung* 71: 461–469.

41 Pepper, J.M., Baylis, P.E.T., and Adler, E. (1959). The isolation and properties of lignins obtained by the acidolysis of spruce and aspen woods in dioxane–water medium. *Canadian Journal of Chemistry* 37: 1241–1248.

42 Zijlstra, D.S., de Santi, A., Oldenburger, B. et al. (2019). Extraction of lignin with high β-O-4 content by mild ethanol extraction and its effect on the depolymerization yield. *J. Visualized Exp.* 14: e58575.

43 Li, M.-F., Sun, S.-N., Xu, F., and Sun, R.-C. (2012). Formic acid based organosolv pulping of bamboo (*Phyllostachys acuta*): comparative characterization of the dissolved lignins with milled wood lignin. *Chemical Engineering Journal* 179: 80–89.

44 Chen, L., Dou, J., Ma, Q. et al. (2017). Rapid and near-complete dissolution of wood lignin at ≤80°C by a recyclable acid hydrotrope. *Science Advances* 3: e1701735.

45 Fan, D., Xie, X., Li, C. et al. (2021). Extraction of noncondensed lignin from poplar sawdusts with *p*-toluenesulfonic acid and ethanol. *Journal of Agricultural and Food Chemistry* 69: 10838–10847.

46 Zhai, Q., Long, F., Hse, C. et al. (2019). Facile fractionation of bamboo wood toward biomass valorization by *p*-TsOH-based methanolysis pretreatment. *ACS Sustainable Chemistry & Engineering* 7: 19213–19224.

47 Zijlstra, D.S., Lahive, C.W., Analbers, C.A. et al. (2020). Mild organosolv lignin extraction with alcohols: the importance of benzylic alkoxylation. *ACS Sustainable Chemistry & Engineering* 8: 5119–5131.

48 Chen, L., Liang, Z., Zhang, X. et al. (2022). A facile and novel lignin isolation procedure–methanolic hydrochloric acid treatment at ambient temperature. *International Journal of Biological Macromolecules* 222: 1423–1432.

49 Lancefield, C.S., Panovic, I., Deuss, P.J. et al. (2017). Pre-treatment of lignocellulosic feedstocks using biorenewable alcohols: towards complete biomass valorisation. *Green Chemistry* 19: 202–214.

50 Zijlstra, D.S., De Korte, J., De Vries, E.P. et al. (2021). Highly efficient semi-continuous extraction and in-line purification of high β-O-4 butanosolv lignin. *Frontiers in Chemistry* 9: 655983.

51 Lan, W., Amiri, M.T., Hunston, C.M., and Luterbacher, J.S. (2018). Protection group effects during α,γ-diol lignin stabilization promote high-selectivity monomer production. *Angewandte Chemie, International Edition* 57: 1356–1360.

52 Lan, W., de Bueren, J.B., and Luterbacher, J.S. (2019). Highly selective oxidation and depolymerization of α, γ-diol-protected lignin. *Angewandte Chemie, International Edition* 58: 2649–2654.

53 Korich, A.L., Fleming, A.B., Walker, A.R. et al. (2012). Chemical modification of organosolv lignin using boronic acid-containing reagents. *Polymer (Guildf)* 53: 87–93.

54 Sun, R., Lawther, J.M., and Banks, W.B. (1996). Effects of extraction time and different alkalis on the composition of alkali-soluble wheat straw lignins. *Journal of Agricultural and Food Chemistry* 44: 3965–3970.

55 Xu, F., Jiang, J.X., Sun, R.C. et al. (2008). Fractional isolation and structural characterization of mild ball-milled lignin in high yield and purity from *Eucommia ulmoides* Oliv. *Wood Science and Technology* 42: 211–226.

56 Chen, W.-J., Zhao, B.C., Cao, X.F. et al. (2018). Structural features of alkaline dioxane lignin and residual lignin from *Eucalyptus grandis* × *E. urophylla*. *Journal of Agricultural and Food Chemistry* 67: 968–974.

57 Zijlstra, D.S., Analbers, C.A., de Korte, J. et al. (2019). Efficient mild organosolv lignin extraction in a flow-through setup yielding lignin with high β-O-4 content. *Polymers (Basel)* 11: 1913.

58 Brandner, D.G., Kruger, J.S., Thornburg, N.E. et al. (2021). Flow-through solvolysis enables production of native-like lignin from biomass. *Green Chemistry* 23: 5437–5441.

59 Zhou, H., Xu, J.Y., Fu, Y. et al. (2019). Rapid flow-through fractionation of biomass to preserve labile aryl ether bonds in native lignin. *Green Chemistry* 21: 4625–4632.

60 Wang, Z., Qin, S., Hirth, K. et al. (2019). Preserving both lignin and cellulose chemical structures: flow-through acid hydrotropic fractionation at atmospheric pressure for complete wood valorization. *ACS Sustainable Chemistry & Engineering* 7: 10808–10820.

61 Kubisa, P. (2005). Ionic liquids in the synthesis and modification of polymers. *Journal of Polymer Science, Part A: Polymer Chemistry* 43: 4675–4683.

62 Lu, J., Yan, F., and Texter, J. (2009). Advanced applications of ionic liquids in polymer science. *Progress in Polymer Science* 34: 431–448.

63 Pu, Y., Jiang, N., and Ragauskas, A.J. (2007). Ionic liquid as a green solvent for lignin. *Journal of Wood Chemistry and Technology* 27: 23–33.

64 Hart, W.E.S., Harper, J.B., and Aldous, L. (2015). The effect of changing the components of an ionic liquid upon the solubility of lignin. *Green Chemistry* 17: 214–218.

65 Rawat, S., Kumar, A., and Bhaskar, T. (2021). Ionic liquids for separation of lignin and transformation into value added chemicals. *Current Opinion in Green and Sustainable Chemistry* 34: 100582.

66 da Costa Lopes, A.M., João, K.G., Morais, A.R.C. et al. (2013). Ionic liquids as a tool for lignocellulosic biomass fractionation. *Sustainable Chemical Processes* 1: 1–31.

67 Szalaty, T.J., Klapiszewski, Ł., and Jesionowski, T. (2020). Recent developments in modification of lignin using ionic liquids for the fabrication of advanced materials–a review. *Journal of Molecular Liquids* 301: 112417.

68 Wen, J.-L., Yuan, T.-Q., Sun, S.-L. et al. (2014). Understanding the chemical transformations of lignin during ionic liquid pretreatment. *Green Chemistry* 16: 181–190.

69 Xu, J.-K., Sun, Y.-C., and Sun, R.-C. (2015). Synergistic effects of ionic liquid plus alkaline pretreatments on eucalyptus: lignin structure and cellulose hydrolysis. *Process Biochemistry* 50: 955–965.

70 Weigand, L., Mostame, S., Brandt-Talbot, A. et al. (2017). Effect of pretreatment severity on the cellulose and lignin isolated from Salix using iono Solv pretreatment. *Faraday Discussions* 202: 331–349.

71 Satlewal, A., Agrawal, R., Bhagia, S. et al. (2018). Natural deep eutectic solvents for lignocellulosic biomass pretreatment: recent developments, challenges and novel opportunities. *Biotechnology Advances* 36: 2032–2050.

72 Smith, E.L., Abbott, A.P., and Ryder, K.S. (2014). Deep eutectic solvents (DESs) and their applications. *Chemical Reviews* 114: 11060–11082.

73 Shen, X.-J., Wen, J.L., Mei, Q.Q. et al. (2019). Facile fractionation of lignocelluloses by biomass-derived deep eutectic solvent (DES) pretreatment for cellulose enzymatic hydrolysis and lignin valorization. *Green Chemistry* 21: 275–283.

74 Liu, Y., Chen, W., Xia, Q. et al. (2017). Efficient cleavage of lignin–carbohydrate complexes and ultrafast extraction of lignin oligomers from wood biomass by microwave-assisted treatment with deep eutectic solvent. *ChemSusChem* 10: 1692–1700.

75 Vigier, K.D.O., Chatel, G., and Jérôme, F. (2015). Contribution of deep eutectic solvents for biomass processing: opportunities, challenges, and limitations. *ChemCatChem* 7: 1250–1260.

76 Loow, Y.-L., New, E.K., Yang, G.H. et al. (2017). Potential use of deep eutectic solvents to facilitate lignocellulosic biomass utilization and conversion. *Cellulose* 24: 3591–3618.

77 Chen, Y. and Mu, T. (2019). Application of deep eutectic solvents in biomass pretreatment and conversion. *Green Energy & Environment* 4: 95–115.

78 Hong, S., Shen, X.J., Xue, Z. et al. (2020). Structure-function relationships of deep eutectic solvents for lignin extraction and chemical transformation. *Green Chemistry* 22: 7219–7232.

79 Chen, Z., Ragauskas, A., and Wan, C. (2020). Lignin extraction and upgrading using deep eutectic solvents. *Industrial Crops and Products* 147: 112241.

80 Liu, Y., Deak, N., Wang, Z. et al. (2021). Tunable and functional deep eutectic solvents for lignocellulose valorization. *Nature Communications* 12: 1–15.

81 Chen, Z., Bai, X., Lusi, A., and Wan, C. (2018). High-solid lignocellulose processing enabled by natural deep eutectic solvent for lignin extraction and industrially relevant production of renewable chemicals. *ACS Sustainable Chemistry & Engineering* 6: 12205–12216.

82 Leskinen, T., Kelley, S.S., and Argyropoulos, D.S. (2015). Refining of ethanol biorefinery residues to isolate value added lignins. *ACS Sustainable Chemistry & Engineering* 3: 1632–1641.

83 Chen, F., Tobimatsu, Y., Havkin-Frenkel, D. et al. (2012). A polymer of caffeyl alcohol in plant seeds. *Proceedings of the National Academy of Sciences of the United States of America* 109: 1772–1777.

84 Li, Y. et al. (2018). An ideal lignin facilitates full biomass utilization. *Science Advances* 4.

85 Wang, S. et al. (2020). Catechyl lignin extracted from castor seed coats using deep eutectic solvents: characterization and depolymerization. *ACS Sustainable Chemistry & Engineering* 8: 7031–7038.

6

Isolation of Lignin in the Biorefinery Process

Qian Xu and Ling-Ping Xiao

Dalian Polytechnic University, Liaoning Key Lab of Lignocellulose Chemistry and BioMaterials, Liaoning Collaborative Innovation Center for Lignocellulosic Biorefinery, College of Light Industry and Chemical Engineering, No. 1 Qinggongyuan, Ganjingzi District, Dalian, 116034, China

6.1 Introduction

Lignocellulosic biomass is composed of cellulose, hemicellulose, and lignin and is considered a renewable resource due to its wide availability and natural carbon properties [1, 2]. Of these, the carbohydrate fractions (cellulose and hemicellulose) can be converted into biofuels and biochemicals, while lignin can be used for the production of high value-added materials [3–5]. The cellulose in lignocellulose is encapsulated by lignin and hemicellulose, and the bonding between cellulose and hemicellulose or lignin molecules relies mainly on oxygen bonding; hemicellulose and lignin are chemically bonded to each other in addition to hydrogen bonding [6, 7].

Lignin is a naturally occurring aromatic polymer with great production value and potential as an important raw material for various types of sustainable materials [8–10]. However, most industrial production, especially the pulp and paper industry, utilizes mainly carbohydrate fractions and lignin is usually incinerated as a by-product [11, 12]. In recent years, the value of lignin has received increasing attention, and researchers have developed a number of methods to increase the added value of lignin. For example, lignin macromolecules are degraded into small molecule monomers, which provide important raw materials for downstream industries. On the other hand, lignin is modified for specific purposes based on its functional groups for the preparation of functional materials. Nevertheless, due to the complex structure and heterogeneity of lignin macromolecule, both processes pose many challenges to the efficient utilization of lignin. Maximizing the retention of the overall structure of native lignin, or altering the structure of lignin by special treatments for its value-added utilization, is important for understanding the reactivity of lignin [13]. The presence of fewer lignin-linked motifs, rare monomeric units, and particle size distributions in lignocellulosic lignin results in significant differences in structure and properties.

Lignin Chemistry: Characterization, Isolation, and Valorization, First Edition.
Edited by Yuhe Liao and Bert F. Sels.

Researchers have continuously developed different pretreatment methods to isolate lignin from lignocellulosic matrices [14–21]. An effective pretreatment is to disrupt the recalcitrance of lignocellulose, enhancing carbohydrate polymers for enzymatic hydrolysis while producing high-quality lignin [22]. However, all the pretreatment methods have their advantages and disadvantages. Therefore, it is of great importance to develop processing methods that enable lignin to be isolated at a high yield in a state that is highly close to the native structure or meets high value-added utilization. This helps lignin to be applied to well-established transformation strategies and facilitates the development of new utilization strategies.

This chapter presents in a concise manner the currently known pretreatment methods for the isolation of lignin in the biorefinery process, such as organosolv process, steam explosion (SE), ammonia-based pretreatment, γ-valerolactone, ionic liquid, deep eutectic solvent (DES), acid treatment, and acidic lithium bromide trihydrate system. The physicochemical properties of the obtained lignin samples were also discussed.

6.2 Methods Used to Isolate Lignin from Lignocellulose

6.2.1 Organosolv Process

The organosolv process method is typically using organic solvents and a combination of acids or bases to achieve the separation of lignocellulosic components, effectively avoiding condensation reactions [23–25] by preserving the β-aryl ether unit [26], thus obtaining lignin fractions with high purity and retaining part of the natural C—O bond structure. Organic acids can ensure the break of ether bonds and the destruction of bonds of lignin–carbohydrate complex (LCC), but may lead to the formation of C—C bonds with higher bond energy, resulting in a more complex lignin structure [27, 28]. This process generally adopts high-temperature conditions, which can promote better dissolution of each component in the solvent. In addition, many organic solvents can be recycled, so the method is environmentally friendly and economical.

In this process, the effects of extraction efficiency are highly related to the organic solvents. Zijlstra et al. [29] used a mild organic alcohol extraction method to obtain lignin with high β-aryl ether retention, and α-alkylation occurred in part of the lignin structure (Figure 6.1). Watkins et al. [30] extracted lignin from nonwoody cellulosic biomass from the treatment of formic acid and peroxyformic acid. The result revealed that the thermal stability of different samples was different. Michelin et al. [31] pretreated corn cob with liquid hot water and ethanol organic solvent to obtain lignin with high purity and high oxidation resistance. Glycerin can be rapidly heated and cooled during the organic solvent treatment and is effective in selectively deconstructing lignocellulosic biomass [32, 33]. Song et al. [34] proposed an anhydrous two-step organic solvent pretreatment process, and the final extraction rate of lignin was as high as 81.5%.

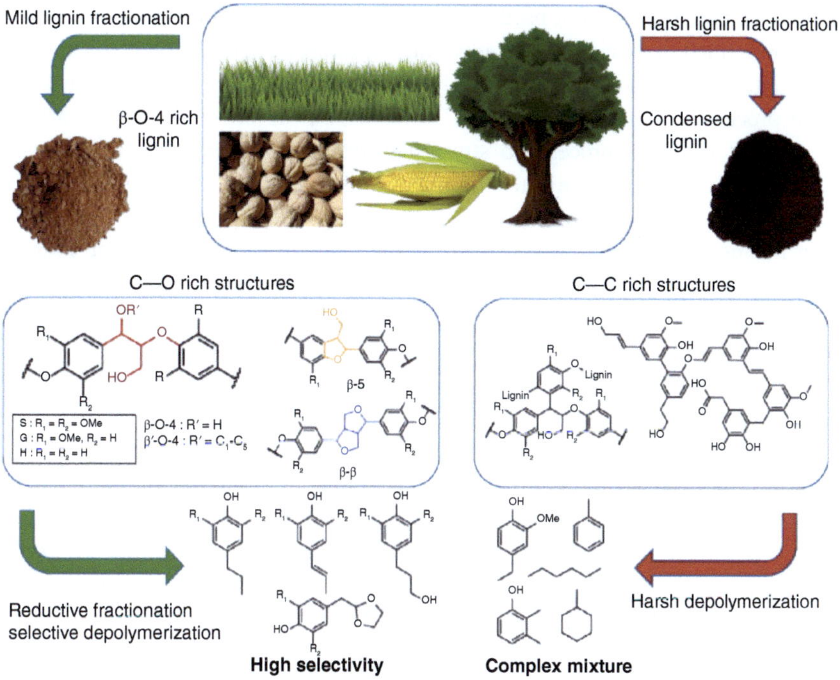

Figure 6.1 Influence of the fractionation conditions on the structural composition of the obtained lignin and possibilities for further depolymerization [29].

6.2.2 Hydrothermal Pretreatment

Hydrothermal pretreatment (HTP) means that water is pressurized at a high temperature to keep it liquid, and the water in this state can effectively hydrolyze the fiber biomass [35]. Using water as the medium, this method has the advantages of fewer by-products, short reaction period, wide applicability, and environmental friendliness [36, 37]. However, the main drawback is that the whole process consumes relatively high water and energy consumption.

HTP process can cleave the *O*-acetyl group attached to hemicellulose, thereby releasing acetic acid and further lowering the pH of the solution. This acidic condition facilitates the cleavage of the LCC linkages and the partial deconstruction of lignin [38, 39]. Similar to other acidic pretreatments, HTP can also decompose monosaccharides into furans at high temperatures through acidic dehydration reactions. The resulting furans can be converted by hydrolysis, ring closure, and dehydration to aromatic hydrocarbons, such as 3,8-dihydroxy-2-methylchromone and 1,2,4-benzenetriol, which can be polymerized to "pseudo lignin" by electrophilic substitution and oxidative polymerization [40, 41]. This pseudo lignin can negatively affect the enzymatic hydrolysis process by blocking the binding sites on the cellulose surface and thereby preventing effective adsorption by the enzyme. Accordingly, it is desirable to minimize the formation of pseudo lignin during the pretreatment process [40].

Microwave radiation is often used to accelerate hydrothermal processes in order to increase extraction efficiency. The advantage of microwave radiation is that it can be heated internally quickly and uniformly, which makes it more popular in hydrothermal processes [42]. Chen et al. [43] investigated the effect of the operating parameters of the HTP process on straw. The result indicated that the optimal parameter was 200 °C for 10 minutes with a 160 ml min^{-1} flow rate, and the lignin was almost completely removed from the straw. Hage et al. [44] used ^{13}C and ^{31}P nuclear magnetic resonance (NMR) to study the lignin after hydrolysis, and found that the addition of an appropriate amount of 2-naphthol during the hydrolysis process not only effectively reduced the repolymerization of lignin but also better recovered hemicellulose sugar.

6.2.3 Steam Explosion

SE destroys fiber cell walls mainly by the instantaneous release of pressure from high-temperature, and high-pressure steam. It is an effective, environmentally friendly, and industrially scalable pretreatment method for the structural destruction of biomass through high-temperature vapor and explosive decompression, respectively [45].

The process of SE for lignin preparation involves passing water vapor at a certain pressure into a reactor, maintaining it for a certain period of time, and then rapidly relieving the pressure [46]. During the SE process, the β-*O*-4′ bonds in lignin are degraded with increasing severity of pretreatment, while the C—C condensed structures and other inter-unit connections (β-1′ and β-5′) formed through free radical coupling reactions increase [47]. Caputo et al. [46] investigated six different pretreatment conditions and found that surface defibrillation and lignin depolymerization increased with increasing levels of steam pretreatment. Nousiainen's group [48] used organic solvents, such as ethanol and formic acid, to technologically fractionate biorefinery lignin obtained after the steam blasting and saccharification processes, producing a more homogeneous lignin fraction with high β-*O*-4′ and aromatic contents and free of residual carbohydrates (Figure 6.2).

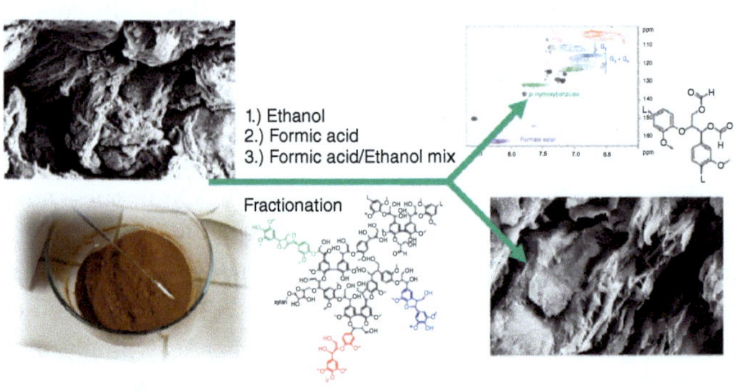

Figure 6.2 Fractionation of vapor blast lignin using ethanol and formic acid [48].

6.2.4 Ammonia-Based Pretreatment

Ammonia fiber expansion (AFEX), which is a traditional ammonia treatment, has been shown to be an effective pretreatment method for lignocellulosic biomass, cleaving ester and LCC but retaining the overall chemical structure of the lignin polymer and retaining most of the lignin in the biomass [49, 50]. During AFEX pretreatment, about 95% of the ammonia can be recovered and reused, and the pretreated biomass is easy to hydrolyze and ferment, making it very economically attractive [51, 52]. Dale's research group has developed a novel liquid ammonia pretreatment method named "Extractive Ammonia" (EA) for the simultaneous conversion of native crystalline cellulose I_β to the highly digestible cellulose III_I isomer and the selective extraction of lignin from lignocellulosic biomass [53]. Moreover, the fermentable sugar yields were higher in EA-pretreated corn stover as compared to the AFEX process (Figure 6.3).

6.2.5 γ-Valerolactone (GVL)

γ-Valerolactone (GVL) is considered as a promising organic solvent for biomass conversion. GVL treatment of lignocellulosic biomass retains more β-ether units and more aliphatic hydroxyl groups, resulting in higher-quality lignin [54–56]. GVL fractionation has been applied to a variety of biomass resources with high lignin yields under mild conditions [57]. Cheng et al. [58] found that the β-*O*-4′ content and the ratio of aliphatic hydroxyl group to phenolic hydroxyl group in the lignin samples extracted by GVL decreased with the increase of the reaction intensity, while the amount of condensed lignin increased and the lignin yield increased. Therefore, the chemical structure of lignin was directly related to the lignin yield. In addition, lignin condensation increased with increasing lignin concentration in the GVL liquid phase. This suggested that lignin underwent a degradation reaction in the liquid phase.

In the actual fractionation process, GVL and water are a binary mixture in which water hydrolyses the hemicellulose while GVL dissolves the lignin fraction. Dumesic and coworkers [55] used a solvent mixture consisting of 80 wt% GVL and 20 wt% water to isolate lignin from corn stover by biomass fractionation, which produced ≥70% of soluble carbohydrates from biomass, and the extracted lignin was similar to that of native lignin. Molecular weight is one of the most important factors affecting industrial lignin utilization (Figure 6.4). Jin et al. [59] adjusted the ratio of water, ethanol, and GVL for the molecular weight-controlled fractionation of industrial alkali lignin at room temperature, which can be fractionated into six grades with progressively controlled molecular weights from low to high.

6.2.6 Ionic Liquids (ILs)

Ionic liquids (ILs), which are organic salts with melt below 100 °C and consist of cation/anion pairs, have attracted increasing attention as environmentally friendly "green" solvents in the pretreatment methods. IL pretreatment also allows for the

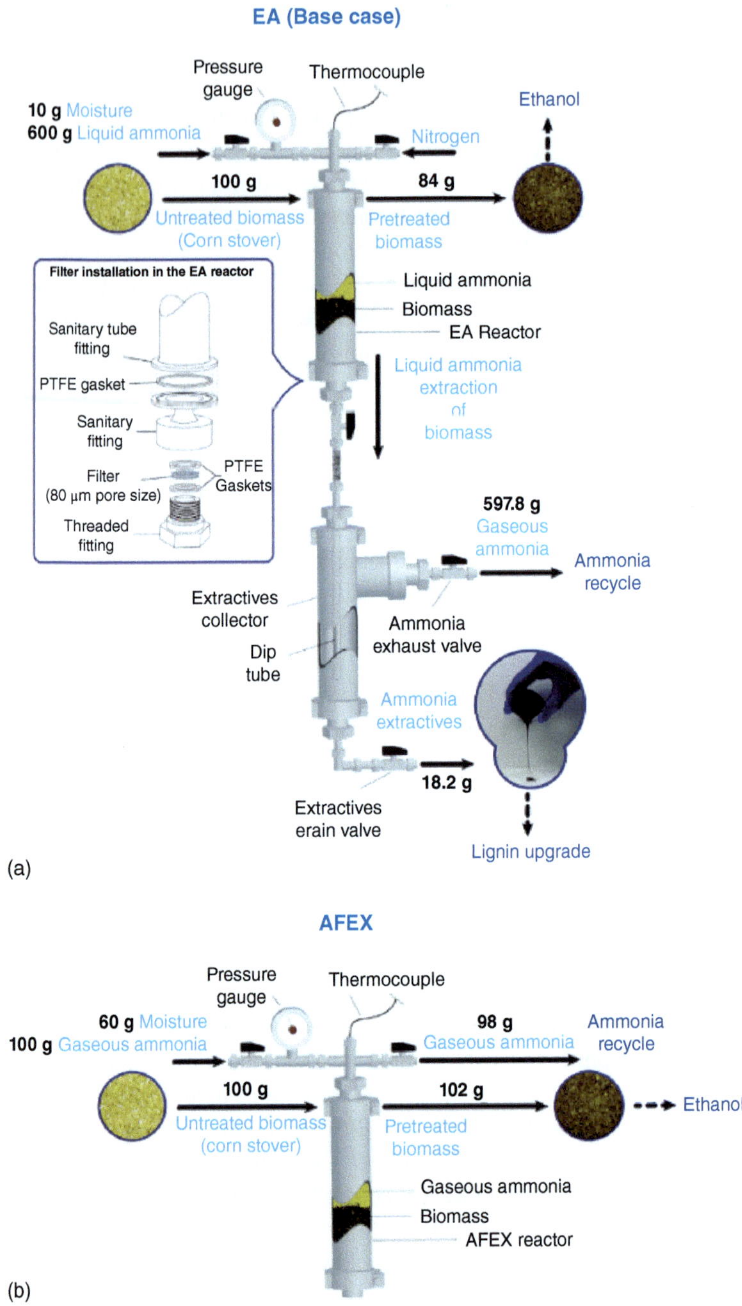

Figure 6.3 Process design differences between EA (a) and AFEX (b) pretreatment [53].

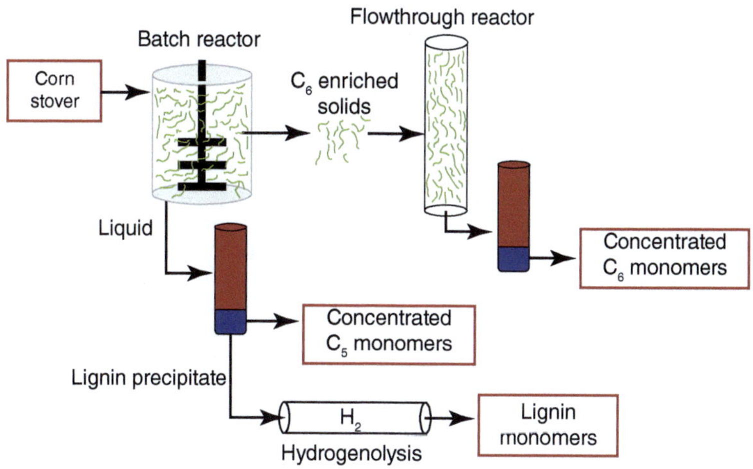

Figure 6.4 Lignin extracted using GVL as a solubilizer was similar to native lignin, which was upgraded using a two-stage hydrogenolysis process over a Ru/C catalyst [55].

collection of partially dissociated lignin. [60, 61] Pu et al. [62] investigated the effect of imidazolium-type ILs on lignin solubility and found that ILs containing methyl sulfate and trifluoro sulfonate anions were effective in dissolving softwood kraft lignin. Hou et al. [63] determined the solubility of kraft lignin in a series of carboxylate anionic ILs containing a choline cation. It was found that the ability of the IL to dissolve the lignin increased with the increase in the length of the anionic alkyl chain.

IL pretreatment leads to structural changes in lignin (cleavage of β-*O*-4′ ether bonds and formation of condensed structures). Wen et al. [64] demonstrated the effects of [C_2mim][OAc] pretreatment at different temperatures and times on the structural characteristics of lignin outside the plant cell wall matrix. The results showed that aliphatic hydroxyl groups in lignin decreased while phenolic hydroxyl groups significantly increased with increasing temperature and time of IL pretreatment. Furthermore, [C_2mim][OAc] pretreatment selectively degraded G-type lignin fragments, and the S-unit was more susceptible to condensation reactions than of the G-unit. Imidazolium acesulfamate ILs have the ability to dissolve wood lignin without dissolving cellulose. Pinkert et al. [65] extracted lignin using 1-ethyl-3-methylimidazolium acesulfamate, which showed a greater average molar mass and a more homogeneous molar mass distribution than that of lignin extracted in the kraft process (Figure 6.5).

6.2.7 Deep Eutectic Solvents (DES)

Deep eutectic solvents (DES) are mixtures of two or more chemicals that act as hydrogen bond donors (HBDs) or hydrogen bond acceptors (HBAs), and may have unique capabilities in solubilizing lignin in plant material [67, 68]. Based on the principle of interaction between HBDs and HBAs, DES may provide a mild

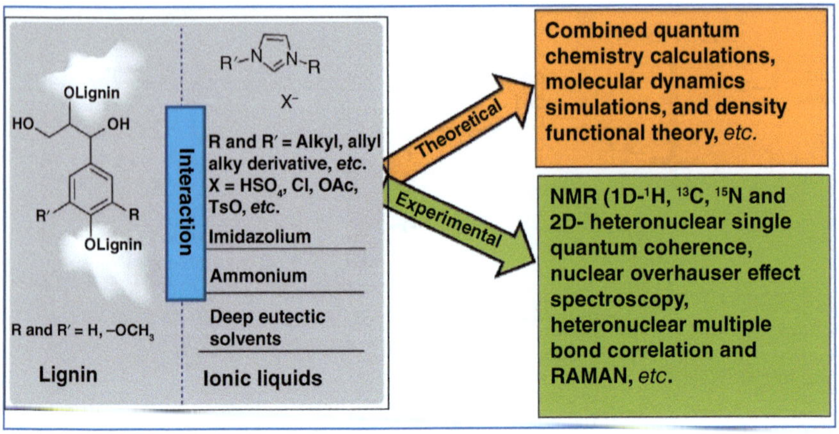

Figure 6.5 Hydrogen bonding, covalent, and $\pi-\pi$ stacking interactions between ionic liquids and lignin or lignin-derived compounds [66].

acid–base catalytic mechanism to induce controlled cleavage of perishable ether bonds between phenylpropane units, leading to lignin depolymerization and/or separation from biomass. Between the phenylpropane units, leading to the depolymerization and/or separation of lignin.

DES is highly selective for the extraction of lignin, which facilitates the high value-added utilization of regenerated lignin and retained cellulose. Yu and coworkers [69, 70] demonstrated that both hydrogen and ether bonds in the LCC could be cleaved during DES pretreatment, which facilitated the selective extraction of lignin (Figure 6.6). Li and coworkers [71] developed an acidic biomass-derived DES (choline chloride/oxalic acid) pretreatment to deconstruct the readhesion of bamboo to enhance lignin fractionation and value (Figure 6.7). Zhang's group reported a method of applying DES to extract lignin from woody biomass with high yield and purity, 78% for poplar and 58% for fir [72]. Shen et al. [73] have developed a biomass-derived DES pretreatment for deconstructing the difficult-to-decompose structure of eucalyptus. The DES consists of biomass-derived chemicals (lactic acid and choline chloride), and the structure of the regenerated lignin (β-*O*-4′ and β-β linkages) is well preserved, and the molecular weight is relatively low and homogeneous (Figure 6.7).

6.2.8 Acid Treatment

Acid treatment is an effective method for separating lignin, and organic acid treatment is increasingly being used because it is cost-effective and environmentally friendly [74]. *p*-Toluenesulfonic acid (p-TsOH) treatment as an emerging treatment has drawn more and more attention to lignin separation. Separation of lignin using *p*-TsOH in a mild reaction condition results in high extraction efficiency (85% of the lignin and 70% of the xylan were solubilized) and excellent treatment recovery [75, 76]. The development of *p*-TsOH treatment provides an efficient and viable

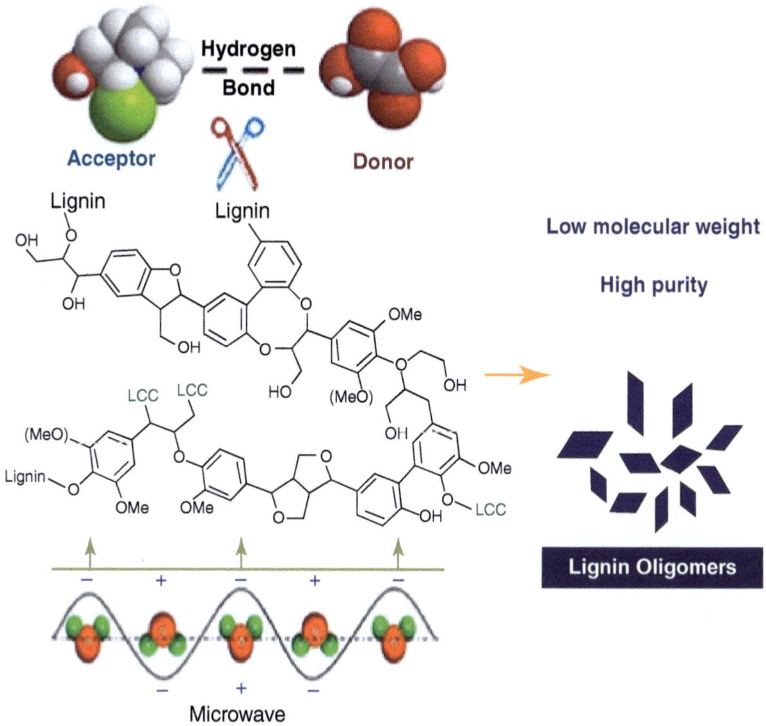

Figure 6.6 A choline chloride/oxalic acid dihydrate deep eutectic solvent (DES) is used for the fractionation of lignocellulose [69].

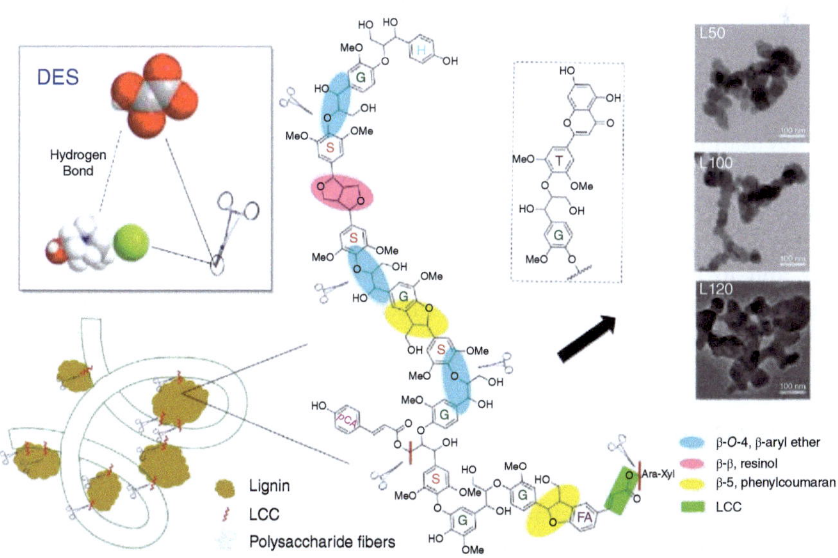

Figure 6.7 Extraction of lignin from lignocellulosic biomass by DES pretreatment and putative cleavage sites for typical intramolecular linkages and LCC benzyl ester structures [71].

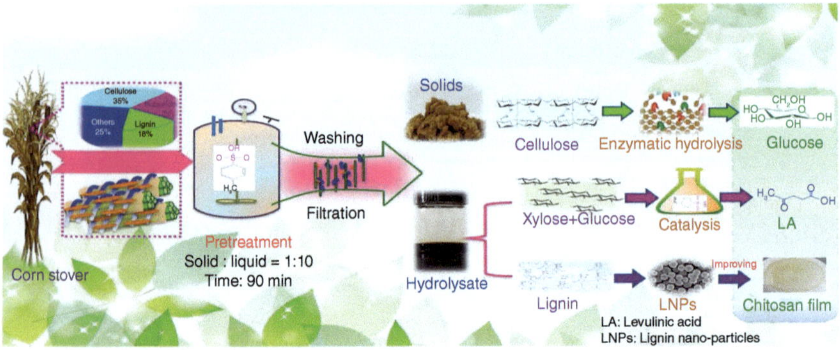

Figure 6.8 Schematic of using *p*-TsOH/water pretreatment to fractionate corn stover to cellulose, xylose, and lignin and their conversions to glucose, levulinic acid (LA), and lignin nanoparticles (LNPs), respectively [78].

method for the high-value utilization of lignocellulosic biomass. Wu and coworkers [77] performed sequential fractionation of woody biomass using *p*-TsOH as the reaction medium and achieved 23.5% biomass solubility and 55.1% xylan removal at 120 °C for two hours. Yu et al. [78] proposed a pathway to fractionate corn stover for the production of glucose, high-quality lignin nanoparticles (LNPs), and levulinic acid (LA) based on *p*-TsOH (Figure 6.8). Ma et al. [79] directly produced LNPs by fractionation poplar wood using *p*-TsOH in an aqueous system at ≤80 °C. The results revealed that LNPs could be directly produced. Zhai et al. [80] treated lignocellulosic biomass in mild methanol using *p*-TsOH as a catalyst, and the extracted lignin was of high purity (>94%) with moderate and uniform molecular weight.

As compared to *p*-TsOH, maleic acid (MA) has a lower acidity and solubility, which facilitates recycling and reduces environmental impact. The esterified lignin obtained by maleic acid hydro fractionation (MAHF) displays a low degree of condensation and very light color. Cai et al. [81] reported that MA was an effective hydrophilic agent that could effectively delignify wood under mild conditions and enhance the lubricating effect of lignin in the fibrillation process (Figure 6.9). Moreover, it was found that MA dissolved lignin and hemicellulose faster and to a greater extent, which were esterified (carboxylate) and dissolved acidic hydrogenated lignin, thus expanding the pathway of lignin functionalization [82].

6.2.9 Acidic Lithium Bromide Trihydrate (ALBTH) System

Cellulose and hemicellulose are easily and efficiently solubilized and digested in the acidic lithium bromide trihydrate (ALBTH) treatment, so that lignin as a solid residue is separated from lignocellulose by centrifugation and the separated lignin structure acquires high activity [83]. Li et al. [84] used ALBTH ($LiBr{\cdot}3H_2O$) as the reaction medium, and the carbohydrate component of the biomass was able to be rapidly dissolved under mild conditions, with the lignin left as an insoluble residual

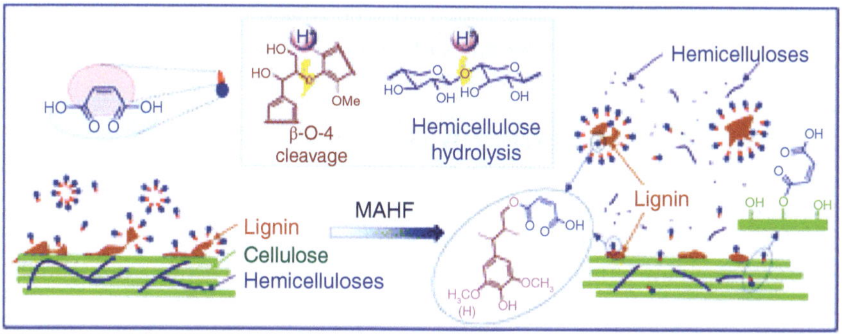

Figure 6.9 The figure shows recyclable MA as an acidic hydrotreating agent for fractionating wood to produce less condensed carboxy lignin, lignin-containing cellulose nanofibers, and furfural [81].

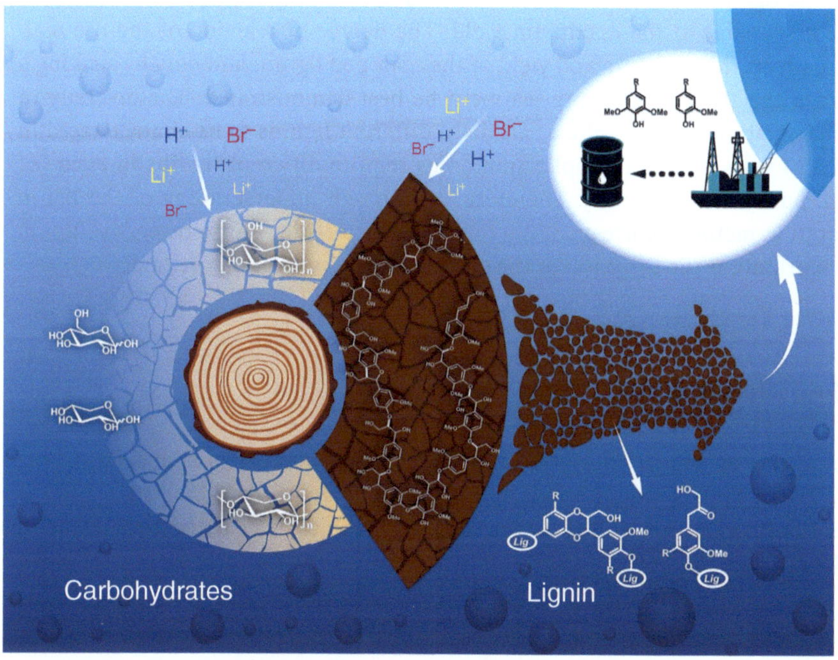

Figure 6.10 High-purity and minimal condensation lignin can be isolated from biomass by dissolving and hydrolyzing cellulose and hemicellulose using an acidic lithium bromide trihydrate (ALBTH) system [85].

solid. Pan's group [85] processed lignocellulose in the ALBTH system to achieve depolymerization and separation of lignin in the solid state after almost quantitative hydrolysis of the carbohydrate fraction to sugars (Figure 6.10). In this process, lignin could be demethylated during ALBTH treatment and the ether bonds in the β-5 and β-β structures were cleaved in addition to those in the diphenyl structure [85, 86]. Yang et al. [87] reported that ALBTH (60% LiBr aqueous solution) could selectively cleave the ether bond of kraft lignin, which led to depolymerization and demethylation of lignin, and improved the reactivity of lignin.

6.3 Conclusion and Perspectives

There is an increasing amount of research on maximizing the utilization of lignocellulosic biomass with lignin-focused biorefinery concepts gradually becoming a hot research topic. In order to fully focus on lignin and convert it into high value-added products, a large number of basic reaction studies are required. Prolonging in plant materials has the potential to deepen our reexploration of lignin structure and reactivity. Functional lignin materials with representative structures are also important for the development of innovative biorefinery methods. Many of the existing pretreatment methods have been developed to separate lignin from the biomass matrix and to obtain lignin as close to its native state as possible at the maximum yield. The degree of reaction of the pretreatment affects the structure and yield of the obtained lignin, and only by striking a balance can the effects of pretreatment be best demonstrated. The diversity of lignin isolation methods allows all these different lignins to be characterized in different ways. Different application areas require different lignins in terms of yield, purity, and solubility. Therefore, a thorough understanding of the lignin separation methods corresponding to the characteristics of the respective lignin is important for the selection of the optimal lignin separation method.

Acknowledgments

This work was supported by the National Natural Science Foundation of China (22278049), the Liaoning Revitalization Talents Program (XLYC2007104), and the Scientific Research Fund Program for Liaoning Provincial Department of Education (LJKZZ20220071).

References

1 Liu, Z., Li, H., Gao, X. et al. (2022). Rational highly dispersed ruthenium for reductive catalytic fractionation of lignocellulose. *Nature Communications* 13: 4716. https://doi.org/10.1038/s41467-022-32451-5.

2 Yu, Y., Cheng, W., Li, Y. et al. (2022). Tailored one-pot lignocellulose fractionation to maximize biorefinery toward versatile xylochemicals and nanomaterials. *Green Chemistry* 24: 3257–3268. https://doi.org/10.1039/D2GC00264G.

3 Wu, Z., Peng, K., Zhang, Y. et al. (2022). Lignocellulose dissociation with biological pretreatment towards the biochemical platform: a review. *Materials Today Bio* 16: 100445. https://doi.org/10.1016/j.mtbio.2022.100445.

4 Ge, H., Zheng, J., and Xu, H. (2023). Advances in machine learning for high value-added applications of lignocellulosic biomass. *Bioresource Technology* 369: 128481. https://doi.org/10.1016/j.biortech.2022.128481.

5 Provost, V., Dumarcay, S., Ziegler-Devin, I. et al. (2022). Deep eutectic solvent pretreatment of biomass: influence of hydrogen bond donor and temperature on lignin extraction with high β-*O*-4 content. *Bioresource Technology* 349: 126837. https://doi.org/10.1016/j.biortech.2022.126837.

6 Gundupalli, M.P., Cheenkachorn, K., Chuetor, S. et al. (2023). Assessment of pure, mixed and diluted deep eutectic solvents on Napier grass (*Cenchrus purpureus*): compositional and characterization studies of cellulose, hemicellulose and lignin. *Carbohydrate Polymers* 306: 120599. https://doi.org/10.1016/j.carbpol.2023.120599.

7 Wang, W., Kong, Y., Peng, J. et al. (2022). Probing the mechanism of green solvent solubilization of hemicellulose based on molecular dynamics simulations. *Industrial Crops and Products* 186: 115159. https://doi.org/10.1016/j.indcrop.2022.115159.

8 Gao, D., Ouyang, D., and Zhao, X. (2022). Electro-oxidative depolymerization of lignin for production of value-added chemicals. *Green Chemistry* 24: 8585–8605. https://doi.org/10.1039/D2GC02660K.

9 Atiwesh, G., Parrish, C.C., Banoub, J., and Le, T.-A.T. (2022). Lignin degradation by microorganisms: a review. *Biotechnology Progress* 38: e3226. https://doi.org/10.1002/btpr.3226.

10 Yao, M., Bi, X., Wang, Z. et al. (2022). Recent advances in lignin-based carbon materials and their applications: a review. *International Journal of Biological Macromolecules* 223: 980–1014. https://doi.org/10.1016/j.ijbiomac.2022.11.070.

11 Beaucamp, A., Muddasar, M., Amiinu, I.S. et al. (2022). Lignin for energy applications – state of the art, life cycle, technoeconomic analysis and future trends. *Green Chemistry* 24: 8193–8226. https://doi.org/10.1039/D2GC02724K.

12 Chauhan, P.S., Agrawal, R., Satlewal, A. et al. (2022). Next generation applications of lignin derived commodity products, their life cycle, techno-economics and societal analysis. *International Journal of Biological Macromolecules* 197: 179–200. https://doi.org/10.1016/j.ijbiomac.2021.12.146.

13 Wang, Z. and Deuss, P.J. (2023). The isolation of lignin with native-like structure. *Biotechnology Advances* 68: 108230. https://doi.org/10.1016/j.biotechadv.2023.108230.

14 Bimestre, T.A., Júnior, J.A.M., Canettieri, E.V., and Tuna, C.E. (2022). Hydrodynamic cavitation for lignocellulosic biomass pretreatment: a review of recent developments and future perspectives. *Bioresources and Bioprocessing* 9: 7. https://doi.org/10.1186/s40643-022-00499-2.

15 Pancholi, M.J., Khristi, A., Ahira, K.M., and Bagchi, D. (2023). Comparative analysis of lignocellulose agricultural waste and pre-treatment conditions with FTIR and machine learning modeling. *Bioenergy Research* 16: 123–137. https://doi.org/10.1007/s12155-022-10444-y.

16 Meraj, A., Singh, S.P., Jawaid, M. et al. (2023). A review on eco-friendly isolation of lignin by natural deep eutectic solvents from agricultural wastes. *Journal of Polymers and the Environment* 31: 3283–3316. https://doi.org/10.1007/s10924-023-02817-x.

17 González-González, R.B., Iqbal, H.M.N., Bilal, M., and Parra-Saldívar, R. (2022). (Re)-thinking the bio-prospect of lignin biomass recycling to meet sustainable development goals and circular economy aspects. *Current Opinion in Green and Sustainable Chemistry* 38: 100699. https://doi.org/10.1016/j.cogsc.2022.100699.

18 Hoang, A.T., Nguyen, X.P., Duong, X.Q. et al. (2023). Steam explosion as sustainable biomass pretreatment technique for biofuel production: characteristics and challenges. *Bioresource Technology* 385: 129398. https://doi.org/10.1016/j.biortech.2023.129398.

19 Liu, B., Liu, L., Deng, B. et al. (2022). Application and prospect of organic acid pretreatment in lignocellulosic biomass separation: a review. *International Journal of Biological Macromolecules* 222: 1400–1413. https://doi.org/10.1016/j.ijbiomac.2022.09.270.

20 Sun, Q., Fu, X., Wang, P. et al. (2023). Selective production of p-hydroxybenzaldehyde in the oxidative depolymerization of alkali lignin catalyzed by copper-containing imidazolium-based ionic liquids. *Journal of Molecular Liquids* 385: 122378. https://doi.org/10.1016/j.molliq.2023.122378.

21 Zhang, B., Ping, T., Mu, L. et al. (2023). Highly selective conversion of alkali lignin into aromatic monomers by pulse dielectric barrier discharge plasma at mild reaction conditions. *Sustainable Materials and Technologies* 36: e00643. https://doi.org/10.1016/j.susmat.2023.e00643.

22 Hassanpour, M., Abbasabadi, M., Strong, J. et al. (2020). Scale-up of two-step acid-catalysed glycerol pretreatment for production of oleaginous yeast biomass from sugarcane bagasse by *Rhodosporidium toruloides*. *Bioresource Technology* 313: 123666. https://doi.org/10.1016/j.biortech.2020.123666.

23 Schutyser, W., Renders, T., Van den Bosch, S. et al. (2018). Chemicals from lignin: an interplay of lignocellulose fractionation, depolymerisation, and upgrading. *Chemical Society Reviews* 47: 852–908. https://doi.org/10.1039/C7CS00566K.

24 Rinaldi, R., Jastrzebski, R., Clough, M.T. et al. (2016). Paving the way for lignin valorisation: recent advances in bioengineering, biorefining and catalysis. *Angewandte Chemie International Edition* 55: 8164–8215. https://doi.org/10.1002/anie.201510351.

25 Ferrini, P. and Rinaldi, R. (2014). Catalytic biorefining of plant biomass to non-pyrolytic lignin bio-oil and carbohydrates through hydrogen transfer reactions. *Angewandte Chemie International Edition* 53: 8634–8639. https://doi.org/10.1002/anie.201403747.

26 Deuss, P.J. and Barta, K. (2016). From models to lignin: transition metal catalysis for selective bond cleavage reactions. *Coordination Chemistry Reviews* 306: 510–532. https://doi.org/10.1016/j.ccr.2015.02.004.

27 Galkin, M.V. and Samec, J.S.M. (2016). Lignin valorization through catalytic lignocellulose fractionation: a fundamental platform for the future biorefinery. *ChemSusChem* 9: 1544–1558. https://doi.org/10.1002/cssc.201600237.

28 Sun, Y. and Cheng, J. (2002). Hydrolysis of lignocellulosic materials for ethanol production: a review. *Bioresource Technology* 83: 1–11. https://doi.org/10.1016/S0960-8524(01)00212-7.

29 Zijlstra, D.S., Lahive, C.W., Analbers, C.A. et al. (2020). Mild organosolv lignin extraction with alcohols: the importance of benzylic alkoxylation. *ACS Sustainable Chemistry & Engineering* 8: 5119–5131. https://doi.org/10.1021/acssuschemeng.9b07222.

30 Watkins, D., Nuruddin, M., Hosur, M. et al. (2015). Extraction and characterization of lignin from different biomass resources. *Journal of Materials Research and Technology* 4: 26–32. https://doi.org/10.1016/j.jmrt.2014.10.009.

31 Michelin, M., Liebentritt, S., Vicente, A.A., and Teixeira, J.A. (2018). Lignin from an integrated process consisting of liquid hot water and ethanol organosolv: physicochemical and antioxidant properties. *International Journal of Biological Macromolecules* 120: 159–169. https://doi.org/10.1016/j.ijbiomac.2018.08.046.

32 Florian, T.D.M., Villani, N., Aguedo, M. et al. (2019). Chemical composition analysis and structural features of banana rachis lignin extracted by two organosolv methods. *Industrial Crops and Products* 132: 269–274. https://doi.org/10.1016/j.indcrop.2019.02.022.

33 Song, G., Madadi, M., Sun, C. et al. (2023). Surfactants facilitated glycerol organosolv pretreatment of lignocellulosic biomass by structural modification for co-production of fermentable sugars and highly reactive lignin. *Bioresource Technology* 383: 129178. https://doi.org/10.1016/j.biortech.2023.129178.

34 Song, G., Sun, C., Hu, Y. et al. (2023). Construction of anhydrous two-step organosolv pretreatment of lignocellulosic biomass for efficient lignin membrane-extraction and solvent recovery. *Journal of Physics: Energy* 5: 014015. https://doi.org/10.1088/2515-7655/acacc7.

35 Xiao, L.-P., Song, G.-Y., and Sun, R.-C. (2017). Effect of hydrothermal processing on hemicellulose structure. In: *Hydrothermal Processing in Biorefineries: Production of Bioethanol and High Added-Value Compounds of Second and Third Generation Biomass* (ed. H.A. Ruiz, M. Hedegaard Thomsen, and H.L. Trajano), 45–94. Cham: Springer International Publishing.

36 Chen, J., Yuan, Z., Zanuso, E., and Trajano, H.L. (2017). Response of biomass species to hydrothermal pretreatment. In: *Hydrothermal Processing in Biorefineries: Production of Bioethanol and High Added-Value Compounds of Second and Third Generation Biomass* (ed. H.A. Ruiz, M. Hedegaard Thomsen, and H.L. Trajano), 95–140. Cham: Springer International Publishing.

37 Peterson, A.A., Vogel, F., Lachance, R.P. et al. (2008). Thermochemical biofuel production in hydrothermal media: a review of sub- and supercritical water technologies. *Energy & Environmental Science* 1: 32–65. https://doi.org/10.1039/B810100K.

38 Yang, B., Tao, L., and Wyman, C.E. (2018). Strengths, challenges, and opportunities for hydrothermal pretreatment in lignocellulosic biorefineries. *Biofuels, Bioproducts and Biorefining* 12: 125–138. https://doi.org/10.1002/bbb.1825.

39 Yao, K., Wu, Q., An, R. et al. (2018). Hydrothermal pretreatment for deconstruction of plant cell wall: part I. Effect on lignin-carbohydrate complex. *AICHE Journal* 64: 1938–1953. https://doi.org/10.1002/aic.16114.

40 Shinde, S.D., Meng, X., Kumar, R., and Ragauskas, A.J. (2018). Recent advances in understanding the pseudo-lignin formation in a lignocellulosic biorefinery. *Green Chemistry* 20: 2192–2205. https://doi.org/10.1039/C8GC00353J.

41 Chuntanapum, A. and Matsumura, Y. (2010). Char formation mechanism in supercritical water gasification process: a study of model compounds. *Industrial & Engineering Chemistry Research* 49: 4055–4062. https://doi.org/10.1021/ie901346h.

42 Amini, N., Haritos, V.S., and Tanksale, A. (2018). Microwave assisted pretreatment of eucalyptus sawdust enhances enzymatic saccharification and maximizes fermentable sugar yield. *Renewable Energy* 127: 653–660. https://doi.org/10.1016/j.renene.2018.05.001.

43 Chen, W.-H., Hsu, M.-H., Wu, A.-Y., and Hwang, W.-S. (2017). Efficient extraction and recovery of xylan and lignin from rice straw using a flow-through hydrothermal system. *Journal of the Taiwan Institute of Chemical Engineers* 79: 103–109. https://doi.org/10.1016/j.jtice.2017.04.021.

44 El Hage, R., Chrusciel, L., Desharnais, L., and Brosse, N. (2010). Effect of autohydrolysis of *Miscanthus x giganteus* on lignin structure and organosolv delignification. *Bioresource Technology* 101: 9321–9329. https://doi.org/10.1016/j.biortech.2010.06.143.

45 Kumari, D. and Singh, R. (2018). Pretreatment of lignocellulosic wastes for biofuel production: a critical review. *Renewable and Sustainable Energy Reviews* 90: 877–891. https://doi.org/10.1016/j.rser.2018.03.111.

46 Caputo, F., Al-Rudainy, B., Naidjonoka, P. et al. (2022). Understanding the impact of steam pretreatment severity on cellulose ultrastructure, recalcitrance, and hydrolyzability of Norway spruce. *Biomass Conversion and Biorefinery* https://doi.org/10.1007/s13399-022-03405-0.

47 Heikkinen, H., Elder, T., Maaheimo, H. et al. (2014). Impact of steam explosion on the wheat straw lignin structure studied by solution-state nuclear magnetic resonance and density functional methods. *Journal of Agricultural and Food Chemistry* 62: 10437–10444. https://doi.org/10.1021/jf504622j.

48 Maltari, R., Kontro, J., Koivu, K. et al. (2022). Fractionation of technical lignin from enzymatically treated steam-exploded poplar using ethanol and formic acid. *ACS Applied Polymer Materials* 4: 9388–9398. https://doi.org/10.1021/acsapm.2c01665.

49 Chundawat, S.P.S., Donohoe, B.S., da Costa Sousa, L. et al. (2011). Multi-scale visualization and characterization of lignocellulosic plant cell wall deconstruction during thermochemical pretreatment. *Energy & Environmental Science* 4: 973–984. https://doi.org/10.1039/C0EE00574F.

50 Liu, Z.-H., Hao, N., Wang, Y.-Y. et al. (2021). Transforming biorefinery designs with 'Plug-In Processes of Lignin' to enable economic waste valorization. *Nature Communications* 12: 3912. https://doi.org/10.1038/s41467-021-23920-4.

51 Pu, Y., Hu, F., Huang, F., and Ragauskas, A.J. (2015). Lignin structural alterations in thermochemical pretreatments with limited delignification. *Bioenergy Research* 8: 992–1003. https://doi.org/10.1007/s12155-015-9655-5.

52 da Costa Sousa, L., Chundawat, S.P.S., Balan, V., and Dale, B.E. (2009). 'Cradle-to-grave' assessment of existing lignocellulose pretreatment technologies. *Current*

Opinion in Biotechnology 20: 339–347. https://doi.org/10.1016/j.copbio.2009.05.003.

53 da Costa Sousa, L., Jin, M., Chundawat, S.P.S. et al. (2016). Next-generation ammonia pretreatment enhances cellulosic biofuel production. *Energy & Environmental Science* 9: 1215–1223. https://doi.org/10.1039/C5EE03051J.

54 Lê, H.Q., Ma, Y., Borrega, M., and Sixta, H. (2016). Wood biorefinery based on γ-valerolactone/water fractionation. *Green Chemistry* 18: 5466–5476. https://doi.org/10.1039/C6GC01692H.

55 Luterbacher, J.S., Azarpira, A., Motagamwala, A.H. et al. (2015). Lignin monomer production integrated into the γ-valerolactone sugar platform. *Energy & Environmental Science* 8: 2657–2663. https://doi.org/10.1039/C5EE01322D.

56 Meng, X., Bhagia, S., Wang, Y. et al. (2020). Effects of the advanced organosolv pretreatment strategies on structural properties of woody biomass. *Industrial Crops and Products* 146: 112144. https://doi.org/10.1016/j.indcrop.2020.112144.

57 Trevorah, R.M., Huynh, T., Brkljača, R., and Othman, M.Z. (2021). Structural and morphological analysis of cellulose pulp produced from the fractionation of *Eucalyptus obliqua* sawdust using γ-valerolactone. *ACS Omega* 6: 4126–4136. https://doi.org/10.1021/acsomega.0c03616.

58 Cheng, F., Liu, S., Karlen, S.D. et al. (2023). Poplar lignin structural changes during extraction in γ-valerolactone (GVL). *Green Chemistry* 25: 336–347. https://doi.org/10.1039/D2GC03446H.

59 Jin, H., Shi, H., Jia, W. et al. (2022). Green solvents-based molecular weight controllable fractionation process for industrial alkali lignin at room temperature. *International Journal of Biological Macromolecules* 207: 531–540. https://doi.org/10.1016/j.ijbiomac.2022.03.049.

60 Kim, J.-Y., Shin, E.-J., Eom, I.-Y. et al. (2011). Structural features of lignin macromolecules extracted with ionic liquid from poplar wood. *Bioresource Technology* 102: 9020–9025. https://doi.org/10.1016/j.biortech.2011.07.081.

61 Wen, J.-L., Sun, S.-L., Xue, B.-L., and Sun, R.-C. (2013). Quantitative structures and thermal properties of birch lignins after ionic liquid pretreatment. *Journal of Agricultural and Food Chemistry* 61: 635–645. https://doi.org/10.1021/jf3051939.

62 Pu, Y., Jiang, N., and Ragauskas, A.J. (2007). Ionic liquid as a green solvent for lignin. *Journal of Wood Chemistry and Technology* 27: 23–33. https://doi.org/10.1080/02773810701282330.

63 Hou, X.-D., Xu, J., Li, N., and Zong, M.-H. (2015). Effect of anion structures on cholinium ionic liquids pretreatment of rice straw and the subsequent enzymatic hydrolysis. *Biotechnology and Bioengineering* 112: 65–73. https://doi.org/10.1002/bit.25335.

64 Wen, J.-L., Yuan, T.-Q., Sun, S.-L. et al. (2014). Understanding the chemical transformations of lignin during ionic liquid pretreatment. *Green Chemistry* 16: 181–190. https://doi.org/10.1039/C3GC41752B.

65 Pinkert, A., Goeke, D.F., Marsh, K.N., and Pang, S. (2011). Extracting wood lignin without dissolving or degrading cellulose: investigations on the use of food additive-derived ionic liquids. *Green Chemistry* 13: 3124–3136. https://doi.org/10.1039/C1GC15671C.

66 Singh, S.K. (2022). Ionic liquids and lignin interaction: an overview. *Bioresource Technology Reports* 17: 100958. https://doi.org/10.1016/j.biteb.2022.100958.

67 Liu, Q.-P., Hou, X.-D., Li, N., and Zong, M.-H. (2012). Ionic liquids from renewable biomaterials: synthesis, characterization and application in the pretreatment of biomass. *Green Chemistry* 14: 304–307. https://doi.org/10.1039/C2GC16128A.

68 Francisco, M., van den Bruinhorst, A., and Kroon, M.C. (2012). New natural and renewable low transition temperature mixtures (LTTMs): screening as solvents for lignocellulosic biomass processing. *Green Chemistry* 14: 2153–2157. https://doi.org/10.1039/C2GC35660K.

69 Liu, Y., Chen, W., Xia, Q. et al. (2017). Efficient cleavage of lignin–carbohydrate complexes and ultrafast extraction of lignin oligomers from wood biomass by microwave-assisted treatment with deep eutectic solvent. *ChemSusChem* 10: 1692–1700. https://doi.org/10.1002/cssc.201601795.

70 Xia, Q., Liu, Y., Meng, J. et al. (2018). Multiple hydrogen bond coordination in three-constituent deep eutectic solvents enhances lignin fractionation from biomass. *Green Chemistry* 20: 2711–2721. https://doi.org/10.1039/C8GC00900G.

71 Li, W.-X., Xiao, W.-Z., Yang, Y.-Q. et al. (2021). Insights into bamboo delignification with acidic deep eutectic solvents pretreatment for enhanced lignin fractionation and valorization. *Industrial Crops and Products* 170: 113692. https://doi.org/10.1016/j.indcrop.2021.113692.

72 Alvarez-Vasco, C., Ma, R., Quintero, M. et al. (2016). Unique low-molecular-weight lignin with high purity extracted from wood by deep eutectic solvents (DES): a source of lignin for valorization. *Green Chemistry* 18: 5133–5141. https://doi.org/10.1039/C6GC01007E.

73 Shen, X.-J., Wen, J.-L., Mei, Q.-Q. et al. (2019). Facile fractionation of lignocelluloses by biomass-derived deep eutectic solvent (DES) pretreatment for cellulose enzymatic hydrolysis and lignin valorization. *Green Chemistry* 21: 275–283. https://doi.org/10.1039/C8GC03064B.

74 Zheng, W., Lan, T., Li, H. et al. (2020). Exploring why sodium lignosulfonate influenced enzymatic hydrolysis efficiency of cellulose from the perspective of substrate–enzyme adsorption. *Biotechnology for Biofuels* 13: 19. https://doi.org/10.1186/s13068-020-1659-5.

75 Zhu, J., Chen, L., and Cai, C. (2021). Acid hydrotropic fractionation of lignocelluloses for sustainable biorefinery: advantages, opportunities, and research needs. *ChemSusChem* 14: 3031–3046. https://doi.org/10.1002/cssc.202100915.

76 Chen, L., Dou, J., Ma, Q. et al. (2017). Rapid and near-complete dissolution of wood lignin at ≤80 °C by a recyclable acid hydrotrope. *Science Advances* 3: e1701735. https://doi.org/10.1126/sciadv.1701735.

77 Wu, X., Zhang, T., Liu, N. et al. (2020). Sequential extraction of hemicelluloses and lignin for wood fractionation using acid hydrotrope at mild conditions. *Industrial Crops and Products* 145: 112086. https://doi.org/10.1016/j.indcrop.2020.112086.

78 Yu, H., Xu, Y., Hou, J. et al. (2020). Efficient fractionation of corn stover for biorefinery using a sustainable pathway. *ACS Sustainable Chemistry & Engineering* 8: 3454–3464. https://doi.org/10.1021/acssuschemeng.9b07791.

79 Ma, Q., Chen, L., Wang, R. et al. (2018). Direct production of lignin nanoparticles (LNPs) from wood using *p*-toluenesulfonic acid in an aqueous system at 80°C: characterization of LNP morphology, size, and surface charge. *Holzforschung* 72: 933–942. https://doi.org/10.1515/hf-2018-0033.

80 Zhai, Q., Long, F., Hse, C.-Y. et al. (2019). Facile fractionation of bamboo wood toward biomass valorization by *p*-TsOH-based methanolysis pretreatment. *ACS Sustainable Chemistry & Engineering* 7: 19213–19224. https://doi.org/10.1021/acssuschemeng.9b05392.

81 Cai, C., Hirth, K., Gleisner, R. et al. (2020). Maleic acid as a dicarboxylic acid hydrotrope for sustainable fractionation of wood at atmospheric pressure and ≤100°C: mode and utility of lignin esterification. *Green Chemistry* 22: 1605–1617. https://doi.org/10.1039/C9GC04267A.

82 Su, C., Hirth, K., Liu, Z. et al. (2021). Acid hydrotropic fractionation of switchgrass at atmospheric pressure using maleic acid in comparison with *p*-TsOH: advantages of lignin esterification. *Industrial Crops and Products* 159: 113017. https://doi.org/10.1016/j.indcrop.2020.113017.

83 Chen, J., Pang, Z., Zhang, Y. et al. (2022). Efficiently conversion of raw lignocellulose to levulinic acid and lignin nano-spheres in acidic lithium bromide-water system by two-step process. *Bioresource Technology* 343: 126130. https://doi.org/10.1016/j.biortech.2021.126130.

84 Li, N., Pan, X., and Alexander, J. (2016). A facile and fast method for quantitating lignin in lignocellulosic biomass using acidic lithium bromide trihydrate (ALBTH). *Green Chemistry* 18: 5367–5376. https://doi.org/10.1039/C6GC01090C.

85 Li, N., Li, Y., Yoo, C.G. et al. (2018). An uncondensed lignin depolymerized in the solid state and isolated from lignocellulosic biomass: a mechanistic study. *Green Chemistry* 20: 4224–4235. https://doi.org/10.1039/C8GC00953H.

86 Li, Z., Sutandar, E., Goihl, T. et al. (2020). Cleavage of ethers and demethylation of lignin in acidic concentrated lithium bromide (ACLB) solution. *Green Chemistry* 22: 7989–8001. https://doi.org/10.1039/D0GC02581J.

87 Yang, X.-H., Li, Z., Li, L. et al. (2021). Depolymerization and demethylation of Kraft lignin in molten salt hydrate and applications as an antioxidant and metal ion scavenger. *Journal of Agricultural and Food Chemistry* 69: 13568–13577. https://doi.org/10.1021/acs.jafc.1c05759.

7

Gasification and Combined Heat and Power (CHP) from Lignin

Subhashini[1], *Michal Jeremias*[2] *and Vineet Singh Sikarwar*[2,3]

[1] *Indian Institute of Technology Ropar, Department of Chemical Engineering, Rupnagar, 140001, Punjab, India*
[2] *Institute of Plasma Physics of the Czech Academy of Sciences, Plasma Chemical Technologies, Za Slovankou 1782/3, Prague, 18200, Czech Republic*
[3] *University of Chemistry and Technology, Department of Power Engineering, Technicka 5, Prague, 16628, Czech Republic*

7.1 Introduction

Energy is vital for the existence of mankind. The majority of energy is obtained from burning fossil fuels like petroleum, coal, and natural gas. It takes millions of years to produce these fossil fuel reserves on Earth to meet our daily energy requirements. Apart from their production, the shortage of fossil fuels is one aspect, but a more pressing issue is the release of carbon dioxide (CO_2) from the stored carbon (C) in the fossil fuels and the related greenhouse effect. This puts pressure to leave fossil fuels (mainly coal) underground by implementing carbon tax and/or carbon trading systems. Therefore, the price of fossil fuels will continually increase and their competitiveness with other energy sources will gradually decrease. Despite the continual increase in their cost, the fossil fuel reserves have been consumed extensively over the past decades, which is triggering a huge foreseen energy crisis across the globe. This rapidly increasing energy crisis has encouraged the researchers and the industry and population to shift from conventional sources of energy and evolve new sustainable alternative techniques of energy generation. Thus, there have been continual efforts to develop more economical, environment-friendly, and advanced alternative fuels to meet the energy demands of the increasing population [1]. Biomass-based materials and energy are sustainable alternatives to fossil fuels, as these are less costly and can be produced from renewable sources (such as plants and waste), releasing much less carbon dioxide and sulfur as compared with conventional fossil fuels. Eventually, the bio-renewable energy sector has become a matter of global importance [2]. It is worth noting that the cost depends on the market. Normally, fossil fuels are cheaper, but their price will increase in time

Lignin Chemistry: Characterization, Isolation, and Valorization, First Edition.
Edited by Yuhe Liao and Bert F. Sels.

because their emissions will be increasingly taxed or traded. For example, in central Europe, the price of biomass increased steeply after the increase in competition between newly built or modified power plants on biomass about 10 years ago. It also reflects the price of fossil fuels. After shortages of fuels at the beginning of 2022 due to the invasion of Russia into Ukraine and due to the sanctions implemented on Russian fuels, the price of fossil fuels increased and the price of biomass-based fuels followed the trend.

Biomass accounts for ~10% of the world's energy consumption [2]. The biomasses can be further categorized into different subsets as follows: wood and woody biomass, herbaceous and agricultural biomass, aquatic biomass, animal and human biomass wastes, and mixture of biomasses from all these varieties [1–4]. Biomass mainly consists of cellulose $(C_6H_{10}O_5)_x$, hemicelluloses $(C_5H_8O_4)_m$, lignin $[C_9H_{10}O_3(COH_3)_{0.9-1.7}]_n$, proteins, fats, sugars, water, ash content, etc. [1]. Cellulose is the main constituent of biomass; it is 35–50% of biomass by weight. Lignin is 15–20% and hemicelluloses are about 20–35% of biomass weight, while the remaining 15–20% is fats, proteins, ash content, and small extractives [5]. Among various biomass species, inedible biomass such as lignocellulosic [6] is the most favorable alternative for producing biofuel to replace conventional energy resources (Figure 7.1).

Over the years, hemicellulose and cellulose have been studied extensively, and some mature technologies have been industrialized for the production of biofuels and important chemicals from the (hemi) cellulose [7]. Whereas, lignin is mostly regarded as a low-value by-product in most biorefinery processes [8]. Although lignin is the largest renewable source of aromatics biopolymer on Earth, its conversion into value-added chemicals and biofuel remains a challenge due to its complex chemical structure [8, 9]. Therefore, we have discussed the various sources, their composition and availability, conventional conversion routes to fuel, and value-added products. Furthermore, the gasification process of lignin is being emphasized due to its various advantages over the other conventional processes of lignin. Along with its combined heat and power generation and utilization methods.

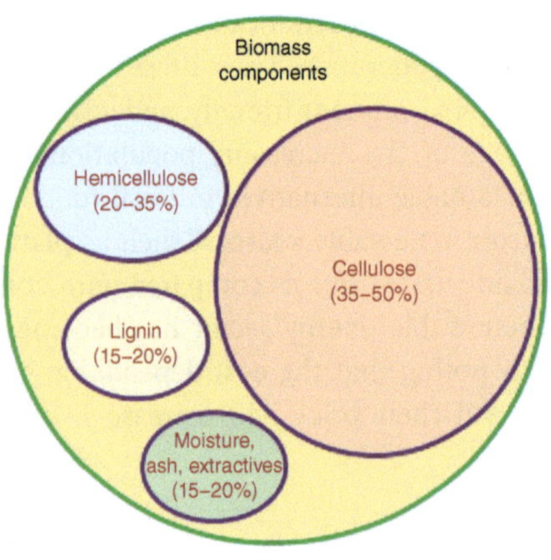

Figure 7.1 Composition of biomass.

7.2 Chemical Structure and Composition of Lignin

Lignin is an irregular and complex three-dimensional amorphous biopolymer synthesized by free radical coupling among three primary methoxylated derivatives of phenols [7]. These primary components present in lignin are also called monolignol units. Three primary monolignol units or so-called phenylpropane monomers that constitute the lignin structure are syringyl (S), guaiacyl (G), and *p*-hydroxyphenyl (H) alcohol (Figure 7.2). These phenylpropane monomers are also named as synapyl, coniferyl, and *p*-coumaryl alcohols, respectively. Depending upon the number of methoxy groups with which these monolignols are attached to an aromatic moiety, differentiates these monolignols from each other. Sinapyl alcohol has two methoxy groups, while coniferyl alcohol has one methoxy group, and *p*-coumaryl alcohol has no methoxy groups [10]. The random polymerization of these three monolignols monomers results in a variety of C—O (ether) and C—C linkages [7, 11], containing linkages such as β-O-4, 5-5′, α-O-4, phenylcoumaran β-5, resinol β-β, 4-O-5′, and β-1 [11]. The cleavage of the β-O-4 bond has been of particular interest in lignin depolymerization strategies since it represents up to 50% of the total linkages in wood materials compared to 5-5′ (4–25%), β-5 (4–12%), α-O-4 (4–8%), and β-β (2–7%) [12].

Lignin along with cellulose and hemicellulose constitutes the different layers in the cell wall of plants. The cell wall exhibits a layer configuration. The first layer, i.e., the outermost layer is called the primary wall, which is made up of pectin. The

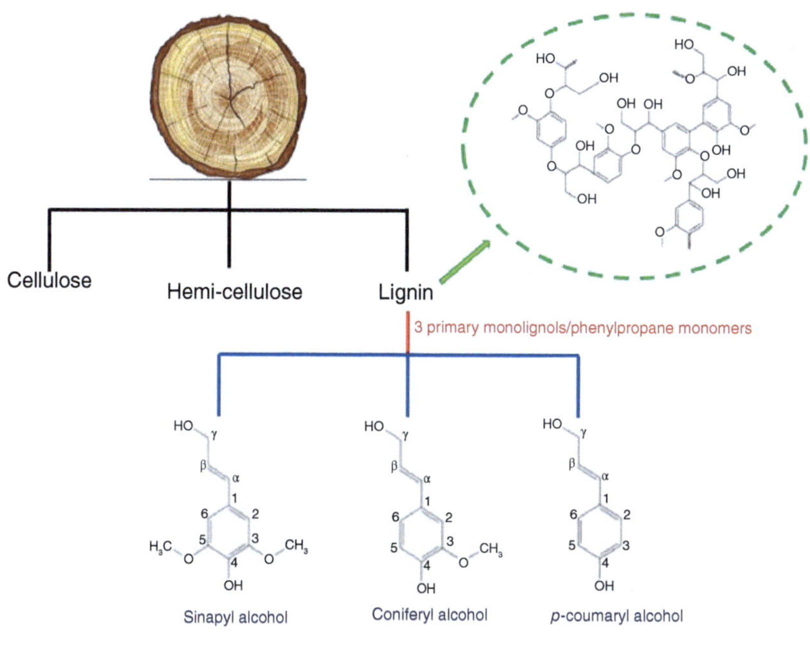

Figure 7.2 Chemical structure of lignin and its monolignol units.

primary wall is followed by the secondary wall, which consists of different laminated filamentous sections. These filamentous sections form different layers subsequently forming the secondary wall of the cell wall [11]. The lignin content concentration changes consistently in these different layers of secondary wall of the cell wall [6]. It varies in an ascending manner from the innermost layer to the outermost layer of the secondary wall of the cell wall. Lignin in the outmost layer serves as a binding agent to hold the adjacent cells together, whereas the lignin within the cell walls gives rigidity through chemical bonding with hemicellulose and cellulose microfibrils [13]. Thereby providing the structural strength, furthermore, it also serves as a sealing of water conduction system between roots and leaves (water impermeability), provides protection against degradation and biological strikes, and provides resistance against UV absorbance [6, 14]. Also, lignin contributes to 40% of the calorific value of lignocellulosic biomass [15]. The heating values of the lignocellulosic biomass species vary linearly with the associated lignin content according to the following equation [16]

$$\text{HHV}\ (\text{MJ kg}^{-1}) = 0.0893\ \left(\text{L\%}\right) + 16.97 \tag{7.1}$$

Here, HHV is the higher heating value expressed in (MJ kg^{-1}) and L represents the corresponding lignin content expressed in (wt%) in the biomass. The lignin content varies in different plants and ends up producing different types of biomass sources. These different biomass sources of lignin are tabulated in Table 7.1, along with their cardinal composition on dry weight % basis. Different biomass sources contain different percentages of each of the monolignols in lignin composition. In softwoods, the coniferyl alcohol reaches 95% of the total monolignols. Whereas in hardwoods, synapylic alcohol content is more significant than coniferylic alcohol by 50%. Herbaceous plants contain a mixture of the three units [25].

7.3 Sources of Lignin and its Characterization Techniques

Generally, biorefineries, pulp and paper mills, and other biomass-processing industries use the lignocellulosic biomass as their feedstock to produce ethanol, sugars, black-liquor, and fuel-grade chemicals, primarily derived from cellulose and in some cases from hemicellulose [9]. A huge amount of lignin content of lignocellulosic biomass remains unutilized or gets discarded during the processes. Hence, it can be concluded that the untreated lignocellulosic biomass in the biorefineries, pulp and paper mills, and other biomass-processing industries serves as the main resource of lignin. In order to determine the total lignin, molecular weight, and functional groups, researchers have been putting a lot of effort into developing an effectual characterization method that can overcome the various limitations

Table 7.1 Different sources of lignin and their composition.

Biomass feed	Cellulose (% w/w)	Hemicellulose (% w/w)	Lignin (% w/w)	References
Switchgrass	37	29	19	[16]
Wheat straw	38	29	15	
Corn stover	38	26	19	
Miscanthus	43	24	19	
Softwood	40–44	20–32	25–35	[17]
Hardwood	40–44	15–35	18–25	
Eucalyptus	42.8	17.1	21.2	[18]
Eucalyptus globulus	42.8	17.1	21.2	
Hemp	50.8	27.7	14.6	[19]
Kenaf core	42.9–43.4	10–13	13–15.5	[20]
Rice straw	32.7	15.6	21.2	[21]
Wheat straw	42.4	20.2	18	[22]
Sugarcane Bagasse	41.6	18.9	21.3	[23]
Cornstalk	36.5	27.9	15.3	[24]

accompanied by lignin, such as its complex structure, inherent variation in the proto-lignin, and significant structural changes occurring by different the isolation methods [6]. Klason procedure is the most widely used method for the determination of the lignin content of a lignocellulosic material [6]. However, techniques like gel permeation chromatography (GPC) performed on acetylated lignin dissolved in THF, matrix-assisted laser desorption ionization/time off light (MALDI-TOF), size exclusion chromatography combined with multi-angel laser light scattering (SEC-MALLS), vapor pressure osmometry (VPO), thioglycolic acid method and acetyl bromide methods are also used for determining the molecular weight and polydispersity of the lignin fragments [6, 26]. To study the complex structure of lignin, spectroscopic methods namely NMR, FTIR, and UV absorption have been very useful tools. Also, degradative techniques are used to quantify the methoxyl, phenolic hydroxyl, and carbonyl functional groups of lignin [6]. The various characterization methods are listed in Figure 7.3.

7.4 Lignin Isolation/Production Process

The heterogeneity of the lignin structure and high polymerization degree (greater than 10,000 units) lignin offers a stiff and recalcitrant structure to chemical attack [7]. Therefore, severe experimental conditions for depolymerization are

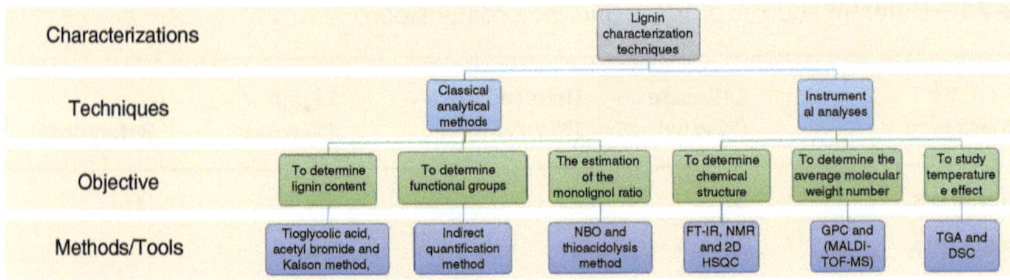

Figure 7.3 Various characterization techniques used to analyze lignin.

required to modify the biopolymer structure [11]. Generally, these depolymerization processes can be classified into two categories [6, 11]:

(i) Processes in which lignin is degraded into soluble fragments and is removed by separating the solid residue from the spent liquor; kraft, sulfite, soda, and organosolv processes and
(ii) Processes that selectively hydrolyze the carbohydrate content (cellulose and hemicellulose) and separate lignin as an insoluble solid residue; dilute acid hydrolysis of lignocellulose to yield sugar monomers, furfural, and levulinic acid.

Employing either of the delignification processes, extensively affects the structure conformation of the lignin as compared with that of the native and also, subsequently causes a substantial change in the behavior of the isolated lignin from the native lignin in wood. The delignification processes such as kraft and sulfite processes, not only bring structural changes but also incorporate the sulfur in the lignin structure. Hence, depending upon the type of the delignification process, lignin is classified into four categories: (i) sulfur-containing lignin, (ii) sulfur-free lignin, (iii) greener solvents, and (iv) other than the first three categories. Figure 7.4 shows different types of lignin obtained from different delignification processes.

Let us now discuss briefly about the various lignin isolation/production processes. There are various well-established processes that produce lignin as one of their major by-products. As we already know that biorefineries, pulp and paper mills, and other biomass-processing units play an important role in handling 150–200 billion tons of lignocellulosic biomass, thereby producing tons of lignin. These industries employ various processes to produce lignin at different stages of their processing. These processes are kraft process, sulfite process, Klason process, soda process, organosolv process, steam explosion process, enzymatic process, and Ionic liquids process. The advantages and disadvantages of these various lignin isolation processes are listed in Table 7.2.

7.5 Lignin Depolymerization and Conversion to Fuel

Biofuel production from lignocellulosic biomass resources is gaining massive attention due to its ability to replace the existing fossil fuels which are considered unsustainable [10]. Industrial biorefineries were acknowledged as one of the best

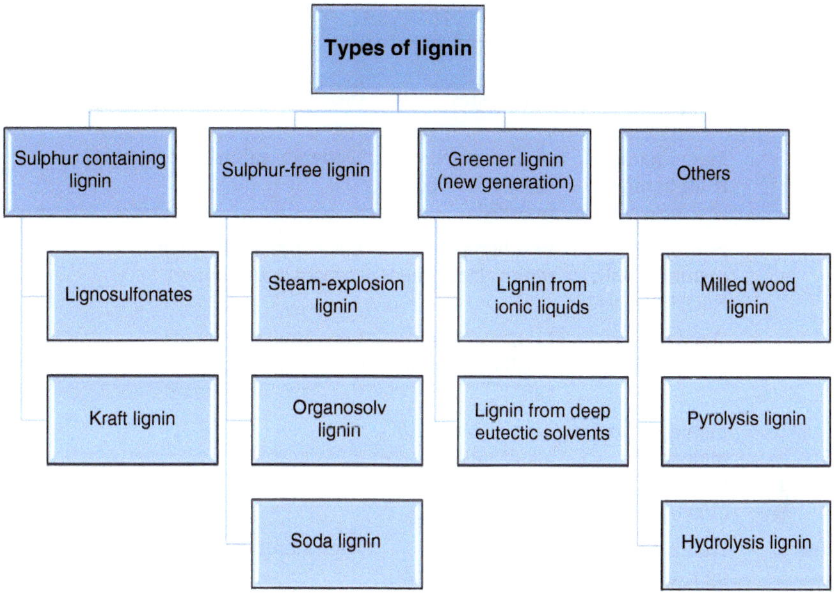

Figure 7.4 Types of lignin obtained after different delignification process.

solutions to mitigate climate change and the demand for fossil fuels, enabling the implementation of sustainable and productive systems [11]. The bioeconomy possibilities based on biochemical products and bioenergy through integrated biorefining systems became a new development strategy in many countries [32]. Hence, the lignin obtained from various isolation processes can be significantly used to serve as an alternative source of energy, if depolymerized effectively. Various forms of renewable fuel production by the use of lignin as feedstock are beneficial as reported by several researchers in recent years [8, 9]. To obtain these fuels, the effective depolymerization of recalcitrant lignin can be achieved either in a single-step process or can be subjected to a series of stepwise techniques. The different types of fuel that can be obtained from lignin feedstock through suitable depolymerization or deoxygenation methods are shown in Figure 7.5. In recent years, the research has been focused on implementing lignin as a platform product that allows valorization through different transformation routes [11]. The depolymerization process generally involves pyrolysis, liquefaction, and other thermochemical catalytic treatments involving a variety of parameters, while the deoxygenation reactions involve several hydrotreating steps (hydrodeoxygenation, hydrogenolysis, etc.) [10].

However, the complex polymeric structure of lignin is the main barrier to overcome and attain selective depolymerization. To attain selective depolymerization and cleave C—C and C—O linkages various reactions, such as hydrolysis, pyrolysis, and hydrogenolysis, are utilized. However, the conventional thermochemical routes such as pyrolysis and gasification may convert 100% lignin, but the desired product selectivity may be low. In contrast, when a catalyst is employed for a particular chemical

Table 7.2 Comparison of different lignin isolation processes.

Isolation type	Process summary	Advantages	Disadvantages	References
Kraft process	Lignin name: Kraft lignin	Implemented mostly for wood materials	Moderate sulfur content	[27]
	Hydrolytic treatment with NaOH and Na2S	Best lignin removal treatment	Black liquor with high carbohydrate content	
	Temperature: 150–180 °C	Lignin with a small content of ash	Long-time treatments (1–2 h)	
	Lignin with mild molecular weight (1000–3000 Da)	Lignin soluble in alkali solutions and organic solvents (polar)	—	
Sulfite	Lignin name: lignosulfonate lignin	Implemented mostly for wood biomass	High sulfur content	[28]
	Hydrolytic treatment with NaOH and Na_2SO_3	Water-soluble lignin	No selective for lignin extraction. Both hemicellulose and lignin are extracted	
	Temperature: 120–180 °C	Soluble in polar organic solvents and amines	High carbohydrates content	
	Lignin with high molecular weight (5700–61,000 Da)	—	High ash content (4–8%)	
Klason	Lignin name: Klason lignin	The standard method for lignin quantification –	An analytical method, not at an industrial scale	[12]
	Concentrated acid hydrolysis with 72% H_2SO_4	High yield isolation	High lignin structure change	
	Temperature: 30 °C	Low carbohydrate content	The high cost of the acid	
	Lignin with mild molecular weight (2000–7000 Da)	—	Severe equipment corrosion	

(Continued)

Table 7.2 (Continued)

Isolation type	Process summary	Advantages	Disadvantages	References
Soda	Lignin name: soda-type lignin	Sulfur free content	Representative cost of the alkali agent	[29]
	Hydrolytic alkali-treatment with NaOH	Low inhibitory compound formation	High alteration on lignin structure	
	Temperature: 90–150 °C	Implemented mostly for non-wood materials	High carbohydrate content at temperatures above 120 °C	
	Lignin with low molecular weight (1000–3000 Da)	Low ash content (<2.8%)	High alkali agent consumption	
Organosolv	Lignin name: organosolv lignin	Sulfur free content	High solvent cost	[9, 14]
	Hydrothermal treatment with solvent	Low alteration on native lignin structure	Require an additional stage for solvent recovery	
	Solvent concentration: 25–75% (vol.)	Lignin with a low molecular weight	Hydrophobic lignin	
	Temperature: 90–210 °C	Soluble in alkali solutions		
	Lignin with mild molecular weight (3000–4500 Da)	Relatively low operation times (<1.5 h)	Pilot/demo production scale	
Steam explosion	Lignin name: steam-exploded lignin	Sulfur free content	Middle lignin changes compared with the native form	[9]
	Hydrothermal treatment	Low or null reagent input requirement	Carbohydrate isolation in the liquor fraction (7–20%)	
	Temperature: 180–230° C	Shortened process time –	High hemicellulose content	
	Pressure: 1.3–3.5 MPa	Low energy requirement	—	
	Lignin with low molecular weight (1000–3000 Da)	Low ash content (<4%)	—	

(Continued)

Table 7.2 (Continued)

Isolation type	Process summary	Advantages	Disadvantages	References
Enzymatic	Lignin name: enzymatic lignin	Sulfur free lignin	Long residence times (several hours or days)	[30]
	Biological treatment	Non-energy intensive processes	Carbohydrate consumption from microbial culture	
	Temperature: 20–45 °C	—	—	
	Bio-reagent: fungi and bacteria	—	—	
	Lignin with mid molecular weight (4500–9500 Da)	—	—	
Ionic liquids	Lignin name: ionic liquid lignin	Sulfur free content	The high cost of the ionic liquid	[31]
	Hydrothermal solvent fractionation	Low lignin structure change	Addition of an antisolvent to biomass regeneration	
	Temperature: 100–170°C	—	Interferences between the residual ionic liquid and the enzyme and microbial activities	
	Lignin with mid molecular weight (1500–5000 Da)	—	Require an additional stage for the ionic liquid recovery	

production, it may not necessarily convert the whole lignin but may yield high desired product selectivity. Thus, it is essential to understand different methods for lignin conversion into fuel and value-added chemicals to select an appropriate process for any industry [9]. In this text, we will be focusing on different thermochemical techniques of lignin and especially on the gasification technique.

7.5.1 Thermochemical Conversion

The thermochemical conversion routes are the most widely used method in industries due to their nil prerequisite requirement of chemical pre-treatment for lignin depolymerization. Exposure to higher temperature is the most common treatment for breaking chemical bonds. However, operating conditions like high temperatures and pressures are the major disadvantage associated with these treatments along with comparatively lower value products.

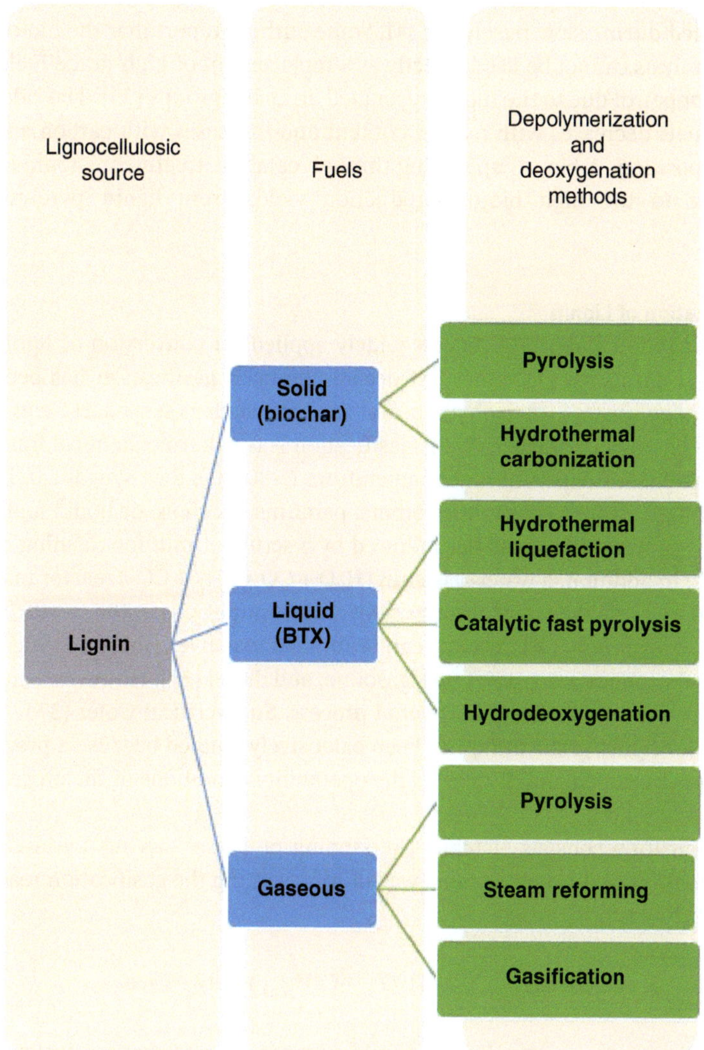

Figure 7.5 Useful fuels obtained from lignin feedstock.

7.5.1.1 Pyrolysis of Lignin

Pyrolysis is the thermal degradation of organic materials without or with limited amounts of oxygen, performed at elevated temperature and pressure conditions. According to the heating rate and residence time, pyrolysis can be classified as fast pyrolysis, intermediate pyrolysis, or slow pyrolysis. The absence of oxygen in pyrolysis allows lignin to be degraded but not converted into CO_2. Most of the final products in this process are bio-oil, combustible gases, and biochar. Lignin degradation in pyrolysis consists of several stages [33]. Unlike fast pyrolysis, where mainly bio-oil is obtained as possible biofuel (16–25 MJ kg^{-1}), high amounts of biochar are expected during slow pyrolysis [34]. Some authors report that the bio-oil from lignin pyrolysis cannot be used directly as a replacement of high-grade fuels, such as diesel or petrol due to the high oxygenated aromatic groups [35]. The calorific value of fuels decreases with oxygen content and increases with carbon and hydrogen composition. A bio-oil upgrading through catalytic treatment should be performed due to the high bio-oil production yields from lignin pyrolysis processes [11].

7.5.1.2 Gasification of Lignin

Gasification is another important process widely applied for conversion of lignin into value-added products. Over the past decade, biomass gasification has been developed to utilize wastes and to obtain useful products such as syngas, H_2, methane (CH_4), and chemical feedstocks [36]. Gasification is the thermochemical treatment through partial oxidation at high temperatures (>800°C). Raw syngas can be used after thorough cleaning to produce ethers, paraffins, alcohols, or liquid fuels. Firstly, lignin is processed in a gasifier followed by a scrubber unit for cleaning of produced gases. In addition, a water–gas-shift ($H_2O + CO \leftrightarrow H_2 + CO_2$) reactor may be employed to increase the concentration of hydrogen (and CO_2) in the produced gas. Thereafter, conventional FT reactors can be implied to convert obtained syngas into value-added products such as jet fuel, gasoline, and diesel [37]. However, leftover residual tar may be a problem for overall process. Supercritical water (374 °C and 22.1 MPa) for lignin gasification has been extensively studied because a previous drying stage is omitted, and it reduces the operability conditions of the process such as temperature [11].

Biomass gasification consists of many overlapping processes: drying, pyrolysis, and partial oxidation [36]. A very simple way of representing the gasification reaction is shown below [2]:

$$\text{Biomass} \rightarrow H_2 + CO + CH_4 + CO_2 + H_2O_{(g)} + C_xH_{y(g)} + \text{Tar}_{(l)} + \text{Char}_{(s)} \tag{7.2}$$

where (g) and (l) mean gaseous or liquid at ambient temperature (after gas cooling). Tar is usually defined as an organic compound with molecule mass higher or equal to 78 (toluene). The high-mass organic molecules are the most troublesome because they condense at relatively high temperatures (200–300 °C).

Gasification is an alternative thermo-chemical route to treat biomass [38], which reduces the various limitations associated with the pyrolysis of lignin. The major limitation associated with the pyrolysis process is that it does not allow the production of directly usable fuels but produces pyrolytic oils requiring further upgrading owing to their high acidity and instability. Steam gasification is an efficient and established method for H_2 production [39–41].

Furthermore, the combined heat and power generation is another advantage of gasification of lignin process. During this process heat is also generated, and as with other combustion processes in recent years, the idea of co-generating heat and power has become highly attractive. Generally, lignin-rich stream is combusted as a fuel in a boiler to produce steam that generates power and heat in a turbine. This is the standard case in paper mills, where Tomlinson boiler is used to combust black liquor. So, it is usually not discarded as waste. This idea provides a useful resource with a profitable secondary market that makes use of what was previously considered to be a waste stream. Recently, thermal plasma has been employed globally to process difficult waste streams because of its several inherent advantages including the potential to produce H_2-rich syngas [42]. Power can be effectively transmitted over long distances with minimal losses. Small decentralized cogeneration plants are established and are successfully in operation across the world [36]. Cogeneration units can provide heat and power to industrial, commercial, and residential buildings. Co-generation of heat and power by biomass combustion is prevalent [43] albeit gasification is better in terms of electrical efficiency and the acceptable range of biomass qualities [44, 45]. However, many cogeneration units for the production of thermal energy with electricity employing gas engines are installed and working successfully around the globe [44]. One of such company is CHEMREC Company, it focuses on producing DME from black liquor, which is deemed a more sophisticated way of black liquor utilization [46]. The black liquor has properties uniquely suitable for entrained flow gasification:

- It is liquid and easily pumped into the pressurized entrained flow gasifier.
- Its HHV is approximately 12–14 $MJ\,kg^{-1}$ [47, 48].
- The liquid state makes it easy to atomize into fine droplets.
- It is highly reactive due to high sodium and potassium content.

These properties make the gasification of black liquor easier and more rapid than of any other biomass feedstock.

A brief description of CHEMREC technology is shown in Figure 7.6.

In this generator (Figure 7.6), gas with $H_2 : CO$ ratio equal to 1 is produced. Such gas is suitable for DME synthesis. According to manufacturers, the initial cost of this device for black liquor processing is only slightly higher than the cost of a new Tomlinson boiler. The payback period is estimated to be 2 years longer than in the case of a new Tomlinson boiler. Bio DME project was started in 2008 as a follow-up to this technology [49].

DME (dimethyl ether) is potentially a very promising fuel for compression ignition (CI) engines. Technologically, it is very similar to LPG and it can be liquified at 0.5 MPa and normal temperature. CI engines need only a simple modification in order to be able to use this type of fuel (much like gasoline engines and LPG); the cetane number of DME is relatively high (specifically 55). The energy conversion of the original material into this product is relatively high as well (Figure 7.7).

The graph shows the distance a heavy truck can drive in a year on fuel produced from on hectare. DME and methanol have the best results.

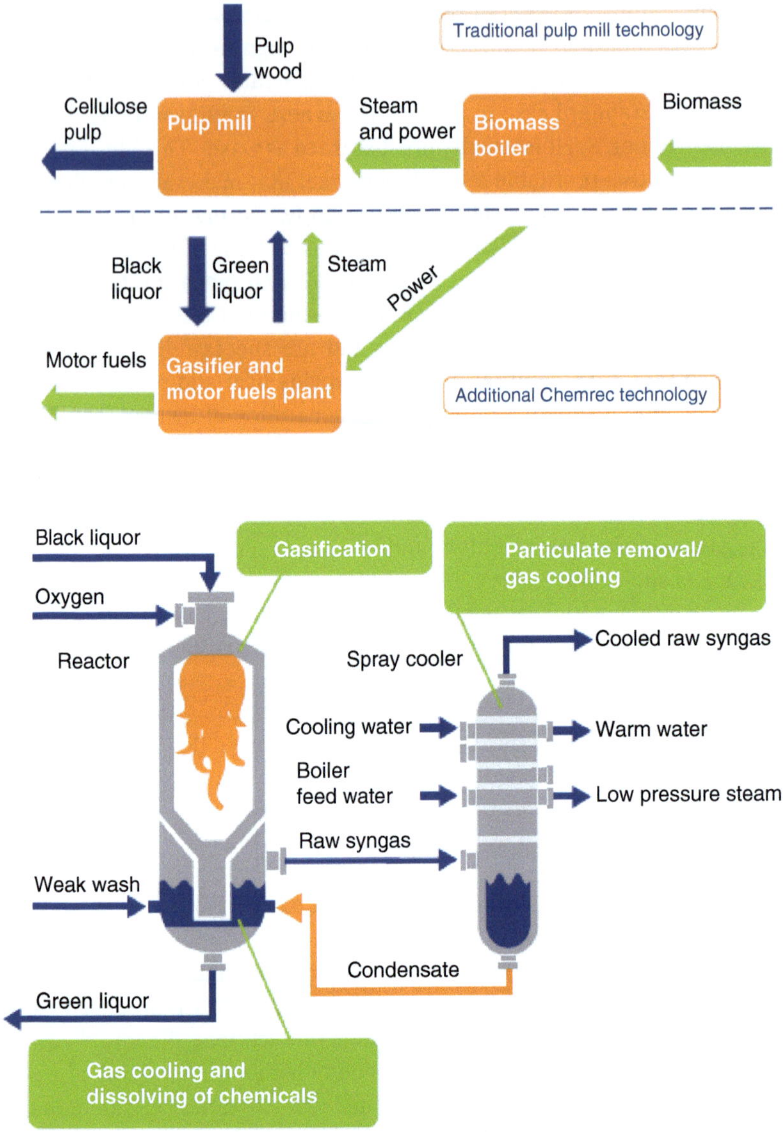

Figure 7.6 Integration into the pulp mill and gasifier scheme [49].

In the year 2008, a European project Bio DME was launched. The project goal is to demonstrate the whole process of transformation of lignocellulosic biomass into DME and use the fuel in a fleet of lorries for testing. The project budget is 28.4 million Euros, out of which 8.2 million Euros originates from European funds and 9.6 from Swedish Energy Agency. The partners of this project are CHEMREC, Delphi, ETC, Haldor Topsoe, Preem, Total, and Volvo. The project was scheduled to end in the year 2013 [50, 51].

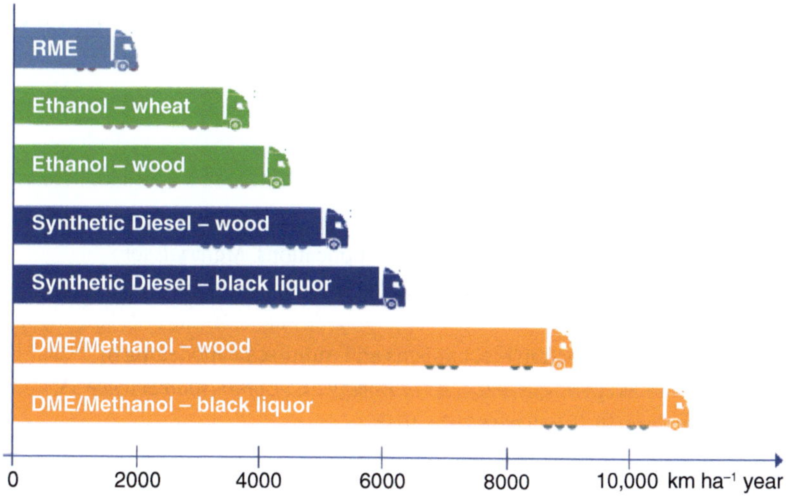

Figure 7.7 Truck distance area using biofuel made from 1 ha of plantation [49]. Source: AB Volvo.

7.6 Conclusion

Lignin as a source of value-added chemicals has potential to make a significant positive impact on the present biorefineries, pulp and paper industries, sugar industries, and agricultural industries. Indeed, lignin-derived value-added chemicals and fuels hold great economic, environmental, and societal value. Implementation of sustainable technologies for lignin conversion may lead to the development of zero-waste biorefineries that in turn will help to save carbon credits for the industries. On the contrary, despite several developments in lignin conversion technologies, their economic and industrial feasibility remains a challenge. Moreover, lignin conversion technologies are limited by feedstock itself. Consequently, lignin structure that plays a fundamental role in deciding the end product are yet to be defined completely, albeit majority of lignin building block units are identified. Similarly, residual lignin composition is dependent on the method of extraction; thus, it is difficult to implement a single technology for the conversion of residual lignin obtained from kraft process, sulfite process, organosolv process, or any other methods. Thus, the lignin obtained from each source needs a different technology for conversion into value-added products and useful chemicals.

On the contrary, thermal conversion methods may be able to process the majority of the lignin obtained from any source. Among thermal conversion methods, gasification overdoes other methods. It is not limited to a particular feedstock and specific product but is flexible toward the treatment of biomass wastes that may be toxic or contaminated, for the generation of diversified usable products. Also, it generates combined heat and power along the process which may lead to advancement of cogeneration units. In this manner, lignin gasification can be used as a sustainable alternative source of energy to replace conventional fossil fuels and can meet the increasing energy demand.

References

1 Kaur, P., Singh, S., and Verma, G. (2018). Facile synthesis of mesoporous carbon material from treated kitchen waste for energy applications. *Materials for Renewable and Sustainable Energy* 7 (2): 1–10. https://doi.org/10.1007/s40243-018-0116-x.

2 Anca-Couce, A. (2016). Reaction mechanisms and multi-scale modelling of lignocellulosic biomass pyrolysis. *Progress in Energy and Combustion Science* 53: 41–79. https://doi.org/10.1016/j.pecs.2015.10.002.

3 Vassilev, S.V., Baxter, D., Andersen, L.K., and Vassileva, C.G. (2010). An overview of the chemical composition of biomass. *Fuel* 89 (5): 913–933. https://doi.org/10.1016/j.fuel.2009.10.022.

4 Basu, P. (2013). *Biomass Gasification, Pyrolysis and Torrefaction: Practical Design and Theory. Biomass Gasification, Pyrolysis and Torrefaction: Practical Design and Theory*, 1–530. Elsevier.

5 Collard, F.X. and Blin, J. (2014). A review on pyrolysis of biomass constituents: mechanisms and composition of the products obtained from the conversion of cellulose, hemicelluloses and lignin. *Renewable and Sustainable Energy Reviews* 38: 594–608. https://doi.org/10.1016/j.rser.2014.06.013.

6 Azadi, P., Inderwildi, O.R., Farnood, R., and King, D.A. (2013). Liquid fuels, hydrogen and chemicals from lignin: a critical review. *Renewable and Sustainable Energy Reviews* 21: 506–523.

7 Cao, Y., Chen, S.S., Zhang, S. et al. (2019). Advances in lignin valorization towards bio-based chemicals and fuels: lignin biorefinery. *Bioresource Technology* 291: 121878.

8 Cao, L., Zhang, C., Chen, H. et al. (2017). Hydrothermal liquefaction of agricultural and forestry wastes: state-of-the-art review and future prospects. *Bioresource Technology* 245 (June): 1184–1193. https://doi.org/10.1016/j.biortech.2017.08.196.

9 Ahmad, E. and Pant, K.K. (2018). Lignin conversion: a key to the concept of lignocellulosic biomass-based integrated biorefinery. In: *Waste Biorefinery: Potential and Perspectives*, 409–444. Elsevier B.V. https://doi.org/10.1016/B978-0-444-63992-9.00014-8.

10 Suresh, S., Viswanathan, V., Angamuthu, M. et al. (2021). Lignin waste processing into solid, liquid, and gaseous fuels: a comprehensive review. *Biomass Conversion and Biorefinery* 13: 4515–4553.

11 Poveda-Giraldo, J.A., Solarte-Toro, J.C., and Cardona Alzate, C.A. (2021). The potential use of lignin as a platform product in biorefineries: a review. *Renewable and Sustainable Energy Reviews* 138: 110688. https://doi.org/10.1016/j.rser.2020.110688.

12 Li, C., Zhao, X., Wang, A. et al. (2015). Catalytic transformation of lignin for the production of chemicals and fuels. *Chemical Reviews* 115 (21): 11559–11624.

13 Shmulsky, R. and Jones, D.P. (2011). *Forest Products & Wood Science An Introduction*, 6e. Wiley 477 p.

14 Behling, R., Valange, S., and Chatel, G. (2016). Heterogeneous catalytic oxidation for lignin valorization into valuable chemicals: what results? What limitations? What trends? *Green Chemistry* 18 (7): 1839–1854.

15 Zhu, G., Qiu, X., Zhao, Y. et al. (2016). Depolymerization of lignin by microwave-assisted methylation of benzylic alcohols. *Bioresource Technology* 218: 718–722. https://doi.org/10.1016/j.biortech.2016.07.021.

16 Demirbaş, A. (2001). Relationships between lignin contents and heating values of biomass. *Energy Conversion and Management* 42 (2): 183–188.

17 Staley, B.F. and Barlaz, M.A. (2009). Composition of municipal solid waste in the United States and implications for carbon sequestration and methane yield. *Journal of Environmental Engineering* 135 (10): 901–909.

18 López, F., García, M.T., Feria, M.J. et al. (2014). Optimization of furfural production by acid hydrolysis of eucalyptus globulus in two stages. *Chemical Engineering Journal* 240: 195–201.

19 Wawro, A., Batog, J., and Gieparda, W. (2019). Chemical and enzymatic treatment of hemp biomass for bioethanol production. *Applied Sciences* 9 (24): 1–11.

20 Wi, S.G., Kim, S.B., Lee, D.S. et al. (2015). A comparative study on enzymatic hydrolysis of kenaf from two different harvest time-points, with- and without pretreatment. *Industrial Crops and Products* 76: 237–243. https://doi.org/10.1016/j.indcrop.2015.06.054.

21 Sindhu, R., Kuttiraja, M., Prabisha, T.P. et al. (2016). Development of a combined pretreatment and hydrolysis strategy of rice straw for the production of bioethanol and biopolymer. *Bioresource Technology* 215: 110–116. https://doi.org/10.1016/j.biortech.2016.02.080.

22 Tsegaye, B., Balomajumder, C., and Roy, P. (2019). Alkali pretreatment of wheat straw followed by microbial hydrolysis for bioethanol production. *Environmental Technology* 40 (9): 1203–1211. https://doi.org/10.1080/09593330.2017.1418911.

23 Andrade, M.F. and Colodette, J.L. (2014). Dissolving pulp production from sugar cane bagasse. *Industrial Crops and Products* 52: 58–64. https://doi.org/10.1016/j.indcrop.2013.09.041.

24 Jiao, Y., Guo, C., Wang, S. et al. (2020). Enhancement of converting corn stalk into reducing sugar by ultrasonic-assisted ammonium bicarbonate pretreatment. *Bioresource Technology* 302 (January): 122878. https://doi.org/10.1016/j.biortech.2020.122878.

25 Calvo-Flores, F.G. and Dobado, J.A. (2010). Lignin as renewable raw material. *ChemSusChem* 3 (11): 1227–1235.

26 Zevallos Torres, L.A., Lorenci Woiciechowski, A., de Andrade Tanobe, V.O. et al. (2020). Lignin as a potential source of high-added value compounds: a review. *Journal of Cleaner Production* 263: 121499.

27 Carvajal, J.C., Gómez, Á., and Cardona, C.A. (2016). Comparison of lignin extraction processes: economic and environmental assessment. *Bioresource Technology* 214: 468–476. https://doi.org/10.1016/j.biortech.2016.04.103.

28 José Borges Gomes, F., de Souza, R.E., Brito, E.O., and Costa Lelis, R.C. (2020). A review on lignin sources and uses. *Journal of Applied Biology & Biotechnology* 7: 100–105.

29 Lehto, J. and Alén, R. (2013). Alkaline pre-treatment of hardwood chips prior to delignification. *Journal of Wood Chemistry and Technology* 33 (2): 77–91.

30 Sindhu, R., Binod, P., and Pandey, A. (2016). Biological pretreatment of lignocellulosic biomass – an overview. *Bioresource Technology* 199: 76–82. https://doi.org/10.1016/j.biortech.2015.08.030.

31 Yinghuai, Z., Yuanting, K.T., and Hosmane, N.S. (2013). Applications of ionic liquids in lignin chemistry. In: *Ionic Liquids – New Aspects for the Future* (ed. J. Kadokawa), 35–346. IntechOpen: Futur. First https://doi.org/10.5772/57353.

32 Moncada, B.J., Aristizábal, M.V., and Cardona, A.C.A. (2016). Design strategies for sustainable biorefineries. *Biochemical Engineering Journal* 116: 122–134. https://doi.org/10.1016/j.bej.2016.06.009.

33 Zhou, S., Pecha, B., van Kuppevelt, M. et al. (2014). Slow and fast pyrolysis of Douglas-fir lignin: importance of liquid-intermediate formation on the distribution of products. *Biomass and Bioenergy* 66: 398–409. https://doi.org/10.1016/j.biombioe.2014.03.064.

34 Marathe, P.S., Westerhof, R.J.M., and Kersten, S.R.A. (2019). Fast pyrolysis of lignins with different molecular weight: experiments and modelling. *Applied Energy* (236): 1125–1137. https://doi.org/10.1016/j.apenergy.2018.12.058.

35 Zhang, X., Zhang, Q., Wang, T. et al. (2016). Efficient upgrading process for production of low quality fuel from bio-oil. *Fuel* 179: 312–321. https://doi.org/10.1016/j.fuel.2016.03.103.

36 Sikarwar, V.S., Zhao, M., Clough, P. et al. (2016). An overview of advances in biomass gasification. *Energy & Environmental Science* 9 (10): 2939–2977. https://doi.org/10.1039/C6EE00935B.

37 Lin, Y.C. and Huber, G.W. (2009). The critical role of heterogeneous catalysis in lignocellulosic biomass conversion. *Energy & Environmental Science* 2 (1): 68–80.

38 Sikarwar, V.S., Peela, N.R., Vuppaladadiyam, A.K. et al. (2022). Thermal plasma gasification of organic waste stream coupled with CO_2 sorption enhanced reforming employing different sorbents for enhanced hydrogen production. *RSC Advances 12* (10): 6122–6132.

39 Nipattummakul, N., Ahmed, I.I., Gupta, A.K., and Kerdsuwan, S. (2011). Hydrogen and syngas yield from residual branches of oil palm tree using steam gasification. *International Journal of Hydrogen Energy* 36 (6): 3835–3843. https://doi.org/10.1016/j.ijhydene.2010.04.102.

40 De Lasa, H., Salaices, E., Mazumder, J., and Lucky, R. (2011). Catalytic steam gasification of biomass: catalysts, thermodynamics and kinetics. *Chemical Reviews* 111 (9): 5404–5433.

41 Sikarwar, V.S., Pfeifer, C., Ronsse, F. et al. (2022). Progress in in-situ CO_2 sorption for enhanced hydrogen production. *Progress in Energy and Combustion Science 91*: 101008.

42 Sikarwar, V.S., Reichert, A., Pohorely, M. et al. (2021). Equilibrium modeling of thermal plasma assisted co-valorization of difficult waste streams for syngas production. *Sustainable Energy & Fuels 5* (18): 4650–4660.

43 International Energy Agency Report. http://www.ieabcc.nl/meetings/32_03_meeting_SaltLakeCity.pdf, .

44 Ahrenfeldt, J., Thomsen, T.P., Henriksen, U., and Clausen, L.R. (2013). *Applied Thermal Engineering* 50: 1407–1417.

45 Negro, S.O., Suurs, R.A.A., and Hekkert, M.P. (2008). *Technological Forecasting & Social Change* 75: 57–77.

46 Jeremias, M. (2014). *Intensification of Fluidized-Bed Gasification of Biomass.* Doctoral thesis. Prague: Institute of Chemical Technology.

47 Rikard Gebart, B., Wiinikka, H., Marklund, M. et al. (2011). Recent advances in the understanding of pressurized black liquor gasification. *Cellulose Chemistry and Technology* 45 (7–8): 521–526.

48 Lindblom, M. (2007). *"Gasification of Black Liquor and Sulphur Cycle in the Paper Industry", in Topsoe Catalysis forum.* Denmark: Havreholm Slot.

49 "Chemrec – The Pulp Mill Biorefinery", promotional material, (2010).

50 Spath P. L. and Dayton D. C. (2003). "Preliminary screening-technical and economic assessment of synthesis gas to fuels and chemicals with emphasis on the potential for biomass-derived syngas", technical report NREL/TP-510-34929.

51 Rostrup-Nielsen, T., Nielsen, P. E. H., Joensen, F. and Madsen, J. "Polygeneration – Integration of Gasoline Synthesis and IGCC Power Production Using Topsoe's TIGAS Process", technical paper Risø-R-1608(EN).

8

Enzymatic and Microbial Bioconversion of Lignin to Renewable Chemicals

Timothy D.H. Bugg, Victoria Sodré, Awatif Alruwaili and Goran M.M. Rashid

University of Warwick, Department of Chemistry, Coventry, CV4 7AL, United Kingdom

8.1 Lignin Conversion by Enzymes

Enzymes for lignin conversion have been most extensively studied in Basidiomycete fungi, which secrete extracellular oxidative enzymes to attack lignin. Since 2010, there has been renewed interest in bacterial enzymes for lignin depolymerization by selected soil bacteria, which will then be discussed. Both lignin-oxidizing fungi and bacteria require accessory enzymes in order to generate hydrogen peroxide for peroxidase enzymes and to prevent repolymerization of radical intermediates in lignin depolymerization.

8.1.1 Extracellular Peroxidases from White-rot Fungi

White-rot fungi such as *Phanerochaete chrysosporium* can degrade lignin, hemicellulose, and cellulose components of plant cell wall lignocellulose and are especially well adapted for degrading wood in the natural environment, since they grow on solid substrates and produce mycelia that can penetrate the surface of the wood [1, 2]. White-rot fungi produce an arsenal of extracellular peroxidase and laccase enzymes that can attach lignin [1–3]. There are three types of fungal peroxidase that attack lignin: (i) lignin peroxidase (LiP), a high redox potential peroxidase that can attack non-phenolic lignin units; (ii) manganese peroxidase (MnP) that oxidizes Mn^{2+} to Mn^{3+}, which then acts as a diffusible oxidant to attack the lignin polymer, usually chelated by organic acids; and (iii) versatile peroxidase (VP), reported in *Pleurotus* and other fungi, which possesses both types of activities [4].

The general catalytic cycle of each class of peroxidase enzymes is shown in Figure 8.1 [3]. The resting state of the enzyme as purified contains heme cofactor with a central ferric (Fe^{3+}) ion, which reacts with hydrogen peroxide to form an

Lignin Chemistry: Characterization, Isolation, and Valorization, First Edition.
Edited by Yuhe Liao and Bert F. Sels.

iron–oxo intermediate known as compound I, in which the iron cofactor has a +4 oxidation state, with a radical cation formed on the heme cofactor. If compound I reacts via one-electron oxidation, then a second intermediate compound II is formed, in which the iron cofactor formally has a +4 oxidation state. In lignolytic peroxidases, the environment of the heme cofactor confers a high redox potential to the compound I intermediate, due to the positioning of N_3 of the side chain of the proximal histidine residue [5]. LiP contains specific binding sites for aromatic substrates [5], whereas MnP contains a manganese ion binding site close to the heme edge [6].

The active site of *P. chrysosporium* MnP consists of a proximal histidine ligand (His-173) that binds the iron cofactor, which is hydrogen-bonded to Asp-242 [6]. Generation of intermediate compound I is catalyzed by proton donor His-46, and compound I is stabilized by a nearby arginine residue (Arg-42). The reaction of *P. chrysosporium* MnP with β-aryl ether lignin model compounds has been studied, generating reaction products via Cα–Cβ bond cleavage, aryl-Cα bond cleavage, and a phenoxy radical intermediate [7].

The genome of *P. chrysosporium* contains 10 LiP genes, encoding at least six isozymes [8]. The catalytic cycle of LiP is similar to that of MnP, but due to its high redox potential, LiP can also oxidize non-phenolic lignin units [1–3]. The active site cavity of LiP is only accessible to small phenolic substrates [5]; however, long-range electron transfer from the heme cofactor to the lignin polymer can take place via a Trp-171 residue on the protein surface (see Figure 8.2), which can form a stable radical intermediate via heme oxidation [9]. Trp-171 is conserved in all amino acid sequence alignments of LiP enzymes, and site-directed mutagenesis experiments suggest that Trp-171 is used as a redox-active residue in LiP [9]. Reaction of *P. chrysosporium* LiP with β-aryl ether

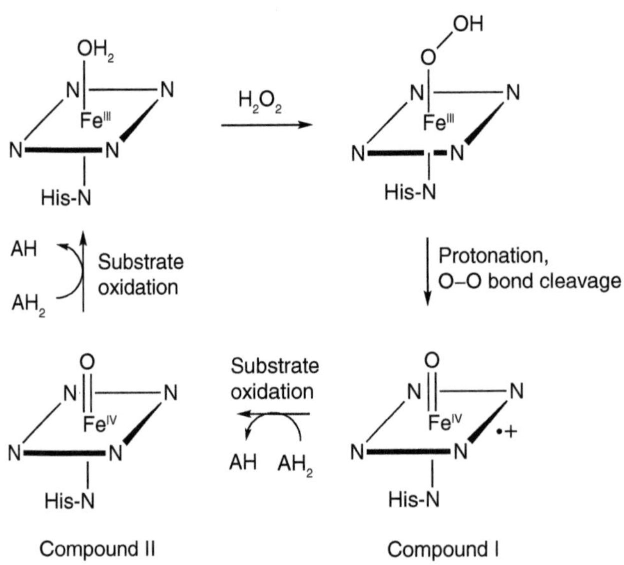

Figure 8.1 Catalytic cycle for lignolytic peroxidases.

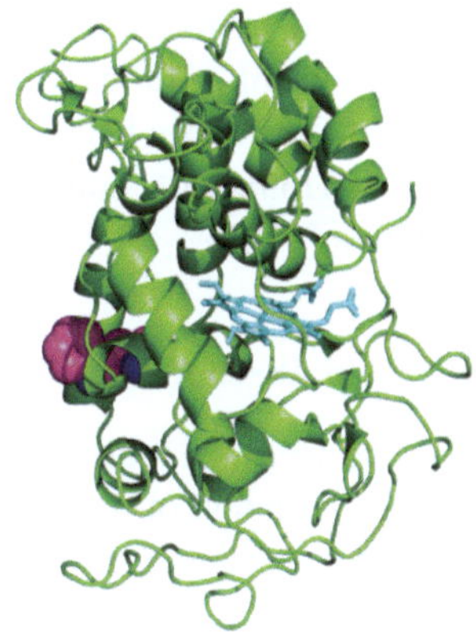

Figure 8.2 Structure of *Phanerochaete chrysosporium* lignin peroxidase, showing location of heme cofactor (cyan) and Trp171 (magenta). Figure prepared using Pymol software, from PDB file 1LGA.

lignin model compounds generates arising products from Cα–Cβ bond cleavage, which is believed to proceed via a radical mechanism [10].

8.1.2 Bacterial Dye-Decolorizing Peroxidases (DyP)

Although activity for degradation of lignin in bacteria was first observed in the 1980s in *Streptomyces viridosporus* [11], no gene or enzyme responsible was identified until after 2010. Lignin degradation activity was reported for *Rhodococcus jostii* RHA1 and *Pseudomonas putida* KT2440, two well-studied aromatic-degrading bacteria [12, 13]. Bioinformatic analysis of the genome of *R. jostii* RHA1 led to the identification of two dye-decolorizing peroxidases (DyPs) found in other lignin-degrading microbes, and gene deletion of the dypB gene led to loss of lignin degradation activity [14]. Recombinant *R. jostii* RHA1 DypB was catalytically active for oxidative cleavage of a beta-aryl ether lignin model compound, and for oxidation of Mn^{2+}, and in the presence of Mn^{2+}, DypB was found to break down lignocellulose [14].

Four classes of dye-decolorizing peroxidases (DyPs) have been identified based on sequence analysis, of which the B-type and C-type enzymes show activity for Mn^{2+} oxidation [15]. *Pseudomonas fluorescens* Dyp1B has also been reported to have activity for Mn^{2+} oxidation and polymeric lignin oxidation [16]. A C-type peroxidase from *Amycolatopsis* sp. 75iv2 has been reported, which has higher activity for Mn^{2+} oxidation, approaching that of fungal MnP [17].

The active sites of DyP-type peroxidases contain an active site aspartate residue, which is thought to act as a proton donor to generate compound 1 from hydrogen peroxide, whose mutation leads in most cases to loss of catalytic activity [18]. There is also a nearby arginine residue, which is thought to stabilize the iron–oxo intermediate [18]. Adjacent to the heme cofactor in *R. jostii* RHA1 DypB is a manganese ion binding site, comprising Glu-156, Glu-215, Glu-239, and a heme propionate [19]. Some protein engineering studies have been reported for DyP-type peroxidases. Replacement of active site Asn-246 in *R. jostii* DypB by Ala increases the k_{cat} for Mn^{2+} oxidation by 80-fold [19] (Figure 8.3). Directed evolution of *Pseudomonas putida* DyP via error-prone PCR led to the identification of three mutations of surface residues E188K, A142V, and H125Y, which together improved activity toward 2,6-dimethoxyphenol by 100-fold [20]. Saturation mutagenesis of residues close to the heme cofactor in *P. fluorescens* Dyp1B led to the isolation of mutant H169L with enhanced activity for 2,4-dichlorophenol and Kraft lignin oxidation [21].

8.1.3 Multi-Copper Oxidases

Laccases are multi-copper oxidase enzymes containing four copper atoms that catalyze the oxidation of a wide variety of substrates, including phenolic compounds, using dioxygen as an oxidant, which is reduced to water via a four-electron reduction. Laccases can also oxidize non-phenolic compounds in the presence of low molecular weight mediators, such as 2,20-azinobis(3-ethylbenzothiazoline-6-sulfonic acid) (ABTS), or 1-hydroxybenzotriazole (HBT) [22]. In plants, laccases are involved in the biosynthesis of lignin, via phenolic oxidation of monolignol precursors, whereas in fungi, laccases are thought to play a role in the degradation of lignin, although the precise role is uncertain and is debated in the literature [23–25].

The four copper atoms are arranged into a mononuclear Type 1 (T1) copper center, which is responsible for the characteristic blue color of laccases (λ_{max} near 600 nm), and a trinuclear cluster consisting of a Type 2 (T2) copper center and two Type 3 (T3) copper centers. Substrate oxidation is thought to occur at the mononuclear T1 copper center, via successive one-electron transfers, and electrons are then transferred to the trinuclear T2/T3 centers, where oxygen is reduced to water [26].

Several lignin-degrading white rot fungi, such as *Trametes versicolor* and *Phlebia radiata*, produce extracellular laccases, which have been characterized [27, 28]. The

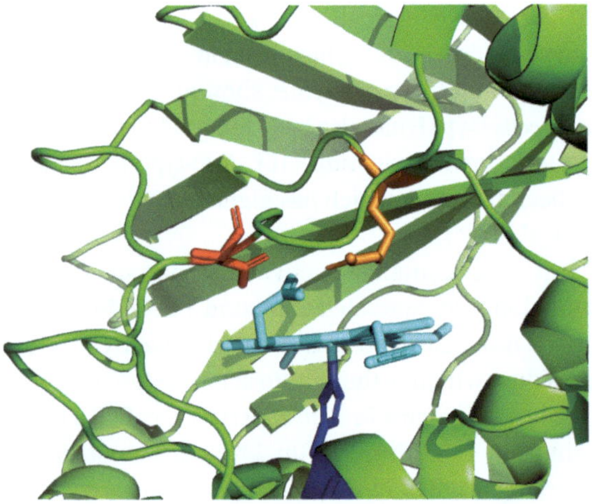

Figure 8.3 Structure of *R. jostii* RHA1 DypB active site, showing heme cofactor (cyan) and active site residues Asp-153 (red) and Arg-244 (orange). Figure prepared using Pymol software, from PDB file 3QNR.

lignin degradation activity of a laccase-deficient mutant of *Sporotrichum pulverulentum* was lost [29], implying a role for laccase in lignin degradation.

In bacteria, laccase enzymes from *Streptomyces coelicolor* A3(2), *Streptomyces lividans* TK24, *Streptomyces viridosporus* T7A, and *Amycolatopsis* sp. 75iv2 have been expressed in recombinant form and characterized kinetically [30]. Using lignin model compounds as substrates, they were found to catalyze Cα oxidation of the β-aryl ether unit [30]. Gene deletion of the laccase gene in *S. coelicolor* A3(2) was found to greatly reduce the production of acid-precipitable lignin (APPL) from lignocellulose, consistent with a role in lignin oxidation; however, treatment *in vitro* of ethanosolv lignin led to the formation of higher molecular weight products, due to repolymerization of phenoxy radical intermediates [30]. Overexpression of *Amycolatopsis* laccase has been shown to increase the release of low molecular weight products arising from aryl-Cα cleavage and Cα-Cβ cleavage from poplar lignocellulose [31]. A blue laccase CueO has been characterized from a lignin-degrading *Ochrobactrum* sp. and was shown to oxidize lignin model compounds and lignosulfonate [32]. The multi-copper oxidase CopA has been identified as a pseudolaccase, requiring the addition of exogenous Cu^{2+} for activity *in vitro*, in *Pseudomonas stutzeri* [33], *Pseudomonas putida* KT2440 [34], and *Pseudomonas fluorescens* Pf-5 [34]. Production of low molecular weight products from polymeric lignin has been observed with CopA [33, 34] but also competing repolymerization of low molecular weight phenols [34].

8.1.4 Beta-Etherase Enzymes for Lignin Ether Cleavage

Degradation of the β-aryl ether lignin dimer in *Sphingobium* SYK-6 (formerly *Sphingomonas paucimobilis* SYK-6) has been shown to proceed via the pathway shown in Figure 8.4. The α-hydroxyl group is first oxidized to the corresponding ketone by an NAD-dependent dehydrogenase LigD [35], which is followed by glutathione-dependent cleavage of the ether linkage by enzymes LigE and LigF [36–38] followed by removal of the glutathione substituent by a further equivalent of reduced glutathione and catalyzed by LigG [37]. Enzymes LigE and LigF have been shown to be specific for opposite enantiomers of the β-ether substrate [38, 39], and crystal structures of these enzymes have been determined [39]. Using a combination of recombinant LigE and LigF enzymes with glutathione lyase LigG was found to be highly effective for bioconversion of lignin models [40].

Until recently, the beta-etherase enzymes had only been shown to act on dimeric lignin model compounds, which may be lignin degradation intermediates. However, there have been some reports of catalytic activity *in vitro* of β-etherase enzymes on polymeric lignin substrates. A combination of β-etherases LigEG from *Sphingobium* SYK-6 and LigF from *Novosphingobium aromaticivorans* was found to convert enzymatically oxidized beech wood lignin to give a bio-oil containing low molecular weight aromatic compounds [41]. A combination of β-etherases LigDEFN and NaGSTNU, and cofactor recycling enzyme glutathione reductase, was reported to release low molecular weight products from polymeric lignin [42].

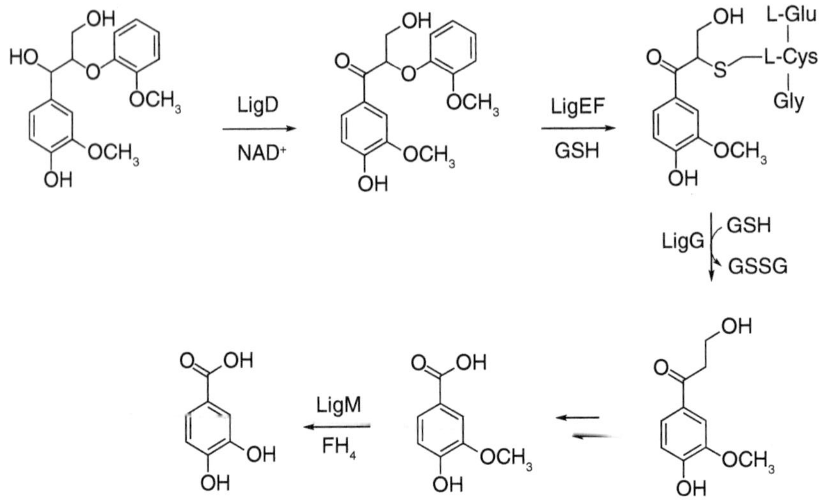

Figure 8.4 β-aryl ether degradation pathway in *Sphingobium* SYK-6.

8.1.5 Other Lignin-Oxidizing Enzymes

Two novel lignin-oxidizing enzymes have been discovered from proteomic analysis of the extracellular fraction of *Sphingobacterium* sp. T2 [43]. Surprisingly, there were neither known lignin-oxidizing genes such as DyP-type peroxidase and laccases in the genome of this organism nor was there the α-ketoadipate pathway for oxidation of aromatic monomers. Both enzymes were found to be extracellular manganese-dependent superoxide dismutases, namely, SpMnSOD1 and SpMnSOD2. The recombinant protein of both enzymes was able to oxidize polymeric lignin organoslov and industrial Kraft lignin. Oxidation of organosolv lignin with SpMnSOD1 generated multiple degradation products, including 2-methoxyhydroquinone, 5-hydroxyvanillic acid, vanillin, guaiacol, and catechol. From these oxidation products, it was concluded that SpMnSOD1 and SpMnSOD2 can exhibit several catalytic activities, including O-demethylation, cleavage of both aryl-Cα and aryl-Cα-Cβ bonds, and hydroxylation of aromatic compounds [43].

Incubation of SpMnSOD1 with lignin model compound guiaiacylglycerol-β-guaiacyl ether using ^{18}O labeling showed that the enzyme uses superoxide or peroxide to insert oxygen atoms, and further mechanistic studies provided evidence that the reactive oxidant was hydroxyl radical [44]. Quantitative ^{31}P NMR spectroscopy showed 20–40% increases in phenolic and aliphatic OH content when alkali lignin was treated by SpMnSOD1, indicating that the major activity of this enzyme is demethylation of polymeric lignin. Site-directed mutagenesis of SpMnSOD1 showed that the enzyme loses demethylation activity when Ala-31 or Tyr-27 close to the enzyme active site were replaced; it is thought that these residues provide a channel for water to access the Mn(II) center to generate hydroxyl radical [44].

8.1.6 Accessory Enzymes for Lignin Oxidation

8.1.6.1 Fungal Accessory Enzymes

White-rot fungi produce several extracellular peroxidase enzymes that are essential for the breakdown of lignocellulosic biomass, and which require hydrogen peroxide as co-substrate, so fungi also produce oxidase enzymes that generate hydrogen peroxide, such as aryl-alcohol oxidase, which acts an accessory enzyme for fungal lignin degradation. These fungal oxidases are flavin-adenine-dinucleotide (FAD)-containing enzymes, also known as aromatic alcohol oxidases (AAOs), veratryl-alcohol oxidases, or benzyl-alcohol oxidases, which catalyze the oxidation of aromatic and aliphatic allylic primary alcohols to the corresponding aldehydes while reducing molecular oxygen to H_2O_2 [45]. The AAO catalytic cycle occurs through two half-reactions: in the first reductive half-reaction, the donor alcohol is oxidized by FAD via two-electron oxidation, to yield an aldehyde product and $FADH_2$. In the second oxidative half-reaction, molecular oxygen is reduced by $FADH_2$ to produce H_2O_2 and regenerate oxidized FAD [46, 47]. Glyoxal oxidase also generates hydrogen peroxide for fungal lignin-degrading enzymes, via oxidation of glyoxal, which is released during lignin breakdown [48]. Fungal glyoxal oxidase enzymes have been characterized from *P. chrysosporium* [49, 50] and *Pycnoporus cinnabarinus* [51].

Cellobiose dehydrogenase is also used as an extracellular enzyme accessory enzyme by the wood-degrading fungus *P. chrysosporium*. This enzyme oxidizes soluble cellulose oligosaccharides produced in the decay of wood. This enzyme has a small electron-transferring cytochrome moiety (CYT) fused to a sugar-oxidizing flavodehydrogenase (DH) domain [52]. The reduction of aromatic radicals produced by laccases and peroxidases during lignin degradation, to prevent their repolymerization, is one of several potential physiological functions, including conversion of cellobiose to cellobionic acid; chelation of manganese ions by cellobionic acid; generation of ferrous ions for involvement in Fenton chemistry; and activation of lytic polysaccharide monooxygenase (LPMO) [52].

8.1.6.2 Bacterial Accessory Enzymes

Dimerization or repolymerization is often observed when recombinant lignin-oxidizing enzymes are used for *in vitro* biotransformation of lignin substrates, because the enzymes generate phenoxy radicals, which can spontaneously recombine. This suggests the existence of further accessory enzymes capable of one-electron reduction of phenoxy radicals (see Figure 8.5). A highly expressed extracellular dihydrolipoamide dehydrogenase from *Thermobifida fusca* has recently been discovered for this activity, as it has been shown to prevent the dimerization of a lignin model compound *in vitro* and to alter the profile of low molecular weight products produced [53, 54].

Two bacterial oxidase enzymes that can produce hydrogen peroxide as a co-substrate for lignin-oxidizing peroxidase enzymes have also been identified. A copper-dependent oxidase enzyme from *Thermobifida fusca* has been demonstrated

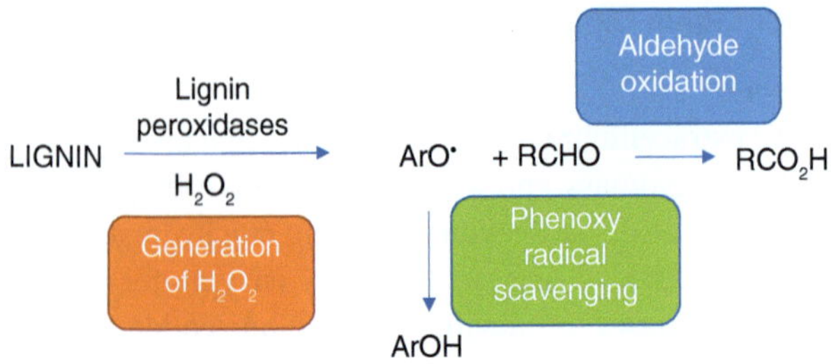

Figure 8.5 Roles of accessory enzymes in lignin degradation.

to decrease lignin content in sugarcane bagasse and generate dilignol products [55]. Another FMN-dependent oxidase enzyme that can oxidize aldehyde intermediates and generate hydrogen peroxide has been demonstrated to be involved in a pathway for the metabolism of aryl-C2 lignin fragments in *R. jostii* RHA1 [56].

8.1.7 Lignin Conversion *In Vitro* by Enzymes

Given that there are several classes of enzymes described above that can attack lignin, it may at first seem surprising that there are no well-established methods for using these enzymes to produce low molecular weight products from lignin breakdown. However, there are several practical problems for such applications, as follows:

1. Solubility. Most native-like lignins such as organosolv lignins are largely insoluble in water, which makes them difficult for processing by enzymes that are active in aqueous media, although there are some examples of enzymes that can attack such lignins.
2. Repolymerization vs. depolymerization. Lignin-oxidizing enzymes generally proceed via radical intermediates (similar to those used in lignin biosynthesis), which can repolymerize (or recondense) to form higher molecular weight products. For many peroxidase and laccase biotransformations of lignin, it is quite common that the majority of products have higher molecular weight than the starting material. Hence there is interest in accessory enzymes that can trap radical intermediates to solve this problem (see Section 8.1.6).
3. Complex mixtures of products. Since lignin is itself heterogeneous in structure, and oxidative cleavages occur at various sites, it is not surprising that mixtures of products are obtained, some more complex than others. One advantage of using a microbial biotransformation for lignin bioconversion is that microbial aromatic degradation pathways are convergent, hence the microbe is able to funnel multiple substrates into a small number of central degradation pathways.

The enzymatic processing of a set of seven different lignin preparations was studied using *Pseudomonas fluorescens* Dyp1B or *Sphingobacterium* sp. T2 manganese superoxide dismutase [57]. The highest yield of small-molecule products was observed using organosolv lignin (alkaline, ammonia, and dioxasolv) containing high β-O-4 content, implying that this is an important determinant for lignin processing by enzymes [57]. To metabolize such a complex and heterogeneous substrate like lignin, a group of enzymes may be needed for complete degradation, including accessory enzymes mentioned in Section 8.1.6. A combination of β-etherase enzymes LigE, LigF, and LigG was found to be effective on beech wood lignin treated with *Trametes versicolor* laccase lcc2, giving enhanced yields of a bio-oil containing low molecular weight aromatic compounds [41]. A further study using a combination of β-etherase enzymes LigD, LigE, LigF, LigN, and NaGSTNU, together with cofactor recycling enzymes, was found to be effective in releasing G, S, and tricin units from polymeric lignin substrates [42].

In order to test enzyme combinations in a more systematic way, two colorimetric assays have been developed to detect degradation products released upon incubation of polymeric lignin with combinations of lignin-oxidizing enzymes and accessory enzymes that can be applied in a microtiter plate format [58]. The assays showed that combinations of oxidizing with accessory enzymes lead to a 4–10-fold enhancement in release of low molecular weight product from technical lignins. The highest enhancement in generation of lignin-degradation products was observed upon combinations of *Agrobacterium* LigE with DyP-type peroxidases from *Pseudomonas fluorescens*, *Comamonas testosteroni,* and *Agrobacterium* sp. [58].

8.2 Pathways for Microbial Lignin Degradation

Microbial lignin degradation is believed to be closely linked with degradation pathways for key intermediates vanillic acid, protocatechuic acid (PCA), and syringic acid [59], since (i) the genomes of bacterial lignin degraders commonly contain the β-ketoadipate pathway for degradation of PCA [60]; (ii)vanillic acid has been commonly observed as a metabolite in bacterial lignin degraders [59], and those degraders can utilize vanillic acid for growth on minimal media [61]; (iii) metabolic engineering studies on bacteria such as *Pseudomonas putida* KT2440 and *Rhodococcus jostii* RHA1 (to be discussed in Section 8.3) have yielded bioproducts that are consistent with these pathways.

The pathways by which polymeric lignin is converted into monocyclic aromatic compounds are not fully understood, but studies on LiPs and multi-copper oxidases (Section 8.1) indicate that oxidative cleavage of polymeric lignin occurs extracellularly to give oxidized monomers, dimers, and oligomers containing the different substructures found in polymeric lignin, as shown in Figure 8.6. These lignin fragments must then be imported into the cytoplasm and converted into aromatic monomers. Extensive studies on the catabolic pathways for dimeric lignin model compounds by *Sphingobium* SYK-6 have led to the elucidation of several biochemical pathways in this organism [62], including a pathway for β-aryl ether cleavage involving β-etherase enzymes [35–40], shown in Figure 8.4, and an oxidative cleavage pathway for biphenyl dicarboxylic acid DDVA, shown in Figure 8.6. *Rhodococcus*

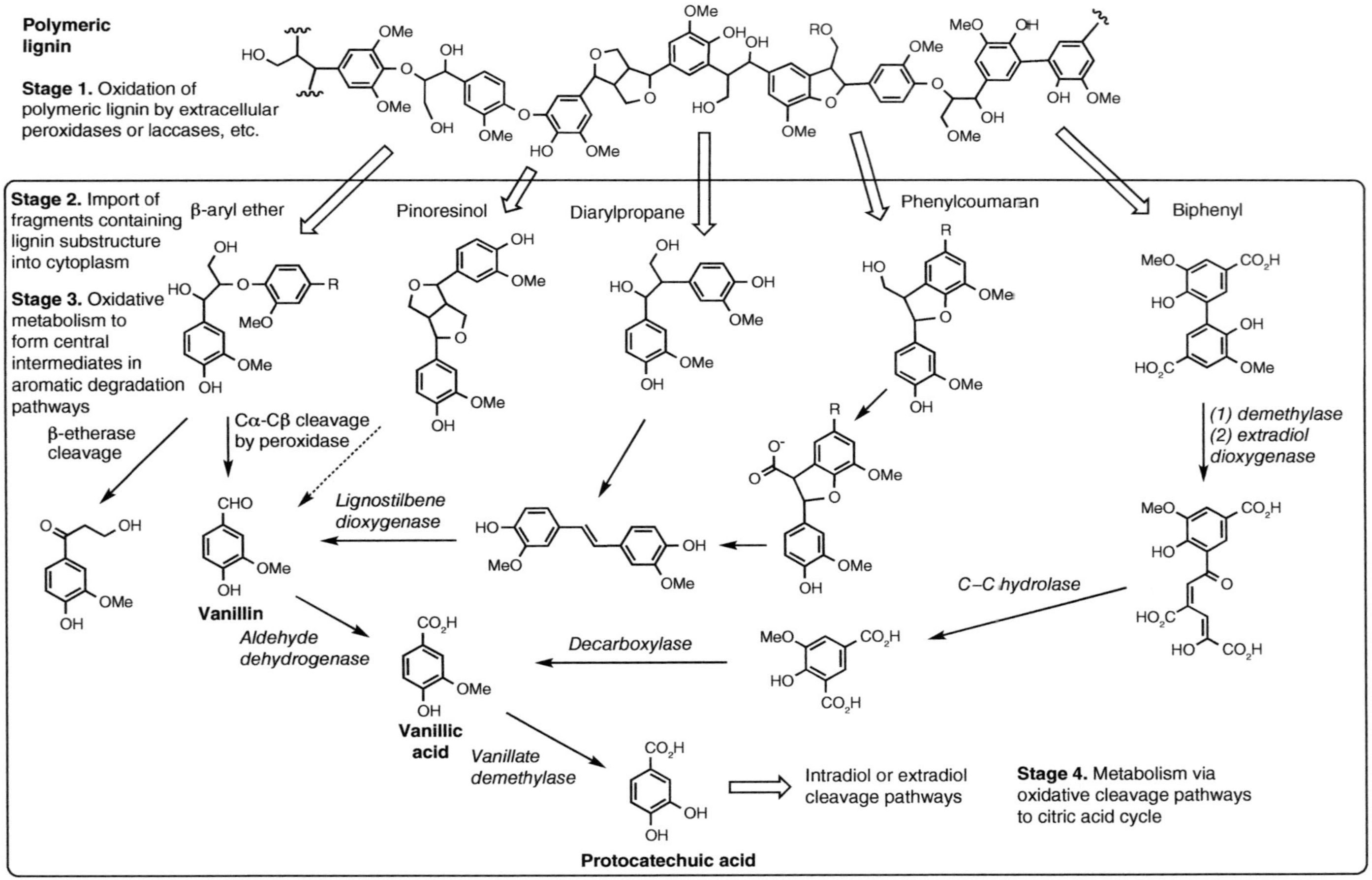

Figure 8.6 Microbial pathways for conversion of polymeric lignin, via oxidized lignin fragments, into vanillic acid and protocatechuic acid.

jostii RHA1 DypB is known to cleave a β-aryl ether lignin model compound via oxidative Cα–Cuβ cleavage [14], generating vanillin, which is oxidized in *R. jostii* RHA1 to vanillic acid, then demethylated to form PCA [63]. The phenylcoumaran (β-5) lignin fragment is metabolized by *Sphingobium* SYK-6 via oxidation of the γ-alcohol functional group to a carboxylic acid, followed by decarboxylation, generating lignostilbene, which is converted to vanillin by non-heme iron-dependent lignostilbene dioxygenase [64]. The diarylpropane (β-1) fragment can also be converted to lignostilbene in *Novosphingobium aromaticivorans* by a novel lyase enzyme LsdE, which generates formaldehyde as a by-product [65]. Metabolism of the pinoresinol (β-β) lignin dimer occurs via reductive cleavage of the tetrahydrofuran ring in *Sphingobium* SYK-6 [66], but via oxidation of the α-position and ring-opening to a benzylic ketone in *Pseudomonas* sp. SG-MS2 [67].

The majority of bacterial lignin degraders then metabolize PCA via the β-keto-adipate pathway [68], which commences via intradiol oxidative cleavage, catalyzed by protocatechuate 3,4-dioxygenase, as shown in Figure 8.7, and which then leads to the citric acid cycle. The majority of actinobacteria and α- and β-proteobacteria utilize the β-ketoadipate pathway [59], for which a gene cluster can be readily identified in their genomes [60]. However, *Sphingobium* SYK-6 and some other aromatic degraders utilize protocatechuate 4,5-dioxygenase to catalyze extradiol cleavage of PCA, followed by a meta-cleavage pathway, shown in Figure 8.7 [59].

8.3 Microbial Lignin Bioconversion by Engineered Bacteria

8.3.1 Microbial Hosts for Lignin Bioconversion

A promising strategy for valorization of lignin with relatively low costs and mild process conditions is the use of microbial lignin degraders, and further modifying them through genetic and metabolic engineering [69]. In nature, white-rot fungi are among the most conspicuous lignin degraders and are capable of secreting enzymes with high catalytic efficiencies to promote lignin degradation. However, metabolic engineering of this group remains a challenge due to their long cultivation times, lack of available molecular biology tools, low transformation and recombination efficiencies, complex genomes, and a still elusive understanding of their complete metabolic pathways [70]. Efforts to develop genetic tools and model organisms for metabolic engineering of white-rot fungi have been reviewed in a recent article [71].

Among bacterial lignin degraders, species belonging to the *Pseudomonas* genus are well known for their ability to degrade aromatic compounds, including lignin-derived monomers such as hydroxycinnamates, vanillate, benzoate, coniferyl alcohol, and phenylacetate, among others [72]. In addition, species such as *P. putida* are also capable of degrading polymeric lignin [73], although there seems to be a preference to consume lower molecular weight lignin first [74]. A recent study has shown that *P. putida* secretes an exoproteome rich in enzymes related to aromatic

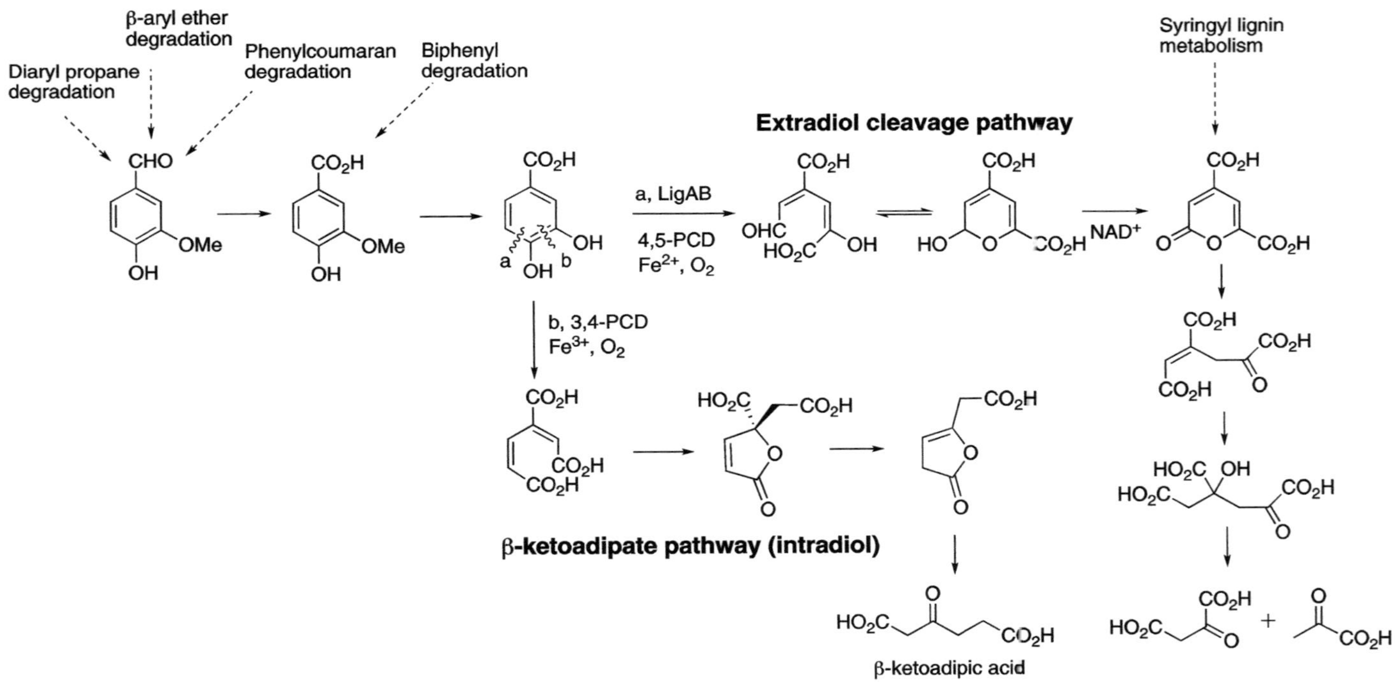

Figure 8.7 β-Ketoadipate pathway (intradiol) and extradiol cleavage pathway for metabolism of protocatechuic acid.

compounds catabolism via outer membrane vesicles, thus providing evidence for extracellular lignin catabolism in this species [75]. A widely used strain is *P. putida* KT2440, which is derived from the toluene-degrading strain *P. putida* mt-2 via deletion of the catabolic TOL plasmid pWW0 [76]. The *P. putida* KT2440 strain is a strong degrader of lignin and aromatic compounds, possesses inherently high tolerance toward oxidative stress, and, importantly, is a "generally regarded as safe" (GRAS) microorganism, favoring its use as a host for genetic engineering. Many high-value compounds that can be produced from lignin or lignin-derived monomers by *Pseudomonas* spp. are intermediates of the β-ketoadipate pathway, notably *cis,cis*-muconic acid (MA), as will be discussed in Section 8.3.2.4. The well-developed genetic toolbox in *P. putida* also allows using biosensors for screening or pathway regulation purposes. Using promoter and protein evolution, Jha et al. adapted a PCA-catechol biosensor from *Escherichia coli* for use in *P. putida*, allowing direct monitoring of lignin-derived aromatic degradation [77]. Similarly, a 4-hydroxybenzoate (4-HB) biosensor was established in *P. putida* and used to monitor the conversion of chorismate to 4-HB by chorismate pyruvate-lyase (UbiC) variants, which had been engineered to overcome product inhibition [78].

The *Rhodococcus* genus contains Gram-positive actinobacteria, with a coccus or rod-like cell morphology, that possess a remarkably versatile metabolism: their large genomes encode a myriad of catabolic pathways, which are highly redundant and regulated by sophisticated mechanisms, many of which are still unknown [79]. *Rhodococcus* strains have the capacity to degrade a wide range of compounds and to withstand harsh environmental conditions, which place these bacteria as ideal candidates for applications in bioremediation and industrial processes [80–82]. Since it was first identified as a lignin degrader, molecular genetics and biochemicals studies with the species *R. jostii* RHA1 have elucidated some of the enzymes involved in lignin degradation, such as LiPs belonging to the DyPs family, accessory enzymes, and at least 3 multi-copper oxidases [14]. DyPB from *R. jostii* was the first bacterial enzyme to be identified and characterized for oxidation of lignin model compounds and polymeric lignin. Notably, DypB is of fundamental importance for the degradation of lignin due to lack of this activity in *R. jostii* RHA1 Δ*dypB* mutants [14]. The degradation of lignin-derived aromatics is relatively well understood due to the availability of annotated genome sequences [83] as well as experimental elucidation of key pathways, such as the catabolism of hydroxycinnamates, hydroxyquinol, and aryl C2 lignin fragments [84, 85]. Compared to *Pseudomonas*, there are fewer molecular biology tools available for metabolic engineering of *Rhodococcus* species. Some plasmids have been developed for use in *Rhodococcus*, many of them *E. coli*-*Rhodococcus* "shuttle-vectors" capable of replicating in both species; for instance, the pTip/pNit expression vector series, which contain either constitutive (pNit) or thiostrepton-inducible promoters (pTip) [86]. Genetic modification of *Rhodococcus* is hindered by low transformation and recombination efficiencies.

In addition to *Pseudomonas* and *Rhodococcus*, *Amycolatopsis* is another lignin-degrading bacterial genus with potential for metabolic engineering toward lignin valorization. Genome analysis of *Amycolatopsis* sp. 75iv2 (ATCC 39116) revealed the presence of 3 DyPs (named Dyp1-3) and a large set of heme-containing peroxidases and multicopper laccases [87, 88]. The capacity of *Amycolatopsis* sp. 75iv2 to convert aromatic monomers was also shown for the bioproduction of vanillin from ferulic acid – by combining the effects of vanillin dehydrogenase (*vdh*) knockout and constitutive expression of feruloyl-CoA synthetase and enoyl-CoA aldolase/lyase, a final vanillin concentration of 22.3 $g\,l^{-1}$ was achieved in a 2-L bioreactor using a fed-batch strategy [89]. *Amycolatopsis* strains have been used to produce *cis,cis*-MA from lignin hydrolysate, as will be discussed in Section 8.3.2.4.

The catabolism of lignin-derived dimers in *Sphingobium* sp. SYK-6 has been extensively characterized by Masai and coworkers, and pathways for the degradation of lignin dimers representative of several lignin linkages have been identified [62]. Genetically modified *Sphingobium* sp. SYK-6 has been used for valorization of chemically depolymerized softwood lignin to *cis,cis*-MA via vanillic acid [90].

Corynebacterium glutamicum is an industrial powerhouse that has been studied for over 60 years, when it was identified as a natural L-glutamate producer [91]. While not capable of depolymerizing lignin, it can assimilate wide array of aromatic compounds, such as 4-hydroxybenzoate, benzoate, benzyl alcohol, 2,4-dihydroxybenzoate, ferulate, *p*-coumarate, caffeate, 4-hydroxyphenylpropionic acid, gentisate, naphthalene, phenol, protocatechuate, and resorcinol [92, 93]. *C. glutamicum* has been successfully applied for valorization of catechol and depolymerized lignin into *cis,cis*-MA and 4-vinylphenol (4-VP), which will be discussed in Section 8.3.2.4.

Escherichia coli, the archetypical bacterial expression system, has also been applied to produce aromatics from lignin-derived compounds, despite its inability to degrade lignin and limited aromatic catabolic pathways. Two studies have applied *E. coli* cells overexpressing phenolic acid decarboxylases (PAD) from *Bacillus* sp. to convert ferulic acid into 4-vinyl guaiacol [94, 95]. In addition, *E. coli* cells overexpressing an aldo-keto reductase from *Coptotermes gestroi* (Cg-AKR) and a carboxylic acid reductase from *Nocardia iowensis* (Ni-CAR) were used to produce coniferyl alcohol from pure and wheat straw-derived ferulic acid [96].

8.3.2 Conversion of Polymeric Lignin and Depolymerized Lignin Hydrolysates into High-Value Products

8.3.2.1 Vanillin

Vanillin is a highly valuable aroma molecule with wide applications in food, flavoring, and cosmetic industries [97]. Naturally extracted from the cured pods of orchid *Vanilla planifolia*, it can also be produced via biotransformation, the product of which is also labeled as "natural vanillin." While vanillin can be produced enzymatically from ferulic acid, there are several advantages associated with the use of engineered lignin-degrading microbes for consolidated bioprocessing.

Genetically engineered *Rhodococcus jostii* RHA1 has been applied for consolidated bioproduction of vanillin from milled wheat straw [98]. In a previous work, the vanillin dehydrogenase gene *vdh* (RHA1_ro02986) was identified as responsible for the oxidation of vanillin into vanillic acid, which is then demethylated by *vanA* to PCA, followed by ring fission toward the central metabolism [99]. Sainsbury and colleagues showed that the Δ*vdh* knockout strain, RHA045, was unable to grow on 1 mM vanillin as sole carbon source but could still grow on 1 mM vanillic acid as well as on minimal media supplemented with 1% milled wheat straw lignocellulose [98]. Cultivation of RHA045 in 2.5% milled wheat straw led to an accumulation of 96 mg l^{-1} of vanillin after six days, in addition to 53 mg l^{-1} benzaldehyde (after five days), consistent with the degradation of both G- and H-lignin units present in wheat straw lignocellulose (see Figure 8.8). Ferulic acid (23–86 mg l^{-1}, four to six days) and vanillic acid (5–120 mg l^{-1}, four to six days) were also detected as major metabolites in the cultivations containing the mutant strain, but not the wild type, although the reasons for this are not entirely clear. Further studies showed that RHA045 accumulated as much as 95 and 243 mg l^{-1} of syringaldehyde upon cultivation in minimal media supplemented with different hardwood organosolv lignins, denoting the capacity of this microorganism to process S-lignin units as well [57].

8.3.2.2 Polyhydroxyalkanoates

Medium-chain-length polyhydroxyalkanoates (*mcl*-PHAs) are naturally produced by some bacteria, notably Pseudomonads, as storage molecules. Using acetyl-CoA as substrate, these compounds are generated via the fatty acid biosynthesis pathway and are considered high-value polymers due to their use as bioplastics and adhesives. In addition, they can be further valorized after depolymerization to chemical precursors and methyl-ester-based fuels, such as biodiesel. Linger and collaborators demonstrated the production of *mcl*-PHAs from lignin-derived carbon by *Pseudomonas putida* KT2440 [100]. Alkaline pretreated lignin (APL) from corn stover was used as substrate, which consists mainly of a mixture of *p*-coumaric acid, ferulic acid, vanillic acid, acetate, and trace amounts of glucose. As much as 32% cell dry weight (CDW; 0.252 g l^{-1}) of *mcl*-PHAs were accumulated from APL, which were then extracted and thermally depolymerized to generate alkenoic acids and oxygenated to produce fuel-rate hydrocarbons [100].

A further study sought to further optimize *P. putida* KT2440 to increase production of *mcl*-PHAs from lignin [101]. *P. putida* was genetically modified to favor *mcl*-PHA accumulation by deleting the PHA depolymerase gene (*phaZ*) as well as β-oxidation genes *fadBA1* and *fadBA2*, using counter-selection vector pK18mobsacB. Furthermore, carbon flux was increased toward *mcl*-PHA formation by overexpressing the PHA-biosynthesis gene *phaG*, *alkK*, *phaC1*, and *phaC2*, under the control of the strong constitutive P_{tac} promoter, and inserting the overexpression cassette into the chromosome (in place of *phaZ*). The final strain, named *P. putida* AG2162, was grown in depolymerized lignin liquor, which was composed primarily of *p*-coumaric acid and smaller amounts of ferulic acid, as well as some oligomeric lignin, and the production of *mcl*-PHAs was measured. Compared to the wild-type, the AG2162 strain had a 100% increase in *mcl*-PHA yield, and 3.3-fold increase in titer (116 ± 35 mg l^{-1}) as well as an increased lignin consumption [101].

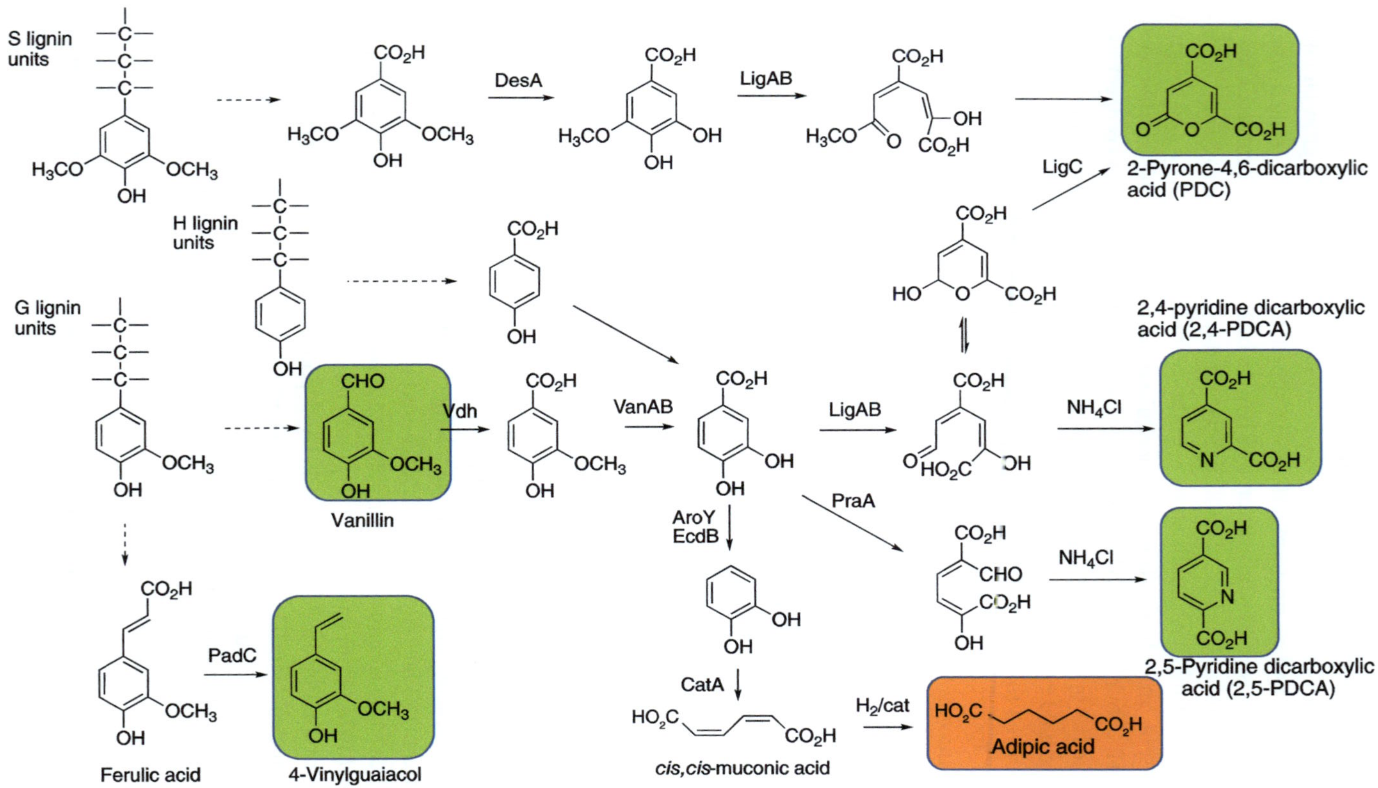

Figure 8.8 Bioproducts formed by microbial bioconversion of polymeric lignin or chemically pretreated lignin. Aromatic products are shown in green, and nonaromatic product are shown in orange.

8.3.2.3 Triacylglycerol Lipids

Similar to PHAs, triacylglycerol lipids can be produced by bacteria as internal carbon storage compounds from acetyl-CoA. Among oleaginous lignin-degrading bacteria, *Rhodococcus* sp. have been extensively studied due to their capacity to accumulate high amounts of lipids from lignin and lignin-derived aromatics. In the past decade, studies have focused on bacterial conversion of aromatic monomers, as well as lignin streams, into triacylglycerols (TAGs). Kosa and Ragauskas showed that *R. opacus* strains DSM 1069 and PD630 were capable of accumulating TAGs from 4-hydroxybenzoic acid (4-HBA) and vanillic acid [102]. *R. opacus* PD630 accumulated a maximum of 0.59 and 0.37 g lipids/g cells from 4-HBA and vanillic acid, respectively, after 12 hours, while strain DSM 1069 accumulated 0.44 and 0.67 g lipids/g cells from 4-HBA and vanillic acid, after 36 and 12 hours, respectively [102]. A further study showed that *R. opacus* DSM 1069 accumulated 67 $mg\,l^{-1}$ of TAGs after 36 hours growth from oxygen-pretreated Kraft lignin [103]. Another *R. opacus* strain, named NRRL B-3311, accumulated up to 32 $mg\,l^{-1}$ lipids after 72 hours, while consuming as much as 20% of the total ammonia fiber expansion corn stover lignin used as substrate [104].

8.3.2.4 *Cis,cis*-Muconic Acid (MA)/Adipic Acid

Cis,cis-MA is a product of catechol catabolism and presents high commercial relevance both as platform molecule for terephtalic and adipic acid productionandas a polymer building block itself [105]. Accumulation of MA from hydroxycinnamic and vanillic acids was achieved in *Pseudomonas putida* KT2440, as shown in Figure 8.8, via integration of the PCA branch by heterologous expression of a PCA decarboxylase (*aroY*) and knocking out *pcaHG*, which promotes the degradation of PCA [106]. To date, the highest MA titer obtained from ferulic and *p*-coumaric acids was 50 $g\,l^{-1}$, using a *P. putida* KT2440 strain that also overexpressed two *aro Y*-associated proteins (EcdBD), and had the global regulator of carbon catabolite repression Crc knocked out [107]. An issue remained that *P. putida* KT2440 was sensitive to high concentrations of catechol. Another study set out to address this bottleneck by overexpressing the native catechol 1,2-dioxygenases *catA* and *catA2*, responsible for converting catechol in MA, under the control of ultra-strong promoter P_{cat} [108]. This led to increased catechol tolerance due to prompt conversion into MA, which in turn acted as an activator of promoter P_{cat}, promoting positive feedback of the system. The authors introduced the phenol hydroxylase genes *dmpKLMNOP* from *P. putida* CF600 into the genome of their novel *P. putida* strain, thus further broadening the substrate range for MA production to include phenol and cresols. Catechol tolerance and strain robustness were further improved by optimizing bioreactor fermentations, and closely monitoring catechol/MA levels, cell growth, and ATP levels. The optimized process resulted in an impressive MA production of 64.2 $g\,l^{-1}$ from catechol in 1-L bioreactor, and the conditions were adapted to a pilot-scale 72-L bioreactor, leading to MA production in the kilogram scale. Furthermore, production of MA was demonstrated from a hydrothermally depolymerized pine wood lignin, which contained mostly catechol, phenol, and cresols; overall, 13 $g\,l^{-1}$ of MA were accumulated as well as small amounts of 3-methyl MA (from *o*- and *m*-cresols), which were then hydrogenated to adipic acid/ 3-methyladipic acid, respectively. Finally, these compounds were polymerized, thus establishing the first process chain from a lignin hydrolysate to nylon [108].

MA has also been accumulated in *Amycolatopsis* sp., by gene deletion of two genes encoding muconate cycloisomerases, leading to a final *cis,cis*-MA concentration of 3.1 $g\,l^{-1}$ from guaiacol. Furthermore, the study also demonstrated the conversion of a lignin hydrolysate (softwood lignin that was submitted to hydrothermal pretreatment), containing mainly guaiacol, catechol, phenol, and *o*-cresol, into 0.18 mM MA with a 72 % molar yield by the mutant *Amycolatopsis* sp. strain [109]. *Corynebacterium glutamicum* has been successfully applied for valorization of catechol and depolymerized lignin into MA, via deletion of *catB* (muconate cycloisomerase) and overexpression of *catA* (catechol 1,2-dioxygenase) [110]. This led to an impressive MA production from catechol of 85 $g\,l^{-1}$ after 60 hours as well as 1.8 $g\,l^{-1}$ of MA from a hydrothermally depolymerized softwood lignin [110].

8.3.2.5 Pyridine- and Pyrone-Dicarboxylic Acids

Pyridine-dicarboxylic acids (PDCAs) were first produced in *Rhodococcus jostii* RHA1 via 2,3- or 4,5-*meta*-cleavage of protocatechuate by *praA* or *ligAB* gene products, followed by cyclization of the extradiol ring fission product with ammonium chloride, as shown in Figure 8.8 [111]. Mycroft and collaborators have achieved the first consolidated bioproduction of 3,4- and 4,5-PDCA directly from wheat straw and Kraft lignin by heterologous overexpression of *ligAB* (protocatechuate 4,5-dioxygenase from *Sphingomonas paucimobilis*) or *praA* (protocatechuate 3,4-dioxygenase from *Paenibacillus* sp. JJ-1b) using the thiostrepton-inducible plasmid pTip-QC2. The ligAB-expressing strain presented a total 2,4-PDCA production of 125 $mg\,l^{-1}$ from 1% wheat straw after nine days, and 53 $mg\,l^{-1}$ from 0.5% Kraft lignin after seven days, in 2.5-l bioreactor cultivations. Similarly, 2,5-PDCA was produced to a maximum concentration of 106 $mg\,l^{-1}$ after 9 days from 1% wheat straw [111].

A further study considerably improved 2,4-PDCA titers by further genetic modification of *R. jostii* RHA1 [112]. Spence and collaborators promoted protocatechuate accumulation and re-routing to *meta*-cleavage pathway by knocking out the *ortho*-cleavage degradation genes *pcaHG*. Furthermore, the *ligAB* gene was inserted into the chromosome rather than relying on plasmid expression, and lignin degradation was enhanced by heterologous expression of the *dyp2* peroxidase from *Amycolatopsis* sp. 75iv2 using the pTip-QC2 vector. The final engineered strain presented considerably higher and faster 2,4-PDCA production compared to the previous study, reaching a maximum titer of 330 $mg\,l^{-1}$ after 40 hours from minimal media containing wheat straw, and 240 $mg\,l^{-1}$ from Green Value Protobind lignin (GVPL), a real soda lignin obtained from wheat straw/sarkanda grass biomass [112]. Synthetic pyridine-dicarboxylic acids have been successfully converted into PDCA-containing plastics, which show similar physical properties to conventional oil-based plastics containing terephthalic acid [113].

Pyrone dicarboxylic acids can also be obtained from lignin via protocatechuate degradation. In *Novosphingobium aromaticivorans* DSM12444, aromatic compounds derived from G- and H-lignin can be degraded by 4,5-*meta*-cleavage of protocatechuate, catalyzed by extradiol dioxygenase LigAB, leading to production of a semialdehyde precursor (4-carboxy-2-hydroxy-*cis,cis*-muconate-6-semialdehyde, CHMS). CHMS undergoes spontaneous cyclization into the hemiacetal, followed by oxidation by dehydrogenase LigC into 2-pyrone 4,6-dicarboxylate (PDC) [114]. PDC is then hydrolyzed by LigI into 4-oxalomesaconate (OMA), which is further processed toward acetyl-CoA and pyruvate [114]. Syringic acid is demethylated to 3-methylgallate, which in turn is also cleaved by LigAB to yield 4-carboxy-2-hydroxy-6-methoxy-6-oxohexa-2,4-dienoate (CHMOD). CHMOD can be directly converted to OMA via sequential action of a *cis,*trans-isomerase (DesD) and methylesterase (DesC). Perez and collaborators have harnessed this system to produce PDC from S-, G-, and H-lignin units by knocking out *ligI*, *desC*, and *desD*, leading to the accumulation of PDC precursors, as shown in Figure 8.8. Interestingly, PDC formation from CHMOD was observed, although it remains unclear whetherby non-enzymatic means or an undisclosed pathway. The engineered strain was able to accumulate 0.49 mM PDC from a chemically depolymerized poplar lignin stream [114].

8.3.2.6 Substituted Styrenes

Styrene derivatives such as 4-vinyl guaiacol (4-VG) are high-value molecules with wide applications in pharmaceutical and food industries [115]. 4-VG can be obtained enzymatically by the decarboxylation of ferulic acid, via cofactor-dependent or independent routes. Likewise, *p*-coumaric and caffeic acids can also be decarboxylated to their corresponding styrenes. A recent work has shown the application of a cofactor-independent phenolic acid decarboxylase (PadC) from *Bacillus subtilis* for production of 4-VG and 4-vinylcatechol (4-VC) from hydroxycinnamic acids derived from the biological degradation of lignocellulose and lignin, as shown in Figure 8.8 [116]. The host of choice was *Pseudomonas putida* KT2440 due to its lignin-degrading capacities and ease for genetic modification. Ferulic acid catabolism in *P. putida* goes through a CoA-dependent β-oxidative pathway, with the sequential action of a feruloyl-CoA synthetase – generating feruloyl-CoA – (encoded by *fcs*) and an enoyl-CoA hydratase/lyase (encoded by *ech*), yielding vanillin and acetyl-CoA. The operon is regulated by a MarR-type repressor encoded by *ferC*, which is derepressed upon feruloyl-CoA binding.

To promote accumulation of ferulic acid from biomass and lignin, the *ech* gene was knocked out, and the PadC gene was inserted in the same genomic location – effectively disrupting the pathway and simultaneously promoting inducible expression of PadC upon ferulic acid release and feruloyl-CoA formation [116]. Cultivation of *P. putida* Δ*ech* in minimal media supplemented with various lignocellulosic biomasses, such as wheat straw, sugarcane bagasse, sugarcane straw, and miscanthus as well as soda lignin (GVPL) led to effective accumulation of ferulic and *p*-coumaric acids, with GVPL presenting the highest contents of released hydroxycinnamic acids. Various concentrations of GVPL (5–30%) were tested as substrate for cultivation with *P. putida* Δ*ech*::PadC, leading to a maximum production of $62.4 \pm 1.9\,mg\,l^{-1}$ 4-VG from 10% GVPL after 18 hours. The 4-VG and 4-VC bioproducts were extracted from the culture medium and used as substrate for polymerization catalyzed by *Trametes*

Table 8.1 Bioproducts from lignin breakdown using engineered microbial strains.

Microbe	Genetic modifications	Feedstock	Bioproduct	Titre (mg l^{-1})	Ref
Rhodococcus jostii RHA1	Δ*vdh*	Wheat straw	vanillin	96	98
	pcaHG::ligAB dyp2	Wheat straw	2,4-PDCA	330	112
	ligAB	Wheat straw	2,5-PDCA	106	111
Rhodococcus opacus DSM1069	wild-type	Kraft lignin	Triglyceride lipids	67	103
Amycolatopsis sp.	Δ*catB12*	HPT softwood	*cis,cis*-muconate	26	109
Corynebacterium glutamicum	*catA* Δ*catB*	HPT softwood	*cis,cis*-muconate	1,800	110
Pseudomonas putida KT2440	wild-type	APL	PHAs	252	100
	aroY ecdBD Δ*pcaGH* Δ*crc*	HPT pine	*cis,cis*-muconate	13,000	108
	Δ*ech::padC*	Soda lignin	4-vinylguaiacol	62	116
Novosphingobium aromaticivorans	Δ*ligI* Δ*desC* Δ*desD*	CPT poplar	PDC	90	114

PDCA, pyridine-dicarboxylic acid; PDC, pyrone-dicarboxylic acid; APL, alkali pretreated lignin; HPT, hydrothermally pretreated; CPT, chemically pretreated.

versicolor laccase, forming "biopolystyrenes" which could have further industrial applications [116]. Table 8.1 summarizes the bioproducts formed from polymeric or chemically pretreated lignin from different engineered microbes.

In conclusion, there has been considerable progress in the understanding of bacterial enzymes and metabolic pathways for lignin degradation in the past 10 years. Metabolic engineering of selected bacterial hosts for bioconversion of polymeric lignin or chemically pretreated lignin into high-value products has been successful, and efforts to develop these microbial biotransformations into processes that could be used commercially are in progress.

Acknowledgments

Research in the author's research group has been supported by research grants BB/T010622/1 and BB/P01738X/1 from BBSRC, and a PhD studentship (to A.A.) funded by Northern Border University, Saudi Arabia.

References

1 Kirk, T.K. and Farrell, R.L. (1987). Enzymatic 'combustion': the microbial degradation of lignin. *Annual Review of Microbiology* 41: 465–505.

2 Ten Have, R. and Teunissen, P.J.M. (2001). Oxidative mechanisms involved in lignin degradation by white-rot fungi. *Chemical Reviews* 101: 3397–3413.

3 Wong, D.W.S. (2009). Structure and action mechanism of lignolytic enzymes. *Applied Biochemistry and Biotechnology* 157: 174–209.

4 Martinez, A.T., Speranza, M., Ruiz-Duenas, F.J. et al. (2005). Biodegradation of lignocellulosics: microbial, chemical, and enzymatic aspects of the fungal attack of lignin. *International Microbiology* 8: 195–204.

5 Edwards, S.L., Raag, R., Wariishi, H. et al. (1993). Crystal structure of lignin peroxidase. *Proceedings of the National Academy of Sciences of the United States of America* 90: 750–754.

6 Sundaramoorthy, M., Kishi, K., Gold, M.H., and Poulos, T.L. (1994). The crystal structure of manganese peroxidase from *Phanerochaete chrysosporium* at 2.06 Å resolution. *The Journal of Biological Chemistry* 269: 32759–32767.

7 Tuor, U., Wariishi, H., Schoemaker, H.E., and Gold, M.H. (1992). Oxidation of phenolic arylglycerol beta-aryl ether lignin model compounds by manganese peroxidase from *Phanerochaete chrysosporium*: oxidative cleavage of an alpha-carbonyl model compound. *Biochemistry* 31: 4986–4995.

8 Martinez, D., Larrondo, L.F., Putnam, N. et al. (2004). Genome sequence of the lignocellulose degrading fungus *Phanerochaete chrysosporium* strain RP78. *Nature Biotechnology* 22: 695–700.

9 Doyle, W.A., Blodig, W., Veitch, N.C. et al. (1998). Two substrate interaction sites in lignin peroxidase revealed by site-directed mutagenesis. *Biochemistry* 37: 15097–15105.

10 Hammel, K.E., Tien, N., Kalyanaraman, B., and Kirk, T.K. (1985). Mechanism of oxidative Cα-Cβ cleavage of a lignin model dimer by *Phanerochaete chrysosporium* ligninase: stoichiometry and involvement of free radicals. *The Journal of Biological Chemistry* 260: 8348–8353.

11 Crawford, D.L., Pometto, A.L., and Crawford, R.L. (1983). Lignin degradation by *Streptomyces viridosporus*: isolation and characterization of a new polymeric lignin degradation intermediate. *Applied and Environmental Microbiology* 45: 898–904.

12 Ahmad, M., Taylor, C.R., Pink, D. et al. (2010). Development of novel assays for lignin degradation: comparative analysis of bacterial and fungal lignin degraders. *Molecular BioSystems* 6: 815–821.

13 Salvachua, D., Karp, E.M., Nimlos, C.T. et al. (2015). Towards lignin consolidated bioprocessing: simultaneous lignin depolymerization and product generation by bacteria. *Green Chemistry* 17: 4951–4967.

14 Ahmad, M., Roberts, J.N., Hardiman, E.M. et al. (2011). Identification of DypB from *Rhodococcus jostii* RHA1 as a lignin peroxidase. *Biochemistry* 50: 5096–5107.

15 Roberts, J.N., Singh, R., Grigg, J.C. et al. (2011). Characterization of dye-decolorizing peroxidases from *Rhodococcus jostii* RHA1. *Biochemistry* 50: 5108–5119.

16 Rahmanpour, R. and Bugg, T.D.H. (2015). Characterisation of Dyp-type peroxidases from *Pseudomonas fluorescens* Pf-5: oxidation of Mn(II) and polymeric lignin by Dyp1B. *Archives of Biochemistry and Biophysics* 574: 93–98.

17 Brown, M.E., Barros, T., and Chang, M.C.Y. (2012). Identification and characterization of a multifunctional dye peroxidase from a lignin-reactive bacterium. *ACS Chemical Biology* 7: 2074–2081.

18 Sugano, Y., Muramatsu, R., Ichiyanagi, A. et al. (2007). DyP, a unique dye-decolorizing peroxidase, represents a novel heme peroxidase family. *The Journal of Biological Chemistry* 282: 36652–36658.

19 Singh, R., Grigg, J.C., Qin, W. et al. (2013). Improved manganese-oxidizing activity of DypB, a peroxidase from a lignolytic bacterium. *ACS Chemical Biology* 8: 700–706.

20 Brissos, V., Tavares, D., Sousa, A.C. et al. (2017). Engineering a bacterial DyP-type peroxidase for enhanced oxidation of lignin-related phenolics at alkaline pH. *ACS Catalysis* 7: 3454–3465.

21 Pour, R.R., Ehibhatiomhan, A., Huang, Y. et al. (2019). Protein engineering of *Pseudomonas fluorescens* peroxidase Dyp1B for oxidation of phenolic and polymeric lignin substrates. *Enzyme and Microbial Technology* 123: 21–29.

22 Camarero, S., Garcia, O., Vidal, T. et al. (2004). Efficient bleaching of non-wood high-quality paper pulp using laccase-mediator system. *Enzyme and Microbial Technology* 35: 113–120.

23 Arora, D.S. and Sharma, R.K. (2010). Ligninolytic fungal laccases and their biotechnological applications. *Applied Biochemistry and Biotechnology* 372: 1760–1788.

24 Giardina, P., Faraco, V., Pezzella, C. et al. (2009). Laccases: a never-ending story. *Cellular and Molecular Life Sciences* 67: 369–385.

25 Munk, L., Sitarz, A.K., Kalyani, D.C. et al. (2015). Can laccases catalyze bond cleavage in lignin? *Biotechnology Advances* 33: 13–24.

26 Bento, I., Silva, C., Chen, Z. et al. (2010). Mechanisms underlying dioxygen reduction in laccases. Structural and modelling studies focusing on proton transfer. *BMC Structural Biology* 10: 28–42.

27 Jönsson, L., Sjöström, K., Häggström, I., and Nyman, P. (1995). Characterization of a laccase gene from the white-rot fungus *Trametes* versicolor and structural features of basidiomycete laccases. *Biochimica et Biophysica Acta, Protein Structure and Molecular Enzymology* 1251: 210–215.

28 Niku-Paavola, M.L., Karhunen, E., Kantelinen, A. et al. (1990). The effect of culture conditions on the production of lignin modifying enzymes by the white-rot fungus *Phlebia radiata*. *Journal of Biotechnology* 13: 211–221.

29 Ander, P. and Eriksson, K.-E. (1976). The importance of phenol oxidase activity in lignin degradation by the white-rot fungus *Sporotrichum pulverulentum*. *Archives of Microbiology* 109: 1–8.

30 Majumdar, S., Lukk, T., Solbiati, J.O. et al. (2014). Roles of small laccases from *Streptomyces* in lignin degradation. *Biochemistry* 53: 4047–4058.

31 Singh, R., Hu, J., Regner, M.R. et al. (2017). Enhanced delignification of steam-pretreated poplar by a bacterial laccase. *Scientific Reports* 7: 42121.

32 Granja-Travez, R.S., Wilkinson, R.C., Persinoti, G.F. et al. (2018). Structural and functional characterisation of a multi-copper oxidase CueO from lignin-degrading bacterium *Ochrobactrum* sp. reveal its activity towards lignin model compounds and lignosulfonate. *The FEBS Journal* 285: 1684–1700.

33 Strachan, C.R., Singh, R., VanInsberghe, D. et al. (2014). Metagenomic scaffolds enable combinatorial lignin transformation. *Proceedings of the National Academy of Sciences of the United States of America* 111: 10143–10148.

34 Granja-Travez, R.S. and Bugg, T.D.H. (2018). Characterisation of multi-copper oxidase CopA from *Pseudomonas putida* KT2440 and *Pseudomonas fluorescens* Pf-5: involvement in bacterial lignin oxidation. *Archives of Biochemistry and Biophysics* 660: 97–107.

35 Masai, E., Kubota, S., Katayama, Y. et al. (1993). Characterization of the Cα-dehydrogenase gene involved in the cleavage of β-aryl ether by *Pseudomonas paucimobilis*. *Bioscience, Biotechnology, and Biochemistry* 57: 1655–1659.

36 Masai, E., Katayama, Y., Nishikawa, S. et al. (1989). Detection and localization of a new enzyme catalyzing the β-aryl ether cleavage in the soil bacterium (*Pseudomonas paucimobilis*) SYK-6. *FEBS Letters* 249: 348–352.

37 Masai, E., Katayama, Y., Kubota, S. et al. (1993). A bacterial enzyme degrading the model lignin compound β-etherase is a member of the glutathione-*S*-transferase superfamily. *FEBS Letters* 323: 135–140.

38 Masai, E., Ichimura, A., Sato, Y. et al. (2003). Roles of the enantioselective glutathione *S*-transferases in cleavage of β-aryl ether. *Journal of Bacteriology* 185: 1768–1775.

39 Gall, D.L., Kim, H., Lu, F. et al. (2014). Stereochemical features of glutathione-dependent enzymes in the *Sphingobium* sp. strain SYK-6 β-aryl etherase pathway. *The Journal of Biological Chemistry* 289: 8656–8667.

40 Picart, P., Sevenich, M., Domínguez de María, P., and Schallmey, A. (2015). Exploring glutathione lyases as biocatalysts: paving the way for enzymatic lignin depolymerization and future stereoselective applications. *Green Chemistry* 17: 4931–4940.

41 Picart, P., Liu, H., Grande, P.M. et al. (2017). Multi-step biocatalytic depolymerization of lignin. *Applied Microbiology and Biotechnology* 101: 6277–6287.

42 Gall, D.L., Kontur, W.S., Lan, W. et al. (2018). In vitro enzymatic depolymerization of lignin with release of syringyl, guaiacyl, and tricin units. *Applied and Environmental Microbiology* 84: e02076–e02017.

43 Rashid, G.M., Taylor, C.R., Liu, Y. et al. (2015). Identification of manganese superoxide dismutase from Sphingobacterium sp. T2 as a novel bacterial enzyme for lignin oxidation. *ACS Chemical Biology* 10: 2286–2294.

44 Rashid, G.M., Zhang, X., Wilkinson, R.C. et al. (2018). Sphingobacterium sp. T2 manganese superoxide dismutase catalyzes the oxidative demethylation of polymeric lignin via generation of hydroxyl radical. *ACS Chemical Biology* 13: 2920–2929.

45 Urlacher, V.B. and Koschorreck, K. (2021). Pecularities and applications of aryl-alcohol oxidases from fungi. *Applied Microbiology and Biotechnology* 105: 4111–4126.

46 Viña-Gonzalez, J. and Alcalde, M. (2020). Directed evolution of the aryl-alcohol oxidase: beyond the lab bench. *Computational and Structural Biotechnology Journal* 29: 1800–1810.

47 Hernandez-Ortega, A., Ferreira, P., and Martinez, A.T. (2012). Fungal aryl-alcohol oxidase: a peroxide-producing flavoenzyme involved in lignin degradation. *Applied Microbiology and Biotechnology* 93: 1395–1410.

48 Daou, M. and Faulds, C.B. (2017). Glyoxal oxidases: their nature and properties. *World Journal of Microbiology and Biotechnology* 33: 1–11.

49 Kersten, P.J. and Kirk, T.K. (1987). Involvement of a new enzyme, glyoxal oxidase, in extracellular H_2O_2 production by *Phanerochaete chrysosporium*. *Journal of Bacteriology* 169: 2195–2201.

50 Whittaker, M.M., Kersten, P.J., Cullen, D., and Whittaker, J.W. (1999). Identification of catalytic residues in glyoxal oxidase by targeted mutagenesis. *The Journal of Biological Chemistry* 274: 36226–36232.

51 Daou, M., Piumi, F., Cullen, D. et al. (2016). Heterologous production and characterization of two glyoxal oxidases from *Pycnoporus cinnabarinus*. *Applied and Environmental Microbiology* 82: 4867–4875.

52 Kracher, D. and Ludwig, R. (2016). Cellobiose dehydrogenase: an essential enzyme for lignocellulose degradation in nature – a review. *Journal of Land Management Food and Environment* 67: 145–163.

53 Bugg, T.D.H., Williamson, J.J., and Rashid, G.M.M. (2020). Bacterial enzymes for lignin depolymerisation: new biocatalysts for generation of renewable chemicals from biomass. *Current Opinion in Chemical Biology* 55: 26–33.

54 Rahmanpour, R., King, L.D.W., and Bugg, T.D.H. (2017). Identification of an extracellular bacterial flavoenzyme that can prevent repolymerisation of lignin fragments. *Biochemical and Biophysical Research Communications* 482: 57–61.

55 Lee, C.C., Chen, H.S., Yang, C.H. et al. (2016). Modification of lignin in sugarcane bagasse by a monocopper hydrogen peroxide-generating oxidase from *Thermobifida fusca*. *Process Biochemistry* 51: 1486–1495.

56 Wei, Z., Wilkinson, R.C., Rashid, G.M.M. et al. (2019). Characterization of thiamine diphosphate dependent 4-hydroxybenzoylformate decarboxylase enzymes from *Rhodococcus jostii* RHA1 and *Pseudomonas fluorescens* Pf-5 involved in degradation of aryl C2 lignin degradation fragments. *Biochemistry* 58: 5281–5293.

57 Lancefield, C.S., Rashid, G.M., Bouxin, F. et al. (2016). Investigation of the chemocatalytic and biocatalytic valorization of a range of different lignin preparations: the importance of β-O-4 content. *ACS Sustainable Chemistry & Engineering* 4: 6921–6930.

58 Rashid, G.M. and Bugg, T.D.H. (2021). Enhanced biocatalytic degradation of lignin using combinations of lignin-degrading enzymes and accessory enzymes. *Catalysis Science & Technology* 11: 3568–3577.

59 Bugg, T.D.H., Ahmad, M., Hardiman, E.M., and Rahmanpour, R. (2011). Pathways for degradation of lignin in bacteria and fungi. *Natural Product Reports* 28: 1883–1896.

60 Granja-Travez, R.S., Persinoti, G.F., Squina, F.M., and Bugg, T.D.H. (2020). Functional genomic analysis of bacterial lignin degraders: diversity in mechanisms of lignin oxidation and metabolism. *Applied Microbiology and Biotechnology* 104: 3305–3320.

61 Taylor, C.R., Hardiman, E.M., Ahmad, M. et al. (2012). Isolation of bacterial strains able to metabolize lignin from screening of environmental samples. *Journal of Applied Microbiology* 113: 521–530.

62 Masai, E., Katayama, Y., and Fukuda, M. (2007). Genetic and biochemical investigations on bacterial catabolic pathways for lignin-derived aromatic compounds. *Bioscience, Biotechnology, and Biochemistry* 71: 1–15.

63 Chen, H.-P., Chow, M., Liu, C.C. et al. (2012). Vanillin catabolism in *Rhodococcus jostii* RHA1. *Applied and Environmental Microbiology* 113: 521–530.

64 Takahashi, K., Kamimura, N., Hishiyama, S. et al. (2014). Characterization of the catabolic pathway for a phenylcoumaran-type lignin-derived biaryl in *Sphingobium* sp. strain SYK-6. *Biodegradation* 25: 735–745.

65 Presley, G.N., Werner, A.Z., Katahira, R. et al. (2021). Pathway discovery and engineering for cleavage of a β-1 lignin-derived biaryl compound. *Metabolic Engineering* 65: 1–10.

66 Fukuhara, Y., Kamimura, N., Nakajima, M. et al. (2013). Discovery of pinoresinol reductase genes in sphingomonads. *Enzyme and Microbial Technology* 52: 38–43.

67 Shettigar, M., Balotra, S., Cahill, D. et al. (2018). Isolation of the (+)-pinoresinol-mineralizing *Pseudomonas* sp. strain SG-MS2 and elucidation of its catabolic pathway. *Applied and Environmental Microbiology* 84: e02531–e02517.

68 Harwood, C.S. and Parales, R.E. (1996). The beta-ketoadipate pathway and the biology of self-identity. *Annual Review of Microbiology* 50: 553–590.

69 Weiland, F., Kohlstedt, M., and Wittmann, C. (2022). Guiding stars to the field of dreams: metabolically engineered pathways and microbial platforms for a sustainable lignin-based industry. *Metabolic Engineering* 71: 13–41. https://doi.org/10.1016/j.ymben.2021.11.011.

70 Kijpornyongpan, T., Schwartz, A., Yaguchi, A., and Salvachúa, D. (2022). Systems biology-guided understanding of white-rot fungi for biotechnological applications: a review. *iScience* 25 (7): 104640. https://doi.org/10.1016/j.isci.

71 Bugg, T.D.H., Williamson, J.J., and Alberti, F. (2021). Microbial hosts for metabolic engineering of lignin bioconversion to renewable chemicals. *Renewable and Sustainable Energy Reviews* 152: 111674.

72 Jiménez, J.I., Miñambres, B., García, J.L., and Díaz, E. (2002). Genomic analysis of the aromatic catabolic pathways from *Pseudomonas putida* KT2440. *Environmental Microbiology* 4 (12): 824–841.

73 Salvachúa, D., Karp, E.M., Nimlos, C.T. et al. (2015). Towards lignin consolidated bioprocessing: simultaneous lignin depolymerization and product generation by bacteria. *Green Chemistry* 17 (11): 4951–4967.

74 Zong, Q.J., Xu, T., Liu, H. et al. (2022). Microbial valorization of lignin to bioplastic by genome-reduced *Pseudomonas putida*. *Frontiers in Microbiology* 13.

75 Salvachúa, D., Werner, A.Z., Pardo, I. et al. (2020). Outer membrane vesicles catabolize lignin-derived aromatic compounds in *Pseudomonas putida* KT2440. *Proceedings of the National Academy of Sciences of the United States of America* 117 (17): 9302–9310.

76 Bagdasarian, M., Lurz, R., Rückert, B. et al. (1981). Specific-purpose plasmid cloning vectors II. Broad host range, high copy number, RSF 1010-derived vectors, and a host-vector system for gene cloning in *Pseudomonas*. *Gene* 16 (1–3): 237–247.

77 Jha, R.K., Bingen, J.M., Johnson, C.W. et al. (2018). A protocatechuate biosensor for *Pseudomonas putida* KT2440 via promoter and protein evolution. *Metabolic Engineering Communications* 6: 33–38.

78 Jha, R.K., Narayanan, N., Pandey, N. et al. (2019). Sensor-enabled alleviation of product inhibition in chorismate pyruvate-lyase. *ACS Synthetic Biology* 8 (4): 775–786.

79 Yam, K.C., van der Geize, R., and Eltis, L.D. (2010). Catabolism of aromatic compounds and steroids by Rhodococcus. In: *Biology of Rhodococcus, Microbiology Monographs 16*, 1e (ed. H.M. Alvarez and A. Steinbüchel), 133–169. Berlin, Germany: Springer-Verlag.

80 Larkin, M.J., Kulakov, L.A., and Allen, C.C.R. (2005). Biodegradation and *Rhodococcus* – Masters of catabolic versatility. *Current Opinion in Biotechnology* 16 (3): 282–290.

81 Martínková, L., Uhnáková, B., Pátek, M. et al. (2009). Biodegradation potential of the genus *Rhodococcus*. *Environment International* 35 (1): 162–177.

82 van der Geize, R. and Dijkhuizen, L. (2004). Harnessing the catabolic diversity of rhodococci for environmental and biotechnological applications. *Current Opinion in Microbiology* 7 (3): 255–261.

83 McLeod, M.P., Warren, R.L., Hsiao, W.W.L. et al. (2006). The complete genome of *Rhodococcus* sp. RHA1 provides insights into a catabolic powerhouse. *Proceedings of the National Academy of Sciences of the United States of America* 103 (42): 15582–15587.

84 Otani, H., Lee, Y.E., Casabon, I., and Eltis, L.D. (2014). Characterization of *p*-hydroxycinnamate catabolism in a soil actinobacterium. *Journal of Bacteriology* 196 (24): 4293–4303.

85 Spence, E.M., Scott, H.T., Dumond, L. et al. (2020). The hydroxyquinol degradation pathway in *Rhodococcus jostii* RHA1 and *Agrobacterium* species is an alternative pathway for degradation of protocatechuic acid and lignin fragments. *Applied and Environmental Microbiology* 86 (19): 1–11.

86 Nakashima, N. and Tamura, T. (2004). Isolation and characterization of a rolling-circle-type plasmid from *Rhodococcus erythropolis* and application of the plasmid to multiple-recombinant-protein expression. *Applied and Environmental Microbiology* 70 (9): 5557–5568.

87 Brown, M.E., Walker, M.C., Nakashige, T.G. et al. (2011). Discovery and characterization of heme enzymes from unsequenced bacteria: application to microbial lignin degradation. *Journal of the American Chemical Society* 133 (45): 18006–18009.

88 Davis, J.R., Goodwin, L.A., Woyke, T. et al. (2012). Genome sequence of *Amycolatopsis* sp. strain ATCC 39116, a plant biomass-degrading actinomycete. *Journal of Bacteriology* 194 (9): 2396–2397.

89 Fleige, C., Meyer, F., and Steinbüchel, A. (2016). Metabolic engineering of the actinomycete *Amycolatopsis* sp. strain ATCC 39116 towards enhanced production of natural vanillin. *Applied and Environmental Microbiology* 82 (11): 3410–3419.

90 Sonoki, T., Takahashi, K., Sugita, H. et al. (2018). Glucose-free *cis,cis*-muconic acid production via new metabolic designs corresponding to the heterogeneity of lignin. *ACS Sustainable Chemistry & Engineering* 6 (1): 1256–1264.

91 Kinoshita, S., Udaka, S., and Shimono, M. (2004). Studies on the amino acid fermentation. Part 1. Production of L-glutamic acid by various microorganisms. *The Journal of General and Applied Microbiology* 50 (6): 331–343.

92 Shen, X.H., Zhou, N.Y., and Liu, S.J. (2012). Degradation and assimilation of aromatic compounds by *Corynebacterium glutamicum*: another potential for applications for this bacterium? *Applied Microbiology and Biotechnology* 95 (1): 77–89.

93 Kallscheuer, N., Vogt, M., Kappelmann, J. et al. (2016). Identification of the *phd* gene cluster responsible for phenylpropanoid utilization in *Corynebacterium glutamicum*. *Applied Microbiology and Biotechnology* 100 (4): 1871–1881.

94 Chen, Y., Li, L., Long, L., and Ding, S. (2018). High cell-density cultivation of phenolic acid decarboxylase-expressing *Escherichia coli* and 4-vinylguaiacol bioproduction from ferulic acid by whole-cell catalysis. *Journal of Chemical Technology & Biotechnology* 93 (8): 2415–2421.

95 Li, L., Long, L., and Ding, S. (2019). Bioproduction of high-concentration 4-vinylguaiacol using whole-cell catalysis harboring an organic solvent-tolerant phenolic acid decarboxylase from *Bacillus atrophaeus*. *Frontiers in Microbiology* 10.

96 Tramontina, R., Galman, J.L., Parmeggiani, F. et al. (2020). Consolidated production of coniferol and other high-value aromatic alcohols directly from lignocellulosic biomass. *Green Chemistry* 22 (1): 144–152.

97 Ciriminna, R., Fidalgo, A., Meneguzzo, F. et al. (2019). Vanillin: the case for greener production driven by sustainability megatrend. *ChemistryOpen* 8 (6): 660–667.

98 Sainsbury, P.D., Hardiman, E.M., Ahmad, M. et al. (2013). Breaking down lignin to high-value chemicals: the conversion of lignocellulose to vanillin in a gene deletion mutant of *Rhodococcus jostii* RHA1. *ACS Chemical Biology* 8 (10): 2151–2156.

99 Chen, H.P., Chow, M., Liu, C.C. et al. (2012). Vanillin catabolism in *Rhodococcus jostii* RHA1. *Applied and Environmental Microbiology* 78 (2): 586–588.

100 Linger, J.G., Vardon, D.R., Guarnieri, M.T. et al. (2014). Lignin valorization through integrated biological funneling and chemical catalysis. *Proceedings of the National Academy of Sciences of the United States of America* 111 (33).

101 Salvachúa, D., Rydzak, T., Auwae, R. et al. (2020). Metabolic engineering of *Pseudomonas putida* for increased polyhydroxyalkanoate production from lignin. *Microbial Biotechnology* 13 (1): 290–298.

102 Kosa, M. and Ragauskas, A.J. (2012). Bioconversion of lignin model compounds with oleaginous Rhodococci. *Applied Microbiology and Biotechnology* 93 (2): 891–900.

103 Wei, Z., Zeng, G., Huang, F. et al. (2015). Bioconversion of oxygen-pretreated Kraft lignin to microbial lipid with oleaginous *Rhodococcus opacus* DSM 1069. *Green Chemistry* 17 (5): 2784–2789.

104 Wang, Z., Li, N., and Pan, X. (2019). Transformation of ammonia fiber expansion (AFEX) corn stover lignin into microbial lipids by *Rhodococcus opacus*. *Fuel* 240: 119–125.

105 Khalil, I., Quintens, G., Junkers, T., and Dusselier, M. (2020). Muconic acid isomers as platform chemicals and monomers in the biobased economy. *Green Chemistry* 22 (5): 1517–1541.

106 Vardon, D.R., Franden, M.A., Johnson, C.W. et al. (2015). Adipic acid production from lignin. *Energy & Environmental Science* 8 (2): 617–628.

107 Salvachúa, D., Johnson, C.W., Singer, C.A. et al. (2018). Bioprocess development for muconic acid production from aromatic compounds and lignin. *Green Chemistry* 20 (21): 5007–5019.

108 Kohlstedt, M., Starck, S., Barton, N. et al. (2018). From lignin to nylon: cascaded chemical and biochemical conversion using metabolically engineered *Pseudomonas putida*. *Metabolic Engineering* 47: 279–293.

109 Barton, N., Horbal, L., Starck, S. et al. (2018). Enabling the valorization of guaiacol-based lignin: Integrated chemical and biochemical production of cis,cis-muconic acid using metabolically engineered *Amycolatopsis* sp ATCC 39116. *Metabolic Engineering* 45: 200–210.

110 Becker, J., Kuhl, M., Kohlstedt, M. et al. (2018). Metabolic engineering of *Corynebacterium glutamicum* for the production of *cis,cis*-muconic acid from lignin. *Microbial Cell Factories* 17 (1).

111 Mycroft, Z., Gomis, M., Mines, P. et al. (2015). Biocatalytic conversion of lignin to aromatic dicarboxylic acids in *Rhodococcus jostii* RHA1 by re-routing aromatic degradation pathways. *Green Chemistry* 17 (11): 4974–4979.

112 Spence, E.M., Calvo-Bado, L., Mines, P., and Bugg, T.D.H. (2021). Metabolic engineering of *Rhodococcus jostii* RHA1 for production of pyridine-dicarboxylic acids from lignin. *Microbial Cell Factories* 20 (1).

113 Pellis, A., Comerford, J.W., Weinberger, S. et al. (2019). Enzymatic synthesis of lignin derivable pyridine based polyesters for the substitution of petroleum derived plastics. *Nature Communications* 10 (1): 1762.

114 Perez, J.M., Kontur, W.S., Alherech, M. et al. (2019). Funneling aromatic products of chemically depolymerized lignin into 2-pyrone-4-6-dicarboxylic acid with *Novosphingobium aromaticivorans*. *Green Chemistry* 21 (6): 1340–1350.

115 Mishra, S., Sachan, A., Vidyarthi, A.S., and Sachan, S.G. (2014). Transformation of ferulic acid to 4-vinyl guaiacol as a major metabolite: a microbial approach. *Reviews in Environmental Science and Biotechnology* 13 (4): 377–385.

116 Williamson, J.J., Bahrin, N., Hardiman, E.M., and Bugg, T.D.H. (2020). Production of substituted styrene bioproducts from lignin and lignocellulose Using engineered *Pseudomonas putida* KT2440. *Biotechnology Journal* 15 (7).

9

Approaches to the Oxidative Depolymerization of Lignin

Elena Subbotina[1,2], Shannon S. Stahl[3], Paul Anastas[4,5,6,7] and Joseph S.M. Samec[8,9]

[1]*Department of Fibre and Polymer Technology, KTH Royal Institute of Technology, Teknikringen 56, 100 44, Stockholm, Sweden*
[2]*Center for Green Chemistry and Green Engineering, Yale University, New Haven, CT, USA*
[3]*Department of Chemistry, University of Wisconsin–Madison, Madison, WI, USA*
[4]*Yale School of the Environment, Yale University, New Haven, CT, USA*
[5]*Center for Green Chemistry and Green Engineering, Yale University, New Haven, CT, USA*
[6]*Department of Chemistry, Yale University, 225 Prospect St, New Haven, CT, USA*
[7]*School of Public Health, Yale University, New Haven, CT, USA*
[8]*Department of Organic Chemistry, Stockholm University, 10691 Stockholm, Sweden*
[9]*Department of Chemistry, Faculty of Science, Chulalongkorn University, Pathumwan, Bangkok 10330, Thailand*

9.1 Introduction

Lignin is a polymer found in terrestrial plants and some aquatic plants (e.g. seagrass). Its role in plants together with hemicellulose, is to provide a structure through binding of the fibrils together in the cell wall and thus affects the strength of the plant. Lignin is formed primarily through radical couplings of phenolic propenols to give various interunit linkages, common examples of which are illustrated in Figure 9.1. The ratio of these linkages varies with the plant species, but the β-O-4′ unit is the most common and is present in all plants. Thus, most types of lignin depolymerization methods target this bond. Besides, the β-O-4′ and the 4-O-5′ structures, the others depicted in Figure 9.1 contain ether bonds (C—O) and carbon—carbon (C—C) bonds, both of which need to be cleaved in order to liberate an aromatic monomer. While the ether bonds are easily cleaved during pulping, the C—C bonds generally stay intact.

During industrial treatment of wood when the C—O bonds in lignin are cleaved, the lignin is deconstructed into fragments. While lignin itself is nucleophilic due to electron-rich aromatic rings, these lignin fragments are usually electrophilic due to quinone methide formation. This leads to the C—C bond couplings between depolymerized lignin fragments and leads to formation of condensed polymers, often referred to as "technical lignin" and possess a different structure than native

Lignin Chemistry: Characterization, Isolation, and Valorization, First Edition.
Edited by Yuhe Liao and Bert F. Sels.

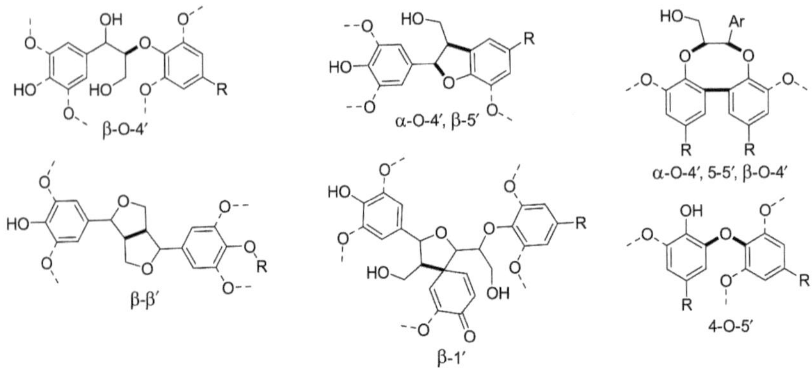

Figure 9.1 Examples of interlinkages in native lignin.

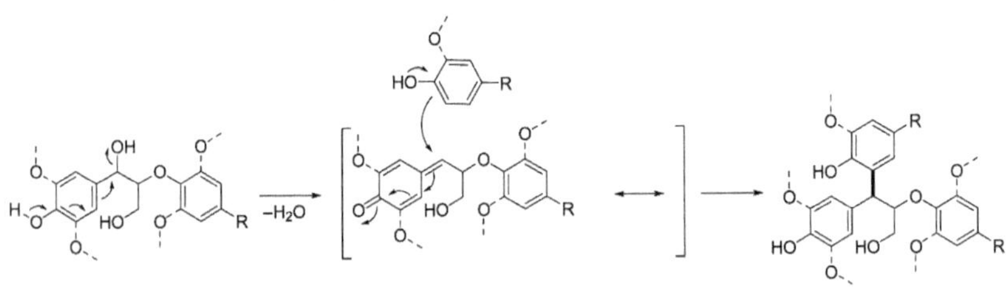

Scheme 9.1 Example of a reaction pathway to form condensed lignin.

lignins [1, 2]. An illustrative example of such a coupling is depicted in Scheme 9.1, where the formed quinone methide is attacked by another lignin polymer through a Friedel–Crafts type alkylation.

Pulping is an industrial process to produce cellulose pulp from wood usually performed by heating the wood in a liquor until approximately 80% delignification is attained [3]. Further heating is detrimental to the carbohydrate matrix, therefore the pulping processes usually utilize an additional step to remove the lignin that still remains bonded to the carbohydrate matrix. This additional step is termed bleaching and is commonly a combination of various oxidative processes. Many of these bleaching steps involve cleavage of C—C bonds in the aromatic structure or other double bonds that have been formed during the pulping. Scheme 9.2 shows the action of oxygen, a common bleaching agent used in pulping, on the aromatic structure of lignin.

In the 1990s, much of the pulping industry transformed from using chlorine to accomplish bleaching to more environmentally benign procedures. This transformation made lignin oxidation an important research area. In recent decades, the research field has evolved from an exclusive focus on technologies to remove lignin from pulp to strategies that also seek to valorize lignin. There are many reviews concerning lignin oxidation, including several recent examples focused on: lignin bleaching [4], heterogeneous catalysis [5], alkaline oxidation [6], reaction mechanisms [7], methods employing lignin models and isolated lignin

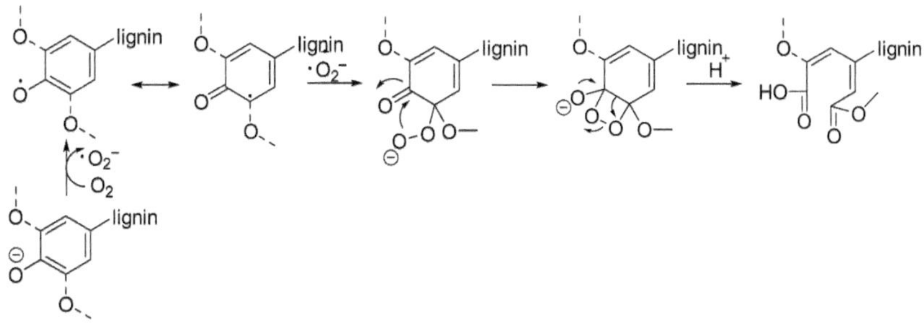

Scheme 9.2 Example of lignin oxidation in bleaching.

streams [8], sequential oxidation/depolymerization [9], TEMPO-mediated oxidation [10], and electrochemical oxidation of lignin model compounds [11].

The intention of this chapter is to summarize the most significant contributions relevant to selective oxidative depolymerization methods of native or technical lignins with a specific focus on the various types and yields of the obtained products by these methods. The first part of the chapter comprises metal-free oxidations, while the second part deals with transition metal-catalyzed oxidations, and the third part addresses electrochemical oxidation of lignin. Although the reactivity of lignin model compounds is outside of the scope of this chapter, we have included selected mechanistic studies of model compounds when they provide relevant insights into the reaction pathway for lignin depolymerization.

9.2 Metal-Free Oxidative Depolymerization of Lignin

The metal-free oxidation of lignin encompasses a wide spectrum of technologies ranging from aerobic nitrobenzene and alkaline oxidations to TEMPO-mediated oxidations. In certain methods, the oxidant is recycled *in situ* either directly by oxygen or electrochemically, and represents good examples of more sustainable approaches. As is well-known with all oxidation chemistry on lignin, selectivity is a major challenge. In order to achieve selectivity, it is useful to understand the reaction mechanism of the transformations. Mechanistic details for both well-understood pathways and novel methodologies provide useful insights that can help to improve oxidative processing of lignin.

9.2.1 Nitrobenzene Oxidation

Nitrobenzene oxidation (NBO) [12] in addition to thioacidolysis [13] is the main analytical method used to identify monomers derived mostly from β-ether linkage motif in lignin. The nitrobenzene procedure consists of heating lignin in a sodium hydroxide solution with more than stoichiometric quantities of nitrobenzene, and it provides near quantitative yields of monophenolic compounds from the cleavage of all β-ether linkages [14]. Even though NBO has no significant degree of industrial application, the method is an important benchmark for determining the theoretical maximum yield of monophenols from oxidative depolymerizations of lignin and is often used to evaluate an efficiency of a given oxidative method.

9.2.2 Ozone Mediated Depolymerization of Lignin

Ozone is a very powerful reagent and will readily react with any double bond, including aromatic structures resulting in the formation of products of the oxidative degradation of the aromatic ring (Scheme 9.3). Ozonolysis of lignin has been used in the pulp and paper industry to remove residual lignin from the carbohydrate matrix [15]. More recently, it has been evaluated for lignin depolymerization into aromatic molecules.

Recent studies have shown that ozononlysis may be used to convert lignin into various low molecular weight products, including aromatics and dicarboxylic acids. The composition of the product mixture can be tuned. For example, Chuck et al. performed ozonolysis in ethanol on alkaline lignin (lignin derived from soda pulping) and found that shorter reaction times gave aromatic compounds and longer reaction times favored dicarboxylic acids [16]. Heeres and Deuss have studied ozonolysis of lignin pyrolysis products [17, 18] and kraft lignin [19]. The yields of oils and dissolved lignin were high where only 26 wt% of kraft lignin remains as residue after 120 minutes. Alcell lignin (obtained via organosolv pulping of wood) gave even less residue where the major products were dicarboxylic acids. Bala and coworkers performed ozonolysis using a spray reactor where grass lignins were dissolved in a mixture of acetic acid and formic acid to yield monophenols, including vanillin in 5 wt% [20]. The same authors also reported ozonolysis using a continuously stirred tank reactor set up to achieve 7 wt% aromatic aldehydes from grass lignins [21].

9.2.3 Other Organic Oxidants

Oxidations can also be promoted by organic compounds such as 2,2,6,6-tetramethylpiperidin-1-yl)oxyl (TEMPO) and their derivatives, 2,3-dichloro-5,6-dicyano-1,4-benzoquinone (DDQ), or *N*-Hydroxyphthalimide (NHPI). Advantages stem from the possibility of selectively transforming the lignin. Challenges include the reoxidation of the reduced form of the oxidant and the separation of reagents after completion of the reaction.

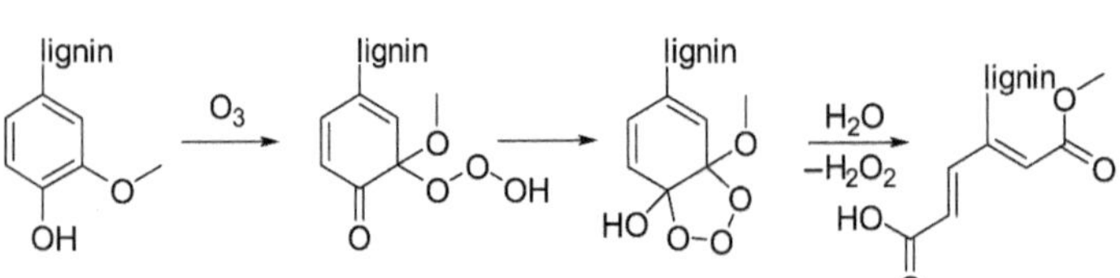

Scheme 9.3 A possible pathway for the ozonolysis of lignin [15].

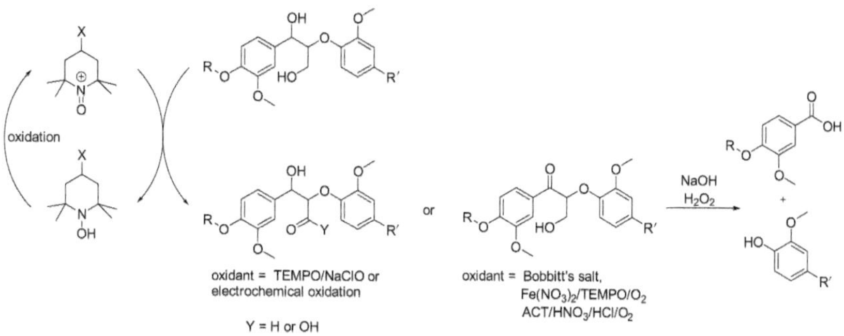

Scheme 9.4 Nitroxyl-mediated oxidation of either the α- or γ-position of the β-O-4′ motif (Nitroxyl species: TEMPO, X = H; 4-acetamido-TEMPO or ACT, X = AcNH).

Stahl and coworkers significantly contributed to the development of the field of TEMPO-promoted oxidation of lignin with the development of selective oxidation of either the α or γ positions of the β-O-4′ interlinkage [22, 23]. An extensive screening of the oxidation conditions resulted in an aerobic system consisting of 4-acetoamido-TEMPO (AcNH-TEMPO or ACT) in combination with $HCl/HNO_3/O_2$. The resulting system allowed for selective oxidation of benzylic alcohol in lignin model compounds as well as in native aspen lignin [22]. In this work, the authors did not perform a further depolymerization of the oxidized lignin but probed a possible pathway on model compounds via H_2O_2/NaOH transformation giving a carboxylic acid and phenol. Regeneration of the oxoammonium species ACT+ using electrochemistry was also developed by the the Stahl group, *vide infra* (X = AcNH in Scheme 9.4) [22].

Bolm and coworkers developed a mechanochemical depolymerization of lignin using a TEMPO/KBr/Oxone system [24]. Studies on lignin model compounds showed that while non-phenolic substrates undergo selective benzylic oxidation, compounds containing free phenolic groups undergo oxidative cleavage to produce quinones and phenols (guaiacol or syringol). Application of the method to organosolv beech lignin led to depolymerization with total monomer yield of 16 wt%. The composition of the reaction mixture was complex and included quinones, phenols, aromatic ketones, and aldehydes, along with some products of bromination of aromatic rings. In another work by the same authors, a selective oxidation of primary alcohols in lignin using TEMPO/diacetoxyiodobenzene was followed by proline catalyzed retro-aldol reaction [25]. The yields of monomers obtained from lignin were not reported.

A different approach was developed by Yan and coworkers, where metal-free oxidation of lignin was catalyzed by nitrogen-containing graphene material (LCN) with tert-butyl hydroperoxide (TBHP) as a terminal oxidant [26]. Nitrogen was incorporated into pyrolytic graphene oxide via a chemical vapor deposition method with acetonitrile as a nitrogen source. Studies on model compounds revealed the formation of a complex mixture of products, including oxidized aromatics and products from the decomposition of the aromatic rings. A radical mechanism was proposed involving oxidation of the benzylic alcohol of the β-O-4′ linkage. The authors applied the developed method to organosolv birch lignin (24 hours, 12 equiv. TBHP, 140 °C)

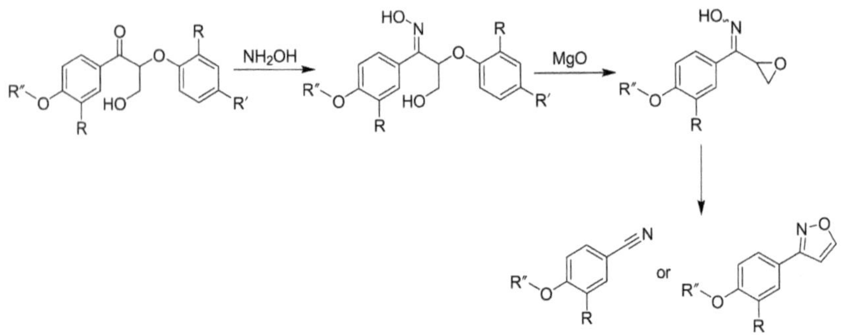

Scheme 9.5 Cleavage of oxidized lignin via oxime formation.

and observed the formation of an oil fraction and a solid fraction. The fractions were subjected to acid hydrolysis and monomeric products were quantified. However, due to a complex mixture of products, the yield of individual compounds was very low (0.13 wt% of diethyl succinate and 0.75 wt% of diethyl phthalate).

There are number of metal-free methods using oxidation of lignin as a pretreatment method, followed by depolymerization of the oxidized lignin by other means. In the following paragraphs, we will briefly highlight the most revelant studies. Westwood and coworkers studied the oxidation of lignin using DDQ/t-BuONO/O_2 system. Similar to the system reported by Stahl and coworkers, the oxidation proceeded selectively at the benzylic position. This reactivity was successfully translated from lignin model compounds to dioxasolv birch lignin. The depolymerization of the preoxidized lignin was performed via reductive cleavage of β-O-4′ bonds with Zn. Using this DDQ/t-BuONO/O_2 system, Wang and coworkers developed a method for generation of isoxazoles and nitriles from lignin via oximation of the benzylic ketone [27]. The proposed mechanistic pathway is shown in Scheme 9.5. The authors proposed that the formation of oxime weakens C—O bond leading to the formation E and Z isomers of oxirane oximes, which result in the formation of the nitrile and isoxazole, respectively. However, when the system was applied to oxidized dioxasolv birch lignin, the depolymerization did not proceed. A further modification of the system (addition of triflic or hydroxylamine-O-sulfonic acid) allowed for the formation of the target nitriles and isoxazoles, but in significantly lower yields (2–3 wt%) compared to model compounds. This reflects a general trend in lignin chemistry, where the reactivity of realistic feedstock does not match with the reactivity of lignin model compounds.

Luterbacher and coworkers developed a three-step procedure to generate aromatic diketones from lignin in high selectivity. The first step is a formaldehyde-assisted fractionation of biomass to yield a lignin comprising an acetal involving α and γ hydroxyl groups within the β-O-4′ motif [28]. The second step is a deprotection of the acetal and subsequent oxidation of lignin using DDQ in catalytic amounts (HNO_3 co-catalyst, O_2). And finally depolymerization of the oxidized lignin using formic acid to yield diketones in 31 mol% [29].

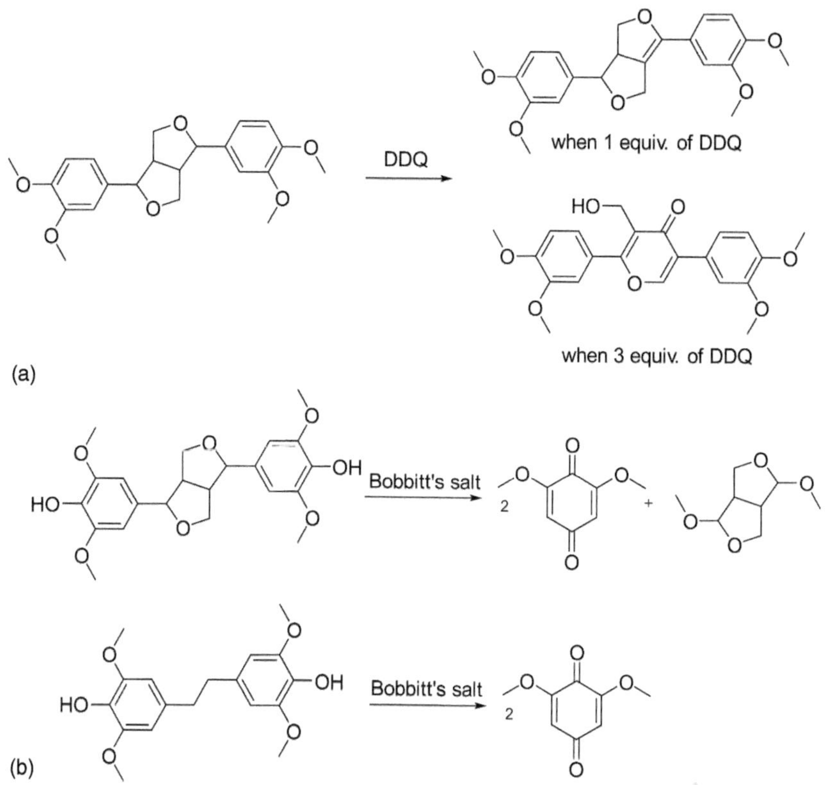

Scheme 9.6 (a) Oxidation of the β−β′ motif using DDQ; (b) Oxidative cleavage of β−β′ and β−1′ motifs using Bobbitt's salt.

In another work from Westwood and coworkers, DDQ oxidation of β-β′ bonds in kraft lignin was studied by HSQC. No cleavage was observed, but oxidation of the linkage to pyran-4-one had occurred (Scheme 9.6a) [30]. A similar approach was demonstrated by Wang and coworkers, where $NaNO_2$/DDQ/NHPI system was used to obtain α ketone structures from lignin, and depolymerization was also performed via reductive cleavage. The authors performed an oxidation/reduction sequence for birch wood and were able to generate up to 32 wt% of monophenolic compounds [31].

Cleavage of C—C bonds in lignin motifs rather than β-O-4′ is important for valorization of lignin streams arising from reductive catalytic fractionation. These streams consist of monomeric fraction (originated from the near quantitative cleavage of C—O bonds in β-O-4′ structures) and a large (ca. 60 wt%) fraction of lignin dimers and oligomers, lacking β-O-4′ bonds and containing a significantly lower amount of α and γ hydroxyl groups [32–34]. Thus, oxidative depolymerization of lignin dimers and oligomers cannot rely on the oxidation of α and γ hydroxyl groups (as in the methodologies described above). To cleave these C—C bonds in lignin requires other strategies. An example of a different approach is work by Samec and coworkers, who used Bobbitt's salt (a common name used for the oxoammonium species, ACT+; see Scheme 9.4) was used in the C—C bond cleavage of the dimeric and oligomeric fraction of a reductive catalytic fractionation mixture to produce 2,6-dimethoxybenzoquinone (Scheme 9.6b) [35]. The reduced form of the oxidant was recycled by electrochemistry. In this way, the total monomer yield from lignin was increased by 32% and thus exceeds the theoretical maximum yield based on only cleavage of beta-ether bonds. Two models from the RCF oil were studied, β-β′ and β-1′ motifs, to shed light on the reaction pathway. The reaction mechanism starts with an oxidative dearomatization to form a carbocation in position 1 of the generated quinone derivative. Attack by first methanol, then exchange with water gives a tertiary alcohol in this position. A second oxidation by Bobbitt's salt gives dimethoxybenzoquinone (DMB) product by a C—C bond cleavage.

9.2.4 Alkaline Oxidation of Lignin

Alkaline oxidative depolymerization of lignin is one of the oldest technologies used to deconstruct lignin into valuable compounds, such as vanillin [6]. Alkaline oxidation of lignosulfonates to vanillin catalyzed by copper is the only existing method for commercial valorization of lignin into aromatic monomers [36–40]. The reaction was developed in the middle of the last century when many patent applications were filed [41, 42]. The reactions are either uncatalyzed or performed in the presence of a transition metal with a low redox potential, such as copper. We include both approaches (metal-free and metal-catalyzed) in this section as it is challenging to separate the technologies, vide infra.

Alkaline oxidative depolymerizations are typically performed at 150–180 °C in 2 M of NaOH and the reactions are run for less than one hour using an oxygen pressure below 5 bar. The yield of depolymerized monophenols such as aldehydes strongly depends on the experimental parameters [43, 44]. For example, more dilute reactions perform better than concentrated ones [37]. It has been shown that the reaction is very sensitive and product formation is followed by product degradation. Thus, quenching the reaction at the right time is crucial for the results. This makes comparison of different technologies challenging as yields will depend on both reaction conditions but also how well the reactions are monitored [45]. In addition, the lignin source also affects the yields of monophenols. In contrast to NBO, alkaline oxidative depolymerizations target other motifs in the lignin besides the β-O-4′ and α-O-4′ bonds. For instance, higher yields were obtained with kraft lignin, which has relatively low beta ether content, than with isolated organosolv lignin [46]. Uncatalyzed alkali oxidation reactions can deliver products in reasonable yields. For example, Rodrigues and coworkers have shown that 10% yield of vanillin can be obtained from softwood kraft lignin when performing the alkaline reactions at 130 °C for 35 minutes [43, 47]. Yields of monophenolics as high as 58 wt% were reported for the aerobic oxidation of stem exploded corn stover lignin at 130 °C at low pressures of oxygen [48]. However, the quantification should be evaluated with care, because final products possess higher molecular weight due to the added oxygen atoms.

The role of the catalyst employed, most commonly a transition metal with low reduction potential, is not fully understood. It has been observed that the presence of a catalyst makes the reaction less sensitive to the reaction parameters (i.e. oxygen pressure, reaction temperature and reaction time). By careful screening of the oxidative depolymerization, the groups of Román-Leshkov and Beckham found that the same yields were obtained in the presence and absence of catalysts, but only when the reaction was carefully monitored [45]. The yields of the uncatalyzed reactions dropped rapidly after reaching a maximum. For instance, the uncatalyzed reaction performed at 175 °C reached 30% yield of monophenolic compounds after 10 minutes. After 60 minutes, less than 10% remained. For the catalyzed version, however, the yield dropped to around 20%.

Stahl and coworkers reported the alkaline aerobic oxidation of lignin derived from poplar wood using $CuSO_4$ as a catalyst, and they were able to achieve 30.5 and 28.8 wt% yield of monomers, respectively [66]. Notably, the process showed high selectivity towards aldehydes. Vanillin and syringaldehyde were formed in 19.8 wt% and 19.2 wt%, respectively. The reactions were performed at 160 °C using 25 bar of air. By implementing centrifugal partition chromatography they were able to isolate individual monophenols from the product mixture in good purity.

In some cases, a synergistic effect of combining different metals has been observed. Xiang and Lee studied aerobic alkaline oxidation of precipitated hardwood lignin (produced from acid treatment of biomass) catalyzed by $CuSO_4$ and $FeCl_3$ at 160–180 °C [49]. The process generated vanillin and syringaldehyde in a total yield of 15 wt%, and aromatic ketones (4 wt%) were also isolated. The authors found that Cu^{2+} and Fe^{3+} ($Cu^{2+}/Fe^{3+} = 10/1$) had a synergistic effect on the selectivity of the process, allowing for the highest yield of aldehydes and ketones. This synergy was observed in previous studies, which suggested that Cu^{2+} catalyzed one-electron oxidation of phenolate lignin moieties to phenoxy species, while Fe^{3+} act as an oxygen carrier [50]. A similar example of oxidation using $CuO/Fe_2(SO_4)_3/NaOH$ system with H_2O_2 as a terminal oxidant was shown by Qiu and coworkers where the yield of monomers reached 18 wt% running the reaction at 150 °C in a methanol/water mixture. The authors found that the solvent composition is crucial for the transformation. Water facilitates NaOH dissolution, but MeOH can play a crucial role by preventing repolymerization reactions via methoxylation reactions [51]. Other examples of using hydrogen peroxide as the terminal oxidant have also been reported for alkaline oxidations [52, 53]. It should be noted that the synergistic effect of using different metals is not always observed [45].

Liu and coworkers tested a Co-based perovskite catalyst ($LaCoO_3$) for alkaline oxidation of steam-exploded lignin from cornstalk. The total yield of aromatic aldehydes was improved to 10 wt% (vs. <5 wt% for uncatalyzed reaction) [54]. The catalytic activity of $LaCoO_3$ was attributed to the formation of surface-active sites ($Co(surf)^{3+}O_2^-$), confirmed by O 1s XPS. These surface-active species facilitate the formation of quinone methide radicals and peroxides. In addition, the authors found that the catalyst exhibited no drop in activity after five consecutive runs. In another work, the authors investigated the influence of Cu substitution on $LaCo_{1-x}Cu_xO_3$ performance in catalytic wet oxidation of steam-exploded lignin from cornstalk [55]. $LaCo_{0.8}Cu_{0.2}O_3$ exhibited the best performance, to give *p*-hydroxybenzaldehyde, syringaldehyde, and vanillin in 20 wt% yield. An introduction of Cu-doping into the catalyst resulted in the formation of anion vacancies,

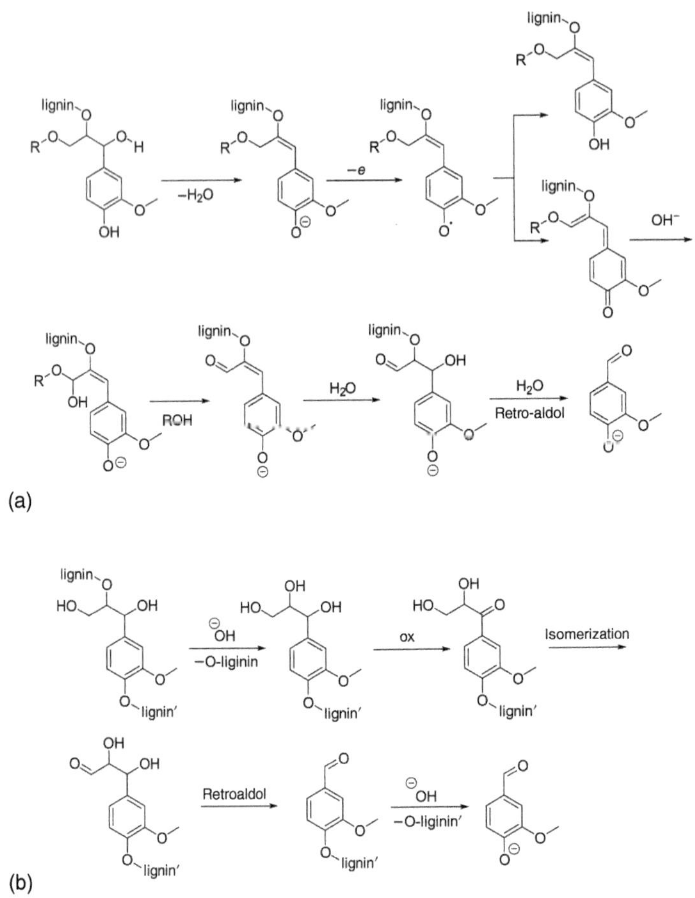

Scheme 9.7 Mechanistic proposals for the alkaline aerobic oxidative depolymerizations of the β-O-4′ bond: (a) radical mechanism; (b) initial triol formation.

which in turned led to the increased amount of adsorbed oxygen surface active site species [$Co(surf)^{3+}O_2^-$] and [$Cu(surf)^{2+}O_2^-$]. Other combinations of metals were also studied [56, 57]. The groups of Románn-Leshkov and Beckham also confirmed the importance of copper in the perovskite-based catalysts, vide infra [45].

Au and Pt nanoparticles supported on TiO_2 were evaluated for oxidative depolymerization of organosolv and kraft lignin under relatively mild alkaline reaction conditions (T: 150 °C, P: 20 bar, c(NaOH): $10\,g\,l^{-1}$, 1 hour) resulting in total yields of monoaromatic products of 4.3 wt% [58]. Pt/TiO_2 demonstrated higher yields of targeted aromatic products compared to Au, and was used to build a kinetic model of the oxidation. The authors proposed a model where lignin depolymerization into oligomeric species, depolymerization into small aromatics and into aliphatic products proceed in parallel. In addition, condensation reactions between monoaromatic products occur. Pt increases the rate of lignin depolymerization, but also of recondensation reactions between aromatic units, which overall leads to a decrease of ratio between aromatic and aliphatic products.

From a mechanistic point of view, different reaction pathways have been suggested [59, 60, 61]. Several proposals have been made for cleavage of the β-O-4′ bond. Radical pathways have been proposed by Tarabanko where the phenoxy anion is central for the reactivity. Radical formation from the vinyl ether intermediate generates a species that disproportionate into a reduced form and an oxidized form (Scheme 9.7a). Michael addition of water to the γ position results in formation of coniferyl aldehyde. Subsequent addition of water, followed by a retro-aldol reaction affords the final product [61]. In a recent study, Hosoya and coworkers investigated the possibility of the aliphatic triol as a key intermediate from an initial nucleophilic substitution reaction by the OH^- (Scheme 9.7b). Benzylic oxidation, followed by tautomerization would give rise to the aldehyde at the γ position [62, 63]. This intermediate can undergo a sequence of two retro-aldol reactions to give the aldehyde that can undergo attack by a hydroxide at the [Aryl-O]–C position to give the final product. This mechanism would not require a radical, or a phenoxy anion, and it also explains the formation of aryl ketones where the retro-aldol would occur from benzylic oxidation.

As compared to NBO oxidations, the alkaline oxidations use oxygen as a terminal oxidant and thus represent greener methodologies. The reactions can be performed in the presence or absence of a transition metal with low reduction potential. The mechanism is still under debate and role of the transition metal catalyst is not fully understood.

9.2.5 Oxidative Catalytic Fractionation of Wood (OCF)

Oxidative catalytic fractionation (OCF) of wood is a complementary methodology to the reductive catalytic fractionation (RCF) approach [32, 34, 64, 65] where the fractionation and depolymerization of lignin occur in situ. As in the case of RCF, the major benefits of OCF is that lignin fragments released into the solution during the treatment have more native-like structure, and contain less C—C bonds which make the material more amenable for further transformations. One of the advantages of OCF is that the obtained oxygenated products generally possess a better solubility in water and thus are more suitable for microbial funneling. One of the examples of OCF was already discussed, vide supra, in the alkaline oxidative depolymerization of lignin in poplar wood by the groups of Román-Leshkov and Beckham [45]. The study disclosed that the use of wood rather than isolated lignin can significantly improve the yield of monomeric products. Yields of monophenols, mostly syringylaldehyde and vanillin of 31 wt% was obtained and benchmarked to NBO where 35 wt% was observed.

The groups of Sels and Wang have developed an OCF process using CuO nanoparticles. In addition to a high yield of lignin-derived monomers (45.6 wt%), a high-quality cellulose was preserved in up to 81% [67]. In another contribution from the same group, the authors optimized OCF of pine (7.5 wt% NaOH, 1 MPa O_2, 400 rpm, 160 °C). In this process vanillin, a main lignin-derived monomer, was obtained in 21.1 wt%, and the total yield of monomers constituted 29 wt% based on the lignin content, which is a very high yield for softwood lignin. The hemicellulose fraction was converted to a mixture of short di- and monocarboxylic acids in 63.2 wt%. And the cellulose was recovered in 45.4 wt% of 95.5% purity. The authors found that the stirring rate had a great impact on the outcome of the retention of cellulose and the yield of lignin-derived monomers caused by increased overoxidation. Thus, the authors determined the stirring rate of 400 rpm to be optimal. With regards to other substrates (grasses and hardwood) the authors found that harsher reaction conditions were required for an efficient depolymerization of lignin. In the case of hardwood, the authors posited that the decreased acidity of phenol of syringyl unit compared to guaiacyl unit in softwood to be responsible for less efficient lignin oxidation given that the deprotonation of a phenol moiety is necessary for the oxidation to occur [68].

Stahl and coworkers developed an OCF method using acetone as a solvent and various heterogeneous catalyst under non-alkaline conditions: metal oxides (Co, Mn, Fe, Cu), carbon-supported transition metals (Ru, Pd, Pt), and metal-containing nitrogen-doped carbon catalysts (Co, Fe). Among the tested catalysts, Co-based ones (Co-PANI-C and Co-phen-C) afforded the highest monomer yield of 15 wt%. The product mixture consisted of vanillin, vanillic acid, syringaldehyde, syringic acid, and *p*-hydroxy benzoic acid as the main components [69]. Thus, the method allowed for the generation of a set of products complementary to those obtained from RCF. In addition, non-alkali conditions allow for a preservation of cellulose pulp, prone to degradation at high pH via peeling reactions.

Deng and coworkers developed an OCF process with wood catalyzed by $H_3PMo_{12}O_{40}$ [70]. The developed system is a two-step/one-pot OCF of poplar sawdust, where the wood is first treated at 100 °C to extract α-OH alkylated lignin into the solution, followed by the separation of solid cellulose residue and the oxidative depolymerization of lignin at higher temperature (140 °C) into low molecular weight syringol and guaiacol derived products in total of 45.9 wt%.

In general, OCF results in higher yields of lignin-derived monomers, in comparison to oxidation of technical lignin, which is in line with previous work on RCF and other lignin-first concepts.

9.3 Transition Metal Catalysis for Oxidative Cleavage of Lignin

Transition metal catalysis in lignin oxidation includes diverse methods, ranging from the use of homogeneous salts (e.g. $FeCl_3$, $CuSO_4$) to improve alkaline oxidation of lignin, vide supra, to complex systems mimicking lignin peroxidase and manganese dependent peroxidase (e.g. metalloporphyrins). Both homogeneous and heterogeneous versions of metal catalysis were studied with respect to lignin oxidation. While offering multiple advantages, e.g. allowing to tune redox properties of a metal–organic complex via changes in the ligand structure, transition metal catalysis can pose environmental and economic problems for the application of a given process on an industrial scale, and thus needs to be thoughtfully designed. In this section, we will cover the main transition metal-catalyzed systems developed for the oxidation of lignin. A large variety of metal-catalyzed methods for lignin oxidation makes them difficult to be classified. Well-established systems (Re, Co, and V-based) will be addressed initially followed by an overview of various complementary methods.

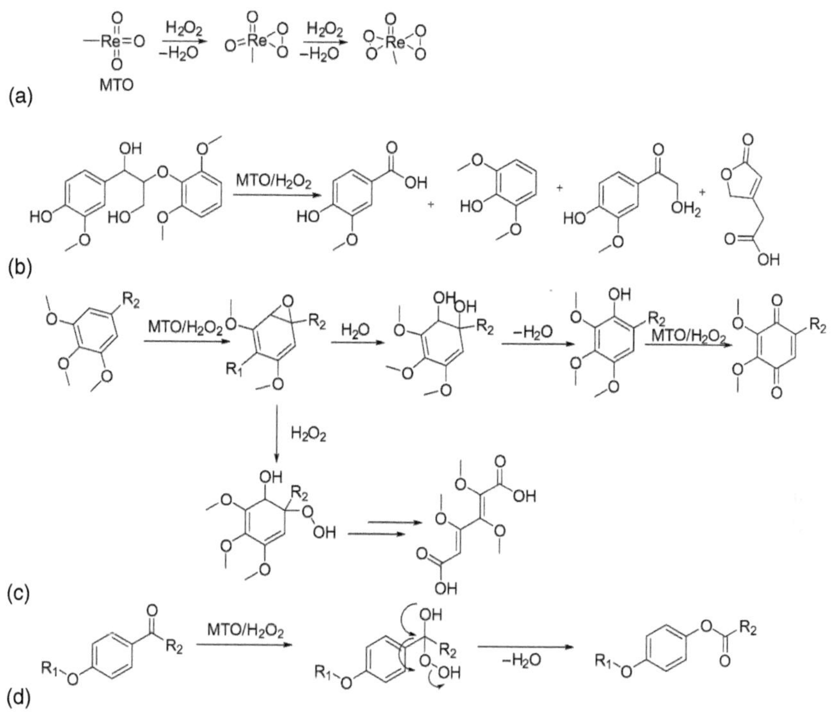

Scheme 9.8 (a) MTO and bisperoxo η_2-Re-complex; (b) an example of product mixtures obtained from lignin model compounds with MTO/H_2O_2; (c) a mechanism of oxidative cleavage of the aromatic ring by MTO/H_2O_2; (d) Dakin-type oxidation pathways.

9.4 Rhenium-Based Catalysts

Methyltrioxorhenium (MTO) was extensively studied for the oxidation of both lignin model compounds and lignin. Crestini and coworkers strongly contributed to the understanding of mechanistic aspects and to the development of systems applicable to real substrates. The reactivity of the MTO/H_2O_2 system is attributed to the formation of mono- and bis-peroxo η_2-complexes (Scheme 9.8a) [71]. Studies on lignin model compounds are well summarized by Crestini [72, 73], and include monomeric compounds, such as vanillin alcohol [74], β-O-4′ dimers [74], β—β′ dimers [75], and diphenylmethane neolignans [74]. These studies reveal that MTO-catalyzed oxidation leads to the formation of complex product mixtures via ring opening, side chain oxidation and C—O and C—C bond cleavage in lignin (Scheme 9.8b). There are several proposed mechanistic pathways of oxidation, including epoxidation of an aromatic ring and Dakin-type oxidation (Scheme 9.8c,d) [74]. In addition, MTO can act as a Lewis acid leading to the products of hydrolytic cleavage of C—O bonds in lignin models [74, 76].

Studies on MTO oxidation of lignin itself were mainly conducted by Crestini and coworkers [74, 77]. In these works the authors assessed the reactivity of several types of lignins (hydrolytic sugarcane lignin (SCL), a red spruce kraft lignin (RSL) and a hardwood organosolv lignin (OSL), representing *p*-hydroxyphenyl–guaiacyl, guaiacyl and guaiacyl–syringyl lignin with homogeneous and heterogeneous versions of the catalyst, respectively. Reactions were performed in glacial acetic acid at room temperature. The structural changes were accessed by ^{31}P NMR, allowing estimation of the amount of aliphatic, phenolic (condensed and non-condensed), and carboxylic acid structures. In the case of homogeneous MTO, all samples demonstrated a decrease of aliphatic hydroxyl groups, which was the lowest for OSL (14%) and comparable for SCL and RSL (43% and 67%, respectively), and an increase of carboxylic group content indicative of side chain oxidation. The reason for the observed difference in aliphatic hydroxyl groups was not determined. An additional finding was the poor reactivity of *p*-hydroxy phenol units in SCL compared to guaiacyl and syringyl units, which probably reflects a higher propensity of electron-rich aromatic rings to undergo oxidation. Crestini and coworkers also studied the reactivity of MTO immobilized on a heterogeneous support (various polyvinylpyridines and polystyrenes (PS)) [77]. These catalysts were applied for the oxidation of RSL and SCL lignin, and the results were compared to the homogeneous conditions. While the heterogeneous catalysts still exhibited oxidation activity, the homogeneous reaction conditions were found to be significantly more efficient (e.g. 5% decrease in aliphatic hydroxyl groups for PS supported MTO vs. 46% for the homogeneous MTO in case of SCL). In addition, a different reactivity of the heterogeneous version of MTO was revealed. MTO on a solid support was found to be more prone to catalyze a Dakin-type reaction on alfa-carbonyl units vs. aromatic ring oxidation. No information regarding the molecular weight distribution of the products and their chemical structures was reported.

Thus, while MTO is one of the most widely studied metal catalysts for lignin oxidation, the efficiency of MTO toward production of valuable monomers has not yet been fully explored; and the compositions of the mixtures obtained from real feedstock were not fully characterized in terms of yields of monomeric products and molecular weights of the fractions.

9.5 Cobalt-Based Catalysts

Cobalt salen (Co-salen)/O_2 represents another system widely applied for the oxidation of lignin. Co-salen is known for its ability to bind oxygen and form Co-peroxy species, which possess catalytic activity for oxidation of phenolic substrates (Scheme 9.9a) [78]. The ability of the Co-salen complex to bind oxygen is largely improved when pyridine or imidazole are used as ligands giving rise to pentacoordinated Co complexes [79]. Studies with monomeric and dimeric lignin model compounds representing S and G units revealed that while oxidation of the S unit proceeds readily, the oxidation of the G unit is drastically less efficient. However, the reaction was improved upon addition of a bulky base, which facilitates the oxidation of the substrate via phenol deprotonation, and, due to steric reasons, does not coordinate to the metal center and thus does not compromise its catalytic

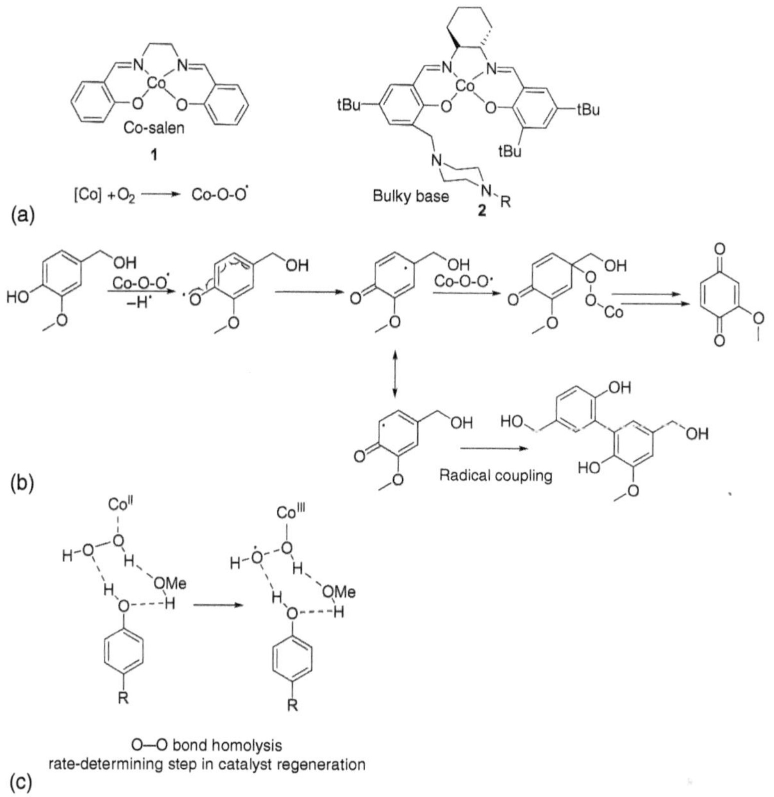

Scheme 9.9 (a) Examples of Co-salen type complexes; (b) mechanistic pathway for oxidation of guaiacyl moeites with Co-salen type complexes; (c) rate-determining step in Co-salen catalyst regeneration pathway [80].

perfomance [79]. Studies on dimeric lignin model compounds by Bozell and coworkers showed that cleavage of phenolic lignin fragments can proceed via C—C bond cleavage and leads to the formation of benzoquinones, where the formation of methoxybenzoquinone from G lignin unit is compromised by the radical coupling of phenoxy radicals at position 5 of the aryl (Scheme 9.9b). A combination of computational and experimental studies on Co-salen oxidation of lignin model compounds was recently published [80]. The authors identified the rate-limiting step and the corresponding transition state with the highest energy in the catalyst regeneration pathway. In case of S lignin models this transition state has significantly lower energy (24.9 $kcal\,mol^{-1}$) vs. 33.1 and 40.9 $kcal\,mol^{-1}$ in case of G and H units, respectively. It was also found that the energy of the transition state cannot be lowered by addition of a stronger or bulkier base. As a possible solution, the authors propose the use of a co-catalyst to promote O—O bond homolysis in the intermediate leading to the regenerated catalyst (Scheme 9.9c).

Given the mechanism of the reaction, the presence of free phenolic groups is a prerequisite for the successful oxidative cleavage of lignin. While native lignin possesses a limited number of free phenolic groups, industrial lignins, such as kraft lignin [1], have a higher content of phenolics, rendering them amenable for the oxidative cleavage with Co-salen complexes. Bozell and coworkers performed depolymerization of poplar organosolv lignin at room temperature using 10 mol% of Co-salen complex bearing a bulky heterocyclic nitrogen base as a substituent (reaction conditions: 50 psi O_2, in MeOH/DMSO, room temperature) [81]. A total monomer yield of 3.5% was observed. The product mixture consisted of syringaldehyde, vanillin, 2,6-dimethoxy benzoquinone and 2,5-dimethoxy benzoquinone (formed from methoxy benzoquinone). The authors attributed the rather low yield of monomers to low content of free phenolic groups in lignin.

Heterogeneous versions of the Co-salen catalyzed aerobic oxidation of lignin have also been reported. Zhou studied depolymerization of kraft lignin using Co-salen and Co-tetrahydrosalen complexes encapsulated in NaY zeolites in combination with peracetic acid [82]. The encapsulation was performed to prevent catalyst degradation via the formation of bridged dimeric species. Guaiacol was found to be the major product. Although the authors did not propose any reaction pathway, the formation of this product and absence of quinones suggests that zeolite catalyzed acidolysis was probably operating.

Another oxidative approach using cobalt was developed by Bolm and coworkers, where the system $Co(acac)_2/NHPI/O_2$ initially developed by Ishii and coworkers [83] was employed as an oxidant for kraft lignin (Sigma) and beech organosolv lignin [84]. The main observed reactivity was an oxidation of the benzylic position with minor formation of products resulting from the cleavage.

Similar to the previously described MTO system, Co-salen type catalyzed oxidation of lignin is well studied on model compounds, via both computational and experimental methods. However, with respect to the realistic substrates (wood and isolated lignin) experimental data are significantly more scarce and protocols are less optimized, once again demonstrating the challenge of the transferring reactivity from lignin model compounds to a biomass feedstock.

9.6 Vanadium-Based Catalysts

A number of vanadium catalysts were developed and applied to the oxidative cleavage of lignin models. Examples include oxidative cleavage of non-phenolic [85] and phenolic lignin model compounds [86] as well as aerobic redox neutral cleavage of lignin models [87–89] using homogeneous vanadium complexes. These studies revealed important mechanistic insights of vanadium catalyzed oxidative cleavage of interunit linkages. Silks and coworkers discovered that catalysts **3** and **4** (Figure 9.2) exhibit a different reactivity towards phenolic dimeric lignin model compounds, where bis(8-oxyquinolate) coordinated vanadium complex **3** led to the formation of the products of C—C bond cleavage, while the Schiff base vanadium complex **4** yielded products of the C—O bond scission, when the reaction was performed under air in the presence of a base [86]. The mechanistic rationalization of this reactivity via computation was provided by Fu and coworkers [88].

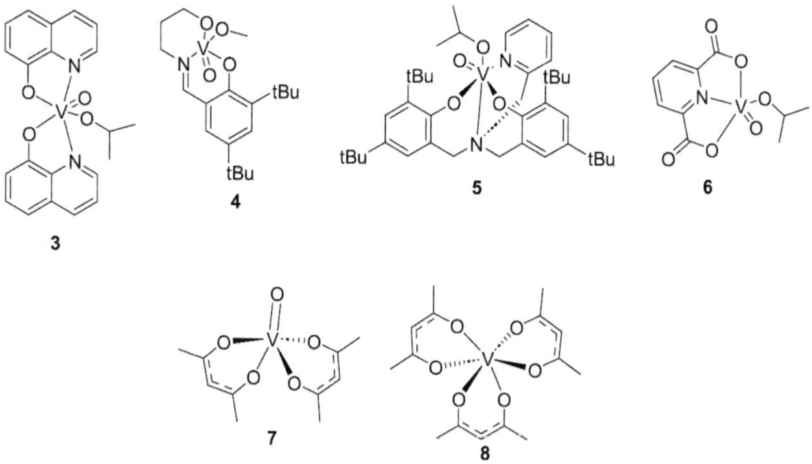

Figure 9.2 Examples of vanadium-based catalyst.

These catalysts were also applied to the depolymerization of realistic lignin substrates. In their later work, Hanson and Baker studied the translation of the reactivity of lignin model compounds to organosolv lignin [90]. Catalysts **3**, **5**, and **6** (Figure 9.2) were tested, and the reaction was assessed by GPC, HSQC, and ^{31}P NMR experiments. The authors found that catalyst **3** was the most efficient, allowing for a significant decrease in the molecular weight and the amount of the interunit linkages, which was evaluated by NMR. No analysis of the monomeric products was presented. Catalyst **4** was also probed by Silks and coworkers for the redox-neutral depolymerization of Miscanthus giganteus [91]. A decrease in molecular weight and cleavage of interunit linkages was evident, however, the yield of monomeric products (vanillin, vanillic acid, syringaldehyde, syringic acid, etc.) only reached 3 wt%.

With respect to the vanadium-catalyzed oxidation of lignin, the reactivity of kraft lignin under alkaline oxidation conditions in the presence of vanadium-based catalyst $VO(acac)_2$ was studied [92]. Vanillin, vanillic acid, acetovanillone were obtained as the main products using $VO(acac)_2$ in combination with $Cu(OAc)_2$, reaching maximum 7 wt% yield of monomers. Optimization of reaction conditions revealed an optimal V : Cu ration of 0.75 and optimal O_2 pressure (5 bar). It is worth mentioning that the same products albeit in a lower yield (5.5 wt%) were generated under control conditions in the absence of a catalyst. Another work contributing to the translation of the reactivity of model compounds to lignin is a study by Baker and coworkers, where a number of vanadium and copper catalysts were tested for the depolymerization of organosolv lignin [93]. The authors accessed the depolymerization of lignin by GPC and HSQC NMR experiment. Catalyst **3** in the presence of a base and **5** containing a tethered base demonstrated the best reactivity in terms of molecular weight decrease (from c. 2500 Da to 500 Da) and interunit bonds cleavage (β-O-4′, phenylcoumaran, pinoresinol). Oxidation of organosolv poplar lignin using $V(acac)_2O$ complexes was performed by the groups of Zhang and Huang (reaction conditions, MeCN/H_2O, 140 °C, 48 hours, 1.5 MPa O_2) [94]. Model compounds studies showed the formation of derivatives of phenol and benzoic acid, that supported C—C and C—O bond cleavage. No yield of monomers obtained from lignin was reported, however a significant drop in Mw of lignin from 5720 to 1140 Da was observed.

Heterogeneous vanadium catalysts are more common in conversion of lignin. An example by Bolm and coworkers, is where copper–vanadium double-layered hydrotalcites were applied as catalysts for oxidative cleavage of organosolv beech lignins and kraft lignin. The authors did not report yields of individual monomeric products, however reported a significant decrease in Mw of lignin reaching 300 Da and cleavage of β-O-4′ and resinol structures [95]. Patience and coworkers studied the conversion of softwood kraft lignin to acids (lactic, formic acid) and maleic anhydride over vanadium pyrophosphate and Al-V-Mo catalysts (with predominant $(VO)_2P_2O_7$ and $(VO)_2P_4O_{12}$ as predominant phases) in fluidized bed reactor at temperatures of 327–427 °C [96]. The highest yield obtained (6 wt%) of lactic acid was observed with Al-V-Mo, while vanadium pyrophosphate produced malonic acid and maleic anhydride mixture in total yields around 2 wt%. In another report from the same authors, oxidation directly on black liquor instead of the isolated lignin was performed. Using vanadium pyrophosphate/ZSM-5 zeolites gave a mixture of malonic acetic formic, lactic, acrylic, maleic, and butyric acids with a total monomer yield of 5.5 wt% [97].

9.7 Other Metals

This part of the chapter provides an overview of other types of metal catalysts applied to lignin oxidation.

Copper: Copper, in a form of copper oxide or copper salts, has been widely used as a catalyst for lignin oxidation. The use of copper was widely explored for alkaline oxidation of lignin (vide supra). Some additional notable examples of copper-catalyzed oxidative upgrading of lignin are presented below.

Zhang and coworkers developed ($CuFeS_2$) catalyzed method for the production of dicarboxylic acids from diluted-acid corn stover lignin and steam-exploded spruce lignin via an oxidative ring opening of aromatic monomers with H_2O_2 as a terminal oxidant [98]. In the proposed mechanistic pathway, lignin is depolymerized via a side chain substitution to produce low molecular weight products which subsequently get oxidized to quinones that get converted into fumaric, maleic and muconic acids. The process gave a mixture of monomers mainly consisting of malonic, succinic, and malic acids in total 11–14 wt% yield. The origin of succinic acid is not clear; the authors propose that formic acid generated in the process acts as a reducing agent.

Another example of copper-based oxidative catalytic cleavage is $CuCl_2$/polybenzoxazine composite catalysts for C—C and C—O bond cleavage of lignin using H_2O_2 as a terminal oxidant at room temperature [99]. The catalyst was prepared from $CuCl_2$ and benzoxazine monomer (based on melanie/formaldehyde/phenol). The main benefits of the system are mild operational conditions and the heterogeneous nature of the catalyst, allowing for the catalyst recycling. After the optimization of the reaction on model substrates, the authors performed an oxidation of alkali lignin. While the GPC analysis revealed a significant decrease in Mw (from c. 2600 to 500 Da after 8 hours and 395 after 24 hours), the mixture of products was complex and consisted of trimeric and dimeric acids, aldehydes, and phenols.

Palladium: Pd-catalyzed systems are present in literature, but scarcer and less systematically investigated. Pd/Al_2O_3 was used in several studies. A continuous oxidation of sugar-cane bagasse lignin over Pd/Al_2O_3 was carried out by Pereira and coworkers [100]. The maximum yield of aldehydes observed was 12 wt% using 4 wt% catalyst loading. In this work, the authors built kinetic models for the production of *p*-hydroxy benzaldehyde, vanillin, and syringaldehyde. Another example is oxidation of alkali lignin with Pd/Al_2O_3 in a batch and flow reactor under 2–10 bar of oxygen. In the batch process, vanillin and syringaldehyde were obtained at 3.5 wt%, while under flow conditions the yield was improved to 6 wt% [101].

Wang and coworkers performed oxidation of organosolv lignin using Pd/CeO_2 system. When the system was applied to lignin models, a complex mixture of products resulting from benzylic oxidation, C—C and C—O bond cleavage was observed. The products from organosolv lignin included vanillin (5.2 wt%), guaiacol (0.87 wt%), and 4-hydroxybenzaldehyde (2.4 wt%) [102].

As can be concluded from the examples above, the currently developed Pd-based systems do not offer a significant improvement in product yields or selectivity, while significantly increasing the cost of the process.

Gold: A noteworthy example of gold-catalyzed oxidative cleavage of lignin was reported Crocker and coworkers where the authors studied an oxidative depolymerization of kraft (pine) and GVL extracted (maple) lignin on Au nanoparticles supported on a basic Li–Al LDH (layered double hydroxide) [103]. Oxidation of lignins was performed in DMF, followed by hydrolysis with NaOH. The monomer yield for kraft and GVL lignin was 10 and 40 wt%, respectively. The product mixtures consisted of various aromatic acids, aldehydes, and alcohols. The mechanism of the oxidation was probed with model compounds. The authors found that the oxidation begins with the benzylic alcohol, resulting in the corresponding ketone, which in turn can undergo either dehydration or loss of the $C\gamma$ via a retro-aldol reaction. $C\alpha$-$C\beta$ bond cleavage can take place via either radical or ionic pathways to result in the corresponding phenol and carboxylic acid, which in turn react to form an ester. The formed esters are hydrolytically cleaved during the NaOH treatment of the oxidized lignin. As the key features of the catalytic system, the authors identified the size of Au nanoparticles and the charge transfer from the basic support to Au atoms, leading to efficient oxygen activation.

Cobalt/Manganese: Partenheimer applied a mixture of Co/Mn/Zr/Br (salts of metals and HBr) for oxidative cleavage of various types of lignin to produce hydroxyl benzaldehydes and carboxylic acids in acetic acid/water mixture at 180 °C under aerobic conditions, where the maximum yield achieved was 10.9 wt% with syringic acid as a main product [104]. The authors discovered that the presence of free phenolic group retarded the oxidative cleavage due to the oxidation of the phenolic functionality to quinone, which subsequently degraded to CO_2. However, acetylation prevented this reaction and allowed the oxidative cleavage to proceed. Thus, acetic acid or acetic anhydride were good choices for the reaction medium.

Polyoxometalates: Polyoxometalates (POMs) represent another class of metal-containing catalysts for lignin oxidation. POMs were widely applied to lignin degradation in pulp bleaching processes and water treatment processes [105–108]. An early example is work by Hill and coworkers where the authors demonstrated the potential of POM ($[AlV^VW_{11}O_{40}]^{6-}$) in wood delignification given their enhanced stability compared to metal catalysts and internal pH management [109]. POMs were used for the production of targeted monomers via oxidative depolymerization of kraft lignin. Using $H_3PMo_{12}O_{40}$ (9.125 wt% related to lignin), at 170 °C and 5 bar of O_2 led to improvement in the yield from 2.8 to 5.2 wt% [110]. The authors found that addition of alcohols (MeOH and EtOH) significantly improved the monomer yield, presumably via suppressing repolymerization pathways. In the follow-up work, the authors demonstrated that this process can be applied for the production of vanillin and methyl vanillate on a laboratory scale to achieve 7 wt% yield. The products were not isolated and were obtained in a mixture with oligomers [111]. After the separation of solid residue the reaction medium containing $H_3PMo_{12}O_{40}$ was reused in five consecutive runs without a significant drop in yields.

9.8 Biomimetic Oxidations

Biomimetic oxidation of lignin focuses on mimicking enzymes from white-rot basidiomycetes, lignin peroxidase (LiP) and manganese-dependent peroxidase (MnP), which can selectively degrade lignin in the presence of cellulose and hemicellulose. Metalloporphyrins and phthalocyanines are two major classes of catalysts for biomimetic lignin degradation with the possibility to tune redox properties and their higher stability toward oxidants compared to enzyme-based counterparts [112]. Crestini and coworkers studied oxidation of lignin model compounds and kraft lignin by Fe and Mn porphyrin complexes [113]. The products of the oxidation of 5-5′ and diphenylmethane lignin models with Fe and Mn porphyrins in the presence of H_2O_2 led to the formation of quinone structures and products that were hydroxylated at the benzylic position. Mn-based catalyst demonstrated superior activity compared to Fe-based. The oxidation of kraft lignin was also assessed and it was found that Fe-based complexes showed increase in condensed structures, presumably due to the coupling reaction occurring during the oxidation of lignin by generated HO· species. On the contrary, Mn-based catalysts did not exhibit this reactivity, which probably reflects the difference in the generated active species from hydrogen peroxide which in the case of Fe results from the homolytic and in the case of Mn from heterolytic cleavage. Molecular weights and monomeric product distribution were not examined [113].

Self-decomposition of metalloporphyrins via the formation of μ-oxo species is an issue, which limits the potential of this type of catalysts in lignin oxidation. In naturally occurring enzymes this problem is overcome by the protection of the metal center by the polypeptide envelope. Crestini and coworkers mimic this via the preparation of heterogeneous manganese meso-tetrakis(tetramethylpyridinio)porphyrin pentaacetate immobilized on the smectite clay montmorillonite. The authors tested the catalyst for oxidation of lignin model compounds. The catalyst was active for several hours, which was evident by the continuous increase in conversion when the reaction time was prolonged [114]. As an extension of the heterogeneous version of the catalytic system, the same authors examined the influence of low molecular weight mediators (veratryl alcohol (VA) and 1-hydroxybenzotriazole (HBT)) [115]. Use of low molecular weight mediators in lignin oxidation under heterogeneous conditions is beneficial, since it helps to overcome kinetic problems for the reactivity between lignin and heterogeneous catalyst. In this work, the authors studied oxidation of Black Spruce milled wood and residual kraft lignins with cationic porphyrin, [Mn(TMePyP)(OAc)$_5$], supported on the smectic clay montmorillonite, [Mn(TMePyP)/clay] in an aqueous solution of H_2O_2. The oxidized lignin was analyzed by ^{31}P NMR which allowed for assessing the content of aliphatic, condensed, and non-condensed phenolic and carboxylic groups. While homogeneous version of the catalyst and a heterogeneous version of the catalyst were not active in lignin oxidation in the absence of the mediator, [Mn(TMePyP)/clay] in the presence of VA and especially HBT demonstrated a significant decrease in aliphatic hydroxyl groups and concomitant increase in carboxylic groups. Importantly, the authors did not observe a significant increase in condensed lignin structures, showing that recondensation reactions are prevented under the applied conditions.

Oxidation using Fenton reagents represents another biomimetic approach to lignin depolymerization. Tong and coworkers developed a two-step process for lignin depolymerization where organosolv hardwood lignin was first treated with Fenton reagent (Fe^{3+}/H_2O_2) at room temperature with H_2O_2/Fe^{3+}ratios from 1.98 to 7.92 followed by depolymerization in supercritical ethanol (7 MPa and 250 °C) [116]. The authors studied the structure of lignin after the treatment with Fenton reagent, which is known to generate reactive oxygen species HO· and HOO· and cause demethoxylation [117] and partial ether bond cleavage in lignin, via Py–GC–MS. Decreased relative content of S and G units to H units confirmed the demethoxylation pathway. The lignin was depolymerized into a liquid oil in a 66.0 ± 8.5 wt% yield and char formation constituted (20.0 ± 4.4) wt%. The resulting mixture consisted of a complex mixture of aromatic monophenolics and ring oxidation products (dicarboxylic acids and their esters). When untreated organolsolv lignin was subjected to supercritical ethanol a higher char formation ((36 ± 2.8)%) and lower yield of liquid oil fraction (50.0 ± 5.7 wt%) were observed. The authors proposed that the formation of chelated Fe-lignin complexes suppresses the recondensation reactions and thus decreases the char formation.

Biomimetic oxidation of lignin certainly has potential benefits, such a possibility to perform the reaction under milder conditions. At its current state, this method possesses a higher applicability to delignification (pulp bleaching or water treatment) rather than for the production of targeted products, given the complex reaction mechanism often involving reactive oxygen species that results in a poor selectivity.

9.9 Electrochemical Oxidative Cleavage of Lignin

9.9.1 General Aspects

Electrochemical oxidation of lignin is an attractive approach for lignin valorization as it aligns with a large number of green chemistry principles: [118] (i) electrochemical reactions can allow for milder reaction conditions compared to thermal reactions; (ii) electrooxidation avoids the need for stoichiometric oxidants; (iii) oxidation of a substrate at an anode may be coupled to useful processes at the cathode such as hydrogen gas production or other reductive reactions; and (iv) product identity and/or selectivity often may be tuned by varying the applied potential.

Lignin depolymerization via electrochemical oxidation (reaction at the anode) can be divided into several classes: direct oxidation, indirect oxidation, and electrochemical-chemical combination reactions [11]. In direct electrochemical oxidation, the electron transfer from lignin to the anode takes place directly at the surface of the electrode. In the case of indirect electrooxidation, lignin is oxidized by a homogeneous mediator (e.g. TEMPO or phthalimide N-oxyl (PINO)) in a solution. The mediator is then regenerated via electrochemical oxidation at the anode. Electrochemical–chemical combination reactions are more complicated and feature both direct oxidation of lignin at the surface of the electrode and by electrogenerated reactive species, such as H_2O_2, $O_2-^{\cdot}$, and OH·).

9.10 Direct Electrochemical Oxidation of Lignin

In direct electrochemical oxidations, lignin is ideally fully dissolved in the electrolyte to maximize contact with the surface of the electrode, a consideration that influences the choice of electrolyte. Many reactions have been conducted under alkaline aqueous conditions [119–122]; however, an aqueous reaction medium limits the magnitude of applied potentials, due to the possibility of competing water oxidation at high potentials. Indulin AT lignin was oxidized in alkaline solution at 0.8 V (vs. SCE) with simultaneous production of hydrogen gas at the cathode [123]. In work by Caravaca and coworkers, oxidation of lignin was coupled to hydrogen gas production in a continuous-flow mode at the overall cell potential of 0.45 V [124]. To address the potential limits in aqueous media, ionic liquids (ILs) have been evaluated as an alternative electrolyte for electrochemical oxidation of lignin to expand the potential range [125].

The mechanism of direct electrochemical oxidation of lignin model compounds has been the subject of several studies [119, 126, 127]. Mechanistic studies were performed on β-O-4′ model compounds. Utley and coworkers studied several phenolic and non-phenolic lignin model compounds oxidized in 1.5–2 M NaOH solution at temperatures of 150–160 °C, using a Ni gauze anode [119]. It is crucial to mention that structural changes in lignin occur at elevated temperatures under alkaline conditions in the absence of applied potential, resulting in the formation of vinyl ethers, vide supra. The authors concluded that vinyl ether intermediates were mainly converted into polymeric by-products rather than monomeric species. The

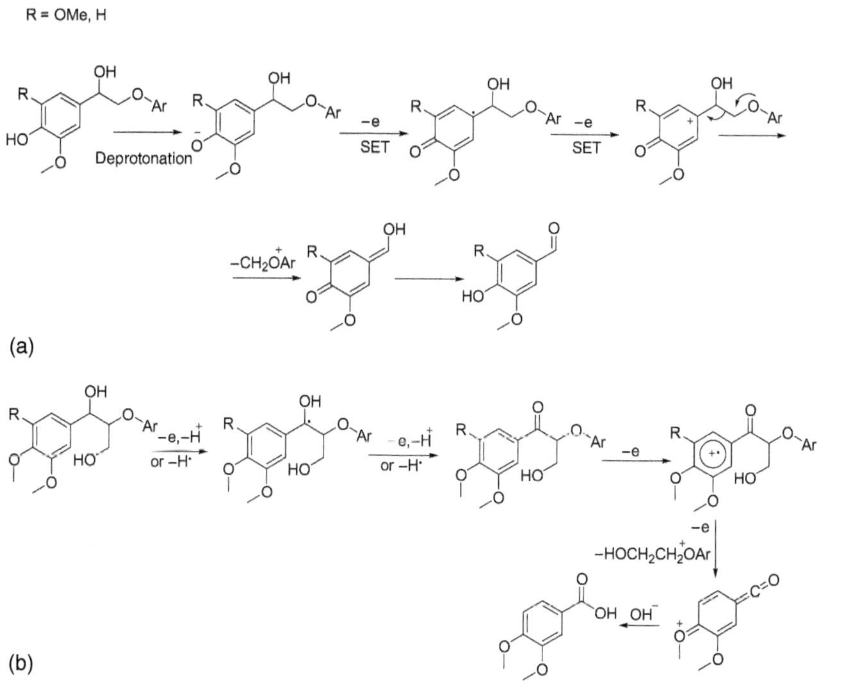

Scheme 9.10 (a) Proposed mechanism for the direct oxidation of phenolic lignin model compounds on Ni anode; (b) proposed mechanism for the direct oxidation of non-phenolic lignin model compounds on Ni anode.

oxidation occurred at the Ni-oxide electrode surface. Under direct electro-oxidative conditions of dimeric lignin models bearing free phenols, the authors observed C—C bond cleavage giving rise to aromatic aldehydes, acids, and phenols as the main products (Scheme 9.10a). When the free phenol was protected as a methoxyl group, cleavage was observed only when the γ-alcohol function was present (possibly reflecting coordination of the γ-alcohol the Ni surface) or the hydroxyl group at Cα position was oxidized to a ketone. Based on these observations, the authors proposed mechanisms for the direct (Ni anode) electrochemical oxidation of lignin (Scheme 9.10a). Alkaline conditions generate phenolates in lignin that can undergo single electron transfer (SET) at the anode surface, followed by a second SET, and subsequent Cα—Cβ bond scission generates the corresponding aldehydes. Cleavage of the lignin models lacking free phenolic groups in addition to the formation of aldehydes resulted in acids. For these models, the authors proposed a mechanistic pathway involving formation of ketones generated via hydrogen atom abstraction or proton coupled electron transfer at the α position (Scheme 9.10b). A subsequent SET and C—C bond scission followed by a nucleophilic attack of OH^- delivers acid as an oxidation product.

A choice of the anode material is governed by its catalytic activity for lignin oxidation and deactivation stability. Anodic corrosion and deposition of polymeric by-products of lignin degradation, formed via radical coupling of monomeric phenols are the main deactivation pathways [120, 128]. One of the most common materials used for anodes is Ni, due to its low price and good performance [121, 129–132]. Metal alloys are also commonly used as anode materials (Ni, Fe, Co, Ti, Mo, W, Al, V) for the direct electrooxidation of lignin [121, 132]. Metal oxides or dimensionally stable anodes [133] (e.g. SnO_2 [134, 135], IrO_2 [136, 137], PbO_2 [122, 138–141], etc.) are especially attractive anodic materials for industrial applications due to their high corrosion stability, and they have been widely studied for electrochemical oxidative depolymerization of lignin.

Lignin oxidation generally leads to complex mixtures of products, however there are systems that give high selectivity toward vanillin. For example, Waldvogel and coworkers obtained vanillin in 0.5–1.8 wt% yields in a mixture of minor amounts of acetovanillone from kraft lignin carrying out electrolysis at 80 °C in 3M NaOH solution with various metal alloys as anodes [121, 142]. Interestingly, the authors observed an improved performance of Ni foam electrode in consecutive runs. They attributed this to the depositions of organic matter on the surface of the electrode, which facilitate the oxidative degradation of lignin by enhancing lipophilicity of the electrode surface and change in the electrode potential. Another example of a selective process was demonstrated by Utley and coworkers, where vanillin was formed in 5–7 wt% yield with acetovanillone as a by-product. The electrolysis was carried out at 145 °C/500 kPa/3 M NaOH with applied potentials in a range of 1.3–1.6 V under flow conditions [129]. Utley and coworkers were able to obtain vanillin in a direct oxidation of kraft lignin at Ni and Co anodes at 80 °C and a constant current density of 1.9 mA cm^{-2} [131]. While giving high selectivity, the process resulted in moderate yields, ranging from 0.7 to 1.8 wt%. In this work, the authors also designed a product separation technique using basic anion exchange resins for the removal of vanillin from the solution. Apart from vanillin and acetovanillone, products of direct electrooxidation of lignin may include syringaldehyde, vanillic acid, syringic acid, acetovanillone, guaiacol and syringol. This product mixture was obtained by direct oxidation of alkali lignin as a 5 wt% solution in protic ionic liquid (PIL, triethylammonium methanesulfonate) on $Ru_{0.25}V_{0.05}Ti_{0.7}O_x$ electrode at 110 °C. The highest total monomer yield (6 wt%) was obtained at a potential of 1.5 V vs. an Ag pseudo reference electrode [143]. Relatively high yields of monomeric products (11.5 wt%) were achieved using Ir-based electrodes ($Ru_{0.2}Mn_{0.2}Ir_{0.6}O_x$ mixed oxide) in triethylammonium methanesulfonate at 70 °C when the applied potential was set to 1.8 V vs. an Ag-quasi-reference electrode (reaction time: 3 hours, lignin content: 10 wt%) [137]. However, the observed selectivity of the process was not high, and the product mixture consisted of 18 compounds, including vanillin, ferulic acid, vanillic and homovanillic acid, sinapyl aldehyde, among others.

9.11 Indirect Electrooxidation of Lignin

In the indirect electrooxidation, lignin is oxidized in a homogeneous fashion by a mediator (e.g. NHPI, TEMPO, violuric acid (VLA)), which in turn is regenerated to its active form at the anode. The mediator needs to exhibit chemical stability in both oxidized and reduced forms and possess a lower redox potential than a

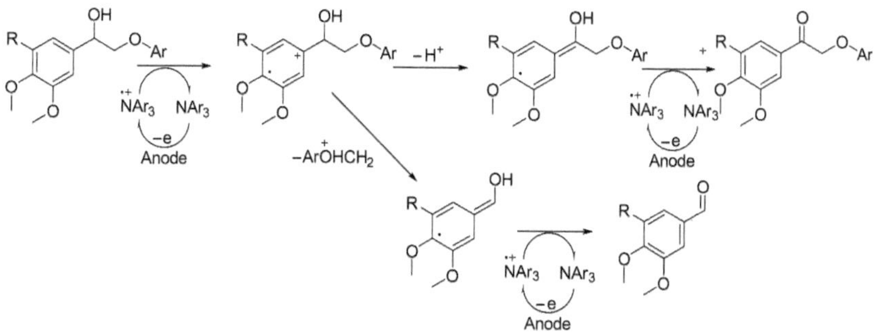

Scheme 9.11 Proposed mechanism for the mediated oxidation of dimeric lignin model compounds.

substrate. Electro-mediated reactions allow performance of reactions at lower potentials and control over the number of transferred electrons in individual oxidation events. The mechanism of oxidation is often specific for each particular oxidative mediator [144, 145]. Utley and coworkers proposed a mechanism for tris(4-bromphenyl)amine mediated oxidation of dimeric lignin model compounds in the presence of 2,6-lutidine (Scheme 9.11) in $MeCN/CH_2Cl_2$ solvent mixture containing $LiClO_4$ and Na_2CO_3 [126].

The reaction begins with SET from the substrate to the oxidized form of the mediator, followed by the deprotonation followed by a second SET/deprotonation event resulting in formation of the benzylic ketone or C—C bond cleavage leading to the formation of the aldehyde. In this mediated system (unlike the direct oxidation) the cleavage of lignin model compounds did not require the presence of a free phenolic group or γ-alcohol moiety.

TEMPO and NHPI are among the most widely studied electrochemical mediators [146, 147]. In the case of TEMPO, the oxidation takes place via an ionic mechanism where oxoammonium cation is formed by the anodic oxidation of TEMPO. Stahl and coworkers developed a system where organosolv lignin was oxidized in an indirect electrochemical process with 4-AcNH-TEMPO as a mediator [23]. This process did not directly result in the formation of monomeric products. A subsequent depolymerization step was achieved under acidic conditions. This two-step procedure was applied to poplar lignin, where electrochemical oxidation was followed by acid treatment to yield the aldehyde and other monomeric and dimeric compounds in up to 29 wt% yields.

Another example of mediated electrooxidation of lignin without depolymerization is the NHPI-mediated oxidation of lignin developed by Stephenson and coworkers [148]. Lignin was selectively oxidized at the benzylic position, and the depolymerization of the oxidized lignin was achieved in a photocatalytic process. Other mediators were also used for electrochemical oxidation of lignin. For example, VLA-mediated electrochemical delignification of pulp as an alternative to chlorine was applied in pulp bleaching processes [149]. POM-promoted electrooxidation of lignin was realized by Deng and coworkers. The authors performed depolymerization of kraft lignin into monomeric products in 14 wt% yield, with phosphomolybdic acid ($H_3PMo_{12}O_{40}$) or Fe^{3+} as mediator using a graphite anode at

100 °C [150]. This indirect approach eliminated the necessity of a noble metal anode, since lignin oxidation proceeded in the solution. Electroenzymatic oxidation of lignin is yet another type of indirect electrooxidation of lignin. Lignin peroxidase [151] and manganese peroxidase [152] mediated processes were developed and applied to lignin model compounds.

9.12 Lignin Oxidation via Electrochemical Combination Reactions

During this type of electrooxidation, several different types of oxidation events are occurring. In addition to direct oxidation of lignin at the anode, oxidation by reactive oxygen species (ROS) generated either at the cathode or anode, when the reaction is performed in an undivided cell. ROS can be generated at an anode from water via either one-electron (OH·), two-electron (H_2O_2) or four-electron (O_2) processes. ROS can also be generated at the cathode via one-electron reduction of oxygen to superoxide ($O_2^{-\cdot}$) or two-electron reduction to H_2O_2 [153]. An example of the latter is the lignin oxidation in cylindrical electrolytic cell with RuO_2–IrO_2/Ti anode, whereby 59.2% of lignin was converted into low molecular weight products after one hour of electrolysis at 80 °C. The product mixture consisted of guaiacol- and syringol-derived aldehydes and aromatic ketones and acids [154].

Lignin oxidation involving ROS can lead to the formation of products resulting from the cleavage of the aromatic rings. An example is an electrolysis of lignin using a Pb/PbO_2 anode and copper sheet as an auxiliary electrode, where the decomposition of the aromatic rings of lignin occurred by superoxide species ($O_2^{-\cdot}$) generated via water oxidation followed by the reduction of the oxygen to superoxide [140]. When hydroxyl radicals (OH·) are involved in the oxidation process, lower selectivity is observed. Lignin depolymerization by OH· generated at the Pb/PbO_2 anode, followed by their hydrogenation at the Cu/Ni–Mo–Co cathode was demonstrated by the groups of Fan and Li and resulted in 24 different products, including *trans*-ferulic acid, vanillin, 3-hydroxy-4-methoxyphenyl-ethanone, syringaldehyde, acetosyringone and 4-methoxy-3-methyl-phenol [155]. The same authors studied the depolymerization of cornstalk lignin in a similar process combining OH· assisted oxidative depolymerization followed by hydrogenation at the Ni cathode. The monomeric products were obtained in 3.5 wt% including toluene, *m*-xylene, *o*-xylene and anisole [156]. ROS assisted lignin depolymerization in ILs was also reported. Volmer and coworkers studied the oxidative depolymerization of lignin in ionic liquids (1-ethyl-3-methylimidazolium trifluoromethanesulfonate). In the presence of water, hydroperoxyl radicals promoted lignin depolymerization into low molecular weight products. Complex mixtures were obtained, and the products were not quantified [157].

While electrochemical oxidation of lignin is a potentially promising method for lignin upgrading, it shows several limitations that need to be addressed to improve its feasibility. The main challenges are low selectivity, high working potentials required for lignin oxidation and limited options for electrolytes. Thus, future efforts must be directed toward improving products separation methods, development of novel catalysts that work at lower potentials and design of alternative electrolyte systems.

9.13 Concluding Remarks and Outlook

The research field of oxidative depolymerization of lignin has evolved from technologies to degrade lignin that remain after pulping to selective depolymerization of lignin to give monophenolic compounds or quinones. Benign methods using air as a terminal oxidant or electrochemical oxidation have advantages in terms of environmental and economic sustainability. Even though a lot of research has been performed, aerobic alkaline oxidation still gives the highest yields of monophenols. In fact, aerobic alkaline oxidation gives close to and even beyond the theorctical maximum yield when benchmarked to nitrobenzene oxidation. The reaction mechanism is still under debate, and especially the role of the catalyst is not understood.

The major technical challenge with oxidative depolymerizations is selectivity. The over-oxidation of the final products and side reactions (e.g. radical coupling, Friedel–Crafts alkylation, etc.) of the formed intermediates or products limit yields of the targeted products. The presence of several sites for the oxidation (α hydroxyl group, γ hydroxyl group, aromatic ring) imposes an additional challenge. In this chapter, several strategies disclose very selective oxidation protocols. For instance, the procedures developed by Stahl et al. show that oxidation can be steered to either the α-position or γ-position, and that the γ-position can be oxidized to either aldehyde or carboxylic acid, illustrating that selectivity issues can be addressed.

A challenge for this research direction is a discrepancy in reporting yields and protocols. A guideline on how to report yields and correlate these yields to the feedstock would be helpful for the community. For analysis of feedstock and correlation to yields, the guidelines written for 'lignin-first' could be used [34]. A report by Roman-Leshkov and Beckham is a nice example of a thorough study that correlate yields to feedstock and compare to NBO oxidations [45]. Furthermore, reporting mass balances and quality of all feedstock is important. Another challenge for the whole biomass transformation community is the lack of standardized feedstocks [158]. If such feedstocks were present, it would be more straight forward to compare different methodologies.

Acknowledgments

ES thanks Marie Skłodowska-Curie foundation for funding. SSS is supported by the Great Lakes Bioenergy Research Center, U.S. Department of Energy, Office of Science, Office of Biological and Environmental Research under Award Number DE-SC0018409. JS thanks Swedish Foundation for Strategic Environmental Research (Mistra: project Mistra SafeChem, project number 2018/11). The authors thank Ciaran Lahive and Gregg Beckham for fruitful discussions.

References

1 Crestini, C., Lange, H., Sette, M., and Argyropoulos, D.S. (2017). *Green Chemistry* 19: 4104–4121.

2 Lancefield, C.S., Wienk, H.L.J., Boelens, R. et al. (2018). *Chemical Science* 9: 6348–6360.

3 Sixta, H., Potthast, A., and Krotschek, A.W. (2006, https://doi.org/10.1002/9783527619887.ch4a). *Handbook of Pulp*, 109–229. Wiley.

4 Ma, R., Xu, Y., and Zhang, X. (2015). *ChemSusChem* 8: 24–51.

5 Behling, R., Valange, S., and Chatel, G. (2016). *Green Chemistry* 18: 1839–1854.

6 Costa, C.A.E., Vega-Aguilar, C.A., and Rodrigues, A.E. (2021). *Molecules* 26: 4602.

7 Tarabanko, V.E. and Tarabanko, N. (2017). *International Journal of Molecular Sciences* 18: 2421.

8 Vangeel, T., Schutyser, W., Renders, T., and Sels, B.F. (2020, https://doi.org/10.1007/978-3-030-00590-0_3). *Lignin Chemistry* (ed. L. Serrano, R. Luque, and B.F. Sels), 53–68. Cham: Springer International Publishing.

9 Cui, Y., Goes, S.L., and Stahl, S.S. (2021). *Advances in Inorganic Chemistry*, vol. vol. 77 (ed. P.C. Ford and R. van Eldik), 99–136. Academic Press.

10 Gharehkhani, S., Zhang, Y., and Fatehi, P. (2019). *Progress in Energy and Combustion Science* 72: 59–89.

11 Du, X., Zhang, H., Sullivan, K.P. et al. (2020). *ChemSusChem* 13: 4318–4343.

12 Erdtman, H., Leopold, B., Linderholm, H. et al. (1948). *Acta Chemica Scandinavica* 2: 1358–1374.

13 Rolando, C., Monties, B., and Lapierre, C. (1992, https://doi.org/10.1007/978-3-642-74065-7_23). *Methods in Lignin Chemistry* (ed. S.Y. Lin and C.W. Dence), 334–349. Berlin, Heidelberg: Springer.

14 Chen, C.L. (1992, https://doi.org/10.1007/978-3-642-74065-7_21). *Methods in Lignin Chemistry* (ed. S.Y. Lin and C.W. Dence), 301–321. Berlin, Heidelberg: Springer.

15 Sixta, H., Süss, H.-U., Potthast, A. et al. (2006, https://doi.org/10.1002/9783527619887.ch7a). *Handbook of Pulp*, 609–708.

16 Chuck, C.J., Parker, H.J., Jenkins, R.W., and Donnelly, J. (2013). *Bioresource Technology* 143: 549–554.

17 Figueirêdo, M.B., Deuss, P.J., Venderbosch, R.H., and Heeres, H.J. (2019). *ACS Sustainable Chemistry & Engineering* 7: 4755–4765.

18 Figueirêdo, M.B., Keij, F.W., Hommes, A. et al. (2019). *ACS Sustainable Chemistry & Engineering* 7: 18384–18394.

19 Figueirêdo, M.B., Heeres, H.J., and Deuss, P.J. (2020). *Sustainable Energy & Fuels* 4: 265–276.

20 Silverman, J.R., Danby, A.M., and Subramaniam, B. (2019). *Reaction Chemistry & Engineering* 4: 1421–1430.

21 Danby, A.M., Lundin, M.D., and Subramaniam, B. (2018). *ACS Sustainable Chemistry & Engineering* 6: 71–76.

22 Rahimi, A., Azarpira, A., Kim, H. et al. (2013). *Journal of the American Chemical Society* 135: 6415–6418.

23 Rafiee, M., Alherech, M., Karlen, S.D., and Stahl, S.S. (2019). *Journal of the American Chemical Society* 141: 15266–15276.
24 Dabral, S., Wotruba, H., Hernández, J.G., and Bolm, C. (2018). *ACS Sustainable Chemistry & Engineering* 6: 3242–3254.
25 Dabral, S., Hernández, J.G., Kamer, P.C.J., and Bolm, C. (2017). *ChemSusChem* 10: 2707–2713.
26 Gao, Y., Zhang, J., Chen, X. et al. (2014). *ChemPlusChem* 79: 825–834.
27 Li, H., Wang, M., Liu, H. et al. (2018). *ACS Sustainable Chemistry & Engineering* 6: 3748–3753.
28 Shuai, L., Amiri, M.T., Questell-Santiago, Y.M. et al. (2016). *Science* 354: 329–333.
29 Lan, W., de Bueren, J.B., and Luterbacher, J.S. (2019). *Angewandte Chemie, International Edition* 58: 2649–2654.
30 Tran, F., Lancefield, C.S., Kamer, P.C.J. et al. (2015). *Green Chemistry* 17: 244–249.
31 Zhang, C., Li, H., Lu, J. et al. (2017). *ACS Catalysis* 7: 3419–3429.
32 Galkin, M.V. and Samec, J.S.M. (2016). *ChemSusChem* 9: 1544–1558.
33 Renders, T., Van den Bossche, G., Vangeel, T. et al. (2019). *Current Opinion in Biotechnology* 56: 193–201.
34 Abu-Omar, M.M., Barta, K., Beckham, G.T. et al. (2021). *Energy & Environmental Science* 14: 262–292.
35 Subbotina, E., Rukkijakan, T., Marquez-Medina, M.D. et al. (2021). *Nature Chemistry* 13: 1118–1125.
36 Fache, M., Boutevin, B., and Caillol, S. (2016). *ACS Sustainable Chemistry & Engineering* 4: 35–46.
37 Araújo, J.D.P., Grande, C.A., and Rodrigues, A.E. (2010). *Chemical Engineering Research and Design* 88: 1024–1032.
38 Hocking, M.B. (1997). *Journal of Chemical Education* 74: 1055.
39 Tomlinson, G.H. and Hibbert, H. (1936). *Journal of the American Chemical Society* 58: 345–348.
40 Tomlinson, G.H. and Hibbert, H. (1936). *Journal of the American Chemical Society* 58: 348–353.
41 *Unites States Pat.*, US2516827, 1950.
42 *United States Pat.*, US2692291, 1954.
43 Fargues, C., Mathias, Á., and Rodrigues, A. (1996). *Industrial and Engineering Chemistry Research* 35: 28–36.
44 Santos, S.G., Marques, A.P., Lima, D.L.D. et al. (2011). *Industrial and Engineering Chemistry Research* 50: 291–298.
45 Schutyser, W., Kruger, J.S., Robinson, A.M. et al. (2018). *Green Chemistry* 20: 3828–3844.
46 Rodrigues Pinto, P.C., Borges da Silva, E.A., and Rodrigues, A.E. (2011). *Industrial and Engineering Chemistry Research* 50: 741–748.
47 Pinto, P.C.R., Costa, C.E., and Rodrigues, A.E. (2013). *Industrial and Engineering Chemistry Research* 52: 4421–4428.
48 Lyu, G., Yoo, C.G., and Pan, X. (2018). *Biomass and Bioenergy* 108: 7–14.
49 Xiang, Q. and Lee, Y.Y. (2000). Oxidative cracking of precipitated hardwood lignin by hydrogen peroxide. *Appl. Biochem. Biotechnol.* 84: 153–162.

50 Wu, G. and Heitz, M. (1995). *Journal of Wood Chemistry and Technology* 15: 189–202.

51 Ouyang, X., Ruan, T., and Qiu, X. (2016). *Fuel Processing Technology* 144: 181–185.

52 Yuan, Z., Klinger, G.E., Nikafshar, S. et al. (2021). *ACS Sustainable Chemistry & Engineering* 9: 1118–1127.

53 Das, A., Rahimi, A., Ulbrich, A. et al. (2018). *ACS Sustainable Chemistry & Engineering* 6: 3367–3374.

54 Deng, H., Lin, L., Sun, Y. et al. (2009). *Energy & Fuels* 23: 19–24.

55 Deng, H., Lin, L., and Liu, S. (2010). *Energy & Fuels* 24: 4797–4802.

56 Zhang, J., Deng, H., and Lin, L. (2009). *Molecules* 14: 2747–2757.

57 Deng, H., Lin, L., Sun, Y. et al. (2008). *Catalysis Letters* 126: 106.

58 Cabral Almada, C., Kazachenko, A., Fongarland, P. et al. (2021). *Catalysts* 11: 467.

59 Gierer, J., Imsgard, F., and Norén, I. (1977). *Acta Chemica Scandinavica* 31b: 561–572.

60 Tarabanko, V.E., Fomova, N.A., Kuznetsov, B.N. et al. (1995). *Reaction Kinetics and Catalysis Letters* 55: 161–170.

61 Tarabanko, V.E., Petukhov, D.V., and Selyutin, G.E. (2004). *Kinetics and Catalysis* 45: 569–577.

62 Hirano, Y., Izawa, A., Hosoya, T., and Miyafuji, H. (2022). *Reaction Chemistry & Engineering* https://doi.org/10.1039/D2RE00036A.

63 Hosoya, T., Yamamoto, K., Miyafuji, H., and Yamada, T. (2020). *RCS Advances* 10: 19199–19210.

64 Rinaldi, R., Jastrzebski, R., Clough, M.T. et al. (2016). *Angewandte Chemie, International Edition* 55: 8164–8215.

65 Renders, T., Van den Bosch, S., Koelewijn, S.F. et al. (2017). *Energy & Environmental Science* 10: 1551–1557.

66 Alherech, M., Omolabake, S., Holland, C.M. et al. (2021). *ACS Central Science* 7: 1831–1837.

67 Lu, L., Liao, Y., Lv, W. et al. (2021). *ChemRxiv* https://doi.org/10.26434/chemrxiv.14558814.v1.

68 Zhu, Y., Liao, Y., Lv, W. et al. (2020). *ACS Sustainable Chemistry & Engineering* 8: 2361–2374.

69 Luo, H., Weeda, E.P., Alherech, M. et al. (2021). *Journal of the American Chemical Society* 143: 15462–15470.

70 Du, X., Tricker, A.W., Yang, W. et al. (2021). *ACS Sustainable Chemistry & Engineering* 9: 7719–7727.

71 Herrmann, W.A., Fischer, R.W., Scherer, W., and Rauch, M.U. (1993). *Angewandte Chemie International Edition in English* 32: 1157–1160.

72 Lange, H., Decina, S., and Crestini, C. (2013). *European Polymer Journal* 49: 1151–1173.

73 Crestini, C., Crucianelli, M., Orlandi, M., and Saladino, R. (2010). *Catalysis Today* 156: 8–22.

74 Crestini, C., Pro, P., Neri, V., and Saladino, R. (2005). *Bioorganic and Medicinal Chemistry* 13: 2569–2578.

75 Saladino, R., Fiani, C., Crestini, C. et al. (2007). *Journal of Natural Products* 70: 39–42.

76 Harms, R.G., Markovits, I.I.E., Drees, M. et al. (2014). *ChemSusChem* 7: 429–434.

77 Crestini, C., Caponi, M.C., Argyropoulos, D.S., and Saladino, R. (2006). *Bioorganic and Medicinal Chemistry* 14: 5292–5302.

78 Zombeck, A., Drago, R.S., Corden, B.B., and Gaul, J.H. (1981). *Journal of the American Chemical Society* 103: 7580–7585.

79 Cedeno, D. and Bozell, J.J. (2012). *Tetrahedron Letters* 53: 2380–2383.

80 Cooper, C.J., Alam, S., Nziko, V.D.P.N. et al. (2020). *ACS Sustainable Chemistry & Engineering* 8: 7225–7234.

81 Biannic, B. and Bozell, J.J. (2013). *Organic Letters* 15: 2730–2733.

82 Zhou, X.-F. (2014). *Journal of Applied Polymer Science* 131.

83 Ishii, Y. (1997). *Journal of Molecular Catalysis A: Chemical* 117: 123–137.

84 Rinesch, T. and Bolm, C. (2018). *ACS Omega* 3: 8386–8392.

85 Hanson, S.K., Baker, R.T., Gordon, J.C. et al. (2010). *Inorganic Chemistry* 49: 5611–5618.

86 Hanson, S.K., Wu, R., and Silks, L.A.P. (2012). *Angewandte Chemie, International Edition* 51: 3410–3413.

87 Son, S. and Toste, F.D. (2010). *Angewandte Chemie, International Edition* 49: 3791–3794.

88 Jiang, Y.-Y., Yan, L., Yu, H.-Z. et al. (2016). *ACS Catalysis* 6: 4399–4410.

89 Sedai, B., Díaz-Urrutia, C., Baker, R.T. et al. (2011). *ACS Catalysis* 1: 794–804.

90 Hanson, S.K. and Baker, R.T. (2015). *Accounts of Chemical Research* 48: 2037–2048.

91 Chan, J.M.W., Bauer, S., Sorek, H. et al. (2013). *ACS Catalysis* 3: 1369–1377.

92 Walch, F., Abdelaziz, O.Y., Meier, S. et al. (2021). *Catalysis Science & Technology* 11: 1843–1853.

93 Díaz-Urrutia, C., Chen, W.-C., Crites, C.-O. et al. (2015). *RCS Advances* 5: 70502–70511.

94 Zhao, X., Zhang, Y., Yang, M. et al. (2018). *Polymers* 10: 907.

95 Mottweiler, J., Puche, M., Räuber, C. et al. (2015). *ChemSusChem* 8: 2106–2113.

96 Lotfi, S., Boffito, D.C., and Patience, G.S. (2015). *ChemSusChem* 8: 3424–3432.

97 Lotfi, S., Bahrpaima, K., Boffito, D.C., and Patience, G.S. (2020). *Energy & Fuels* 34: 9683–9696.

98 Ma, R., Guo, M., and Zhang, X. (2014). *ChemSusChem* 7: 412–415.

99 Ren, X., Wang, P., Han, X. et al. (2017). *ACS Sustainable Chemistry & Engineering* 5: 6548–6556.

100 Sales, F.G., Abreu, C.A.M., and Pereira, J.A.F.R. (2004). *Brazilian Journal of Chemical Engineering* 21.

101 Sales, F.G., Maranhão, L.C.A., Filho, N.M.L., and Abreu, C.A.M. (2007). *Chemical Engineering Science* 62: 5386–5391.

102 Deng, W., Zhang, H., Wu, X. et al. (2015). *Green Chemistry* 17: 5009–5018.

103 Song, Y., Mobley, J.K., Motagamwala, A.H. et al. (2018). *Chemical Science* 9: 8127–8133.

104 Partenheimer, W. (2009). *Advanced Synthesis and Catalysis* 351: 456–466.

105 Bujanovic, B., Reiner, R.S., Ralph, S.A., and Atalla, R.H. (2011). *Journal of Wood Chemistry and Technology* 31: 121–141.

106 Sonnen, D.M., Reiner, R.S., Atalla, R.H., and Weinstock, I.A. (1997). *Industrial and Engineering Chemistry Research* 36: 4134–4142.

107 Gaspar, A.R., Gamelas, J.A.F., Evtuguin, D.V., and Pascoal Neto, C. (2007). *Green Chemistry* 9: 717–730.
108 Gaspar, A.R., Evtuguin, D.V., and Neto, C.P. (2004). *Industrial and Engineering Chemistry Research* 43: 7754–7761.
109 Weinstock, I.A., Barbuzzi, E.M.G., Wemple, M.W. et al. (2001). *Nature* 414: 191–195.
110 Voitl, T. and Rudolf von Rohr, P. (2008). *ChemSusChem* 1: 763–769.
111 Voitl, T. and Rohr, P.R.V. (2010). *Industrial and Engineering Chemistry Research* 49: 520–525.
112 Li, C., Zhao, X., Wang, A. et al. (2015). *Chemical Reviews* 115: 11559–11624.
113 Crestini, C., Saladino, R., Tagliatesta, P., and Boschi, T. (1999). *Bioorganic and Medicinal Chemistry* 7: 1897–1905.
114 Crestini, C., Pastorini, A., and Tagliatesta, P. (2004). *Journal of Molecular Catalysis A: Chemical* 208: 195–202.
115 Crestini, C., Pastorini, A., and Tagliatesta, P. (2004). *European Journal of Inorganic Chemistry* 2004: 4477–4483.
116 Zeng, J., Yoo, C.G., Wang, F. et al. (2015). *ChemSusChem* 8: 861–871.
117 Hildén, L., Johansson, G., Pettersson, G. et al. (2000). *FEBS Letters* 477: 79–83.
118 Anastas, P.T. and Warner, J.C. (1998). *Green Chemistry: Theory and Practice.* Oxford [England]; New York: Oxford University Press.
119 Pardini, V.L., Smith, C.Z., Utley, J.H.P. et al. (1991). *The Journal of Organic Chemistry* 56: 7305–7313.
120 Parpot, P., Bettencourt, A.P., Carvalho, A.M., and Belgsir, E.M. (2000). *Journal of Applied Electrochemistry* 30: 727–731.
121 Zirbes, M., Schmitt, D., Beiser, N. et al. (2019). *ChemElectroChem* 6: 155–161.
122 Chen, A., Wen, Y., Han, X. et al. (2019). *Environmental Progress & Sustainable Energy* 38: 13117.
123 Lalvani, S.B. and Rajagopal, P. (1993). *Holzforschung* 47: 283–286.
124 Caravaca, A., Garcia-Lorefice, W.E., Gil, S. et al. (2019). *Electrochemistry Communications* 100: 43–47.
125 Chatel, G. and Rogers, R.D. (2014). *ACS Sustainable Chemistry & Engineering* 2: 322–339.
126 Pardini, V.L., Vargas, R.R., Viertler, H., and Utley, J.H.P. (1992). *Tetrahedron* 48: 7221–7228.
127 Shiraishi, T., Takano, T., Kamitakahara, H., and Nakatsubo, F. (2012). *Holzforschung* 66: 303–309.
128 Ežerskis, Z. and Jusys, Z. (2001). *Journal of Applied Electrochemistry* 31: 1117–1124.
129 Smith, C.Z., Utley, J.H.P., and Hammond, J.K. (2011). *Journal of Applied Electrochemistry* 41: 363–375.
130 Di Marino, D., Jestel, T., Marks, C. et al. (2019). *ChemElectroChem* 6: 1434–1442.
131 Schmitt, D., Regenbrecht, C., Hartmer, M. et al. (2015). *Beilstein Journal of Organic Chemistry* 11: 473–480.
132 Ghahremani, R. and Staser, J.A. (2018). *Holzforschung* 72: 951–960.
133 Trasatti, S. (2000). *Electrochimica Acta* 45: 2377–2385.

134 Jia, Y., Wen, Y., Han, X. et al. (2018). *Catalysis Science & Technology* 8: 4665–4677.

135 Shao, D., Liang, J., Cui, X. et al. (2014). *Chemical Engineering Journal* 244: 288–295.

136 Tolba, R., Tian, M., Wen, J. et al. (2010). *Journal of Electroanalytical Chemistry* 649: 9–15.

137 Rauber, D., Dier, T.K.F., Volmer, D.A., and Hempelmann, R. (2018). *Zeitschrift für Physikalische Chemie* 232: 189–208.

138 Pan, K., Tian, M., Jiang, Z.-H. et al. (2012). *Electrochimica Acta* 60: 147–153.

139 Hao, X., Quansheng, Y., Dan, S. et al. (2015). *Journal of Hazardous Materials* 286: 509–516.

140 Zhang, Y.-M., Peng, Y., Yin, X.I. et al. (2014). *Journal of Chemical Technology & Biotechnology* 89: 1954–1960.

141 Liu, M., Wen, Y., Qi, J. et al. (2017). *ChemistrySelect* 2: 4956–4962.

142 Zirbes, M., Quadri, L.L., Breiner, M. et al. (2020). *ACS Sustainable Chemistry & Engineering* 8: 7300–7307.

143 Reichert, E., Wintringer, R., Volmer, D.A., and Hempelmann, R. (2012). *Physical Chemistry Chemical Physics* 14: 5214–5221.

144 Shiraishi, T., Takano, T., Kamitakahara, H., and Nakatsubo, F. (2012). *Holzforschung* 66: 311–315.

145 Sannami, Y., Kamitakahara, H., and Takano, T. (2017). *Holzforschung* 71: 109–117.

146 Yang, C., Maldonado, S., and Stephenson, C.R.J. (2021). *ACS Catalysis* 11: 10104–10114.

147 Nutting, J.E., Rafiee, M., and Stahl, S.S. (2018). *Chemical Reviews* 118: 4834–4885.

148 Bosque, I., Magallanes, G., Rigoulet, M. et al. (2017). *ACS Central Science* 3: 621–628.

149 Mickel, M., Kim, H.-C., and Hampp, N. (2003). *Green Chemistry* 5: 8–14.

150 Du, X., Liu, W., Zhang, Z. et al. (2017). *ChemSusChem* 10: 847–854.

151 Ko, M., Pham, L.T.M., Sa, Y.J. et al. (2019). *Nature Communications* 10: 5123.

152 Rouco, L., Fernández-García, M.I., González-Noya, A.M. et al. (2019). *Applied Sciences* 9: 3430.

153 Jiang, H., Xue, A., Wang, Z. et al. (2020). *Sustainable Chemistry* 1: 345–360.

154 Zhu, H., Wang, L., Chen, Y. et al. (2014). *RCS Advances* 4: 29917–29924.

155 Cai, P., Fan, H., Cao, S. et al. (2018). *Electrochimica Acta* 264: 128–139.

156 Lan, C., Fan, H., Shang, Y. et al. (2020). *Sustainable Energy & Fuels* 4: 1828–1836.

157 Dier, T.K.F., Rauber, D., Durneata, D. et al. (2017). *Scientific Reports* 7: 5041.

158 Galkin, M.V., Di Francesco, D., Edlund, U., and Samec, J.S.M. (2017). *Faraday Discussions* 202: 281–301.

10

Photocatalytic Conversion of Lignin

Xuejiao Wu[1,2], Shunji Xie[1] and Ye Wang[1]

[1] *Xiamen University, State Key Laboratory of Physical Chemistry of Solid Surfaces, Collaborative Innovation Center of Chemistry for Energy Materials, National Engineering Laboratory for Green Chemical Productions of Alcohols, Ethers and Esters, College of Chemistry and Chemical Engineering, 422 Siming S Rd, Xiamen, 361005, China*
[2] *KU Leuven, Centre for Sustainable Catalysis and Engineering, Faculty of Bioscience Engineering, Celestijnenlaan 200F, Heverlee, 3001, Belgium*

10.1 Introduction

Solar light is by far the largest renewable power source on earth. However, the diffuse and intermittent nature of sunlight exerts difficulty in its direct use. Photocatalysis, in which photocatalysts absorb proton energy and then drive chemical transformations, has been recognized as an effective approach to directly harnessing the clean, abundant, and inexhaustible solar energy [1–3]. Recently, substantial research progress in photocatalysis has been achieved in the depolymerization of biomass, a process that can be thought of as reverse photosynthesis [4–7]. This emerging field has gained a lot of attention in the scientific community, with much research focusing on the selective conversion of lignin [8–10].

Lignin is an irregular aromatic polymer that is composed of various methoxylated phenylpropanoid units connected by C—O (e.g. β-O-4 and α-O-4) bonds and C—C (e.g. β-5 and 5-5) bonds [11]. As the most abundant resource of renewable aromatics, the conversion of lignin by selective scission of the cross-linked C—O and/or C—C bonds provides a feasible and sustainable approach to preparing aromatic chemicals [12]. Notably, aromatics in lignin are characterized by their functionalized nature, including phenolic, methoxyl, and propanoic groups. Aromatics with high functionality, which is commonly found in specialties and fine chemicals, often imply high values. Currently, these value-added aromatics are mainly obtained from the functionalization of petroleum-based "BTX" (benzene, toluene, and xylenes), and these processes involve tedious synthetic procedures [13]. Therefore, it is highly desirable to obtain functionalized aromatics directly from lignin. To achieve selective depolymerization of lignin and produce functionalized aromatics,

Lignin Chemistry: Characterization, Isolation, and Valorization, First Edition.
Edited by Yuhe Liao and Bert F. Sels.

it is essential to use mild reaction conditions. High temperatures or pressures can result in unwanted side reactions such as de-functionalization or humin formation. Moreover, the production of valuable, reactive products like aldehydes or acids is difficult to achieve under severe conditions. Therefore, photocatalysis is a promising approach for the selective conversion of lignin, as it enables selective chemical transformation under mild conditions [8–10].

Generally, photocatalysis is initiated by the light absorption of photocatalysts, transforming photocatalysts into their excited state. Photocatalysts involved in the conversion of lignin mainly refer to semiconductors and molecular photocatalysts (MPC). The excitation of semiconductors can produce charge carriers (electrons and holes), and the excitation of MPC results in excited chemicals with redox ability. The photoinduced charge carriers or excited chemicals can directly activate lignin (Direct approach in Figure 10.1a and b) or result in the formation of active species to activate lignin (Indirect approach in Figure 10.1a and b), generating redox intermediates. In other words, charge carriers and excited chemicals, which have redox ability, may replace the stoichiometric reductant (e.g. H_2) and/or oxidant (e.g. H_2O_2) that are usually required in thermocatalytic systems for lignin conversion. The

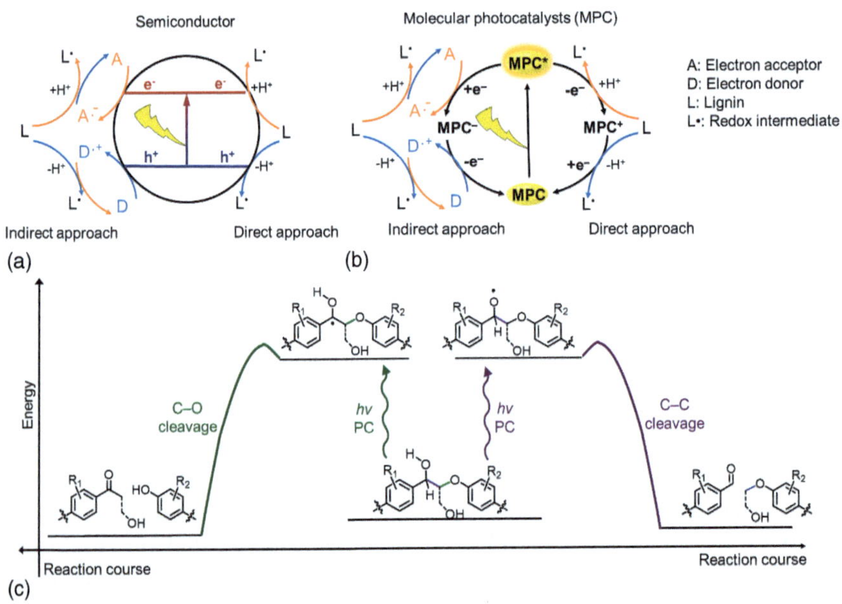

Figure 10.1 Fundamentals of photocatalytic lignin conversion. Schematic illustration of photocatalytic conversion of lignin over (a) semiconductors and (b). MPC. Semiconductors and MPC are the most investigated catalysts for photocatalytic lignin conversion, both the direct and indirect approaches are illustrated. The blue and orange lines in (a) and (b) denote oxidative and reductive transformations, respectively. (c) An example of lignin β-O-4 scission through C_β—O or C_α—C_β bond cleavage, which is dependent on the formed redox intermediates. Both redox radical intermediates are of high energy, they can facilely undergo following bond cleavage with a low energy barrier. Note that the energy scale here is only a schematic illustration. Source: Ye Wang.

redox intermediates from lignin activation, which mainly refers to radical species and is characterized by their active and energetic nature, can facilely undergo sequent transformations with a very low or even zero energy barrier, allowing bond cleavage to be performed under mild conditions (Figure 10.1c) [4]. The mild and environmental-benign conditions that are commonly employed in the photocatalytic conversion of lignin enable the cleavage of specific C—O/C—C bonds with other functional groups remaining intact.

In addition to the mild reaction conditions, the distinctive reaction patterns of photocatalysis involving photo-generated charge carriers or photoexcited species have provided extra opportunities beyond the accessibility of conventional catalysis. For instance, photocatalysis allows overall redox-neutral reactions to take place in one-pot thanks to the concurrent presence of reductive species/electrons and oxidative species/holes [2]. In contrast, other approaches usually require multiple steps because the required stoichiometric chemical oxidants and reductants are often incompatible with each other [12]. The cleavage of the C—O bond in lignin β-O-4 motif represents one example to illustrate the uniqueness of photocatalysis. The two-step strategy for the cleavage of β-O-4 motif, that is peroxidation of C_α—OH to a ketone intermediate that weakened adjacent C—O bond followed by reductive cleavage of the C—O bond, has been proven to be effective for the depolymerization of lignin [14]. Interestingly, the two-step process can be realized concurrently by the photogenerated electrons and holes over semiconductor photocatalysts in one step [15]. This redox mechanism will be unfolded in the following text.

Furthermore, charge transfer leads to the generation of active intermediates, majorly radical species, which possess fundamentally different properties from their ground states. These photoinduced active intermediates can participate in the following transformations rapidly, allowing kinetically difficult or high energy-barrier transformations to be performed with high reactivity [16]. Typical examples are the generation of carbon or oxygen radical intermediates that facilitate the efficient cleavage of the C—O or C—C bonds (Figure 10.1c) [7]. Access to these intermediates using other approaches is typically challenging, the addition of hazardous/toxic additives (e.g. BEt_3, Bu_3SnH) or high temperatures is usually required [17]. It is important to note that the unique reactivity of radicals cannot be harnessed productively under severe conditions. For instance, carbon radical intermediates generated during lignin acidolysis at high temperatures typically undergo unselective C—C re-polymerization, which limits the selectivity of the process.

The present book chapter is devoted to provide an overview of major progress in the photocatalytic conversion of lignin to produce aromatic products. The emphasis of this contribution is placed on the unique photocatalytic chemistry for the selective C—O and C—C bond cleavage. The unselective degradation of lignin without generating useful organic products by photocatalytic or photochemical approaches is out of this chapter's scope. Due to the irregular and complex structures of lignin polymers, relatively simple model compounds with typical lignin bonding patterns are often used to develop efficient photocatalytic systems and to investigate reaction mechanisms; some photocatalytic systems were also employed for the conversion of

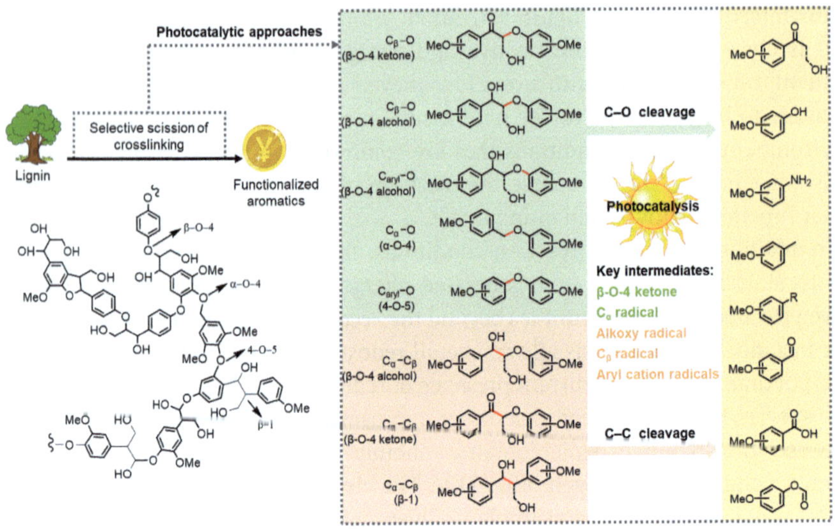

Figure 10.2 Photocatalytic conversion of lignin through selective cleavage of C—O and C—C bonds involved in this chapter, which are highlighted in red, and a representative lignin fragment. Source: Ye Wang.

extracted lignin or native lignin. According to the target chemical bonds in lignin or model compounds, distinct photocatalytic methods with different mechanisms have been developed, which usually lead to different products. To this end, we summarized the photocatalytic systems for lignin conversion based on the target bonds to be cleaved, including $C_β$—O bond in β-O-4 linkage, C_{aryl}—O bonds in β-O-4 linkage and 4-O-5 linkage, $C_α$—O bond in α-O-4 linkage, as well as $C_α$—$C_β$ bonds in β-O-4 linkage and β-1 linkage (Figure 10.2).

10.2 Photocatalytic C—O Bond Cleavage

C—O bonds consist of more than two-thirds of the total linkages in native lignin [12], therefore, selective cleavage of C—O bonds is the key to obtaining functionalized aromatic monomers from lignin. β-O-4 motif represents the most frequent C—O bond, varying from 43% to 65% in native lignin [12]. Photocatalytic fragmentation of β-O-4 model compounds, extracted lignin, and native lignin has been widely investigated through the cleavage of $C_β$—O bonds. To simplify, $C_β$—O refers $C_β$—O bond in β-O-4 lignin linkages in the following text. Studies of $C_β$—O cleavage are roughly classified into two categories, i.e. stepwise and direct approaches. Besides the $C_β$—O bond, several photocatalytic systems were developed for the selective cleavage of other lignin C—O bonds, including C_{aryl}—O bond in β-O-4 linkage, $C_α$—O bond in α-O-4 linkage, and C_{aryl}—O bond in 4-O-5 linkage (Figure 10.2).

10.2.1 Stepwise Photocatalytic Cleavage of C_β–O (β-O-4) Bond

The stepwise approach for C_β—O bond cleavage usually involves two individual steps: the first step is the oxidation of the hydroxyl group at C_α position (C_αH—OH) into a carbonyl group (C_α=O), and the second step is the reductive cleavage of the C_β—O (=O) bond. Hereafter, C_β—O (=O) is used to refer C_β—O bond in β-O-4 ketone. The peroxidation of C_αH—OH into C_α=O resulted in the generation of β-O-4 ketone, which possesses a weaker C_β—O (=O) bond than the original β-O-4 alcohol [18], rendering C_β—O bond cleavage easier. Such a pre-oxidation following reduction strategy was first employed in thermocatalytic systems to break C_β—O bonds under relatively mild conditions, and the pre-oxidation and reduction steps are conducted separately in a stepwise manner [14]. This strategy was then adopted by photocatalytic systems. Photocatalytic systems were developed for both the oxidation of C_αH—OH into C_α=O, i.e. the first step to generate β-O-4 ketone intermediates, and the reductive cleavage of C_β—O bond in β-O-4 ketones, i.e. the second step.

10.2.1.1 Oxidation of C_α Hydroxyl into C_α Carbonyl

The first step, oxidation of C_αH—OH into C_α=O, is the premise for the subsequent easy scission of β-O-4 motifs thanks to the weakening of C_β—O bond. Stephenson and coworkers found that the pre-oxidation could be achieved by merging Ir-based molecular photocatalysis and palladium-based catalysis [19]. 2.0 equiv. of $Na_2S_2O_8$ additive was added as a sacrificial agent. The system is successfully employed for the oxidation of a wide range of dimeric lignin models containing C_αH—OH. By light irradiation, the Ir-based molecular photocatalyst (Ir^{III}) was activated to its excited state (Ir^{III*}), which is a strong oxidant, i.e. $E_{1/2}$ Ir^{III*}/Ir^{II} = +1.21 V vs. saturated calomel electrode (SCE). The excited Ir photocatalyst can then oxidize $Pd(OAc)_2$ to generate a Pd^{III} species, turning Ir^{III*} into its reduced form, Ir^{II}. Subsequently, the reduced Ir^{II} photocatalyst ($E_{1/2}$ Ir^{III}/Ir^{II} = −1.37 V vs SCE) could easily react with $S_2O_8^{2-}$ to regenerate the ground-state Ir^{III} photocatalyst, concurrently, $S_2O_8^{2-}$ was converted to SO_4^{2-} and $SO_4^{\bullet-}$. $SO_4^{\bullet-}$ is a potent oxidant with a sufficient potential of +2.2 V vs. SCE that can oxidize Pd^{III} to Pd^{IV}. The high-valent Pd^{IV} was believed to be the active species for the oxidation of C_αH—OH to C=O. After the oxidation reaction, Pd^{IV} could be restored to its original Pd^{II} state (Figure 10.3a).

Phthalimide-N-oxyl (PINO) radical, which is generated from photoinduced proton-coupled electron transfer (PCET) oxidation of N-hydroxyphthalimide (NHPI), was discovered to effectively convert lignin C_αH—OH into C_α=O [20]. For the photocatalytic oxidation of NHPI to PINO, donor-substituted aromatic ketones, e.g. 4,4′-bis(diphenylamino)-benzophenone (DPA-BP), exhibited the highest efficiency among different metal-free photocatalysts under 465 nm LED light irradiation. The strong hydrogen bonding between NHPI and the carbonyl group of photoexcited DPA-BP was proposed to be essential for the PCET process. Control experiments by using strong H-acceptors such as DMF as solvents led to total suppression of the reaction, indicating that hydrogen bonding between DPA-BP and NHPI plays an important role in the generation of PINO. PINO radicals could

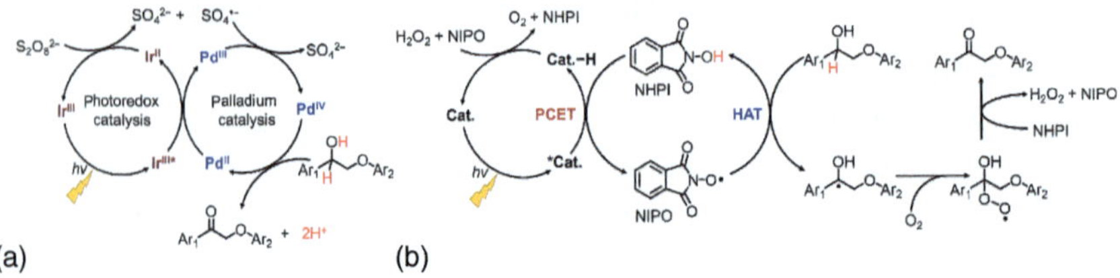

Figure 10.3 Proposed mechanism for photocatalytic oxidation of C_α hydroxyl into C_α carbonyl in β-O-4 motif by (a) merging Ir-based molecular photocatalysis and palladium-based catalysis, (b) merging photocatalysis and hydrogen atom transfer (HAT) catalysis. Source: Refs. [19, 20]. Reproduced with permission of American Chemical Society (ACS), Copyright 2016, and American Chemical Society, Copyright 2016. Figure 10.3a is an unofficial adaptation of an article that appeared in an ACS publication. ACS has not endorsed the content of this adaptation or the context of its use. Further permissions related to the material excerpted should be directed to the ACS, https://pubs.acs.org/doi/full/10.1021/acs.orglett.6b02651.

efficiently abstract the C_α—H bond of β-O-4 alcohols, restore NHPI, and generate the C_α radical, the latte then transformed to β-O-4 ketone by reacting with O_2 (Figure 10.3b). The DPA-BP/NHPI/O_2 photocatalytic system effectively catalyzed a wide range of dimeric lignin β-O-4 models to the corresponding ketones in 81–99% isolated yields at room temperature [20].

The combination of $(NH_4)_2S_2O_8$ and $CuBr_2$ facilitated the oxidation of dimeric β-O-4 alcohols to afford β-O-4 ketones in 26–99% yields under visible light irradiation [21]. Highly oxidative $SO_4^{\bullet-}$ generated from light irradiation of $(NH_4)_2S_2O_8$ was proposed as the oxidant for converting β-O-4 alcohols into C_α radical, which can be further oxidized into β-O-4 ketones by oxidation of $CuBr_2$. It should be noted that $(NH_4)_2S_2O_8$ cannot be decomposed under visible light to generate $SO_4^{\bullet-}$. Control experiments with different copper salts indicate that Cu^{2+} may form charge-transfer complexes *in situ* to promote the photolysis of $(NH_4)_2S_2O_8$ under the irradiation of visible light to produce $SO_4^{\bullet-}$. Unfortunately, the detailed structure of Cu^{2+}-based complexes remains elusive.

Semiconductor δ-MnO_2 was reported to efficiently catalyze the aerobic oxidation of dimeric β-O-4 alcohols selectively to corresponding ketones in 89–90% yields under the irradiation of 470 nm LED light [22]. The feasibility of δ-MnO_2 was further examined in the transformation of technical lignin samples, including organosolv lignin, Kraft lignin, and Alkali lignin. As estimated by the 2D NMR results from different technical lignin samples, over 80% C_α signal belonging to C_αH—OH disappeared after oxidation for all tested samples, which is attributed to the oxidation of β-O-4 alcohols into ketones. The heterogeneous photocatalyst δ-MnO_2 catalyst could be easily recovered and then be reused by heating treatment.

10.2.1.2 Reductive Cleavage of C_β—O Bond in β-O-4 Ketone

In a series of papers published by the group of Stephenson, the pre-oxidation has been achieved by thermocatalysis [23, 24], photocatalysis [19], or electrocatalysis [25], and the authors found that a metal complex photocatalyst, $[Ir(ppy)_2$

(dtbbpy)]PF_6, exhibited excellent performance for the photocatalytic reductive cleavage of C_β—O bonds in β-O-4 ketones obtaining from different pre-oxidation approaches. The reductive cleavage of C_β—O (=O) is conducted under visible-light irradiation and an inert atmosphere in the presence of a sacrificial reagent (diisopropylethylamine, ^{i}Pr$_2$NEt, DIPCA) and a hydrogen source (HCOOH). Quenching experiments and redox potential analyses by electrochemical measurements were employed to study the reaction mechanism [23]. These mechanistic studies suggest that light irradiation results in metal-to-ligand charge transfer (MLCT) and thus the generation of an excited state Ir^{III}* (Figure 10.4a), which has estimated redox potentials of $E_{ox}(Ir^{VI}/Ir^{III*}) = -1.31$ and $E_{red}(Ir^{III*}/Ir^{II}) = +0.96$ V vs. SCE. The excited Ir^{III}* could accept an electron from the sacrificial reagent, i.e. ^{i}Pr$_2$NEt, generating a reduced Ir^{II}, which is a strong reductant with $E_{ox}(Ir^{III}/Ir^{II}) = -1.53$ vs. SCE. Therefore, Ir^{II} can efficiently drive the reduction of dimeric β-O-4 ketones to ketyl radical anions by single electron transfer (SET). The generated ketyl radical anions would easily undergo C_β—O bond scission to afford 3-hydroxy-1-phenylpropan-1-ones and phenols after the HAT and protonation, respectively (Figure 10.4a) [23]. Interestingly, the reductive cleavage of C_β—O (=O) rates can dramatically improve by using a flow instead of batch reaction setup. For example, the cleavage of C_β—O in methoxy groups substituted dimeric β-O-4 ketone increased from 0.050 in batch to 1.8 mmol h^{-1} in flow, even when the catalyst loading was reduced 1.0–0.030 mmol% [24]. The system probably benefited from enhanced light absorption and mass transfer under flow conditions.

In a later work, Hao et al. immobilized the [Ir(ppy)$_2$(bpy)]PF_6 complex onto a mesoporous cellular silica foam (MCF) by a thiol-ene click reaction between prefabricated thiol-functionalized mesoporous cellular silica foams and vinyl-tagged [Ir(ppy)$_2$(bpy)]PF_6 complex. The immobilized molecular heterogeneous photocatalyst exhibited excellent activity and selectivity for the reductive cleavage of C_β—O (=O), which is similar to the homogeneous [Ir(ppy)$_2$(bpy)]PF_6 complex [27]. The heterogeneous photocatalyst can be easily recovered, and only slight decreases in

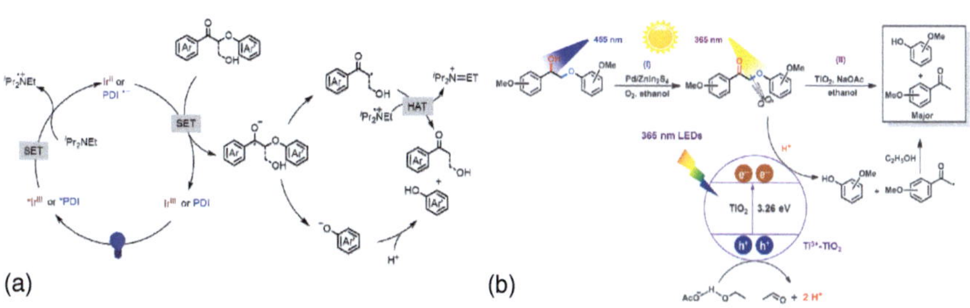

Figure 10.4 Proposed mechanism for photocatalytic fragmentation of dimeric β-O-4 model compounds through the cleavage of C_β—O (=O) (a) by [Ir(ppy)$_2$(bpy)]PF_6 complex or PDI organic photocatalyst and (b) by dual-wavelength switch strategy. Source: Refs. [23, 26]. Reproduced with permission of American Chemical Society, Copyright 2019, and American Chemical Society, Copyright 2016. Figure 10.4a is an unofficial adaptation of an article that appeared in an ACS publication. ACS has not endorsed the content of this adaptation or the context of its use. Further permissions related to the material excerpted should be directed to the ACS, https://pubs.acs.org/doi/10.1021/acscatal.8b04172.

activity were observed for the immobilized molecular photocatalyst after recycling uses to six times.

Metal-free organocatalyst perylene diimide (PDI) was discovered as an efficient catalyst to drive the fragmentation of β-O-4 ketones under visible light illumination, also in the presence of a sacrificial reagent ($^{i}Pr_2NEt$) and a hydrogen source (HCOOH) [28]. Mechanistic studies by radical trapping experiments together with UV–Vis and fluorescence spectroscopy revealed a reductive bond cleavage mechanism similar to the above system catalyzed by the Ir complex. The reaction is proposed to be initiated by photo-excitation of PDI to PDI*, which was quenched by $^{i}Pr_2NEt$ to produce catalytic active species $PDI^{\bullet -}$. SET from the $PDI^{\bullet -}$ to the dimeric lignin β-O-4 ketones produced the ketyl radical anions and concurrently regenerated the ground state PDI. The ketyl radical anions underwent selective C—O bond cleavage, affording phenol and ketone fragments as final products after HAT and protonation (Figure 10.4a). Isoptic experiments revealed that HCOOH was the proton source in the system. The activity of DPI in a homemade continuous-flow reactor was around 10 times higher than that in a batch reactor. Oxidized organosolv lignin was also tested in the continuous-flow reactor catalyzed by PDI, and an 86 wt% yield of low-molecular-mass aromatics, mainly composed of monomers and dimers, was obtained as products [28].

A series of carbazolic copolymers (CzCPs) fabricated by copolymerizing π-conjugated porous organic electron donor (D) and electron acceptor (A) have shown unique performance for the cleavage of C_β—O bond in dimeric β-O-4 lignin model compounds owing to their tunable redox properties [29]. By intentionally altering the ratio of D/A, the redox potentials of CzCPs could be finely manipulated. Interestingly, CzCPs with proper D/A ratios could function for both steps in the stepwise strategy for C_β—O bond cleavage. More specifically, the CzCP100 with a D/A ratio of 0 : 100, which has the strongest oxidative ability, exhibited the best performance for the pre-oxidation photocatalysis to convert dimeric β-O-4 alcohols into ketones in the presence of HAT mediator (NHPI) and oxidant (O_2) under irradiation of white compact fluorescent lamp. On the other hand, the CzCP33 with a D/A ratio of 66 : 33 was the most efficient catalyst for the reductive cleavage of C_β—O (=O) [29]. Similar to the above catalytic systems catalyzed by Ir complex and PDI, reductive cleavage of C_β—O (=O) over CzCP33 also requires the sacrificial reagent ($^{i}Pr_2NEt$), hydrogen source (HCOOH), and inert atmosphere (Ar). The authors proposed a reaction mechanism going through a ketyl radical anion intermediate, similar to the above research based on Ir complex and DPI (Figure 10.4a) [23, 28].

The requirement for the tedious separation of pre-oxidation products from the first step for subsequent reductive bond cleavage is one challenge for the two-step conversion route. Therefore, the development of two tandem catalytic systems for one-pot conversion is preferred over two separate steps. Wang and coworkers developed a one-pot photocatalytic system composed of two photocatalysts, i.e. Pd/$ZnIn_2S_4$ and TiO_2(anatase)-HCOONa, which could catalyze the oxidative dehydrogenation of dimeric β-O-4 models and the reductive cleavage of C_β—O bond in the β-O-4 ketones from the pre-oxidation step, respectively (Figure 10.4b) [26]. These two photocatalysts could work in an intriguing switchable tandem manner due to their different light adsorption spectrum. The reaction was first conducted under 455 nm LED light irradiation, in which Pd/$ZnIn_2S_4$ can catalyze the oxidation of β-O-4 alcohols to ketones, while TiO_2 is a UV-responsive catalyst and thus not activated under such conditions. Subsequently, the light was switched from 455 to 365 nm LED, under which TiO_2 could facilitate $C_β$—O (=O) bond cleavage in the presence of HCOONa (sacrificial reagent) and ethanol (hydrogen donor) (Figure 10.4b). EPR experiments indicated that $C_β$—O (=O) was most likely to be activated and cleaved by the *in situ* formed Ti^{3+}. This conclusion was also supported by metal loading experiments, where the immobilization of metals on Ti^{3+} sites deactivated the TiO_2 photocatalyst. TiO_2 (P25) catalyst was also reported as an effective photocatalyst in another work for the reductive cleavage of $C_β$—O bond in the dimeric β-O-4 ketones [30].

10.2.2 Direct Photocatalytic Cleavage of Lignin $C_β$–O (β-O-4) Bond

Despite the promising performances exhibited by the stepwise approach for $C_β$—O bond cleavage, the separation and purification of ketone intermediates are energy-consuming; and the addition of sacrificial reagents for the separated photocatalytic oxidation and reduction processes are often required. In a direct conversion route, $C_β$—O bond cleavage may proceed in a redox-neutral manner, in which both photo-generated electrons and photo-generated holes participate in the reaction [15], and thus the requirement of separation/purification procedures and the expensive sacrificial reductants/oxidants may be avoided. Currently, two predominate mechanisms, which are distinguished by different intermediates, i.e. β-O-4 ketone and $C_α$ radical, have been proposed for the direct cleavage of $C_β$—O bond by photocatalysis.

10.2.2.1 Direct Approach through β-O-4 Ketone Intermediate

Homogeneous photocatalytic systems composed of a photocatalyst (PC), $[Ir(ppy)_2 dtbpy]PF_6$ or 4CzIPN (an organic dye), and a HAT catalyst, e.g. $HSCH_2CO_2Me$, were found to be effective for the cleavage of $C_β$—O in dimeric lignin models [31]. A dual-cycle mechanism was proposed (Figure 10.5a). In the first cycle, the PC was activated to its excited state (PC*) under the irradiation of blue light, PC* then oxidized thiol, turning it into a thiol radical to drive HAT catalysis in the second cycle. The thiol radical could activate the $C_α$—H to give a ketyl radical, which was further converted to the corresponding β-O-4 ketone by photocatalytic oxidation. The reduced photocatalyst from thiol oxidation gives an electron to the β-O-4 ketone intermediate to produce a ketyl radical anion, which is unstable and thus undergoes $C_β$—O bond cleavage, producing ketones and phenols as final products. Of note, when $[Ir(ppy)_2dtbpy]PF_6$ was the photocatalyst, the reaction was performed in an inert atmosphere, while in the case of using 4CzIPN as the photocatalyst, O_2 was needed to accelerate the reaction. The β-O-4 ketone intermediate was detected by GC

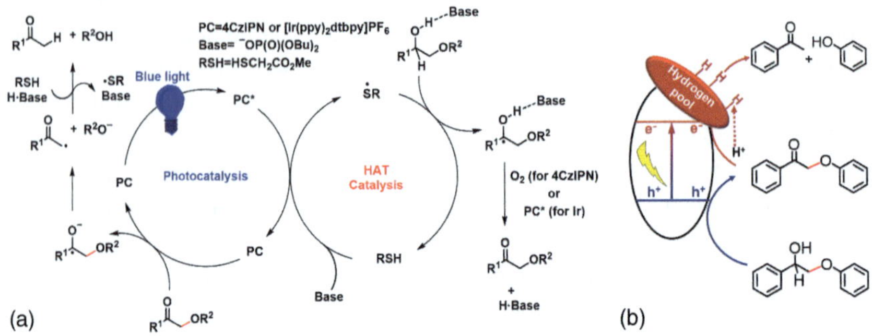

Figure 10.5 Direct fragmentation of dimeric β-O-4 lignin models through β-O-4 ketone intermediate (a) in an integrating system composed of a photocatalyst (PC = $Ir(ppy)_2dtbpy]PF_6$ or 4CzIPN) and a hydrogen abstraction catalyst (HAT = $HSCH_2CO_2Me$), and (b) by self-transfer hydrogenolysis over $ZnIn_2S_4$. Source: Refs. [31, 32]. Reproduced with permission of Royal Society of Chemistry, Copyright 2019, and American Chemical Society, Copyright 2017.

analysis of the crude reaction mixture after four hours of irradiation. As illustrated in the proposed mechanism, the same PC first participated in the oxidation of the reactant and then reduced the intermediate for bond cleavage, indicating a redox-neutral photocatalytic process.

In the above dual catalytic system, a HAT catalyst and a base are required to shuttle the hydrogen. The system can be simplified if one catalyst can accomplish a so-called self-transfer or borrowing-hydrogen hydrogenolysis mechanism [33, 34]. In this case, hydrogen generated in the oxidative dehydrogenation step, which converts $C_\alpha H—OH$ into $C_\alpha{=}O$, can be shuttled by the photocatalyst to perform reductive cleavage of $C_\beta—O$ (=O) bond. Wang and coworkers found that $ZnIn_2S_4$ photocatalyst could fragment lignin β-O-4 models under blue LEDs irradiation via such a self-transfer hydrogenolysis mechanism [32]. Photogenerated holes of $ZnIn_2S_4$ first oxidized the alcoholic group ($C_\alpha H—OH$) of dimeric lignin models to ketone intermediates and protons. Subsequently, these protons were reduced by electrons, generating $ZnIn_2S_4$ surface-adsorbed hydrogen, which then hydrogenolysis the $C_\beta—O$ bond in β-O-4 ketone intermediates (Figure 10.5b). Temperature-programmed-desorption revealed the existence of adsorbed hydrogen species on the $ZnIn_2S_4$. The system was also employed for the conversion of organosolv poplar lignin with isopropanol as the additional hydrogen donor, and around 10 wt% yields of monomeric products after 20 hours irradiation were achieved [32].

The self-transfer hydrogenolysis mechanism was also adopted in a Ni/CdS nanosheet-based system for the fragmentation of β-O-4 lignin model [35]. In this system, 2-phenoxy-1-phenylethanol (PP-ol), a simple β-O-4 model, was converted to the corresponding ketone, 2-phenoxy-1-phenylethanone (PP-one), by the photogenerated holes, under 460 nm LED light irradiation when MeCN alone was solvent. H_2 was formed by photogenerated electron reduction at the same time. Interestingly, when the solvent changed from pure MeCN to a mixture of MeCN and 0.1 M of

KOH in a volume ratio of 2 : 8, C_β—O bond cleavage occurred. In other words, with the addition of KOH, the adsorbed hydrogen participated in breaking the C_β—O bond instead of being reduced to H_2, leading to the generation of acetophenone and phenol as final products [35].

10.2.2.2 Direct Approach through C_α Intermediate

Wang and coworkers discovered a new redox-neutral photocatalytic system for direct cleavage of C_β—O bond, which does not involve ketone intermediate. Zinc blende CdS NPs were found as effective catalysts for the fragmentation of PP-ol, a β-O-4 dimeric model, to produce acetophenone and phenol under visible-light irradiation. CdS quantum dots (QDs) (<6.5 nm) have with high surface area and enhanced redox ability, and thus exhibited higher activity than CdS NPs (20–40 nm). Control experiments have indicated that PP-one, which was commonly assumed as the reaction intermediate in the two-step mechanism and self-transfer hydrogenolysis mechanism, is not the reaction intermediate. Further mechanistic studies with combined experimental and computational results have pointed out a new C_β—O bond cleavage chemistry, which is called electron-hole coupled (EHCO) mechanism [15]. Generally speaking, a photogenerated hole first oxidizes the cleavage of C_α—H bond, forming C_α radical as the reaction intermediate, such as C_α radical possesses a remarkably lower BDE (7.8 kcal mol^{-1}) than PP-ol (55 kcal mol^{-1}) or PP-one (47 kcal mol^{-1}), the low BDE renders facile cleavage of C_β—O bond in C_α radical intermediate, which affords acetophenone and phenol upon accepting a photogenerated electron (Figure 10.6a).

CdS QDs successfully fragmented various dimeric lignin models to corresponding aromatics in 79–95% yields at room temperature under visible-light irradiation [15]. In addition to dimeric models, CdS QDs were also employed for the depolymerization of technical lignins and even native lignin in lignocellulosic biomass through precise cleavage of β-O-4 linkage [15, 36, 37]. Technical lignins refer to lignins that are extracted from lignocellulose with different techniques, such as soda, methanosolv, and dioxasolv. During the extraction, relatively liable β-O-4 linkages could break, producing fragments that undergo unselective coupling to form recalcitrant C—C cross-linked structures. Therefore, β-O-4 linkage contents in different technical lignins change from case to case and are lower than that in the native lignin. For the conversion of both native and technical lignins over CdS QDs, nearly theoretical maximum yields of monomeric aromatics have been obtained [36]. Increasing yields of monomers were observed with increasing β-O-4 contents in lignins, the highest yield (27 wt%) was obtained from the conversion of native lignin in birch woodmeal (Figure 10.6b). For the conversion of native lignin, a reversible aggregation–colloidization strategy was developed to separate CdS QDs catalyst from the liquid lignin oil mixture and the solid carbon hydrogen residue [15]. The carbon–hydrogen residue, which is composed of hemicellulose and cellulose, was well preserved. Stepwise acidolysis and enzymolysis were further performed to convert hemicellulose to xylose and cellulose to glucose, respectively [15]. The conversion of birch woodmeal based on CdS QDs is the first illustration of the selective

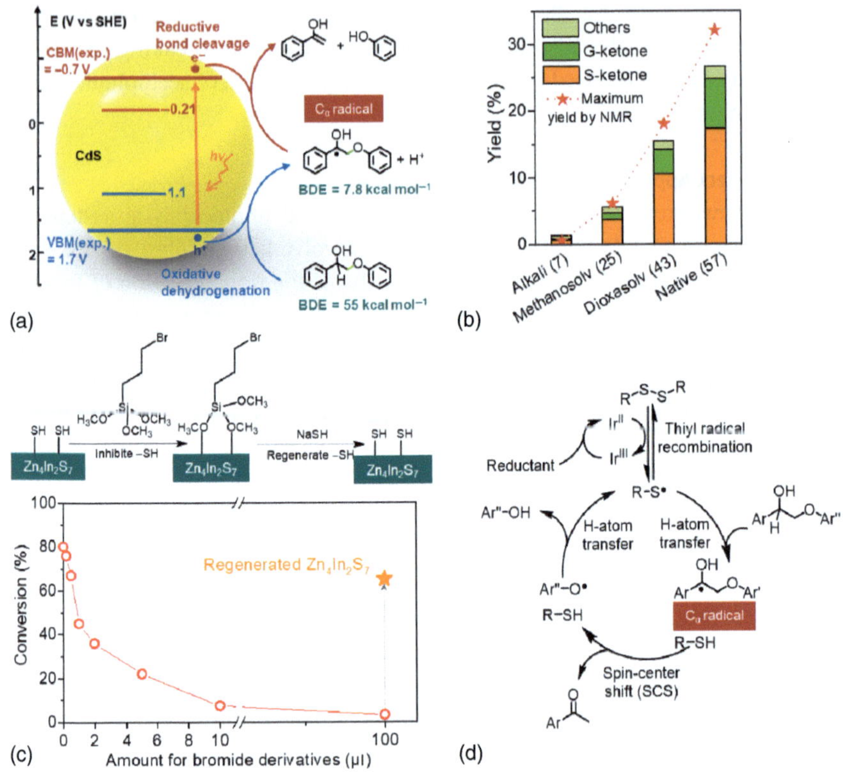

Figure 10.6 (a) Proposed EHCO mechanism for CdS-catalyzed direct cleavage of C_β–O bonds through C_α intermediate. (b) Yields of aromatic monomers from CdS QD-catalyzed conversion of lignins with different β-O-4 contents. The contents of β-O-4 bonds in different lignin are shown in the brackets. (c) Photocatalytic performance of $Zn_4In_2S_7$ after inhibiting and regenerating the surface –SH. (d) Proposed mechanism of merging photocatalyst with disulfide cystine for the cleavage of C_β–O bonds through C_α intermediate. Source: Refs. [15, 36–38]. Reproduced with the permission of Springer-Nature, Copyright 2018, American Chemical Society, Copyright 2019, John Wiley & Sons, Inc, Copyright 2019, and American Chemical Society, Copyright 2021.

depolymerization of native lignin by photocatalysis. The results demonstrate that conducting a lignin-first strategy, i.e. the catalytic valorization of native lignin in biomass in the first step, by photocatalysis at ambient temperature is a promising approach to the production of bio-based functionalized aromatics and the full utilization of all the three components of lignocellulosic biomass.

The authors also found the important roles of CdS QD ligands, which are attached to a large fraction of surface atoms, in the transformation of native lignin in solid lignocellulose [36]. The colloidal stability of QDs, which determines the contact between QDs and native lignin, is controlled by the hydrophilic/hydrophobic property of surface ligands. CdS QDs can remain in a pseudo-homogeneous state in polar CH_3OH/H_2O solution when they are modified with hydrophilic organic ligands, for instance, mercaptoalkanoic acids. The intimate contact between "pseudo homogeneous" QDs and the solid native lignin contributed to their high catalytic efficiency. Further, increasing yields of aromatic monomers were observed upon decreasing the length of alkyl chains of the hydrophilic mercaptoalkanoic acid ligands. Femtosecond transient absorption spectroscopy was performed to study the electron decay kinetics changing with alkyl chain lengths of mercaptoalkanoic acid ligands. The electron decay kinetics illustrated an exponential decrease in charge transfer rate constants with increasing ligand lengths, which suggested an electron-tunneling mechanism for the charge carriers transferring from the QD core to native lignin through organic ligands.

In subsequent work, Wang and coworkers synthesized a series of $Zn_mIn_2S_{m+3}$ (m = 1–6) semiconductors, which were employed for the conversion of various β-O-4 models and organosolv lignin [37]. Among the series catalyst, $Zn_4In_2S_7$ was found to be the most efficient one for the cleavage of C_β—O in PP-ol, producing 80% yield of acetophenone and phenol after 4-hour irradiation of visible light. The photocatalytic system was further used for the conversion of dioxasolv birch lignin, and an 18.4 wt% yield of functionalized aromatic monomers was obtained. Control experiments illustrated that the reaction mainly proceeds by the EHCO mechanism through C_α radical, similar to the mechanism proposed for the CdS-catalyzed system. The self-transfer hydrogenolysis mechanism may also present but only a minor contribution. Moreover, the important role of surface thiol groups (—SH) in the cleavage of β-O-4 linkages was discovered. Intentionally inhibiting and regenerating the surface —SH resulted in decreasing and increasing activity, respectively (Figure 10.6c) [37]. It was proposed that thiyl radical, which results from hole trapping in surface —SH, is the active species for C_α—H bond activation to form C_α radical intermediate and thus determine the activity.

Recently, a variety of metal sulfide solid materials, including Ag^+-exchanged wurtzite CdS, [39] $Cd_xZn_{1-x}S$, [40] Zn/S rich Zn_xInS_y, [41] and $Zn_4In_2S_7$ on graphene oxide [42], were also found to be efficient photocatalysts for the cleavage of C_β—O bond through C_α radical intermediates by different groups. The discovery of these systems further supports the important roles of surface sulfide in the C_α-radical-mediated route for C_β—O bond cleavage.

A homogeneous system merging a photocatalyst, $[Ir(ppy)_2(dtbpy)](PF_6)$, and a HAT reagent, disulfide cystine, was found to be effective for the cleavage of C_β—O bonds in dimeric lignin models and extracted lignin [38]. Electrochemical measurements indicated the reductive dissociation of disulfide cystine was energetically more feasible by Ir^{II} than Ir^{III*}, i.e. $E(Ir^{III}/Ir^{II}) = -1.26$ V vs. normal hydrogen electrode (NHE) while $E(Ir^{IV}/Ir^{III*}$ = −0.71 V vs. NHE. The thiyl radical (BDE (S–H) = 87 kcal mol^{-1}), which is generated from the photocatalytic dissociation of cystine, can induce hydrogen-atom abstraction from the substrate C_α—H (BDE is 85 kcal mol^{-1} in 1-phenylethanol) to produce C_α radical, similar to the proposed chemistry over $Zn_4In_2S_7$ [37]. The author proposed that the generated C_α radical could undergo a spin-center shift (SCS) for C_β—O bond cleavage (Figure 10.6d), which is similar to the bond cleavage chemistry that occurs in the active site of ribonucleotide reductase [38].

10.2.3 Cleavage of C_{aryl}—O (β-O-4, 4-O-5) and C_α—O (α-O-4)

The fragmentation of β-O-4 linkage typically proceeds by the cleavage of C_β—O bond, the cleavage of C_{aryl}—O bond is generally more difficult due to the presence of the resonance effect. In this case, a nucleophilic aromatic substitution strategy was developed to cleave the C_{aryl}—O bond in a pre-oxidized lignin β-O-4 model compound [43]. Upon 455 nm LED light irradiation, excited acridinium photocatalyst with a strong oxidative ability ($E_{Mes\text{-}Acr+^*/Mes\text{-}Acr\bullet}$ = +2.15 V vs. SCE) oxidized the electron-rich arene into cation radical. Nucleophilic attack by imidazole species then took place on *iso*-position, leading to C_{aryl}—O bond cleavage (Figure 10.7a). Notably, this method delivered valuable N-containing aromatics directly in a moderate yield of 51% [43].

Wang and coworkers developed an aryl migration strategy by installing an internal nucleophile in the β-O-4 linkage, and this facilitates the cleavage of C_{aryl}—O bond (Figure 10.7b) [44]. The aryl migration strategy began with stepwise oximation and acylation reactions of β-O-4 ketone, generating a nucleophile as the reactant. This nucleophilic reactant was photooxidized by phenothiazine (PTH), an organic molecular photocatalyst, through a SET mechanism, delivering an iminyl radical as an intermediate. Aryl migration from the O to N atom of iminyl radical was proposed through a five-membered-ring transition state, leading to the cleavage of

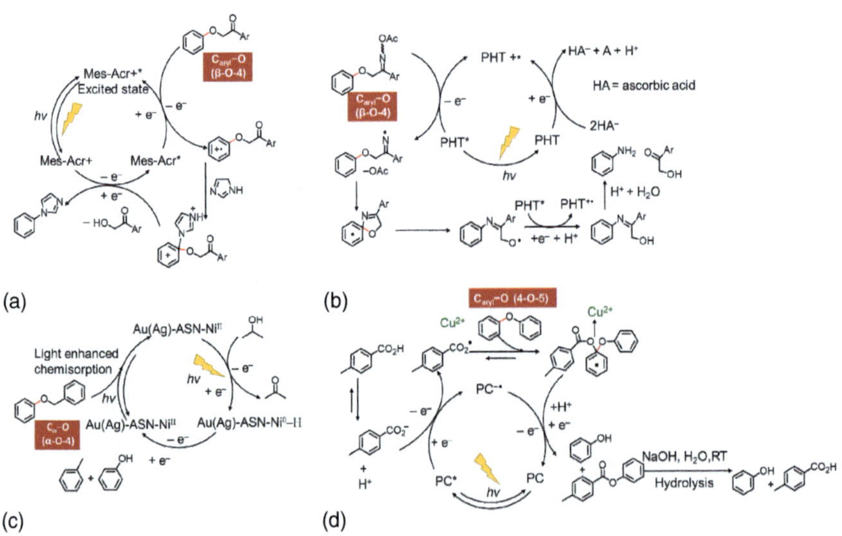

Figure 10.7 (a) Proposed mechanism for the cleavage of C_{aryl}—O bond in β-O-4 linkage by nucleophilic aromatic substitution strategy. (b) Proposed mechanism for the cleavage of C_{aryl}—O bond in β-O-4 linkage by aryl migration strategy to produce primary arylamine. (c) Proposed mechanism for photocatalytic transfer hydrogenolysis of C_α—O bond in benzyl phenyl ether, an α-O-4 lignin model compound, over Au(Ag)-ASN-Ni^{II}. (d) Proposed mechanism for the cleave C_{aryl}—O bond in diphenyl ether, a lignin 4-O-5 model substrate, by acidolysis-hydrolysis strategy. Source: Refs. [43–46]. Reproduced with permission of American Chemical Society, Copyright 2017, American Chemical Society, Copyright 2019, Royal Society of Chemistry, Copyright 2018, and Springer-Nature, Copyright 2020.

C_{aryl}—O bond and the formation of imine, which experienced hydrolysis to produce arylamine as the final product (Figure 10.7b). Notably, the addition of acid is essential for reducing PTH to accomplish the photocatalytic cycle and for donating protons in the hydrolysis of the imine. Isolated yields of arylamine from 14% to 64% were achieved by applying such an aryl migration strategy for the cleavage of C_{aryl}—O bonds in a wide range of lignin model compounds. This aryl migration strategy was also tested for the conversion of Dioxasolv poplar lignin, which resulted in 1.9 wt% arylamine monomers [44].

Pd/TiO_2 (Degussa P25) was discovered as an effective photocatalyst to catalyze the cleavage of C_α—O bond in a typical α-O-4 lignin model compound, benzyl phenyl ether [47]. About 99% yield of toluene and phenol have been obtained after two hours of irradiation under UV light. 2-Propanol is the solvent of the system and also the hydrogen donor. Pd outperformed other noble metals to be the best cocatalyst. 2-Propanol could accept photo-generated electrons to produce reductive hydrogen adsorbed on the Pd surface. The Pd—H species was proposed to be responsible for the reductive cleavage of the C_α—O bond.

Localized surface plasmon resonance (LSPR) occurs on noble-metal surfaces when free electrons of plasmonic nano-metal oscillate in resonance with the light's frequency [48]. LSPR can induce hot-electron generation, heat dissipation, and electric-near-field enhancement. The LSPR effect of noble metals was also found to be applicable to the cleavage of C_α—O bond in α-O-4 linkage [45]. By co-immobilization of plasmonic Au or Ag NPs and Ni(II) complex on silane containing an amino group (ASN) grafted on Al_2O_3 nanofibers, plasmonic-antenna-promoted catalysts Au-ASN-Ni^{II} or Ag-ASN-Ni^{II} photocatalysts were fabricated. Both photocatalysts could catalyze the C_α—O bond cleavage of benzyl phenyl ether to afford phenol and toluene in high yield under visible light at 80–90 °C in the presence of KOH. Isopropanol not only is the sacrificial reagent for photogenerated holes but also the hydrogen source for hydrogenolysis of C_α—O bond. The chemisorption of benzyl phenyl ether on Ni^{II} complex was enhanced by concentrated light irradiation of Au(Ag)-ASN-Ni^{II}. The transfer of plasmonic hot electrons from Au or Ag NPs reduced Ni^{II} to catalytically active Ni^0, and Ni^0—H was proposed as the active species for the reductive cleavage of C_α—O bond (Figure 10.7c) [45].

Among different lignin C—O linkage, the cleavage of lignin 4-O-5 bonding is extremely difficult due to its high BDE (~314 kJ mol^{-1}). A one-pot acidolysis-hydrolysis strategy was adopted to cleave C_{aryl}—O bond in lignin 4-O-5 model compounds under 455 nm LED irradiation [46]. The reaction was initiated by the oxidation of benzoate to produce aryl carboxylic radical by strong oxidative photoexcited acridinium. The generated aryl carboxylic radical is an electrophilic radical species and thus can attack the electron-rich aryl ring to produce a metastable intermediate. Cu$(TMHD)_2$, a Lewis acid catalyst, not only functions as a stabilizer for the metastable intermediate but also facilitates the SET process for C_{aryl}—O bond cleavage to generate ester and phenol products. The following hydrolysis can easily take place in an alkaline aqueous solution, converting benzoate ester into another molecule of phenol and recovering the benzoate (Figure 10.7d). This method was applied to a wide range of 4-O-5 lignin models with different functional groups. In an upscaling reaction, an 80% yield of phenol on a gram-scale was achieved by the conversion of diphenyl ether, a lignin 4-O-5 model substrate, in a flow reactor [46].

10.3 Photocatalytic C—C Bond Cleavage

By solely breaking lignin C—O linkages, the theoretical maximum yield of monomeric aromatics is usually no more than 50% [49]. A higher yield of monomers requires also the cleavage of more robust lignin C—C bonds. Note that one representative type of product from the cleavage of lignin C—C bonds is the aromatic aldehyde, which can easily undergo recondensation with phenolic aromatics in thermocatalytic systems when high temperature is provided. In contrast, aldehydes and phenols may remain under mild photocatalytic conditions. Therefore, photocatalysis can be a useful approach for the cleavage of lignin interlinking C—C bonds to produce a high yield of aromatic monomers. Currently, research focus for C—C bond scission is mainly placed on the cleavage of the C_α—C_β bond of β-O-4 and β-1 linkages. Key intermediates for C_α—C_β bond cleavage include alkoxy radical, C_β radical, and aromatic radical cation.

10.3.1 C_αO—H Bond Activation to Alkoxy Radical Intermediate

As validated by the ring-opening reactions in cyclic alcohols, activation of the O—H bond to form alkoxy radical intermediates could decrease the BDE of the adjacent C—C bond, therefore facilitating C—C bond cleavage [50, 51]. The selective activation of C_αO—H bond in β-O-4 and β-1 motifs is also expected to be an effective protocol to enable the adjacent C_α—C_β bond scission. For the success of this method, the competitive activation of C_α—H bond should be avoided. Many homogenous catalysts including different vanadium [52–54] and cerium complexes [55] were reported as efficient photocatalysts for the cleavage of C_α—C_β bond in β-O-4 or β-1 lignin models. The high efficiency of these systems is contributed to the chelation ability of metal centers to C_αO—H bond, which ensures the selective activation of the C_αO—H bond, instead of C_α—H bond, to produce alkoxy radical intermediates. In these systems, the reaction generally begins with coordination between the catalytic active metal and the C_αO—H group of lignin substrate. Upon light irradiation, ligand-to-metal charge transfer (LMCT) occurs to activate the C_αO—H bond, leading to the generation of alkoxy radical intermediates and then facile cleavage of C_α—C_β bonds (Figure 10.8a and b).

Hirao, Soo, and their coworkers found that a vanadium V^V-oxo complex can effectively cleave C_α—C_β bond in a dimeric β-O-4 lignin model PP-ol under visible-light irradiation ($\lambda > 420$ nm, AM 1.5 solar intensity) [52]. Under optimized conditions, benzyl aldehyde and benzyl formate were produced as major products in high yields of 70 and 61%, respectively. Other products include benzoic acid (24%) from the over-oxidation of benzaldehyde, and phenol (6%) together with formaldehyde (32%) from the decomposition of phenyl formate. The same group has performed

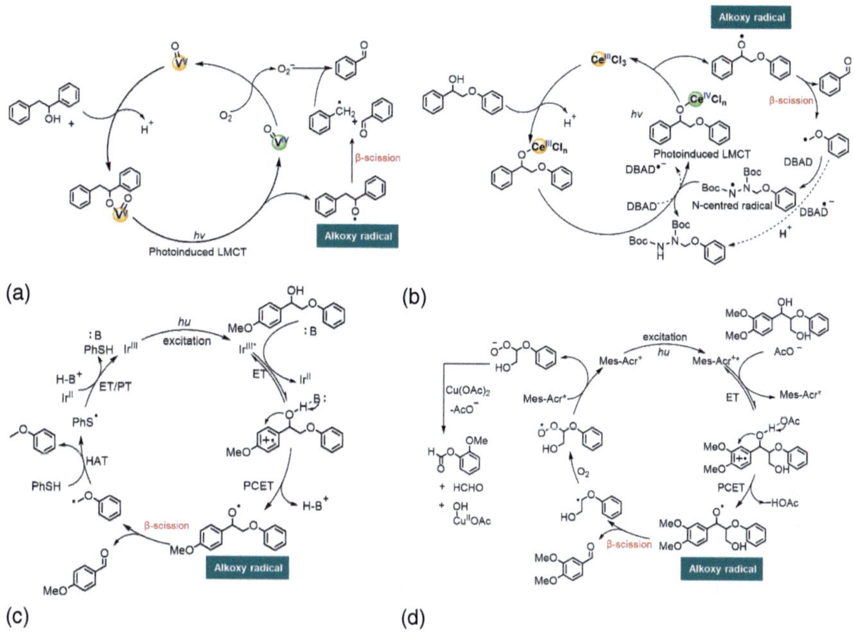

Figure 10.8 Proposed reaction mechanisms for photocatalytic cleavage of C_α–C_β bonds in β-1 or β-O-4 linkages through alkoxy radical intermediates over (a) V complex, (b) Ce complex, (c) Ir complex together with a base cocatalyst, and (d) acridinium salt (Mes-Acr$^+$)/copper/base (AcO$^-$) system. The LMCT mechanism is involved in A and B while the PCET mechanism is involved in C and D. Source: Refs. [54–57]. Reproduced with permission from American Chemical Society, Copyright 2019, Chinese Chemical Society, Copyright 2020, Elsevier B.V., Copyright 2019, and John Wiley & Sons, Inc, Copyright 2018. Figure 10.8B has been published in *CCS Chemistry 2020*; $CeCl_3$-promoted simultaneous photocatalytic cleavage and amination of C_α–C_β bond in lignin model compounds and native lignin is available online at https://doi.org/10.31635/ccschem.020.201900076.

subsequent work to increase the photocatalytic efficiency of V-based photocatalysts for C—C bond cleavage, an enhanced activity that is 2.5–17 times faster than the original V^V oxo complex was achieved by incorporating electron-withdrawing groups at selected sites of the ligand [53].

Phenolic hydroxyl groups in lignin can deactivate the above V^V oxo complexes. A robust $VO(^iPrO)_3$ photocatalyst that is tolerant phenolic hydroxyl group was disclosed by Wang and coworkers for the cleavage of C_α—C_β bond in dimeric β-1 lignin models (Figure 10.8a) [54]. 86% benzaldehyde and 7% benzoic acid were obtained from the photocatalytic conversion of 1,2-diphenylethanol, a β-1 lignin model compound, over $VO(^iPrO)_3$ after six hours of irradiation with blue LEDs ($\lambda = 455$ nm) under O_2. $VO(acac)_2$ exhibited higher efficiency than $VO(^iPrO)_3$ in the cleavage of C_α—C_β bond in β-O-4 lignin models. Depolymerization of different dioxasolv lignin, including birch-, straw- and poplar-derived lignin samples, was performed over $VO(^iPrO)_3$ and $VO(acac)_2$. The monomer yields were still unsatisfactory at this moment, the best result was 0.6 vanillin and 0.79 wt% syringaldehyde from the depolymerization of birch lignin using $VO(acac)_2$ [54].

Cerium(III) chloride ($CeCl_3$) was found to facilitate the scission of C_α—C_β bonds in lignin β-O-4 dimeric models through a mechanism similar to V-based systems [55] The reaction was initiated by deprotonation and the coordination between the metal (Ce^{3+}) active center and the C_αO–H group of lignin substrate. Photoinduced LMCT, which was accelerated by the oxidation of di-*tert*-butyl azodicarboxylate, led to the production of alkoxy radicals. The alkoxy radical then underwent β-scission to produce benzaldehyde and a carbon radical intermediate. Interestingly, the generated carbon radical was able to couple with di-*tert*-butyl azodicarboxylate to produce N-containing products (Figure 10.8b). Various lignin dimeric β-O-4 models were converted into benzaldehydes and N-containing products in yields up to 96 and 95%, respectively. The feasibility of this system was further proved by the depolymerization of pine lignin, and an 11.94 wt% yield of monomers has been achieved.

The relatively strong polarity of O—H bonds suggests that their activation in the presence of a base can be easy, which offers another approach to providing alkoxy radicals intermediate. A combination of Ir complex together with a base cocatalyst could efficiently fragment dimeric β-O-4 models through C_α—C_β bond cleavage. By blue LED irradiation, benzyl aldehydes and phenyl ethers were produced as products under N_2 atmosphere.[56] Among different base additives, collidine ($pK_a = 14.98$) was the optimal choice to afford the highest yields of products from C_α—C_β bond cleavage. As proposed by the authors, the excited Ir^{III*} complex first oxidized the β-O-4 models to form a radical cation, which could be converted to an alkoxy radical intermediate through PCET. The alkoxy radical then underwent a facile β-scission, generating benzaldehyde and anisole as final products. The Ir^{II} complex was restored to its original state (Ir^{III} complex) by electron transfer to the generated PhS$^\bullet$ radical (Figure 10.8c). However, this PCET strategy was inefficient for the direct degradation of dioxasolv lignin. The low efficiency may be due to the presence of abundant phenolic hydroxyl groups in lignin. The phenolic hydroxyl group (ArO—H) can compete with the PCET C_αO—H activation, and thus slow down or even inhibit the C—C bond cleavage. To solve this problem, MeI/K_2CO_3 was added to protect the phenolic hydroxy groups in lignin. The protection of phenolic hydroxy groups led to a yield of ~5 wt% of aromatic monomers from dioxasolv birch lignin under blue LED (100 W) irradiation after 12 hours [56]. A similar PCET system containing an Ir(III) complex, a dialkyl phosphate base, and an aryl thiol H atom donor, was reported to enable selective scission of lignin β-O-4 and β-1 linkages by C_α—C_β bond cleavage [58]. Moreover, the system was compatible with phenolic hydroxyl groups in lignin substrates. Various extracted lignin were directly depolymerized without conducting any pretreatment procedure, generating monomers with yields from 3.4 to 5.1 wt% after irradiation under blue LED for 48 hours. The authors hypothesized that even though the phenolic O—H bonds could also be activated, the resulting phenoxy radical prefers to undergo back electron transfer to the initial state. In this case, the activation of C_αO—H is still possible even in the presence of phenolic hydroxyl groups when the PCET system is efficient enough [58].

In addition to Ir-based photocatalytic systems, an acridinium salt (Mes-Acr$^+$)/copper/base (AcO^-) system was developed for the cleavage of C_α—C_β bonds in both β-O-4 and β-1 models, and a similar PCET mechanism to achieve alkoxy-radical intermediates has been proposed [57]. The radical cation from the oxidation of light-excited acridinium salt (Mes-Acr^{+*}) was deprotonated with the aid of base AcO^-, affording the corresponding alkoxy radical to allow facile C—C bond cleavage. $Cu(OAc)_2$ then catalyzed the transformation of alkoxy anion into phenol formate (Figure 10.8d).

10.3.2 C_β–H Bond Activation to C_β Radical Intermediate

The rupture of C_α—C_β bond in β-1 and β-O-4 linkages can also be achieved through C_β—H bond activation to C_β radical intermediates. Such a C_β radical can react with O_2-based species to give a hydroperoxide intermediate, which undergoes facile C—C bond cleavage to produce benzaldehydes, benzoic acids from β-1 motifs, or benzaldehydes, benzoic acid, and phenol formate derivatives from β-O-4 motifs (Figure 10.9).

Mesoporous graphitic carbon nitride (mpg-C3N4) semiconductor was found as an efficient catalyst to catalyze the fragmentation of both β-O-4 and β-1 lignin models through the cleavage of C_α—C_β bond under the irradiation of 455 nm LED light in the presence of O_2 [59]. Combined yields of benzaldehyde and benzoic acid have reached 66–90% from both the conversion of β-O-4 and β-1 model compounds over C_3N_4. Conversion of β-O-4 model compounds also afforded phenol formates. When C_β—H was replaced by C_β—D, reaction rates for the scission of C—C decreased significantly, which indicated the importance of C_β—H bond activation step for the bond cleavage process. The reaction was completely stopped in the absence of O_2. The authors proposed that the reaction was initiated by C_β—H bond activation to produce a key C_β radical intermediate, which could then undergo C_α—C_β cleavage when reacting with O_2 (Figure 10.9a). Solid-state NMR and DFT calculations suggested the presence of π–π stacking interactions between the mpg-C_3N_4 surface and lignin models, which was proposed as a key reason for the high activity of the mpg-C_3N_4 photocatalyst.

A Z scheme Ag_3PO_4-polymeric carbon nitride (PCN) nanocomposite was fabricated and employed for the conversion of β-1 lignin model compounds [60]. Aromatic aldehydes with high yields of 66–95% were obtained under simulated solar light irradiation at room temperature. Similar to the C_3N_4-based system, the C_β—H bond activation to C_β radical intermediates is a key step in the dominant mechanism under O_2 atmosphere. Further, control experiments with different additives suggested that O_2 was first reduced by an electron into superoxide radical ($O_2^{\bullet}$−), which then reacted with C_β radical intermediate to form a six-membered transitional state that could easily undergo C_α—C_β bond cleavage (Figure 10.9b). The participation of $O_2^{\bullet}$− indicated the conduction band minimum of the catalyst should be more negative than the reduction potential (−0.33 V) for the conversion of

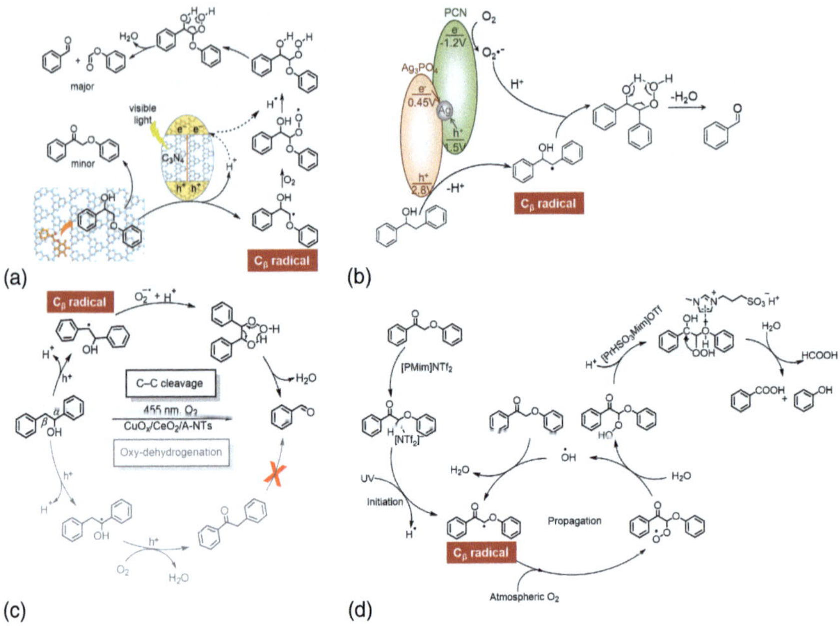

Figure 10.9 Proposed reaction mechanisms for photocatalytic cleavage of $C_\alpha-C_\beta$ bonds in β-1 and β-O-4 linkages through C_β radical intermediates catalyzed by (a) mpg-C_3N_4, (b) Z scheme Ag_3PO_4-PCN, (c) CuO_x/ceria/TiO_2, and (d) ionic liquids [$PrHSO_3$Mim][OTf] and [PMim][NTf_2]. Source: Refs. [59–62]. Reproduced with permission from American Chemical Society, Copyright 2018, Royal Society of Chemistry, Copyright 2021, American Chemical Society, Copyright 2017, and John Wiley & Sons, Inc, Copyright 2019. Figure 10.9C is an unofficial adaptation of an article that appeared in an ACS publication. ACS has not endorsed the content of this adaptation or the context of its use. Further permissions related to the material excerpted should be directed to the ACS, https://pubs.acs.org/doi/10.1021/acscatal.7b00629.

O_2 to $O_2^{\bullet}-$. The construction of the Z-scheme structure, which simultaneously preserves the strong reduction capability of PCN and the strong oxidation capability of Ag_3PO_4, is the key to the EHCO photoredox $C_\alpha—C_\beta$ bond cleavage process [60].

Parallel to the oxidative $C_\beta—H$ bond cleavage, the $C_\alpha—H$ bond cleavage that eventually leads to the pre-oxidation of the $C_\alpha H—OH$ group to $C_\alpha=O$ in oxidation conditions is also possible. However, such a pre-oxidation reaction may increase the difficulty in C—C bond cleavage, taking 1,2-diphenylethanol (β-1 model) as an example, the pre-oxidation will increase the BED of $C_\alpha—C_\beta$ bond from 295 to 354 kJ mol^{-1} [61]. Therefore, an effective photocatalyst for the cleavage of $C_\alpha—C_\beta$ bond in β-1 linkage should facilitate the cleavage of $C_\beta—H$ bond and suppress the oxidation of $C_\alpha—H$. A three-component $CuO_x/CeO_2/TiO_2$ catalyst, which can facilitate the $C_\beta—H$ bond rather than oxy-dehydrogenation of $C_\alpha—H$, was developed for the efficient cleavage of $C_\alpha—C_\beta$ bond in β-1 linkage (Figure 10.9c) [61]. When TiO_2 alone was employed as the photocatalyst, a 16% yield of dipheylethanone, which was from pre-oxidation, and a 22% yield of benzaldehyde, which was from $C_\alpha—C_\beta$ bond cleavage, were obtained under 365 nm UV-light irradiation. The decoration of

TiO_2 with CeO_2 increased the selectivity of benzaldehyde, and a higher yield of benzaldehyde (27%) was achieved. The authors proposed that the defective ceria clusters, which contain Ce^{3+} and oxygen vacancies, were inclined to catalyze the C_β—H bond cleavage, and therefore can promote the scission of C_α—C_β bonds. CuO*x* could increase the density of defective ceria clusters, thereby further improving the selectivity of the photocatalyst for C_α–C_β bond cleavage. The conversion of several dimeric β-1 models under the irradiation of 455 nm LED in the presence of O_2 over the CuO_x/ceria/TiO_2 resulted in high yields of over 90% aromatic aldehydes [61].

The cleavage of the C_α—C_β bond in β-O-4 model ketones, which are generated from the oxidation of β-O-4 model alcohols, can also be achieved through a C_β-radical-mediated pathway. A metal-free photocatalytic method with ionic liquids [$PrHSO_3Mim$][OTf] and [PMim][NTf_2] afforded 87% yield of benzoic acid and 78% yield of phenol from the conversion of dipheylethanone under UV-light irradiation and air atmosphere [62]. It was proposed that the scission of C_α—C_β bond in β-O-4 ketone was initiated by the cleavage of C_β—H bond, assisted by the coordination of [PMim][NTf_2]. The reaction of C_β radical with O_2 generates peroxide radical, which was then oxidized by water, delivering peroxide that experiences C_α—C_β bond cleavage catalyzed by [$PrHSO_3Mim$][OTf] to afford benzoic acid and phenol (Figure 10.9d) [62].

10.3.3 Single Electron Transfer to Radical Cation Intermediate

Some enzymes from wood-rotting fungi, such as lignin peroxidase, depolymerize lignin by C—C bond cleavage through radical cation intermediates [63–65]. Similar to the natural ligninolytic peroxidize process, the cleavage of C—C bond in β-O-4 and β-1 linkages can also go through a radical-cation-mediated pathway by photooxidation of aryl rings. Light excited 9,10-dicyanoanthracene (DCA), which has a strong oxidative ability ($E(DCA^*/DCA^{-\bullet}) = +2.8\,V$), can efficiently oxidize a wide range of aromatics into aryl cation radicals through SET [66, 67]. Under the irradiation of UV light, DCA oxidized dimeric lignin β-O-4 and β-1 models into corresponding aryl cation radicals, resulting in a dramatic decrease in the BDE of C_α—C_β bonds (Figure 10.10) [67]. The cation radical intermediates then experienced C—C bond cleavage, delivering a radical and a carbon cation that can be finally converted to benzaldehydes and phenols.

The biomimetic system based on DCA provides higher rates for β-1 model fragmentation compared to that for β-O-4 model fragmentation, regardless of the substituents at the aryl rings [66, 67]. This may be due to the lower BDEs of C_α—C_β bond in β-1 cation radical intermediates, according to DFT calculations (Figure 10.10). Aryl radical cations derived from less methoxy-substituted dimeric β-O-4 models are easier to undergo C_α—C_β bond cleavage. Despite the validity of this radical-cation-mediated approach for lignin fragmentation, both DCA and ligninolytic peroxidase provided very low yields of the phenolic segments from the conversion of lignin β-O-4 models under highly oxidative conditions.

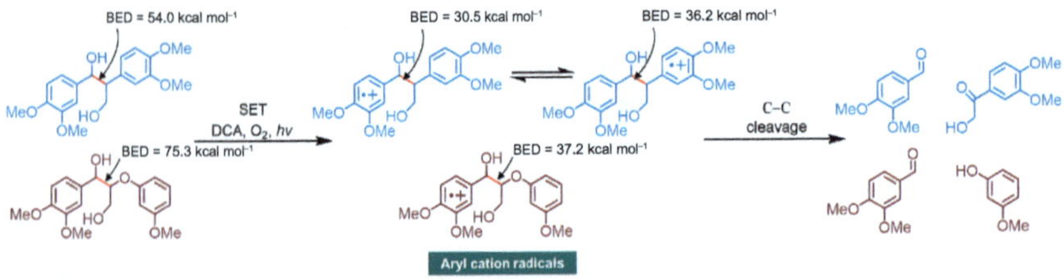

Figure 10.10 Aryl-cation-radical-mediated fragmentation of lignin β-1 and β-O-4 dimeric models triggered by DCA photocatalyst. Source: Cho et al. [67]. Reproduced with permission of American Chemical Society, Copyright 2017.

10.4 Conclusion and Outlook

The development of efficient approaches for the selective conversion of lignin to functionalized aromatics is a highly attractive but challenging research goal. The present chapter has summarized demonstrative photocatalytic systems for precise cleavage of lignin C—O/C—C bonds to produce functionalized aromatics, clearly illustrating that photocatalysis is a promising approach for lignin valorization. The photogenerated charge carriers or photoinduced active species can drive redox reactions without consuming additional reductants or oxidants, enabling environmentally benign transformations. Reactive radicals, which are key intermediates in photocatalytic lignin transformation, are usually with weak C—O or C—C bonds, rendering facile bond cleavage to form desired aromatic products (Figure 10.11).

The cleavage of predominant C—O bonds, in particular the abundant β-O-4 linkages, is essential for lignin valorization. The fragmentation of β-O-4 linkages is intensively investigated through the cleavage of C_β—O bonds. Stepwise approaches for photocatalytic scission of C_β—O bonds (Figure 10.11, Path I), including the first step for oxidation of C_α—OH to C_α=O [19–22] and/or the second step for reductive cleavage of β-O-4 bond in the ketone intermediate [19, 23–30], were investigated for the fragmentation of lignin models or organosolv lignin. In such a stepwise method, photocatalytic oxidation or reduction is performed in a separate manner, and thus sacrificial reagents are required. Redox neural C_β–O bond cleavage, i.e. the direct C_β–O bond scission approach, is more attractive. A few photocatalysts, including organic photocatalysts such as $[Ir(ppy)_2dtbpy]PF_6$ [31] and metal sulfides such as $ZnIn_2S_4$[32] and Ni/CdS, [35] were proposed to drive the direct C_β—O bond cleavage through a β-O-4 ketone intermediate (Figure 10.11, Path II). Recently, many metal-sulfide-based systems have been found to efficiently catalyze the direct C_β—O bond cleavage through a C_α radical intermediate (Figure 10.11, Path III) [15, 36, 37, 39–42]. The CdS QD photocatalyst could also facilitate the conversion of native lignin in solid lignocellulosic biomass into functionalized aromatic monomers with high yield, while cellulose and hemicellulose fractions could be preserved under the mild conditions; the example of CdS QDs demonstrates the potential of photocatalytic lignin-first strategy for full utilization of lignocellulose [15].

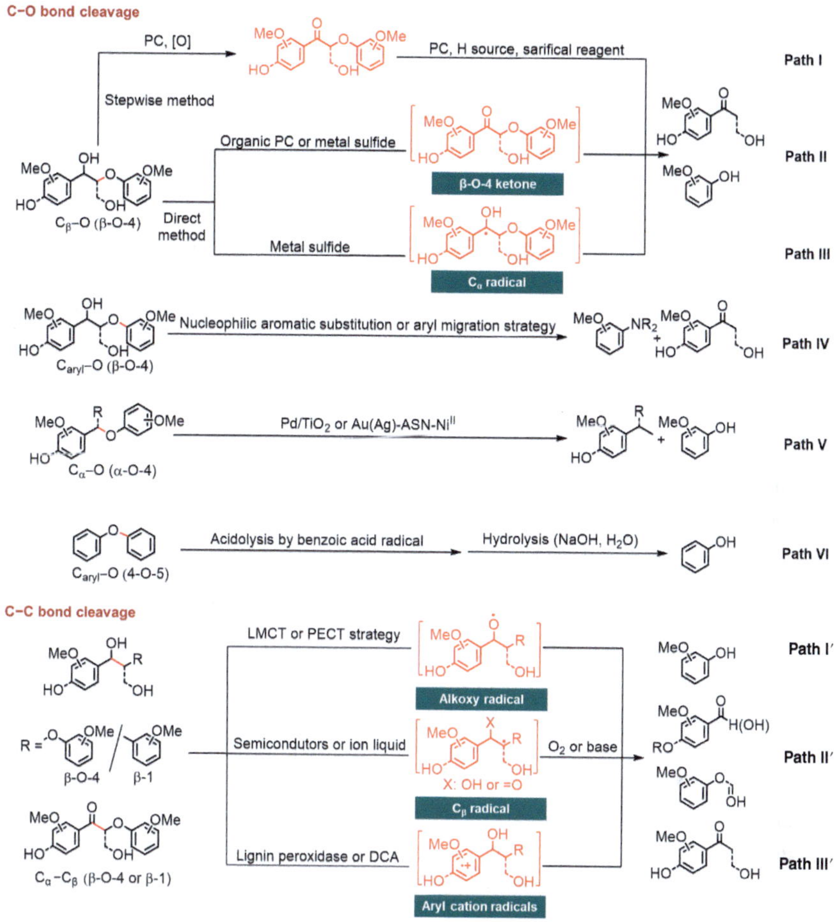

Figure 10.11 Summary of photocatalytic methods in the fragmentation of lignin model compounds. Key intermediates are highlighted in red. Source: Ye Wang.

Photocatalytic systems have also been developed for the cleavage of C_{aryl}—O bond in β-O-4 linkage, C_α—O bond in α-O-4 linkage, and C_{aryl}—O bond in 4-O-5 linkage. A nucleophilic aromatic substitution strategy [43] and an aryl migration strategy [44] were developed to facilitate the cleavage of C_{aryl}—O bond, instead of the more vulnerable C_β—O bond, in β-O-4 linkages (Figure 10.11, Path IV). Pd/TiO_2 or Au(Ag)-ASN-Ni^{II} catalyst catalyzed the cleavage of α-O-4 bond in lignin model compounds by hydrogenolysis, using isopropanol as the hydrogen source (Figure 10.11, Path V) [45, 47]. The cleavage of C_{aryl}—O bond in 4-O-5 linkage is difficult due to its high BED, and a one-pot acidolysis-hydrolysis strategy was adopted to address this challenge (Figure 10.11, Path VI) [46].

Some photocatalytic systems have exhibited high efficiency in the cleavage of C—C bonds, which commonly possess large BEDs and their cleavage usually requires high temperature in thermocatalytic systems. Cleavage of C_α—C_β bonds in β-O-4 and β-1 linkages was achieved by photocatalytic approaches at room temperature. Different radicals, including alkoxy radical, $C_β$ radical, and aromatic radical cation, can undergo facile $C_α$—$C_β$ bond cleavage under mild conditions and are identified as key intermediates for photocatalytic lignin conversion. Metal complexes, including different vanadium [52–54] and cerium complexes [55], and several organic photocatalyst/base-cocatalyst systems [56, 58] could facilitate the $C_α$O—H bond activation to alkoxy radical intermediates through LMTC and PECT strategy, respectively (Figure 10.11, I′). Several semiconductors ruptured $C_α$—$C_β$ bond in β-1 and β-O-4 alcohols through $C_β$—H bond activation to $C_β$ radical intermediates, which can react with O_2-based species to give hydroperoxides to undergo $C_α$—$C_β$ bond cleavage (Figure 10.11, Path II′).[59–61] The cleavage of the $C_α$—$C_β$ bond in β-O-4 model ketones was also achieved through a $C_β$-radical-mediated pathway with the aid of ionic liquids (Figure 10.11, Path II′) [62]. Highly oxidative lignin peroxidase and DCA promoted the $C_α$—$C_β$ bond cleavage through radical cation intermediates (Figure 10.11, Path III′) [63–67].

Although significant progress has been made in the field of photocatalytic valorization of lignin, many important but challenging targets remain unexplored. The future work of photocatalytic lignin valorization can be considered from the following aspects. First, the reported studies by far have only considered the cleavage of one type of C—O or C—C bond, and this would limit the yield of monomeric aromatics. For example, the theoretical maximum yield of aromatic monomer from sole cleavage of $C_β$—O of native lignin in birch is around 30% taking account into the abundance of β-O-4 linkages, therefore, even though CdS QD is highly efficient for the scission of $C_β$—O bond, the highest yield of monomeric aromatics could only achieve 27 wt% [15]. In this case, the development of bifunctional or even multifunctional photocatalysts or integrated photocatalytic systems that enable concurrent cleavage of different linkages is important for increasing the yield of aromatics products.

Designing prototype fluidic reactors is a potential strategy to improve the photocatalytic performance for the conversion of lignin [68]. A boost activity is expected from the increased interaction between macromolecular lignin reactants and photocatalysts due to the enhanced heat- and mass-transfer ability of fluidic systems. Furthermore, higher selectivity is possible by the timely separation of products in fluidic systems to avoid over-oxidation [68]. There are several reports with enhanced activity for the conversion of lignin models [24] and organosolv lignin [28] in flow reactors, while the report for the photocatalytic valorization of native lignin in a fluidic reactor is still lacking. Also, the potential of flow reactors to increase product selectivity remains unexplored. The development of fluidic reactors for effective and selective lignin conversion requires efforts from scientists and engineers.

It should be noted that lignin depolymerization usually results in a mixture of organic products. The mild conditions employed in photocatalysis are beneficial to increase the selectivity of a specific aromatic, still, the separation and purification of products are energy-consuming and costly. An alternative strategy to decrease product complexity is to convert the organic mixture toward chemicals with similar functionality, which is also known as chemical funneling [69]. A typical example of chemical funneling in lignin transformation is hydrodeoxygenation of the depolymerization products to produce phenol [70]. New photocatalytic systems, such as efficient systems for photocatalytic hydrogenolysis of aromatic monomers, should be designed and employed to achieve this purpose.

In-depth insights into the catalyst properties, particularly for heterogenous photocatalysis systems, are still less accumulated. Most of the current mechanistic studies focus on organic transformations induced by photo-initiated electron transfer, and many research efforts were put into tuning the redox potentials of charge carriers or active species. Important questions, such as the specific active sites and the catalyst structure-activity relationships, are not well understood. Advanced characterizations, including ultrafast spectroscopy and *in situ* characterizations, should be applied to monitor the change of catalyst during the reaction. Machine learning could also be used to interpret structure-activity relationships and provide guidelines for the rational design of effective catalysts.

References

1 Nocera, D.G. (2017). Solar fuels and solar chemicals industry. *Accounts of Chemical Research* 50 (3): 616–619.

2 Schultz, D.M. and Yoon, T.P. (2014). Solar synthesis: prospects in visible light photocatalysis. *Science* 343 (6174): 1239176.

3 Yoon, T.P., Ischay, M.A., and Du, J. (2010). Visible light photocatalysis as a greener approach to photochemical synthesis. *Nature Chemistry* 2 (7): 527–532.

4 Wu, X., Luo, N., Xie, S. et al. (2020). Photocatalytic transformations of lignocellulosic biomass into chemicals. *Chemical Society Reviews* 49 (17): 6198–6223.

5 Wu, X., Xie, S., Zhang, H. et al. (2021). Metal sulfide photocatalysts for lignocellulose valorization. *Advanced Materials* 33 (50): 2007129.

6 Liu, X., Duan, X., Wei, W. et al. (2019). Photocatalytic conversion of lignocellulosic biomass to valuable products. *Green Chemistry* 21 (16): 4266–4289.

7 Huang, Z., Luo, N., Zhang, C. et al. (2022). Radical generation and fate control for photocatalytic biomass conversion. *Nature Reviews Chemistry* 6 (3): 197–214.

8 Li, S.-H., Liu, S., Colmenares, J.C. et al. (2016). A sustainable approach for lignin valorization by heterogeneous photocatalysis. *Green Chemistry* 18 (3): 594–607.

9 Xiang, Z., Han, W., Deng, J. et al. (2020). Photocatalytic conversion of lignin into chemicals and fuels. *ChemSusChem* 13 (17): 4199–4213.

10 Xu, J., Zhou, P., Zhang, C. et al. (2022). Striding the threshold of photocatalytic lignin-first biorefining via a bottom-up approach: from model compounds to realistic lignin. *Green Chemistry* 24 (14): 5351–5378.

11 Rinaldi, R., Jastrzebski, R., Clough, M.T. et al. (2016). Paving the way for lignin valorisation: recent advances in bioengineering, biorefining and catalysis. *Angewandte Chemie, International Edition* 55 (29): 8164–8215.

12 Li, C., Zhao, X., Wang, A. et al. (2015). Catalytic transformation of lignin for the production of chemicals and fuels. *Chemical Reviews* 115 (21): 11559–11624.

13 Zimmerman, J.B., Anastas, P.T., Erythropel, H.C. et al. (2020). Designing for a green chemistry future. *Science* 367 (6476): 397–400.

14 Lancefield, C.S., Ojo, O.S., Tran, F. et al. (2015). Isolation of functionalized phenolic monomers through selective oxidation and C–O bond cleavage of the β-O-4 linkages in lignin. *Angewandte Chemie, International Edition* 54 (1): 258–262.

15 Wu, X.J., Fan, X.T., Xie, S.J. et al. (2018). Solar energy-driven lignin-first approach to full utilization of lignocellulosic biomass under mild conditions. *Nature Catalysis* 1 (10): 772–780.

16 Yan, M., Lo, J.C., Edwards, J.T. et al. (2016). Radicals: reactive intermediates with translational potential. *Journal of the American Chemical Society* 138 (39): 12692–12714.

17 Staveness, D., Bosque, I., and Stephenson, C.R.J. (2016). Free radical chemistry enabled by visible light-induced electron transfer. *Accounts of Chemical Research* 49 (10): 2295–2306.

18 Kim, S., Chmely, S.C., Nimlos, M.R. et al. (2011). Computational study of bond dissociation enthalpies for a large range of native and modified lignins. *Journal of Physical Chemistry Letters* 2 (22): 2846–2852.

19 Karkas, M.D., Bosque, I., Matsuura, B.S. et al. (2016). Photocatalytic oxidation of lignin model systems by merging visible-light photoredox and palladium catalysis. *Organic Letters* 18 (19): 5166–5169.

20 Luo, J. and Zhang, J. (2016). Aerobic oxidation of olefins and lignin model compounds using photogenerated phthalimide-N-oxyl radical. *The Journal of Organic Chemistry* 81 (19): 9131–9137.

21 Cao, Y., Wang, N., He, X. et al. (2018). Photocatalytic oxidation and subsequent hydrogenolysis of lignin beta-O-4 models to aromatics promoted by in situ carbonic acid. *ACS Sustainable Chemistry & Engineering* 6 (11): 15032–15039.

22 Dai, J., Patti, A.F., Styles, G.N. et al. (2019). Lignin oxidation by MnO_2 under the irradiation of blue light. *Green Chemistry* 21 (8): 2005–2014.

23 Magallanes, G., Karkas, M.D., Bosque, I. et al. (2019). Selective C–O bond cleavage of lignin systems and polymers enabled by sequential palladium-catalyzed aerobic oxidation and visible-light photoredox catalysis. *ACS Catalysis* 9 (3): 2252–2260.

24 Nguyen, J.D., Matsuura, B.S., and Stephenson, C.R. (2014). A photochemical strategy for lignin degradation at room temperature. *Journal of the American Chemical Society* 136 (4): 1218–1221.

25 Bosque, I., Magallanes, G., Rigoulet, M. et al. (2017). Redox catalysis facilitates lignin depolymerization. *ACS Central Science* 3 (6): 621–628.

26 Luo, N., Wang, M., Li, H. et al. (2016). Photocatalytic oxidation-hydrogenolysis of lignin β-O-4 models via a dual light wavelength switching strategy. *ACS Catalysis* 6 (11): 7716–7721.

27 Hao, Z., Li, S., Sun, J. et al. (2018). Efficient visible-light-driven depolymerization of oxidized lignin to aromatics catalyzed by an iridium complex immobilized on msocellular silica foams. *Applied Catalysis. B, Environmental* 237: 366–372.

28 Li, S., Hao, Z., Wang, K. et al. (2020). Visible light-enabled selective depolymerization of oxidized lignin by an organic photocatalyst. *Chemical Communications* 56 (76): 11243–11246.

29 Luo, J., Zhang, X., Lu, J. et al. (2017). Fine tuning the redox potentials of carbazolic porous organic frameworks for visible-light photoredox catalytic degradation of lignin β-O-4 models. *ACS Catalysis* 7 (8): 5062–5070.

30 Fang, J., Ye, P., Wang, M. et al. (2018). Hydrogenolysis and hydrogenation of β-O-4 ketones by a simple photocatalytic hydrogen transfer reaction. *Catalysis Communications* 107: 18–23.

31 Chen, K., Schwarz, J., Karl, T.A. et al. (2019). Visible light induced redox neutral fragmentation of 1,2-diol derivatives. *Chemical Communications* 55 (87): 13144–13147.

32 Luo, N.C., Wang, M., Li, H.J. et al. (2017). Visible-light-driven self-hydrogen transfer hydrogenolysis of lignin models and extracts into phenolic products. *ACS Catalysis* 7 (7): 4571–4580.

33 Nichols, J.M., Bishop, L.M., Bergman, R.G. et al. (2010). Catalytic C–O bond cleavage of 2-aryloxy-1-arylethanols and its application to the depolymerization of lignin-related polymers. *Journal of the American Chemical Society* 132 (36): 12554–12555.

34 Gao, F., Webb, J.D., Sorek, H. et al. (2016). Fragmentation of lignin samples with commercial Pd/C under ambient pressure of hydrogen. *ACS Catalysis* 6 (11): 7385–7392.

35 Han, G.Q., Yan, T., Zhang, W. et al. (2019). Highly selective photocatalytic valorization of lignin model compounds using ultrathin metal/CdS. *ACS Catalysis* 9 (12): 11341–11349.

36 Wu, X.J., Xie, S.J., Liu, C.X. et al. (2019). Ligand-controlled photocatalysis of CdS quantum dots for lignin valorization under visible light. *ACS Catalysis* 9 (9): 8443–8451.

37 Lin, J., Wu, X., Xie, S. et al. (2019). Visible-light-driven cleavage of C-O linkage for lignin valorization to functionalized aromatics. *ChemSusChem* 12 (22): 5023–5031.

38 Zhu, Q. and Nocera, D.G. (2021). Catalytic C(β)–O bond cleavage of lignin in a one-step reaction enabled by a spin-center shift. *ACS Catalysis* 11 (22): 14181–14187.

39 Yoo, H., Lee, M.-W., Lee, S. et al. (2020). Enhancing photocatalytic β-O-4 bond cleavage in lignin model compounds by silver-exchanged cadmium sulfide. *ACS Catalysis* 10 (15): 8465–8475.

40 Xu, J., Li, M., Qiu, J. et al. (2021). Fine tuning of Cd_xZn_1-X_S for photo-depolymerization of alkaline lignin into vanillin. *International Journal of Biological Macromolecules* 185: 297–305.

41 Shao, S., Wang, K., Love, J.B. et al. (2022). Water promoted photocatalytic $C_β$-O bonds hydrogenolysis in lignin model compounds and lignin biomass conversion to aromatic monomers. *Chemical Engineering Journal* 435: 134980.

42 Xu, J., Li, M., Qiu, J. et al. (2021). Photocatalytic depolymerization of organosolv lignin into valuable chemicals. *International Journal of Biological Macromolecules* 180: 403–410.

43 Tay, N.E.S. and Nicewicz, D.A. (2017). Cation radical accelerated nucleophilic aromatic substitution via organic photoredox catalysis. *Journal of the American Chemical Society* 139 (45): 16100–16104.

44 Li, H., Bunrit, A., Lu, J. et al. (2019). Photocatalytic cleavage of aryl ether in modified lignin to non-phenolic aromatics. *ACS Catalysis* 9 (9): 8843–8851.

45 Han, P., Tana, T., Xiao, Q. et al. (2019). Promoting Ni(II) catalysis with plasmonic antennas. *Chem* 5 (11): 2879–2899.

46 Tan, F.-F., He, X.-Y., Tian, W.-F. et al. (2020). Visible-light photoredox-catalyzed C–O bond cleavage of diaryl ethers by acridinium photocatalysts at room temperature. *Nature Communications* 11 (1): 6126.

47 Imamura, K., Kato, H., Wada, Y. et al. (2018). Photocatalytic chemoselective cleavage of C–O bonds under hydrogen gas- and acid-free conditions. *Chemical Communications* 54 (53): 7298–7301.

48 Baffou, G. and Quidant, R. (2014). Nanoplasmonics for Chemistry. *Chemical Society Reviews* 43 (11): 3898–3907.

49 Renders, T., Van den Bosch, S., Koelewijn, S.F. et al. (2017). Lignin-first biomass fractionation: the advent of active stabilisation strategies. *Energy & Environmental Science* 10 (7): 1551–1557.

50 Ota, E., Wang, H., Frye, N.L. et al. (2019). A redox strategy for light-driven, out-of-equilibrium isomerizations and application to catalytic C–C bond cleavage reactions. *Journal of the American Chemical Society* 141 (4): 1457–1462.

51 Yayla, H.G., Wang, H., Tarantino, K.T. et al. (2016). Catalytic ring-opening of cyclic alcohols enabled by PCET activation of strong O–H bonds. *Journal of the American Chemical Society* 138 (34): 10794–10797.

52 Gazi, S., Hung Ng, W.K., Ganguly, R. et al. (2015). Selective photocatalytic C-C bond cleavage under ambient conditions with earth abundant vanadium complexes. *Chemical Science* 6 (12): 7130–7142.

53 Gazi, S., Đokić, M., Moeljadi, A.M.P. et al. (2017). Kinetics and DFT studies of photoredox carbon–carbon bond cleavage reactions by molecular vanadium catalysts under ambient conditions. *ACS Catalysis* 7 (7): 4682–4691.

54 Liu, H., Li, H., Luo, N. et al. (2019). Visible-light-induced oxidative lignin C–C bond cleavage to aldehydes using vanadium catalysts. *ACS Catalysis* 10 (1): 632–643.

55 Wang, Y., He, J., and Zhang, Y. (2020). $CeCl_3$-promoted simultaneous photocatalytic cleavage and amination of C–C bond in lignin model compounds and native lignin. *CCS Chemistry* 2 (3): 107–117.

56 Wang, Y., Liu, Y., He, J. et al. (2019). Redox-neutral photocatalytic strategy for selective C–C bond cleavage of lignin and lignin models via PCET process. *Scientific Bulletin* 64 (22): 1658–1666.

57 Zhou, W., Nakahashi, J., Miura, T. et al. (2018). Light/copper relay for aerobic fragmentation of lignin model compounds. *Asian Journal of Organic Chemistry* 7 (12): 2431–2434.

58 Nguyen, S.T., Murray, P.R.D., and Knowles, R.R. (2020). Light-driven depolymerization of native lignin enabled by proton-coupled electron transfer. *ACS Catalysis* 10 (1): 800–805.

59 Liu, H.F., Li, H.J., Lu, J.M. et al. (2018). Photocatalytic cleavage of C–C bond in lignin models under visible light on mesoporous graphitic carbon nitride through pi–pi stacking interaction. *ACS Catalysis* 8 (6): 4761–4771.

60 Wu, X., Lin, J., Zhang, H. et al. (2021). Z-scheme nanocomposite with high redox ability for efficient cleavage of lignin C–C bonds under simulated solar light. *Green Chemistry* 23 (24): 10071–10078.

61 Hou, T.T., Luo, N.C., Li, H.J. et al. (2017). Yin and Yang dual characters of CuOx clusters for C–C bond oxidation driven by visible light. *ACS Catalysis* 7 (6): 3850–3859.

62 Kang, Y., Lu, X., Zhang, G. et al. (2019). Metal-free photochemical degradation of lignin-derived aryl ethers and lignin by autologous radicals through ionic liquid induction. *ChemSusChem* 12 (17): 4005–4013.

63 Ten Have, R. and Teunissen, P.J. (2001). Oxidative mechanisms involved in lignin degradation by white-rot fungi. *Chemical Reviews* 101 (11): 3397–3413.

64 Cho, D.W., Latham, J.A., Park, H.J. et al. (2011). Regioselectivity of enzymatic and photochemical single electron transfer promoted carbon-carbon bond fragmentation reactions of tetrameric lignin model compounds. *The Journal of Organic Chemistry* 76 (8): 2840–2852.

65 Tien, M. and Kirk, T.K. (1984). Lignin-degrading enzyme from phanerochaete ehrysosporium: purification, characterization, and catalytic properties of a unique H_2O_2-requiring oxygenase. *Proceedings of the National Academy of Sciences* 81: 2280–2284.

66 Lim, S.H., Nahm, K., Ra, C.S. et al. (2013). Effects of alkoxy groups on arene rings of lignin β-O-4 model compounds on the efficiencies of single electron transfer-promoted photochemical and enzymatic C-C bond cleavage reactions. *The Journal of Organic Chemistry* 78 (18): 9431–9443.

67 Cho, D.W., Parthasarathi, R., Pimentel, A.S. et al. (2010). Nature and kinetic analysis of carbon-carbon bond fragmentation reactions of cation radicals derived from SET-oxidation of lignin model compounds. *The Journal of Organic Chemistry* 75 (19): 6549–6562.

68 Colmenares, J.C., Varma, R.S., and Nair, V. (2017). Selective photocatalysis of lignin-inspired chemicals by integrating hybrid nanocatalysis in microfluidic reactors. *Chemical Society Reviews* 46 (22): 6675–6686.

69 Sun, Z., Fridrich, B., de Santi, A. et al. (2018). Bright side of lignin depolymerization: toward new platform chemicals. *Chemical Reviews* 118 (2): 614–678.

70 Liao, Y., Koelewijn, S.-F., Van den Bossche, G. et al. (2020). A sustainable wood biorefinery for low-carbon footprint chemicals production. *Science* 367 (6484): 1385–1390.

11

Electrochemical Conversion for Lignin Valorization

Qi Zhu[1], Mahlet Garedew[2], Bing Song[3,4], Yuliang Li[5] and Jason Chun-Ho Lam[1]

[1] *School of Energy and Environment, City University of Hong Kong, Tat Chee Avenue, Hong Kong SAR*
[2] *Air Company, 407 Johnson Avenue, Brooklyn, New York 11206, USA*
[3] *Scion, Te Papa Tipu Innovation Park, 49 Sala Street, Private Bag 3020, Rotorua 3046, New Zealand*
[4] *School of Chemical Engineering, University of New South Wales, Sydney, NSW 2052, Australia*
[5] *Key Laboratory of Subsurface Hydrology and Ecological Effects in Arid Region, Ministry of Education, School of Water and Environment, Chang'an University, Xi'an, 710064, P. R. China*

11.1 Introduction

The strategic development of bio-based product manufacturing is urgently needed to satisfy the increasing demand for commodity chemical productions. Over the past several decades, the world has witnessed various climate phenomena that have been resulted from the over-dependence on petroleum resources: it is nearly impossible to identify a product that is completely free-of petroleum precursors. As such, the present industrial commodity fuels and chemicals are dominated by petroleum derivatives [1]. As the second most abundant renewable biopolymer, lignin is an excellent resource to promote a carbon-negative solution for petroleum mitigation [2–4]. However, lignin valorization is no trivial task due to its three-dimensional aromatic structure that has been evolved to provide structural integrity and protection [5]. Decades of research have been devoted to breaking and valorizing lignin and its derived products via homogeneous and heterogeneous catalysis. The most commonly applied physiochemical degradation routes include pyrolysis, gasification, acid catalysis, hydrogenolysis, and some approaches have been developed into commercial-scale applications [1, 6–8]. In these thermo-catalytic processes, combusting temperature conditions at 240–650 °C, and high H_2 pressures of 250–315 bar are typically needed to overcome the various lignin signature bond-cleavage barriers. These severe conditions tend to lead to low-carbon efficiency and reaction rate due to catalyst coking. Indiscriminative thermal treatments are also less likely to produce active and valuable products, which require the cleavage of specific lignin linkages.

Lignin Chemistry: Characterization, Isolation, and Valorization, First Edition.
Edited by Yuhe Liao and Bert F. Sels.

A low-carbon footprint lignin valorization strategy thus greatly appeals to the renewable chemicals and fuel research community. This would require the biomass to be processed with renewable energy input and preferably in a decentralized setting to maximize the sustainability aspect [9, 10]. In light of this rationale, electrocatalysis serves as a protocol to process biomass for renewable chemicals and fuel production. Compared to conventional energy-intense heterogeneous catalysis, electrocatalysis can often achieve a similar catalytic outcome in a milder and greener environment with only a small potential bias [11–13]. There is an increasing number of examples showing the electrocatalytic depolymerization of lignin linkages can be achieved selectively through electrodes and/or reactor design [14, 15]. The advantages of electrocatalysis are elaborated as follows:

1) *Environmentally benign energy source and conditions*: Electrocatalysis can motivate thermodynamically unfavorable reactions by supplying a potential difference generated from renewable energy, such as wind turbines and solar photovoltaics. For instance, hydrogenolysis of diaryl ether C—O bonds in lignin by a hydrothermal process usually requires 80–200 °C in a pressurized hydrogen atmosphere (15–100 bar), but electrocatalysis can accomplish a similar selective transformation at −0.9 V at ambient temperature and pressure [16–19]. Lam and coworkers recently showed that carbon-neutral grid-powered electrocatalysis could reduce greenhouse potential (GHP) by 57% compared to conventional hydrothermal processing scenarios [20].
2) *In situ "reagent" and instant controllability*: The reductants during the electrochemical reaction are via in situ water splitting or proton (or water) reduction. The electrocatalytic activity and selectivity can be regulated by varying the working potential (or current flow), which are instantly responsive. The advantages offer diverse possibilities to control reaction selectivity and even automation.
3) *Convenient catalyst separation and ease of implementation*: Most electrocatalytic reaction operates heterogeneously; electrodes can be easily removed from the reaction medium. Reaction mode can be reconfigured between batch and flow depending on the investigation criteria. Because of the ease of implementation, electrocatalysis is highly suitable to be deployed as a decentralized catalytic treatment strategy. One of the most fitting examples would be localized biomass upgrading, which can address the energy-inefficient aspect of transporting the energy-dilute raw biomass.

The above advantages have promoted great research efforts to promote the electrocatalytic upgrading of lignin in the research community. The general research directions are divided into several areas: energy intensification, chemical transformation, functional groups activation (C_α-carbonylation), and selective cleavage of the inter-monolignol linkages, such as β-O-4, α-O-4, 4-O-5, and C—C bonds of β-1, β-5, 5-5. Besides lignin, it is also worth noting that electrocatalysis is widely applied to the upgrading of cellulose-derived compounds, such as furfural [21–26]. Readers may refer to some of the reviews published recently [27–32]. This chapter will provide the basic concepts of electrocatalysis and discuss the recent advances in the electrocatalytic lignin upgrade with an emphasis on the potential production of high-value chemicals and biofuels.

11.2 The Mechanism of Electrocatalytic Lignin Conversion

Electrocatalytic conversion of organic molecules pertains to the electron transfer (ET) between the electrode surface and the target substrate. Some common electrocatalytic conversions relevant to lignin upgrading are hydrogenation, hydrodeoxygenation, hydrogenolysis, oxidation, and carbonylation [33]. These electrocatalytic transformations have been demonstrated for lignin to biofuels and bulk chemicals production. It is important to note that these transformations are not necessarily described as a single-step transformation; they can be a cascaded reaction where electrocatalytically produced intermediates can interact with the surrounding chemical entity, e.g. solvent molecules, to afford the final product. For example, the electrocatalytic hydrogenation (ECH) of phenol to cyclohexane begins with the hydrogenation of phenol to produce cyclohexanol, which undergoes a solvent-promoted dehydration step to yield a cyclohexene intermediate, followed by another step of ECH of the alkene to yield cyclohexane. The dehydration step occurs favorably in a strongly acidic environment, such as HCl or H_3PO_4 [34]. It is commonly misinterpreted that the final products from an electrocatalytic reaction are solely dictated by the electrochemical parameters alone when external factors, such as solvent composition, temperature, and even pressurize, may influence the yield and selectivity outcome of the reaction [11, 14, 33]. The situation can be even more complicated when dealing with mediated electrochemical reactions commonly found in electrosynthesis studies.

11.2.1 Reaction Mechanism

The essence of electrocatalysis is the ET process between the electrode surface and reactant, which can occur by one of the three possible pathways (Figure 11.1): direct redox, electrocatalytic cascade reaction, and mediated ET [14, 35]. In direct ET, whether it is reduction or oxidation, organic species are adsorbed on the electrode interface, which enables ET between the organic substrate and electrode. The direct ET electrochemical model is restricted to surface catalysis, so the solubility and adsorption of the lignin or its derived species are greatly important to facilitating an efficient ET. For an electrocatalytic cascade reaction, the reactant undergoes a redox reaction and forms a high-energy intermediate prone to producing the final product by rearrangement or dimerization coupling. The success in obtaining the desired intermediate is also subjected to the electrochemical parameters and various reaction conditions. For instance, the electrochemical oxidation of glycerol in a strongly alkaline environment can afford glyceraldehyde, which can be rearranged to yield lactic acid via Cannizzaro rearrangement [36]. However, if the same reaction is

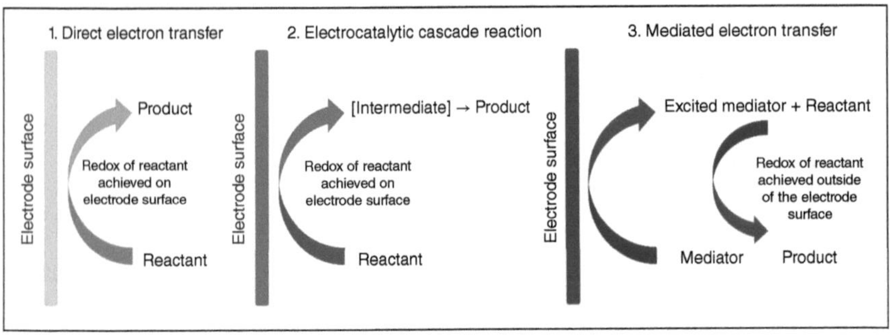

Figure 11.1 General electron transfer modes in electrocatalytic reactions.

performed in a pH-neutral environment in a different current flow, it will produce mainly glyceric acid instead [37].

For mediated ET reaction, an electrochemically active species participates in the electrochemical redox at the surface area. The activated species will engage in the ET with the target organic substrates. The mediator essentially behaves as a homogeneous electrocatalyst. However, unlike direct ET and cascade ET, the use of a mediator will necessitate a post-reaction catalyst removal that is less feasible for low-profit margin products. This is why mediated ET reactions are more commonly explored with electrosynthesis, and in some cases, the mediator may even exert stereo-control on the reaction product [38–40].

On the anode side, electrocatalytic oxidation (ECO) takes place, which is described by reactions Eqs. (11.1)–(11.6). Depending on the electrolyte and the electrodes' properties, different species may be oxidized. In general, ECO of organic substrate in an aqueous environment, which is a common solvent for biomass catalysis, is done preferably with "non-active" anodes that are less active toward water splitting. Typical non-active anodes candidate includes Ti/SnO_2, Ti/SnO_2-Sb, Pb/PbO_2, β-PbO_2, RuO_x–TiO_x, TiO_2–NTs, and different doped materials, such as boron-doped diamond (BDD) and Si/BDD [41–44]. There are several mechanisms by which the dissolved biomass organics can be fragmented by the non-active anodes. The most common mechanism is the generation of hydroxyl radicals from water electrolysis, which begins as active species adsorb to the metal electrode surface (M) first and may interact with the anode to form metal oxides [11]. The main oxidation process features the selective deconstruction of biomass via interaction of the adsorbed organic substrate with the metal oxide and adsorbed/free hydroxyl radicals under mild conditions. Notably, oxygen may form *in situ* via competing for anodic water splitting (oxygen evolution reaction, OER) in this process. Generally accepted ECO reaction equations are:

(1) Hydroxyl radical chemisorption and metal oxide formation

$$H_2O + M \rightarrow (^{\cdot}OH)_{ads}\,M + e^- + H^+ \tag{11.1}$$

$$M(^{\cdot}OH)_{ads} \rightarrow MO + e^- + H^+ \tag{11.2}$$

(2) Substrate oxidation

$$(R)+M\rightarrow(R)_{ads}M \tag{11.3}$$

$$(R)_{ads}M+MO\rightarrow RO+2M \tag{11.4}$$

(3) Oxygen evolution reaction (OER)

$$M(^{\cdot}OH)_{ads}+M(^{\cdot}OH)_{ads}\rightarrow O_2+2M+2e^-+2H^+ \tag{11.5}$$

$$MO+MO\rightarrow O_2+2M \tag{11.6}$$

On the cathode side, electrocatalytic reduction (ECR) takes place on the electrode surface. In a protic environment, the ECR process typically describes the electrocatalytic hydrogenation (ECH) of the unsaturated sites or the reductive cleavage of chemically susceptible sites, such as aryl ether or aromatic benzylic carbon. Similar to ECO, the mechanism of ECH involves a multi-step electrochemical process. In a typical mechanism, surface adsorbed hydrogen (H_{ads}) is generated from the electroreduction of protons, commonly known as Volmer reaction (Eqs. (11.7) and (11.8)). An unsaturated species can adsorb on the electrode surface (Eq. (11.9)), and if the adsorption takes place in close proximity to the H_{ads}, hydrogenation can occur (Eq. (11.10)). In addition to the reaction with ECH, hydrogen evolution reaction (HER) will inevitably occur as a competition by the Heyrovsky (Eq. (11.11)) or Tafel mechanism (Eq. (11.12)). The ECH mechanistic equations are listed as follows:

(1) Hydrogen chemisorption

$$H_2O+e^-+M\rightarrow(H)_{ads}M+OH^- \tag{11.7}$$

$$H_3O^++e^-+M\rightarrow(H)_{ads}M+H_2O \tag{11.8}$$

(2) Substrate reduction

$$(RC=CR)+M\rightarrow(RC=CR)_{ads}M \tag{11.9}$$

$$(RC=CR)_{ads}M+2(H)_{ads}M\rightarrow RHC-CHR+2M \tag{11.10}$$

(3) Hydrogen evolution reaction (HER)

$$(H)_{ads}M+e^-+H_2O\rightarrow H_2+M+OH^-\ \text{(Heyrovsky)} \tag{11.11}$$

$$(H)_{ads}M+(H)_{ads}M\rightarrow H_2+2M\ \text{(Tafel)} \tag{11.12}$$

11.2.2 Faradaic Efficiency

Faradaic efficiency (FE), sometimes termed current efficiency (CE), is of great importance in evaluating the efficiency of any electrochemical reaction. FE evaluates the proportion of electrons that are spent on the desired electrochemical transformation over the side reactions, such as solvent degradation, which is particularly common when conducting aqueous-based electrochemical reactions. Ideally, FE should reach 100%, which mean all the ET occurs to accomplish the desired reactions. It should be emphasized that FE does not directly describe the rate of the reaction. Poor reaction kinetic can still lead to a high FE and vice versa. Also, considering FE is calculated based on the desired product; its calculation can be subjectively interpreted. Readers are often encouraged to pay attention to how authors define the desired product. FE is calculated by dividing the fraction of the electrons passed accomplished the desired target reaction over the total charge passed as described in Eq. (11.13) [11, 45]:

$$\%\mathrm{FE} = \left(\frac{z \cdot n \cdot F}{Q}\right) \times 100\% \tag{11.13}$$

where,

z: Number of electrons transferred for a given product
n: Number of moles of products produced
F: Faraday's constant: 96,485 C mol^{-1}
Q: Total charge passed (C)

Most electrochemical reactions are typically optimized based on yield, selectivity, and FE. The former two terms are applied to assess the reactions outcome, while the latter is used to assess the efficiency of used electricity in desired reactions. FE can be optimized by systematically investigating a number of factors, including but not limited to pH, temperature, electrochemical parameters (working potentials or current density), electrode materials, and electrolyte composition. There are no universal means to optimize the FE. Still, in the authors' research experiences, the electrode materials and working potential tend to have a larger impact than other factors, especially when conducting electrochemical reactions in an aqueous environment.

In brief, when conducting an aqueous-based electrochemical reaction, a high FE can be obtained by suppressing the non-desired electrochemical reactions, such as OER or HER. This is commonly accomplished by employing non-active electrode materials mentioned in the previous section. Another key factor is to control the working potential of the electrodes within or close to the solvent's onset potential. However, the lowering of working potential will decrease the current flow, which represents the rate of ET, and thus will decrease the overall reaction rate. In the authors' research experience, the optimized set of electrochemical parameters is often a compromise between the yield, FE, and product selectivity (when applicable) and generally scores between 30% and 80% for the electrochemical transformation of organics in an aqueous environment.

11.3 Lignin Pretreatment to Enable Efficient Electrocatalytic Upgrading

Due to its poor solubility in various solvents, electrochemcial transformation of raw biomass is challenging. Therefore, it must undergo degradative pretreatment prior to the upgrading process. Several integrated lignin valorization routes were discussed in selected reviews and will be briefly outlined here [2, 46, 47]: (i) partial depolymerization from biomass (cellulose, hemicellulose, and lignin) for solubilization and extraction of lignin by the successful disruption of lignin-carbohydrate complex (LCC) linkages; (ii) cleavage of the C-aryl, C-aryl ether, and C–C linkages in lignin to derive small molecule subunits (monomers, dimers, etc.); (iii) effective recovery and upgrading of depolymerization intermediates to valuable products. More recently, a "lignin-first" strategy that focuses on preserving the lignin's structure integrative during the pretreatment stage has been proposed and investigated by several authors [48, 49]. This process is performed to solubilize/depolymerize the cellulose components while preserving the aromatic moieties that are saved for more elaborated chemical treatments for high-value chemical production.

Figure 11.2 illustrates and summarizes the general pretreatment and degradative strategy for raw biomass. These methods include physical, physicochemical, chemical, and biological approaches. [51, 52]. These strategies are not only developed to maximize the biomass degradation efficiency but also aim to diversify the product outcome of the feedstock. The most commonly known example would be the practice of thermal valorization of lignin via pyrolysis, which produces a variety of commodity liquefied aromatic products, such as phenols (phenol, catechol, guaiacol, syringol, and cresols), aldehydes (vanillin and syringaldehyde), aceton nylon, and polyester [2, 9, 53]. If properly valorized for valuable chemical products, one can improve the economic feasibility of any biorefinery process, which is critically needed in the current petroleum-dominated market.

Ideally, single stream, in-tandem upgrading can be implemented by using the same solvent during pretreatment (assuming solvent is needed) and electrocatalysis. Meanwhile, to achieve a simple, green, and low-cost system for lignin

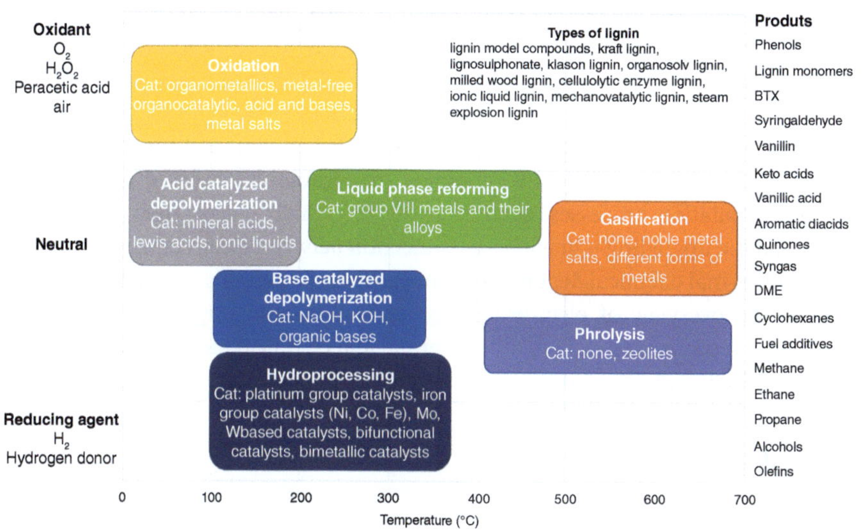

Figure 11.2 A summary and comparison of lignin depolymerization methods. Source: Li et al. [50]/American Chemical Society.

valorization, the pretreatment/depolymerization approach prior to electrocatalysis should avoid known hazardous or nonrecyclable solvents, such as ionic liquid. Thus, thermochemical degradation methods, such as pyrolysis and gasification, are highly advantageous for their convenience in achieving degradation in a solvent-free environment. Nonetheless, chemical-based catalytic degradation protocols, such as hydrothermal liquefaction, acid/base-catalyzed, ionic liquid-assisted, sub- or supercritical fluid-assisted, oxidative, and hydro-processing, are still commonly investigated as the use of solvent provides an additional handle to control yield and selectivity. Beyond degradation, electrocatalysis has also found its role in raw biomass pretreatment. Electro-assisted ionic liquid pretreatment of eucalyptus sawdust and enhance the lignin rate removal under ambient conditions [54].

11.4 Electrocatalytic Cleavage of Lignin-derived Model Compounds

Because of the complex structural makeup of native lignin, model compounds (lignin-relevant monomers and dimers, etc.) with known linkages, such as β-O-4 (β-aryl ether), α-O-4 (α-aryl ether), 4-O-5 (diaryl ether), β-β (resinol), and β-5 (phenylcoumaran), are employed to evaluate the catalytic performance of a catalyst and electrocatalytic system [11]. For example, electrocatalytic conversion for C_α-C_β bond cleavage of the β-O-4 linkage at room temperature with platinum as the anode was achieved with high yields of the aromatic aldehydes and phenols [55]. Cathodic reduction of β-O-4 dimers also afforded over 90% of the corresponding monomeric C-O cleavage products (phenol and acetophenone) in reticulated vitreous carbon (RVC) working cathode in only 1.5 hours [56]. Amongst the known linkages, β-O-4 lignin model dimers are the most explored because of their abundance: β-O-4 linkages account for 43–65% of all lignin linkages. Besides dimer, monomers, such as phenol, guaiacol, and syringol, are also commonly used model compounds for electrocatalytic lignin conversion studies. This bottom-up research approach beginning with the lignin model compounds, makes the mechanistic evaluation easily understood and helps the scientific community benchmark different catalytic systems. Moreover, many model compounds are commonly observed products of thermal degradation processes, such as pyrolysis, which means their successful electrocatalytic upgrading will enable a connection between electrocatalytic studies and other thermal chemical degradation strategies. Although this book chapter only discusses the overview of electrocatalytic degradation and upgrading strategies, readers are encouraged to explore other thermal and chemical methods for effective cleavage of the lignin linkages.

11.4.1 Reductive Approach

11.4.1.1 Monomers

ECH is a common reductive strategy to achieve lignin cleavage through hydrogenation. In terms of catalytic outcome, ECH is similar to conventional heterogeneous hydrogenation that takes place in an elevated temperature and pressure setting, but ECH is typically conducted under ambient conditions. To the best of the authors' knowledge, all biomass-related ECH work is performed below water-boiling temperature and at atmospheric pressure. ECH can achieve the catalytic hydrogenation under mild condition mainly because it can generate the necessary adsorbed hydrogen on the catalytic surface using a small reductive potential bias. In contrast, conventional hydrogenation requires thermal input to achieve the activation of H_2. Li and coworkers reported a Ru-decorated active carbon cloth (Ru/ACC) with 10–20 nm Ru precipitation to achieve reductive cleavage of oxygenated aromatic monomers under constant current electrolysis conditions at 80 °C or less. The reaction yielded cyclohexanol and 2-methoxycyclohexanol as the major products. Two competing reaction pathways were identified in Li's study: the electrocatalytically induced aryl C–O cleavage, which converted the guaiacol to phenol, and the direct hydrogenation of guaiacol, which yielded both cis- and trans-2-methoxycyclohexanol [57]. Li optimized the reaction to favor the production of cyclohexanol because the purpose of the study was to maximize the deoxygenation of the guaiacol model compound. The goal was to raise the C:O ratio, which would improve the energy density, specifically known as specific energy (MJ kg^{-1}) of the substrate. Similarly, Cyr et al. also studied the ECH of guaiacol using a Raney®-nickel cathode, and obtained cyclohexanol as the main product. The result was particularly interesting because (i) the deoxygenation of guaiacol achieved good selectivity, and (ii) the reaction works under ambient conditions without using precious metals [58]. However, both studies only managed to reach cyclohexanol as the final product. The further deoxygenated products, cyclohexene, and cyclohexane, were not detected. It was evident that the electrocatalytically driven deoxygenation was possible, but the efficiency depended on the structural integrity of the reactant.

Nevertheless, the discovery of the mild electrochemical hydrodeoxygenation thus warranted a detailed mechanistic investigation. Lam et al. investigated the selective reduction of guaiacol to cyclohexanol on a Raney®-Nickel cathode [59]. The reaction occurred at 60 °C and atmospheric pressure in a divided cell. Notably, the anode was a cobalt phosphate-coated stainless steel anode instead of a platinum anode to achieve total precious metal-free electrocatalysis. The study observed a hydroxy group-coordinated de-alkoxylation mechanism to induce specific cleavage of aryl-OR ether bond on guaiacol, which produced a phenol intermediate which could further be hydrogenated to cyclohexanol (Figure 11.3). The cleavage occurred with >99% selectivity over the plain hydrogenated product, i.e. 2-methoxycyclohexanol, and the ortho relationship between the hydroxy and the alkoxy group is of great importance to the cleavage selectivity. If the alkoxy group is at a meta- or para-position to the hydroxyl group, the cleavage selectivity dropped to ~50%. The study also proved the phenol intermediate was desorbed from the catalytic site upon its formation in a competition study using *p*-cresol. The mechanism was applicable to syringol, but syringol showed a lower demethoxylation rate, indicating the demethoxylation step could be sensitive to the electron density of the aromatic ring.

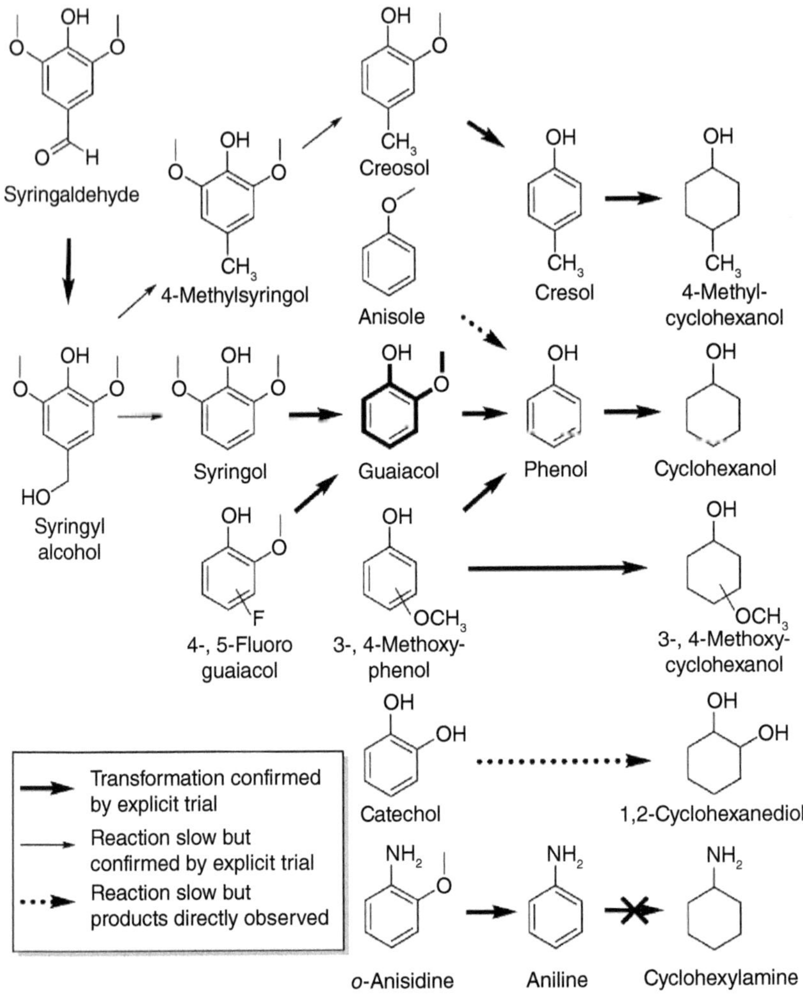

Figure 11.3 Substrates were studied in ECH over Ra–Ni electrodes.

The proposed mechanism by Lam et al. could also explain the demethoxylation of syringol observed by Li et al., who found a product mixture of guaiacol, phenol, cyclohexanol, and methoxycyclohexanols on the Ru-decorated ACC electrodes. Other authors have also explored the ECH of guaiacol on different catalysts. The PtNiB/CMK-3 hybrid electrode with pronounced ECH capability in the conversion of guaiacol was studied by Zhou et al., which exhibited an inspiring high FE of 86.2% [60]. Experimental results and theoretical calculations showed that the B-doping optimized the PtNiB alloy electron structure, simultaneously promoting substrate and intermediate adsorption in the ECH process. This work also confirmed the lab-scale ECH of guaiacol to KA oil (a mixture of cyclohexanone and cyclohexanol) was feasible.

Song et al. investigated the ECH and thermal catalytic hydrogenation (TCH) of phenol using carbon-supported platinum and rhenium catalysts (Pt/C and Rh/C) [61]. While both catalysts were active for the hydrogenation of phenol, Rh/C converted more phenol to cyclohexanol for both ECH and TCH processes via hydrogenation, hydrogenolysis, and hydrolysis mechanisms with similar selectivity. A subsequent study revealed that the electrocatalytic system was effective for the ECH of several substituted phenolic and dialkyl ethers. Sanyal et al. did a follow-up study on Song's work by using Pd/C catalyst for comparison, which showed higher activity and greater FE than Pt/C and Rh/C for the ECH of the non-aromatic fragment (e.g. carbonyl group of benzaldehyde) but was catalytically inactive for the hydrogenation of phenol (Figure 11.4) [18, 62]. Interestingly, in follow-up experiments, the group demonstrated that the reaction turnover frequency (TOF) of phenol ECH was positively correlated with the Pt particle size and lattice plane.

The electrocatalytic reaction pathways and product distributions were also shown to be influenced by synergistic effects among the major variables, such as proton concentration, temperature, and cathode potential in other studies. Recently, Wijaya et al. conducted a comprehensive experimental and theoretical study on the ECH of guaiacol and phenol in a stirred slurry reactor with different acid-base electrolyte combinations [63]. The 5 wt% Pt/C catalysts showed the highest conversion and FE in acid-acid and neutral-acid pair conditions for model compounds. The mechanistic study showed that the reaction occurred through a proton-electron transfer mechanism. In the various electrolytes examined, guaiacol was converted to cyclohexanol, cyclohexanone, 2-methoxycyclohexanol, 2-methoxy-cyclohexanone,

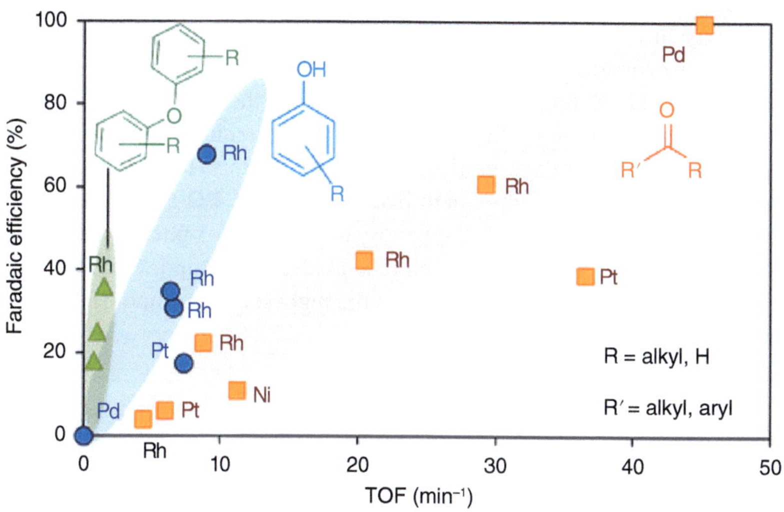

Figure 11.4 Activity and selectivity trends in electrocatalytic hydrogenation by Pd/C, Pt/C and Rh/C electrodes.

and phenol. Subsequently, the performances of different carbon-supported noble metal catalysts (Pt/C, Ru/C, and Pd/C) with the same metal loading of 5 wt% were evaluated for the ECH of guaiacol and phenol. The results suggested that the electrocatalytic upgrading efficiency of the three metal catalysts largely depends on the catholyte-anolyte environment with adsorbed hydrogen atoms as the main active species. This study hence showed the importance of electrocatalyst/electrolyte synergy in the ECH of phenolic compounds.

11.4.1.2 Lignin Dimers

Electrocatalytic reduction of lignin dimers pertains to the reductive cleavage of lignin-relevant linkages (e.g. β-O-4, 4-O-5, and α-O-4) are also well explored. The extent to which intermediates in the ECH of dimers convert to stable products depends on the linkage, the nature of catalysts, and a variety of other operating conditions [11, 12, 14, 15]. One of the earliest studies on lignin model dimers was conducted by Mahdavi et al., where they investigated the ECH of various benzyl phenyl ethers (molecules of α-O-4 linkages) using Raney®-nickel electrode under mild conditions. The absence of aliphatic dimer products indicated selective cleavage of the dimer via hydrogenolysis and subsequent hydrogenation of the aromatic monomer (Figure 11.5) [64]. A more recent in-depth study by Lin et al. demonstrated the cleavage of BPE via an in situ deposited nickel surface in methanol at room temperature [25]. Lin's reaction works exclusively with nickel deposition but not the other divalent metal ions, such as copper, iron, and cobalt ions [132, 133]. The discovery was particularly useful for valorizing C-lignin, which contains mostly α-O-4 linkages [65].

The cleavage of other aryl ether bonds is also important as they account for more than half of the linkages between lignin's phenylpropanoid building blocks. Wu et al. have studied the cleavage of 4-O-5 dimer using Pt in an undivided cell to produce phenol. The study further explored the effect of sodium borohydride ($NaBH_4$) to facilitate high conversion to phenol, presumably through $NaBH_4$ protecting the phenol from subsequent degradation at the anode [66]. They further demonstrated what appears to be a preferential cleavage to produce aromatic monomers with at least one aryl C-O bond. A study by Garedew et al. demonstrated the formation of cyclohexanol via the hydrogenolysis and subsequent hydrogenation of 4-O-5 model dimers (3-phenoxyphenol, 4-phenoxyphenol, 3-phenoxyanisole, and 3-phenoxytoluene) using Ru supported on activated carbon cloth in a divided cell setup [67]. Of all the electrolytes studied, alkaline electrolytes showed the highest cyclohexanol yields for all substrates.

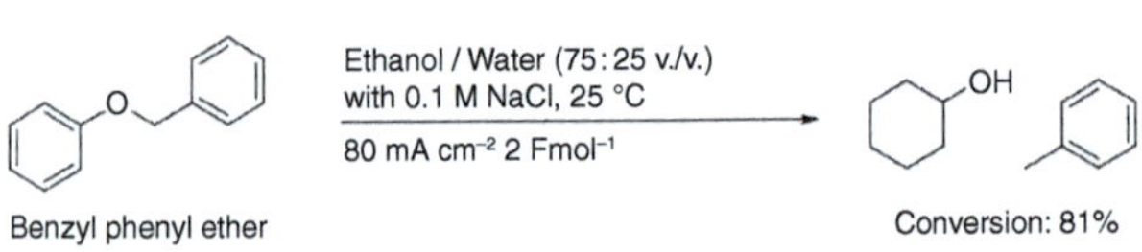

Figure 11.5 ECH of benzyl phenyl ether using Raney®-nickel without aliphatic dimer was observed.

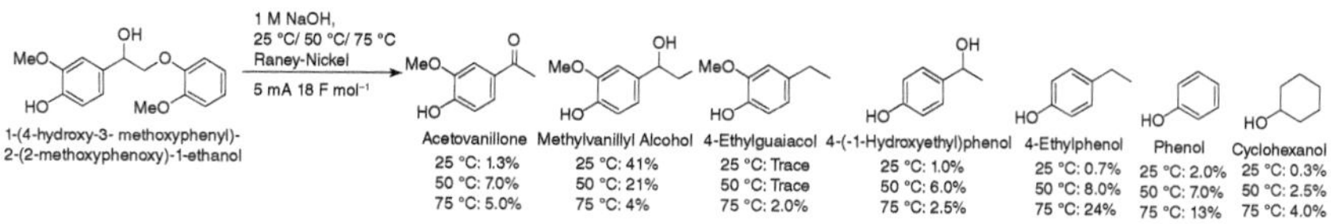

Figure 11.6 Hydrogenation and hydrogenolysis of β-O-4 dimer at different temperatures and the conversion yield.

Cyr et al. explored Pd/C and Raney®-nickel powders circulated in a solution with a reticulated vitreous carbon (RVC) electrode for the hydrogenolysis and hydrogenation of β-O-4 dimer (1-(4-hydroxy-3-methoxyphenyl)-2-(2-methoxyphenoxy)-1-ethanol). Using Pt/C in 1 M NaOH aqueous solution resulted in the formation of mostly hydrogenolysis products, guaiacol and acetovanillone [58].

In contrast, Raney®-nickel catalyst under the same conditions yielded both hydrogenolysis and hydrogenation products such as guaiacol, α-methylvanillyl alcohol, phenol, cyclohexanol, and acetovanillone (Figure 11.6). Another β-O-4 dimer study by Zhou et al. investigated ketone and alcohol dimers and concluded that ketone dimer cleaved roughly twice as fast as the alcohol dimer using Raney®-nickel electrode [68].

11.4.2 Oxidative Approach

11.4.2.1 Monomers

Electrochemical oxidation of lignin is also commonly explored in waste lignin treatment. Some studies have observed the electrochemical oxidation of guaiacol using precious metals and demonstrated the formation of *o*-benzoquinone, methoxybenzoquinone, and polymer [11, 69]. Proposed mechanisms suggest that guaiacol undergoes irreversible 1- or 2-electron transfer to form a phenoxonium cation mesomer [70]. As phenolic compounds are known environmental pollutants found in industrial effluents, electrochemical oxidative degradation of model aromatic compounds has become increasingly important, whether the aromatic substrate showed relevance to biomass valorization or not. Using SnO_2 and PbO_2 catalysts, conversion of guaiacol to the maleic acid intermediates and further to CO_2 and H_2O has been demonstrated [70]. Shao et al. conducted cyclic voltammetry studies to elucidate the mechanism of the electro-oxidative mineralization of guaiacol using Ti/Pb_3O_4 and Ti/Sb–SnO_2. They demonstrated that guaiacol mineralization using Ti/Sb–SnO_2 was likely achieved via the hydroxyl radicals in the bulk electrolyte and not on the surface of the electrode. For Ti/Pb_3O_4, the CV studies suggested that guaiacol degradation likely took place on the electrode surface [71]. In another study, Zhou et al. introduced a new type of Co-based electrocatalyst for selective oxidative cleavage of C(OH)-C or C(O)-C bond of lignin derivatives to carboxylates [72]. Most ECO work on lignin-derived monomers has focused on using this method for complete mineralization of the substrate to CO_2 and H_2O without much regard for selectivity toward the formation of valuable oxidative products. Designing a proper reaction system to maximize the selective partial oxidation of lignin monomers, would provide a valuable route to access bio-based oxygenated commodity chemicals, such as maleic acid and succinic acid.

11.4.2.2 Lignin Dimers

β-O-4 is the most abundant lignin linkage and can account for over half of the linkages found in lignin. To this end, much of the electrochemical lignin valorization efforts have focused on gaining insight into the effective cleavage of these linkages via the study of β-O-4 linked model dimers. Chen et al. studied electro-oxidation of β-O-4 linked model dimers (2-phenoxy-1-phenylethanol, 2-phenoxyacetophenone, and 2-phenoxy-1-phenylethane); the C_α-hydroxyl groups were oxidized to aldehyde. Using in situ tracing method supported by DFT calculations to elucidate the cleavage pattern, they proposed HOMOs delocalization induced mechanism of selective cleavage of either C_β-O or C_α-C_β bonds (Figure 11.7) [73]. Gao et al. investigated iodide-mediated (using NaI) cleavage of various lignin β-O-4 model dimers using graphite and platinum anodes. Cleavage of C_β-O bond was achieved using graphite plate anode in methanol and generated methylated α carbonyl derivatives with up to 30–40% cleavage yield [74]. The proposed mechanism involves the formation of an epoxy ring opened via methoxide to produce methyl acetal on the α carbon.

In another mediated-electrochemical study, Sannami et al. investigated the degradation of Et-1G-derived β-O-4 model compound in a mixed solvent using 2,2,6,6-tetramethylpiperidine-*N*-oxyl (TEMPO) or 4-acetamido-TEMPO (ACT) [75]. The effect of different solvent mixtures on selectivity was investigated. When using CH_3CN:H_2O (95 : 5 molar ratio) in the presence of excess 2,6-lutidine, the selective oxidation of the α alcohol was observed. In contrast, when using dioxane:H_2O (10 : 90) in neutral phosphate buffer resulted in the oxidation of the primary hydroxyl group to carboxylic acid was observed. Based on this and other studies, it appears that a higher concentration of CH_3CN favored the selectivity of the C-α hydroxy group over the primary alcohol [76]. Further, this study demonstrated ACT was more effective in oxidizing the α-hydroxy group to carbonyl than TEMPO. The

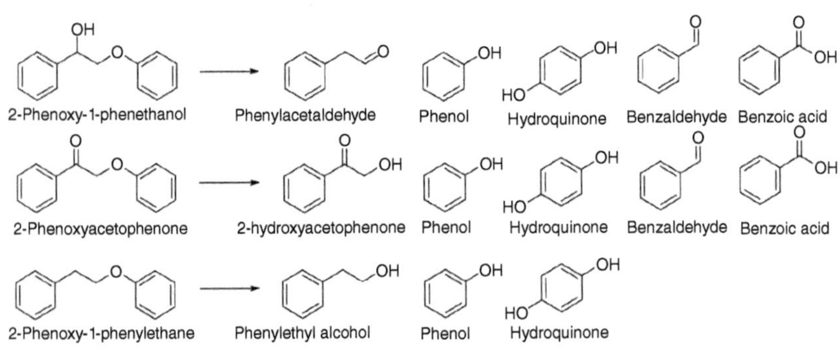

Figure 11.7 ECO of β-O-4 dimers in aqueous environments using a Pt-sheet cathode. Condition: Two Pt-sheet electrodes (20 × 20 × 0.1 mm), and an Ag/Ag^+ nonaqueous reference electrode. Substrate: 0.125 M electrolyte: 0.1 M $LiClO_4/CH_3CN$, room temperature, scan rate: $0.1\,V\,s^{-1}$.

TEMPO and ACT-mediated ECO were also conducted by Rafiee et al., which exhibited high selectivity for oxidation of primary alcohol groups of β-O-4 model compounds. A similar enhanced performance of ACT over TEMPO was also observed. The method was applied to the oxidation of various model compounds, and the primary and secondary alcohols were found to be prone to forming carboxylic acids or ketones. While those compounds bearing β-1 linkages (no ether linkage) undergo cleavage of the aliphatic ether unit to produce a benzaldehyde and an aliphatic aldehyde fragment [77].

11.5 Electrocatalytic Depolymerization of Lignin Polymer

11.5.1 Electrocatalysts for Lignin Depolymerization Using Various Electrodes

11.5.1.1 Iridium Based Anodes

Iridium (Ir)-based electrocatalysts are reported to have high selectivity and stability [78]. More importantly, further modification of IrO_2-based electrodes with other metal oxides can influence their oxidative performance. Tolba et al. investigated ECO of Kraft lignin using modified IrO_2-based anodes ($Ti/Ta_2O_5–IrO_2$, $Ti/SnO_2–IrO_2$, $Ti/RuO_2–IrO_2$ and $Ti/TiO_2–IrO_2$) with Pt as cathode [79]. The electrocatalytic active surface area and consequentially the electrocatalytic activity of the electrodes decreased in the following order $Ti/Ta_2O_5–IrO_2 > Ti/TiO_2–IrO_2 > Ti/SnO_2–IrO_2 > Ti/RuO_2–IrO_2$. Vanillin and vanillic acid were produced from the electro-oxidation of 500 ppm lignin solution under optimized conditions (60 °C, 500 mA cm$-^2$). Rauber et al. reported electrocatalytic conversion of alkali lignin using binary and ternary Ir-based mixed metal oxide (MMO) catalysts ($Ru_{0.4}Ir_{0.6}Ox$, $Ru_{0.2}Mn_{0.2}Ir_{0.6}Ox$, $Ru_{0.2}Pd_{0.2}Ir_{0.6}Ox$, $Ru_{0.2}V_{0.2}Ir_{0.6}Ox$, and $Ru_{0.2}Ti_{0.2}Ir_{0.6}Ox$) [80]. The lignin-degrading activity from these catalysts followed Mn > Pd > V > Ti order, with the ternary catalysts performing better than the binary catalysts. The aromatic derivatives of monolignols were found in the product stream and were most likely formed from the oxidative cleavage of β-O-4 linkage. Additionally, metal composition was found to influence the selectivity of the catalyst as indicated by different product distributions. Specifically, Mn was selective toward forming *p*-coumaric acid, Pd toward forming 4-hydroxy-3,5-dimethoxy cinnamaldehyde, V (vanadium) toward forming 4-hydroxy-3,5-dimethoxy acetophenone, and Ti toward forming vanillin.

11.5.1.2 Lead/Lead Oxide Based Anodes

Lead- and lead oxide-based anodes are one of the more popular anodes that have been studied for oxidative lignin degradation due to their low cost, stability, high conductivity, and high oxygen overpotential [81–90]. Different studies have explored PbO_2 electrodes, modified PbO_2 (e.g. mixed metal oxides), Pb/PbO_2 composite electrodes, and modified Pb/PbO_2 electrodes. Pan et al. studied PbO_2/TiO_2 electrodes for the electrochemical degradation of black liquor Kraft lignin to vanillin and vanillic acid as major products [90]. The study demonstrated that the kinetics of the lignin degradations was mainly affected by reaction temperature and current but independent of initial lignin concentration. Hao et al. used electrodeposition to modify the surface of PbO_2 anode with $[Fe(CN)_6]_3$, which was then used for the depolymerization of alkali lignin [89]. At an optimal reaction rate constant of $0.00609\,min^{-1}$, $50\,mg\,L^{-1}$ lignin substrate was mineralized to 77.3% and the chemical oxygen demand (COD) removal efficiency reached 21.2%.

Li's group explored various Pb/PbO_2 anodes for the redox depolymerization of many types of lignins (black liquor lignin, aspen lignin, bamboo lignin, wheat straw lignin, and rice straw lignin) [81, 82, 84–87]. Many reported Pb/PbO_2 catalysts (referred to herein as composite catalysts) also demonstrate high electrocatalytic conversion activity toward lignin. Zhang et al. synthesized Pb/PbO_2 hybrid electrode for direct degradation of black liquor lignin to form 2,6(1,1-dimethylethyl)-4-methylphenol (BHT) [81]. According to their proposed mechanism, BHT forms mainly through phenolic hydroxyl deprotonation followed by $^{\cdot}O_2-$ formation and selective C-C bonds oxidation, and subsequent cleavage of the side chain of the benzene ring (Figure 11.8). Wang et al. investigated the electrocatalytic redox conversion of aspen lignin to 4-methylanisole using a three-dimensional Pb/PbO_2 anode vs. a stainless-steel cathode at 40 °C [84]. The 4-methylanisole yield was found to increase with increasing current density and temperature. In addition, the choice of the cathodic material was also found to influence the oxidation and reduction efficiency of lignin and product selectivity.

In recent years, there has been more interest in multi-metallic Pb-based catalysts for lignin depolymerization. For instance, $Ti/SnO_2–Sb_2O_3/\alpha\text{-}PbO_2/\beta\text{-}PbO_2$ have been prepared and used by Jia et al., which is the anode for the electrochemical redox of rice straw lignin resulting in a series of products [88].

11.5.1.3 Nickel-, Cobalt-, and Nickel–Cobalt Based Electrodes

The formation of NiOOH layer on the surface of Ni-based electrodes in the alkaline electrolyte was found to enhance the oxidative activity for lignin degradation and avoid further overoxidation of the lignin monomers formed [91]. Moodley et al. studied the electrochemical treatment of complex lignin and organic-extracted mixtures from the Sappi Saiccor pulp mill effluent using a Ni electrode. Syringaldehyde and some vanillin were formed in the first 30 minutes of the reaction but eventually

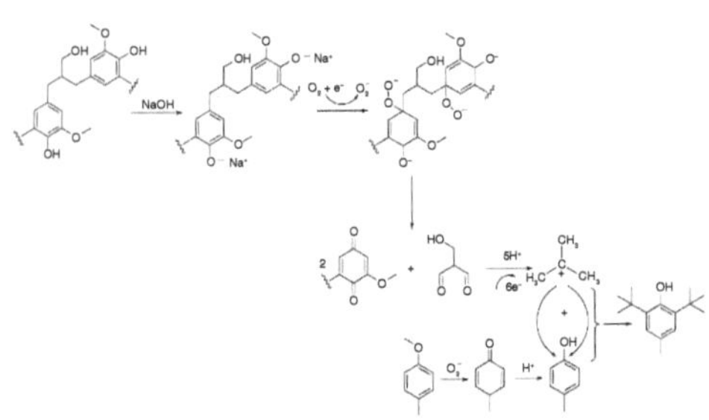

Figure 11.8 Proposed mechanism of reactions for lignin to BHT.

got overoxidized with prolonged reaction time [92]. Schmit et al. used Ni foam electrodes for the oxidation of black liquor lignin that was pretreated with basic anion exchange resins to remove valuable low molecular weight phenolic compounds prior to electrooxidation. Vanillin, acetovanillone, and guaiacol were observed as the main products of the electrochemical degradation of the depleted black liquor lignin [93]. In two consecutive studies, Stiefel et al. reported the electrooxidation of Kraft lignin in an electrochemical membrane reactor (ECMR) with a tubular ceramic membrane using a Ni electrode. They found that the use of a microfiltration membrane helped in controlling product overoxidation and consequently enhanced the recovery of low molecular weight products [94, 95].

The studies by Stiefel et al. have demonstrated that separation of low molecular weight lignin degradation products can help combat their overoxidation. However, the timely removal of products to avoid overoxidation requires a deeper understanding of the complex reaction kinetics. To this end, Bawareth et al. used population balance equations to predict the molecular weight distribution of lignin degradation over time. They modeled and further fitted experimental data of three major lignin-based degradation reactions, namely chain end scission, random recombination, and random scission reactions [96]. The reaction rate coefficients were found to have a linear correlation with the initial polydispersity and average molecular weight of the lignin.

Among the non-precious metal electrode resources, nickel–cobalt (Ni–Co)-based electrodes have aroused extensive interest due to their high activity toward lignin conversion and stability in an alkaline reaction environment [97–101]. Zirbes et al., developed a series of Ni- and Co-based alloy electrodes for the selective transformation of Kraft lignin [101]. Cobalt was susceptible to corrosion but resulted in a relatively high vanillin yield. Interestingly, repeated use of Ni foam electrodes for electrochemical conversion resulted in significantly increased yields of vanillin. This was likely due to a modification phenomenon of the electrode surface during the electrochemical process by forming an adsorption layer on the electrode surface. Movil et al. compared the electrochemical activities of Ni/C, Co/C, and Ni–Co/C via the ECO of lignin [102]. Co/C gave a higher lignin oxidation rate when compared to Ni/C and Ni–Co/C. Additionally, Ni/C and Co/C suffered from reduced lignin oxidation current, presumably from loss of active sites attributed to either lignin adsorption on the electrode surface or oxidative degradation of the electrodes. Interestingly, this phenomenon was not observed for the Ni–Co/C electrodes.

11.5.2 Combination of Electrochemical and Other Process for Lignin Conversion

11.5.2.1 Ionic Liquid and Deep Eutectic Solvent Electrolytes Assisted Approach

Recently there has been increased interest in the use of Ionic liquids (IL) (e.g. $[Et_3NH][MeSO_3]$, ([Emim][OTf]))–) for lignin ECO as ionic liquids possess many favorable attributes, including enhanced conductivity, thermal and chemical stability, non-volatility, greater solubility capacity for lignin substrate, and wider electrochemical window. Reichert et al. studied the electrochemical conversion of alkali lignin to vanillin, guaiacol, syringol, etc., using $Ru_{0.25}V_{0.5}Ti_{0.7}O_x$ anode in $[Et_3NH][MeSO_3]$ at various potentials [103]. The study showed that lower potentials favored the formation of low molecular weight products while higher potentials resulted in the degradation of the ionic liquid solvent. The extraordinary electrocatalytic activity of $Ru_{0.25}V_{0.5}Ti_{0.7}O_x$ toward lignin is likely due to the enhancement of the vanadium-induced single ET efficiency of the catalyst by the proton transporting property of the protic ionic liquid. In a similar study, Deir et al. investigated the ECO of alkali lignin in $[Et_3NH]$ $[MeSO_3]$. In this study, the ionic liquid conveniently facilitated the separation of the products into three distinct layers [104]. Dier et al. also investigated lignin alkali and organosolv lignin depolymerization to low molecular weight products using aprotic ([Emim][OTf]))– and protic (TMS) ionic liquids electrolytes in the presence of water. The presence of water in the electrolyte system promoted H_2O_2 production, which further facilitated lignin degradation and significantly increased the yield of low molecular weight products. Deep eutectic solvents (DES) such as urea-choline chloride (urea-ChCl) or ethylene glycolcholine chloride (EtGly-ChCl) have also been investigated for lignin degradation as they often offer similar advantages as ionic liquids but are generally considered less toxic [105]. Di Marino et al. studied the ECO of Kraft lignin using urea-ChCl or EtGly-ChCl at Ni electrodes. Compared to urea-ChCl and NaOH, EtGly-ChCl showed the best lignin solubility and conductivity; these properties were further enhanced by adding water to the DES [105].

11.5.2.2 Biodegradative Pretreatment

Electrochemical degradation and biodegradation methods have both been explored for lignin wastewater treatment as an alternative to conventional chemical treatment methods [106]. While biodegradation is less energy intensive, it suffers from the need for long incubation periods and inefficient lignin degradation, while electrochemical methods are able to mineralize lignin in a short time but at the cost of high energy consumption [106]. Cui et al. investigated the electrooxidation followed by biodegradation (EO-BD) for the depolymerization of Kraft lignin resulting in a maximum COD removal rate of 70.83% [106]. ECO products such as quinones and short chains formed during the ECO step were completely consumed in the biodegradation step. However, some vanillin and acetosyringone were detected after biodegradation, presumably from bacterial degradation of residual Kraft lignin left after the ECO step. Similarly, Shao and coworkers investigated various anodes for the ECO of lignosulphonate to quinones and short-chain carboxylic acids and their subsequent biodegradation, resulting in 45–60% COD removal efficiency depending on the anode used [107].

11.5.2.3 Photo-Assisted Approach

Photocatalysis is another approach that has emerged as a promising method for selective lignin upgrading under mild conditions [108]. Photoexcitation of catalysts produces charge carriers (electrons and holes) or excited chemicals species with high redox ability that can serve as reductants/oxidants and facilitate the selective transformation of lignin. However, recombination of photogenerated charge carriers is a major barrier observed in photocatalysis and can greatly affect efficiency. Applying external anodic bias is an effective way that can be used to suppress the photogenerated charge carrier recombination. Thus, integrating photocatalysis with electrochemical oxidation can be a promising approach for lignin depolymerization. Tian et al. investigated the degradation of black liquor using a combined photo- and electro-catalytic approach, resulting in the formation of vanillin and vanillic acid as products [109]. The combined photoelectrochemical approach greatly enhanced the lignin degradation rate in comparison to the direct electrochemical-only approach. A reaction rate of 0.021 min^{-1} and a degradation rate of 92% were reported for the combined approach vs. a reaction rate of 0.009 min^{-1} and a degradation rate of 66% for the electrochemical-only approach. Bosque et al. demonstrated selective oxidative cleavage of β-O-4 bonds found in native pine lignin while keeping other linkages intact using a one-pot, two-step photo-electrocatalytic approach [110]. In this study, they demonstrated the integration of photocatalytic, electrocatalytic, and flow approaches, which have promising potential for industrial applications.

11.5.2.4 Mediator-Assisted Approach

ECO degradation of lignin can also be achieved through a mediated approach, where a suitable mediator acts as an oxidazing agent to degrade specific linkages found in lignin. After oxidation, the mediator is regenerated electrochemically to complete the electrocatalytic cycle. Homogeneous oxidants, e.g. polyoxometalates, ferric chloride, laccase mediators, TEMPO, iodine, NHPI, and hydrogen atom transfer (HAT) agents, are found to be effective as proton/electron reservoirs for lignin oxidation. Yao et al. reported on the electrochemical degradation reaction of alkali lignin in a combined laccase-mediator system (LMS) [111]. The synergistic effect of the laccase-mediator promoted charge transfer between the laccase and the mediator by ORR. Meanwhile, Shiraish et al. studied TEMPO/ACT-mediated electrooxidation of non-phenolic lignin model compounds for selective oxidation of the α alcohol [75]. The study showed that a higher concentration of CH_3CN triggered a selective reaction of the C-α hydroxy group over the primary alcohol in the presence of the mediator. Rafiee et al. performed a similar reaction in dioxane: carbonate (10 : 90 molar ratio) buffer with TEMPO or ACT as a mediator. Another electrocatalyic study showed that the ACT oxidized the α-hydroxy group to carbonyl more readily than TEMPO (11.1% over 3.0%, respectively), which was similar to Shiraish's finding [112].

11.5.2.5 H_2O_2-Assisted Approach

Oxidation in H_2O_2 has been identified as an environmentally friendly lignin depolymerization method but suffers from low oxidation capacity. ECO can be a powerful method for lignin degradation, but it suffers from competition for OER and the limited access of the lignin polymer to the electrode surface [113]. Integrating these two methods via in situ anodic generations of H_2O_2 using O_2, one can provide access to a synergistic effect that can enhance the electrocatalytic degradation of lignin. A study by Zhu et al. compared H_2O_2-assisted ECO of crude lignin with anodic oxidation and direct H_2O_2 oxidation. A slew of products (29 total compounds, 16 monomers, and 13 dimers) was generated by the three methods [113]. The integrated (H_2O_2–ECO) method had a powerful depolymerization capacity. It produced C-C bonded dimers (that did not exhibit common lignin structures), which were attributed to possible condensation reactions resulting from the radical formation. In a later study, they further reported that increasing anodic O_2 production by increasing current density further increased in situ H_2O_2 fomentation, but levels off after 20 mins due to its decomposition to form reactive oxygen species ($^{\cdot}OH$, $^{\cdot}O_2-$, $^{\cdot}OOH$) [114].

11.5.2.6 Hydrogen Coproduction Approach

Electrochemical degradation of organic molecules such as lignin has been explored as a low-energy alternative to water electrolysis for the coproduction of hydrogen [115–117]. Some studies have considered black liquor (BL) as a reaction substrate for lignin electrochemical upgrading and simultaneous H_2 hydrogen production [118–120]. Caravaca et al. conducted electrolysis lignin oxidation and hydrogen reduction reactions in continuous flow mode with a maximum H_2 production rate of $0.45\,\mu mol\,s^{-1}$ [121]. NaderiNasrabadi et al. developed a biomass-depolarized electrolyzer that can electrooxidize lignin-rich biorefinery waste to produce high purity (97.6%) H_2 at mild conditions. It was also found that the lignin substrate concentration and the H_2 production rate are positively correlated up to $50\,g\,L^{-1}$ [122].

11.6 Electrocatalytic Valorization of Lignin to Biorefinery Aromatics Products

Lignin is increasingly being applied as a starting material for producing chemicals. Several common technical lignins, i.e. lignosulfonate, Kraft, soda-anthraquinone, enzymatic organosolv, and alcoholysis lignin, can act as suitable feedstocks for producing renewable monomeric aromatic compounds that have relatively high value as renewable raw commodity chemicals for direct use or for building specific polymers. Electrochemical degradation of lignin often occurs in oxidative environments, usually for the purpose of water purification, in which lignin is regarded as a pollutant and is mineralized to CO_2. While electrochemical hydrogenation and deoxygenation are mainly applied to upgrade the various lignin. Accordingly, the mechanism and selective electrochemical conversion of different types of liginins to mono-aromatic specialty chemicals, such as vanillin, syringaldehyde, guaiacol, *trans*-ferulic acid, syringaldehyde, and acetosyringone are widely studied, though the overall yields remain low [2, 9, 123].

11.6.1 Vanillin

Vanillin (3-methoxy-4-hydroxybenzaldehyde) is of particular interest in lignin conversion due to industrial application in food, cosmetics, and tobacco. Various electrode types such as Pt, Au, Ni, Cu, DSA-O_2, PbO_2, boron-doped diamond, Ti/$Ru_{0.1}Sn_{0.2}Ti_{0.7}O_2$, and photo-assisted electrocatalysts as well as mediators ($K_3[Fe(CN)_6]$) have been used under various reaction conditions for the production of value-added vanillin from lignin [124–126]. Ghahremani et al. studied the vanillin production rates (produced from electro-oxidative lignin depolymerization) in an aqueous alkaline media [127]. In the presence of different Sn/Ni ratio catalysts and using different voltages, the yield of vanillin exhibited a volcanic change with the increase of the above two factors. In addition, Singh and Ghatak observed that vanillin generated from electrooxidation of lignin derived from soda pulping of wheat straw and bagasse followed pseudo-first-order kinetics, which produced 6% vanillin from wheat straw and 6.6% vanillin from bagasse in the above process [128].

11.6.2 Syringaldehyde

Syringaldehyde, which has antioxidant, anti-hyperglycemic, and anti-inflammatory activity, can also be produced from lignin through a multi-stage electrochemical approach. Gao et al. reported the example of the NaI-mediated ECO of a series of β-O-4 dimers, producing maximal yields of syringaldehyde (57.30 $g\,kg^{-1}$ lignin) and other chemicals under optimal conditions within two hours [74]. Moodley and coworkers further optimized the reaction system and expanded the substrate scope to native lignin and organic-extracted mixtures from the Sappi Saiccor pulp mill effluent [92]. All the extracted organic components of the effluent, the concentrated effluent, and lignosulfonate provide abundant sources for the chemoselective oxidation of lignin to high yields of syringaldehyde. Interestingly, the authors found that the syringaldehyde products should be collected effectively in the first 30 min to prevent overoxidation. Under different conditions, syringaldehyde and acetovanillone can also be formed at higher current densities of 6.71 and 4.47 $mA\,cm^{-2}$ from wheat straw soda lignin in Singh's work, which is a selective transformation compared to other operating parameters [129].

11.6.3 Guaiacol

Guaiacol produced by chemical transformation is mainly used as an organic synthetic preparation in pharmaceuticals, pesticides, and fragrances, and is commonly used in cosmetics and chemical reagents. Di Marino et al. proposed the mechanism of ECO of Kraft lignin, which involved two electrolytes (urea-choline chloride, urea-ChCl and ethylene glycolcholine chloride, EtGly-ChCl) [105]. In this reaction, water-assisted EtGly-ChCl showed the best lignin solubility and molecular weight transformation rate compared to urea-ChCl with 30–38% guaiacol yield. Du et al. integrated a simultaneous electrooxidation of Kraft lignin and cogeneration of H_2 using polyoxometalate (POM) phosphomolybdic acid and $FeCl_3$ as catalyst and charge transfer mediators, leading to the efficient output of low-molecular-weight guaiacol [116].

11.6.4 *Trans*-Ferulic Acid

Trans-ferulic acid is also one of the important high-value compounds that can be synthesized by the electrochemical reaction of lignin in the presence of a suitable catalytic system. Cai et al. reported the electro-upgrading of lignin-derived oligomers in an electrolytic cell at room temperature and ambient pressure in a high lignin substrate concentration (above $50\,g\,L^{-1}$) [87]. It was found that blockage of active sites on the electrode surface negatively affect redox reactions in the system. This resulted in decreased aromatic compound yield with prolonged processing time, while oxidation products, e.g. *trans*-ferulic acid, were produced with the highest yield of $22.4\,g\,kg^{-1}$ of lignin.

11.6.5 Acetosyringone

Acetosyringone, a natural phenolic compound produced by plants, is a promising substance for the identification of plant-based natural pathogens. Several electrochemical studies have been conducted on the manufacture of acetosyringone from different technical lignins, including wheat straw soda lignin, and Kraft lignin. Cai et al. used Cu/Ni–Mo–Co as cathode and Pb/PbO_2 as anode for the electrocatalytic depolymerization of corn stover lignin to high-value aromatics in sodium hydroxide solution. Fourier transform infrared spectroscopy (FTIR) and high-performance liquid chromatography (HPLC) were used to characterize the functional groups and molecular weight distribution of the lignin residue. By GC–MS analysis, acetosyringone ($6.9\,g\,kg^{-1}$ of lignin) was generated. Cui et al. reported a high-efficiency electrooxidation and biodegradation (EO–BD) integrated system to upgrade oxygenated Kraft lignin under mild conditions [106]. The lignin substrate was first decomposed electrochemically and then reformed into the acetosyringone and vanillin products through biodegradation process using aerobic flora inoculum.

11.7 Summary

As a promising yet under-utilized means of producing aromatics from renewable natural sources, lignin valorization has made a significant progress over the past several decades. However, many challenging research goals remain due to lignin's highly recalcitrant structure that resists mild catalytic depolymerization. The present chapter demonstrates that electrocatalysis integrated with renewable electricity is a promising approach for transforming lignin into high-value chemicals; strategies addressed herein include depolymerization of native lignin and downstream upgrading of lignin-relevant model compounds. Overall, the electrocatalytic approach for lignin valorization currently is relevant for three primary applications: (i) preparation of biorefinery chemicals and fuels, (ii) stabilization of lignin or lignin-derived oils, and (iii) degradation of lignin substrates. Although some advances in electrode fabrication, electrolyte development, and synergistic catalysis have been made, there are still significant advances that need to be made and challenges that must be overcome before commercial application of these technologies.

In the electrocatalytic process, some low molecular weight lignin products are susceptible to mineralization to CO_2 or small volatile acid when the working potential is too high or the oxidation period is not optimized. While there are electrochemical means to valorise volatile acids for fuel product, these conditions can diminish the renewable carbon value of the lignin-derived aromatic products [130, 131]. Although overoxidation can be suppressed by a few in situ product separation methods, including extraction, membrane separation, and nanofiltration, these implementations also increase the operating cost, which will challenge the scaled-up biorefinery application. In order to compete with petroleum products, biorefinery products must be produced in a sustainable and cost-effective manner. Developing novel and effective electrocatalysts that can achieve selective transformation (both oxidative and reductive) will undoubtedly be one of the feasibile solutions.

Though research indicates value-added products (biofuel, biopolymers, aromatic monomers, cosmetics, and pharmaceuticals) can be generated via electrochemical lignin treatment, the low yield and diversity of products hinder the practical application of this process. The in situ monitoring of the generated products and mechanistic studies become increasingly important and can further be leveraged to design selective lignin degradation and catalytic schemes. Higher faradaic efficiencies and current densities are also needed to make this technology profitable for the production of high-value chemicals and fuels.

In conclusion, the electrocatalytic upgrading of lignin shows remarkable promise for converting biomass to industrial chemicals and fuels. This chapter presents an overview of basic concepts, general mechanisms, and features of electrocatalytic lignin conversion with related recent developments in the this field, including experimental and computational insights. Some studies that involve in-depth understanding of the lignin depolymerization mechanism have also been discussed. But scaled-up operation and economic challenges still remain in this field, and we hope that the present chapter can motivate further research on electrocatalytic valorization of lignin resources.

References

1 Corma, A., Iborra, S., and Velty, A. (2007). Chemical routes for the transformation of biomass into chemicals. *Chemical Reviews* 107 (6): 2411–2502.

2 Ragauskas, A.J., Beckham, G.T., Biddy, M.J. et al. (2014). Lignin valorization: improving lignin processing in the biorefinery. *Science* 344 (6185): 1246843.

3 Calvo-Flores, F.G. and Dobado, J.A. (2010). Lignin as renewable raw material. *ChemSusChem* 3 (11): 1227–1235.

4 Wang, X., Li, C., Lam, C.H. et al. (2022). Emerging waste valorisation techniques to moderate the hazardous impacts, and their path towards sustainability. *Journal of Hazardous Materials* 423: 127023.

5 Tolbert, A., Akinosho, H., Khunsupat, R. et al. (2014). Characterization and analysis of the molecular weight of lignin for biorefining studies. *Biofuels, Bioproducts and Biorefining* 8 (6): 836–856.

6 Zhang, C. and Wang, F. (2020). Catalytic lignin depolymerization to aromatic chemicals. *Accounts of Chemical Research* 53 (2): 470–484.

7 Patwardhan, P.R., Brown, R.C., and Shanks, B.H. (2011). Understanding the fast pyrolysis of lignin. *ChemSusChem* 4 (11): 1629–1636.

8 Ye, K., Liu, Y., Wu, S. et al. (2021). A review for lignin valorization: challenges and perspectives in catalytic hydrogenolysis. *Industrial Crops and Products* 172: 114008.

9 Khan, R.J., Lau, C.Y., Guan, J. et al. (2022). Recent advances of lignin valorization techniques toward sustainable aromatics and potential benchmarks to fossil refinery products. *Bioresource Technology* 346: 126419.

10 Lam, C.H., Das, S., Erickson, N.C. et al. (2017). Towards sustainable hydrocarbon fuels with biomass fast pyrolysis oil and electrocatalytic upgrading. *Sustainable Energy & Fuels* 1 (2): 258–266.

11 Garedew, M., Lam, C.H., Petitjean, L. et al. (2021). Electrochemical upgrading of depolymerized lignin: a review of model compound studies. *Green Chemistry* 23 (8): 2868–2899.

12 Du, X., Zhang, H., Sullivan, K.P. et al. (2020). Electrochemical lignin conversion. *ChemSusChem* 13 (17): 4318–4343.

13 Erythropel, H.C., Zimmerman, J.B., de Winter, T.M. et al. (2018). The Green ChemisTREE: 20 years after taking root with the 12 principles. *Green Chemistry* 20 (9): 1929–1961.

14 Garedew, M., Lin, F., Song, B. et al. (2020). Greener routes to biomass waste valorization: lignin transformation through electrocatalysis for renewable chemicals and fuels production. *ChemSusChem* 13 (17): 4214–4237.

15 Wijaya, Y.P., Smith, K.J., Kim, C.S., and Gyenge, E.L. (2020). Electrocatalytic hydrogenation and depolymerization pathways for lignin valorization: toward mild synthesis of chemicals and fuels from biomass. *Green Chemistry* 22 (21): 7233–7264.

16 Gao, F., Webb, J.D., and Hartwig, J.F. (2016). Chemo-and regioselective hydrogenolysis of diaryl ether C-O bonds by a robust heterogeneous Ni/C catalyst: applications to the cleavage of complex lignin-related fragments. *Angewandte Chemie* 128 (4): 1496–1500.

17 Rengshausen, S., Etscheidt, F., Großkurth, J. et al. (2019). Catalytic hydrogenolysis of substituted diaryl ethers by using ruthenium nanoparticles on an acidic supported ionic liquid phase (Ru@ SILP-SO_3H). *Synlett* 30 (04): 405–412.

18 Sanyal, U., Lopez-Ruiz, J., Padmaperuma, A.B. et al. (2018). Electrocatalytic hydrogenation of oxygenated compounds in aqueous phase. *Organic Process Research and Development* 22 (12): 1590–1598.

19 Zhou, Y., Klinger, G.E., Hegg, E.L. et al. (2022). Skeletal Ni electrode-catalyzed CO cleavage of diaryl ethers entails direct elimination via benzyne intermediates. *Nature Communications* 13 (1): 1–13.

20 Tu, Q., Parvatker, A., Garedew, M. et al. (2021). Electrocatalysis for chemical and fuel production: investigating climate change mitigation potential and economic feasibility. *Environmental Science & Technology* 55 (5): 3240–3249.

21 Huang, S., Jin, Y., Zhang, M. et al. (2022). MoS_2-catalyzed selective electrocatalytic hydrogenation of aromatic aldehydes in an aqueous environment. *Green Chemistry*.

22 Huang, S., Gong, B., Jin, Y. et al. (2022). The structural phase effect of MoS_2 in controlling the reaction selectivity between electrocatalytic hydrogenation and dimerization of furfural. *ACS Catalysis* 12 (18): 11340–11354.

23 Zhang, M., Hu, D., Chen, Y. et al. (2022). Electrocatalytic reductive amination and simultaneous oxidation of biomass-derived 5-hydroxymethylfurfural. *Industrial & Engineering Chemistry Research* 61 (4): 1912–1919.

24 Xia, H., Xu, S., Hu, H. et al. (2018). Efficient conversion of 5-hydroxymethylfurfural to high-value chemicals by chemo- and bio-catalysis. *RSC Advances* 8 (54): 30875–30886.

25 Lin, F., Tse, H.-Y., Erythropel, H.C. et al. (2022). Development of a Ni-promoted, selective electrochemical reductive cleavage of the C–O bond in lignin model compound benzyl phenyl ether. *Green Chemistry*.

26 Li, Z., Kelkar, S., Lam, C.H. et al. (2012). Aqueous electrocatalytic hydrogenation of furfural using a sacrificial anode. *Electrochimica Acta* 64: 87–93.

27 Akhade, S.A., Singh, N., Gutiérrez, O.Y. et al. (2020). Electrocatalytic hydrogenation of biomass-derived organics: a review. *Chemical Reviews*.

28 Zhou, H., Li, Z., Ma, L., and Duan, H. (2022). Electrocatalytic oxidative upgrading of biomass platform chemicals: from the aspect of reaction mechanism. *Chemical Communications* 58 (7): 897–907.

29 Li, K. and Sun, Y. (2018). Electrocatalytic upgrading of biomass-derived intermediate compounds to value-added products. *Chemistry - A European Journal* 24 (69): 18258–18270.

30 Du, L., Shao, Y., Sun, J. et al. (2018). Electrocatalytic valorisation of biomass derived chemicals. *Catalysis Science & Technology* 8 (13): 3216–3232.

31 Prabhu, P., Wan, Y., and Lee, J.-M. (2020). Electrochemical conversion of biomass derived products into high-value chemicals. *Matter* 3 (4): 1162–1177.

32 Ge, R., Li, J., and Duan, H. (2022). Recent advances in non-noble electrocatalysts for oxidative valorization of biomass derivatives. *Science China Materials* 1–29.

33 Yang, C., Chen, H., Peng, T. et al. (2021). Lignin valorization toward value-added chemicals and fuels via electrocatalysis: a perspective. *Chinese Journal of Catalysis* 42 (11): 1831–1842.

34 Liu, W., You, W., Gong, Y., and Deng, Y. (2020). High-efficiency electrochemical hydrodeoxygenation of bio-phenols to hydrocarbon fuels by a superacid-noble metal particle dual-catalyst system. *Energy & Environmental Science* 13 (3): 917–927.

35 Yang, C., Maldonado, S., and Stephenson, C.R. (2021). Electrocatalytic lignin oxidation. *ACS Catalysis* 11 (16): 10104–10114.

36 Lux, S., Stehring, P., and Siebenhofer, M. (2010). Lactic acid production as a new approach for exploitation of glycerol. *Separation Science and Technology* 45 (12-13): 1921–1927.

37 Dodekatos, G., Schünemann, S., and Tüysüz, H. (2018). Recent advances in thermo-, photo-, and electrocatalytic glycerol oxidation. *ACS Catalysis* 8 (7): 6301–6333.

38 Osa, T., Kashiwagi, Y., Ono, T. et al. (2014). Design of electrode providing preparative stereoselective synthesis. *International Journal of the Society of Materials Engineering for Resources* 20 (1): 49–53.

39 Elinson, M., Zaimovskaya, T., Vereshchagin, A., and Nikishin, G. (2005). Stereoselective electrocatalytic transformations of malononitrile and aromatic aldehydes into (1R, 5S, 6R)*-4, 4-dialkoxy-2-amino-6-aryl-1, 5-dicyano-3-azabicyclo [3.1.0] hex-2-enes. *Russ. Chemical Bulletin (Beijing, China)* 54 (3): 673–677.

40 Zhang, S., Ruccolo, S., Fryszkowska, A. et al. (2021). Electrochemical activation of galactose oxidase: mechanistic studies and synthetic applications. *ACS Catalysis* 11 (12): 7270–7280.

41 Santos, J.E.L., de Moura, D.C., Cerro-Lopez, M. et al. (2020). Electro-and photo-electrooxidation of 2, 4, 5-trichlorophenoxiacetic acid (2, 4, 5-T) in aqueous media with PbO_2, Sb-doped SnO_2, BDD and TiO_2-NTs anodes: a comparative study. *Journal of Electroanalytical Chemistry* 873: 114438.

42 Berenguer, R., Sieben, J.M., Quijada, C., and Morallón, E. (2014). Pt-and Ru-doped SnO_2-Sb anodes with high stability in alkaline medium. *ACS Applied Materials & Interfaces* 6 (24): 22778–22789.

43 Aquino, J.M., Pereira, G.F., Rocha-Filho, R.C. et al. (2011). Electrochemical degradation of a real textile effluent using boron-doped diamond or β-PbO_2 as anode. *Journal of Hazardous Materials* 192 (3): 1275–1282.

44 Basha, C.A., Sendhil, J., Selvakumar, K. et al. (2012). Electrochemical degradation of textile dyeing industry effluent in batch and flow reactor systems. *Desalination* 285: 188–197.

45 Daiyan, R., Saputera, W.H., Masood, H. et al. (2020). A disquisition on the active sites of heterogeneous catalysts for electrochemical reduction of CO_2 to value-added chemicals and fuel. *Advanced Energy Materials* 10 (11): 1902106.

46 Azadi, P., Inderwildi, O.R., Farnood, R., and King, D.A. (2013). Liquid fuels, hydrogen and chemicals from lignin: a critical review. *Renewable and Sustainable Energy Reviews* 21: 506–523.

47 Wang, H., Pu, Y., Ragauskas, A., and Yang, B. (2019). From lignin to valuable products–strategies, challenges, and prospects. *Bioresource Technology* 271: 449–461.

48 Renders, T., Van den Bosch, S., Koelewijn, S.-F. et al. (2017). Lignin-first biomass fractionation: the advent of active stabilisation strategies. *Energy & Environmental Science* 10 (7): 1551–1557.

49 Van den Bosch, S., Renders, T., Kennis, S. et al. (2017). Integrating lignin valorization and bio-ethanol production: on the role of Ni-Al_2O_3 catalyst pellets during lignin-first fractionation. *Green Chemistry* 19 (14): 3313–3326.

50 Li, C., Zhao, X., Wang, A. et al. (2015). Catalytic transformation of lignin for the production of chemicals and fuels. *Chemical Reviews* 115 (21): 11559–11624.

51 Kumar, P., Barrett, D.M., Delwiche, M.J., and Stroeve, P. (2009). Methods for pretreatment of lignocellulosic biomass for efficient hydrolysis and biofuel production. *Industrial and Engineering Chemistry Research* 48 (8): 3713–3729.

52 Putro, J.N., Soetaredjo, F.E., Lin, S.-Y. et al. (2016). Pretreatment and conversion of lignocellulose biomass into valuable chemicals. *RSC Advances* 6 (52): 46834–46852.

53 Pandey, M.P. and Kim, C.S. (2011). Lignin depolymerization and conversion: a review of thermochemical methods. *Chemical Engineering and Technology* 34 (1): 29–41.

54 Sun, W., Greaves, T.L., and Othman, M.Z. (2020). Electro-assisted pretreatment of lignocellulosic materials in ionic liquid-promoted organic solvents. *ACS Sustainable Chemistry & Engineering* 8 (49): 18177–18186.

55 Ma, L., Zhou, H., Kong, X. et al. (2021). An electrocatalytic strategy for C-C bond cleavage in lignin model compounds and lignin under ambient conditions. *ACS Sustainable Chemistry & Engineering* 9 (4): 1932–1940.

56 Fang, Z., Flynn, M.G., Jackson, J.E., and Hegg, E.L. (2021). Thio-assisted reductive electrolytic cleavage of lignin β-O-4 models and authentic lignin. *Green Chemistry* 23 (1): 412–421.

57 Li, Z., Garedew, M., Lam, C.H. et al. (2012). Mild electrocatalytic hydrogenation and hydrodeoxygenation of bio-oil derived phenolic compounds using ruthenium supported on activated carbon cloth. *Green Chemistry* 14 (9): 2540–2549.

58 Cyr, A., Chiltz, F., Jeanson, P. et al. (2000). Electrocatalytic hydrogenation of lignin models at Raney nickel and palladium-based electrodes. *Canadian Journal of Chemistry* 78 (3): 307–315.

59 Lam, C.H., Lowe, C.B., Li, Z. et al. (2015). Electrocatalytic upgrading of model lignin monomers with earth abundant metal electrodes. *Green Chemistry* 17 (1): 601–609.

60 Zhou, Y., Gao, Y., Zhong, X. et al. (2019). Electrocatalytic upgrading of lignin-derived bio-oil based on surface-engineered PtNiB nanostructure. *Advanced Functional Materials* 29 (10): 1807651.

61 Song, Y., Chia, S.H., Sanyal, U. et al. (2016). Integrated catalytic and electrocatalytic conversion of substituted phenols and diaryl ethers. *Journal of Catalysis* 344: 263–272.

62 Sanyal, U., Song, Y., Singh, N. et al. (2019). Structure sensitivity in hydrogenation reactions on Pt/C in aqueous-phase. *ChemCatChem* 11 (1): 575–582.

63 Wijaya, Y.P., Grossmann-Neuhaeusler, T., Dhewangga Putra, R.D. et al. (2020). Electrocatalytic hydrogenation of guaiacol in diverse electrolytes using a stirred slurry reactor. *ChemSusChem* 13 (3): 629–639.

64 Mahdavi, B., Lafrance, A., Martel, A. et al. (1997). Electrocatalytic hydrogenolysis of lignin model dimers at Raney nickel electrodes. *Journal of Applied Electrochemistry* 27 (5): 605–611.

65 Li, Y., Shuai, L., Kim, H. et al. (2018). An "ideal lignin" facilitates full biomass utilization. *Science Advances* 4 (9): eaau2968.

66 Wu, W.-B. and Huang, J.-M. (2014). Electrochemical cleavage of aryl ethers promoted by sodium borohydride. *The Journal of Organic Chemistry* 79 (21): 10189–10195.

67 Garedew, M., Young-Farhat, D., Bhatia, S. et al. (2020). Electrocatalytic cleavage of lignin model dimers using ruthenium supported on activated carbon cloth. *Sustainable Energy & Fuels* 4 (3): 1340–1350.

68 Zhou, Y., Klinger, G.E., Hegg, E.L. et al. (2020). Multiple mechanisms mapped in aryl alkyl ether cleavage via aqueous electrocatalytic hydrogenation over skeletal nickel. *Journal of the American Chemical Society* 142 (8): 4037–4050.

69 Reddy, S.J. and Krishnan, V. (1978). Electrochemical oxidation of guaiacol at platinum electrode. *Indian Journal of Chemistry* 16A: 684–687.

70 Samet, Y., Abdelhedi, R., and Savall, A. (2002). A study of the electrochemical oxidation of guaiacol. *Physical and Chemical News* 8 (1): 89–99.

71 Shao, D., Chu, W., Li, X. et al. (2016). Electrochemical oxidation of guaiacol to increase its biodegradability or just remove COD in terms of anodes and electrolytes. *RSC Advances* 6 (6): 4858–4866.

72 Zhou, H., Li, Z., Xu, S.M. et al. (2021). Selectively upgrading lignin derivatives to carboxylates through electrochemical oxidative C (OH)-C bond cleavage by a Mn-doped cobalt oxyhydroxide catalyst. *Angewandte Chemie* 133 (16): 9058–9064.

73 Chen, J., Yang, H., Fu, H. et al. (2020). Electrochemical oxidation mechanisms for selective products due to C-O and C-C cleavages of β-O-4 linkages in lignin model compounds. *Physical Chemistry Chemical Physics* 22 (20): 11508–11518.

74 Gao, W.-J., Lam, C.M., Sun, B.-G. et al. (2017). Selective electrochemical CO bond cleavage of β-O-4 lignin model compounds mediated by iodide ion. *Tetrahedron* 73 (17): 2447–2454.

75 Sannami, Y., Kamitakahara, H., and Takano, T. (2017). TEMPO-mediated electro-oxidation reactions of non-phenolic β-O-4-type lignin model compounds. *Holzforschung* 71 (2): 109–117.

76 Shiraishi, T., Sannami, Y., Kamitakahara, H., and Takano, T. (2013). Comparison of a series of laccase mediators in the electro-oxidation reactions of non-phenolic lignin model compounds. *Electrochimica Acta* 106: 440–446.

77 Rafiee, M., Alherech, M., Karlen, S.D., and Stahl, S.S. (2019). Electrochemical aminoxyl-mediated oxidation of primary alcohols in lignin to carboxylic acids: polymer modification and depolymerization. *Journal of the American Chemical Society* 141 (38): 15266–15276.

78 Miller, B. and Chen, A. (2005). Effect of concentration and temperature on electrochemical oscillations during sulfide oxidation on Ti/Ta_2O_5-IrO_2 electrodes. *Electrochimica Acta* 50 (11): 2203–2212.

79 Tolba, R., Tian, M., Wen, J. et al. (2010). Electrochemical oxidation of lignin at IrO_2-based oxide electrodes. *Journal of Electroanalytical Chemistry* 649 (1-2): 9–15.

80 Rauber, D., Dier, T.K., Volmer, D.A., and Hempelmann, R. (2018). Electrochemical lignin degradation in ionic liquids on ternary mixed metal electrodes. *Zeitschrift für Physikalische Chemie* 232 (2): 189–208.

81 Zhang, Y.M., Peng, Y., Yin, X.L. et al. (2014). Degradation of lignin to BHT by electrochemical catalysis on Pb/PbO_2 anode in alkaline solution. *Journal of Chemical Technology and Biotechnology* 89 (12): 1954–1960.

82 Meng, D., Li, G., Liu, Z., and Yang, F. (2011). Study of depolymerization of cotton cellulose by Pb/PbO_2 anode electrochemical catalysis in sulfuric acid solution. *Polymer Degradation and Stability* 96 (7): 1173–1178.

83 Lan, C., Fan, H., Shang, Y. et al. (2020). Electrochemically catalyzed conversion of cornstalk lignin to aromatic compounds: an integrated process of anodic oxidation of a Pb/PbO_2 electrode and hydrogenation of a nickel cathode in sodium hydroxide solution. *Sustainable Energy & Fuels* 4 (4): 1828–1836.

84 Wang, Y.-S., Yang, F., Liu, Z.-H. et al. (2015). Electrocatalytic degradation of aspen lignin over Pb/PbO_2 electrode in alkali solution. *Catalysis Communications* 67: 49–53.

85 Liu, M., Wen, Y., Qi, J. et al. (2017). Fine chemicals prepared by bamboo lignin degradation through electrocatalytic redox between Cu cathode and Pb/PbO_2 anode in alkali solution. *ChemistrySelect* 2 (17): 4956–4962.

86 Chen, A., Wen, Y., Han, X. et al. (2019). Electrochemical decomposition of wheat straw lignin into guaiacyl-, syringyl-, and phenol-type compounds using Pb/PbO_2 anode and alloyed steel cathode in alkaline solution. *Environmental Progress & Sustainable Energy* 38 (4): 13117.

87 Cai, P., Fan, H., Cao, S. et al. (2018). Electrochemical conversion of corn stover lignin to biomass-based chemicals between Cu/NiMoCo cathode and Pb/PbO_2 anode in alkali solution. *Electrochimica Acta* 264: 128–139.

88 Jia, Y., Wen, Y., Han, X. et al. (2018). Electrocatalytic degradation of rice straw lignin in alkaline solution through oxidation on a $Ti/SnO_2\text{-}Sb_2O_3/\alpha\text{-}PbO_2/\beta\text{-}PbO_2$ anode and reduction on an iron or tin doped titanium cathode. *Catalysis Science & Technology* 8 (18): 4665–4677.

89 Hao, X., Quansheng, Y., Dan, S. et al. (2015). Fabrication and characterization of PbO_2 electrode modified with $[Fe(CN)_6]^3-$ and its application on electrochemical degradation of alkali lignin. *Journal of Hazardous Materials* 286: 509–516.

90 Pan, K., Tian, M., Jiang, Z.-H. et al. (2012). Electrochemical oxidation of lignin at lead dioxide nanoparticles photoelectrodeposited on TiO_2 nanotube arrays. *Electrochimica Acta* 60: 147–153.

91 Smith, C.Z., Utley, J.H., and Hammond, J.K. (2011). The electro-oxidative conversion at laboratory scale of a lignosulfonate into vanillin in an FM01 filter press flow reactor: preparative and mechanistic aspects. *Journal of Applied Electrochemistry* 41 (4): 363–375.

92 Moodley, B., Mulholland, D., and Brookes, H. (2012). The chemical oxidation of lignin found in Sappi Saiccor dissolving pulp mill effluent. *Water SA* 38 (1): 1–8.

93 Schmitt, D., Regenbrecht, C., Schubert, M. et al. (2017). Treatment of black liquor (BL) by adsorption on AE resins and a subsequent electrochemical degradation of BL to obtain vanillin. *Holzforschung* 71 (1): 35–41.

94 Stiefel, S., Lölsberg, J., Kipshagen, L. et al. (2015). Controlled depolymerization of lignin in an electrochemical membrane reactor. *Electrochemistry Communications* 61: 49–52.

95 Stiefel, S., Schmitz, A., Peters, J. et al. (2016). An integrated electrochemical process to convert lignin to value-added products under mild conditions. *Green Chemistry* 18 (18): 4999–5007.

96 Bawareth, B., Di Marino, D., Nijhuis, T.A., and Wessling, M. (2018). Unravelling electrochemical lignin depolymerization. *ACS Sustainable Chemistry & Engineering* 6 (6): 7565–7573.

97 Di Marino, D., Jestel, T., Marks, C. et al. (2019). Carboxylic acids production via electrochemical depolymerization of lignin. *ChemElectroChem* 6 (5): 1434–1442.

98 Ghahremani, R. and Staser, J.A. (2018). Electrochemical oxidation of lignin for the production of value-added chemicals on Ni-Co bimetallic electrocatalysts. *Holzforschung* 72 (11): 951–960.

99 Schmitt, D., Regenbrecht, C., Hartmer, M. et al. (2015). Highly selective generation of vanillin by anodic degradation of lignin: a combined approach of electrochemistry and product isolation by adsorption. *Beilstein Journal of Organic Chemistry* 11 (1): 473–480.

100 Movil-Cabrera, O., Rodriguez-Silva, A., Arroyo-Torres, C., and Staser, J.A. (2016). Electrochemical conversion of lignin to useful chemicals. *Biomass and Bioenergy* 88: 89–96.

101 Zirbes, M., Schmitt, D., Beiser, N. et al. (2019). Anodic degradation of lignin at active transition metal-based alloys and performance-enhanced anodes. *ChemElectroChem* 6 (1): 155–161.

102 Movil, O., Garlock, M., and Staser, J.A. (2015). Non-precious metal nanoparticle electrocatalysts for electrochemical modification of lignin for low-energy and cost-effective production of hydrogen. *International Journal of Hydrogen Energy* 40 (13): 4519–4530.

103 Reichert, E., Wintringer, R., Volmer, D.A., and Hempelmann, R. (2012). Electrocatalytic oxidative cleavage of lignin in a protic ionic liquid. *Physical Chemistry Chemical Physics* 14 (15): 5214–5221.

104 Dier, T.K., Egele, K., Fossog, V. et al. (2016). Enhanced mass defect filtering to simplify and classify complex mixtures of lignin degradation products. *Analytical Chemistry* 88 (2): 1328–1335.

105 Di Marino, D., Stöckmann, D., Kriescher, S. et al. (2016). Electrochemical depolymerisation of lignin in a deep eutectic solvent. *Green Chemistry* 18 (22): 6021–6028.

106 Cui, X., Liang, J., and Wang, D. (2015). Optimization of integrated electrochemical oxidation and biodegradation (EO-BD) for Kraft lignin. *Journal of Chemical Technology and Biotechnology* 90 (4): 747–753.

107 Shao, D., Liang, J., Cui, X. et al. (2014). Electrochemical oxidation of lignin by two typical electrodes: Ti/$SbSnO_2$ and Ti/PbO_2. *Chemical Engineering Journal* 244: 288–295.

108 Wu, X., Luo, N., Xie, S. et al. (2020). Photocatalytic transformations of lignocellulosic biomass into chemicals. *Chemical Society Reviews* 49 (17): 6198–6223.

109 Tian, M., Wen, J., MacDonald, D. et al. (2010). A novel approach for lignin modification and degradation. *Electrochemistry Communications* 12 (4): 527–530.

110 Bosque, I., Magallanes, G., Rigoulet, M. et al. (2017). Redox catalysis facilitates lignin depolymerization. *ACS Central Science* 3 (6): 621–628.

111 Yao, B., Kolla, P., Koodali, R. et al. (2017). Enzymatic decomposition and electrochemical study of alkali lignin by laccase (Trametes versicolor) in the presence of a natural mediator (methyl syringate). *New Journal of Chemistry* 41 (3): 958–964.

112 Rafiee, M., Miles, K.C., and Stahl, S.S. (2015). Electrocatalytic alcohol oxidation with TEMPO and bicyclic nitroxyl derivatives: driving force trumps steric effects. *Journal of the American Chemical Society* 137 (46): 14751–14757.

113 Zhu, H., Chen, Y., Qin, T. et al. (2014). Lignin depolymerization via an integrated approach of anode oxidation and electro-generated H_2O_2 oxidation. *RSC Advances* 4 (12): 6232–6238.

114 Zhu, H., Wang, L., Chen, Y. et al. (2014). Electrochemical depolymerization of lignin into renewable aromatic compounds in a non-diaphragm electrolytic cell. *RSC Advances* 4 (56): 29917–29924.

115 Hibino, T., Kobayashi, K., Nagao, M., and Teranishi, S. (2017). Hydrogen production by direct lignin electrolysis at intermediate temperatures. *ChemElectroChem* 4 (12): 3032–3036.

116 Du, X., Liu, W., Zhang, Z. et al. (2017). Low-energy catalytic electrolysis for simultaneous hydrogen evolution and lignin depolymerization. *ChemSusChem* 10 (5): 847–854.

117 Bateni, F., NaderiNasrabadi, M., Ghahremani, R., and Staser, J.A. (2019). Low-cost nanostructured electrocatalysts for hydrogen evolution in an anion exchange membrane lignin electrolysis cell. *Journal of the Electrochemical Society* 166 (14): F1037.

118 Ghatak, H.R. (2006). Electrolysis of black liquor for hydrogen production: some initial findings. *International Journal of Hydrogen Energy* 31 (7): 934–938.

119 Ghatak, H.R., Kumar, S., and Kundu, P. (2008). Electrode processes in black liquor electrolysis and their significance for hydrogen production. *International Journal of Hydrogen Energy* 33 (12): 2904–2911.

120 Nong, G., Zhou, Z., and Wang, S. (2015). Generation of hydrogen, lignin and sodium hydroxide from pulping black liquor by electrolysis. *Energies* 9 (1): 13.

121 Caravaca, A., Garcia-Lorefice, W.E., Gil, S. et al. (2019). Towards a sustainable technology for H_2 production: direct lignin electrolysis in a continuous-flow polymer electrolyte membrane reactor. *Electrochemistry Communications* 100: 43–47.

122 NaderiNasrabadi, M., Rakshit, S.K., Viswanathan, G. et al. (2021). A Techno-economic analysis for integrating an electrochemical reactor into a lignocellulosic biorefinery for production of industrial chemicals and hydrogen. *Applied Biochemistry and Biotechnology* 193 (3): 791–806.

123 Ayub, R. and Raheel, A. (2022). High-value chemicals from electrocatalytic depolymerization of lignin: challenges and opportunities. *International Journal of Molecular Sciences* 23 (7): 3767.

124 Parpot, P., Bettencourt, A., Carvalho, A., and Belgsir, E. (2000). Biomass conversion: attempted electrooxidation of lignin for vanillin production. *Journal of Applied Electrochemistry* 30 (6): 727–731.

125 Lu, Z., Tu, B., and Chen, F. (2003). Electro-degradation of sodium lignosulfonate. *Journal of Wood Chemistry and Technology* 23 (3-4): 261–277.

126 Pelegrini, R., Reyes, J., Duran, N. et al. (2000). Photoelectrochemical degradation of lignin. *Journal of Applied Electrochemistry* 30 (8): 953–958.

127 Ghahremani, R., Farales, F., Bateni, F., and Staser, J.A. (2020). Simultaneous hydrogen evolution and lignin depolymerization using NiSn electrocatalysts in a biomass-depolarized electrolyzer. *Journal of the Electrochemical Society* 167 (4): 043502.

128 Singh, S. and Ghatak, H.R. (2017). Vanillin formation by electrooxidation of lignin on stainless steel anode: kinetics and by-products. *Journal of Wood Chemistry and Technology* 37 (6): 407–422.

129 Singh, S. and Ghatak, H.R. (2018). Process optimization of lignin conversion into value added chemicals by thermochemical pretreatment and electrooxidation on a stainless steel anode. *Holzforschung* 72 (3): 187–199.

130 Wang, C., Liu, K., Jin, Y. et al. (2023). Amorphous RuO_2 catalyst for medium size carboxylic acid to alkane dimer selective Kolbe electrolysis in an aqueous environment**. *ChemSusChem* 16 (16). Portico. https://doi.org/10.1002/cssc.202300222.

131 Lang, L., Li, Y., Lam, J. C.-H. et al. (2022). Enhancing the selectivity of hydrocarbons during the Kolbe electrolysis of biomass-derived short-chain carboxylic acids by anionic surfactants. *Sustainable Energy & Fuels* 6 (11): 2797–2804. https://doi.org/10.1039/d1se02067f.

132 Lin, F., TSE, H.-Y., Erythropel, H. C. et al. (2022). Development of a Ni-promoted, selective electrochemical reductive cleavage of the C–O bond in lignin model compound benzyl phenyl ether. *Green Chemistry* 24 (16): 6295–6305. https://doi.org/10.1039/d2gc01510b.

133 Lin, F., Petrović, P. V., Tse, H.-Y. et al. (2023). Mechanistic investigation of a Ni-catalyzed electrochemical reductive cleavage of the α-O-4 bond in the lignin model compound benzyl phenyl ether. *Green Chemistry* 25 (23): 9720–9732. https://doi.org/10.1039/d3gc01814h.

12

Recent Advances in Thermoset and Thermoplastic Polymeric Materials Produced using Technical and Depolymerized Native Lignins

Kirk M. Torr[1], Oliver J. Driscoll[2], Daniel J. van de Pas[1] and Elias Feghali[2,3]

[1] *Chemistry and Physics Research Group, Scion, 49 Sala Street, Rotorua 3046, New Zealand*
[2] *Flemish Institute for Technological Research (Vito N.V.), Sustainable Polymer Technologies Team, Boeretang 200, Mol 2400, Belgium*
[3] *Notre Dame University-Louaize, Chemical Engineering Program, Zouk Mosbeh 1211, Lebanon*

12.1 Introduction

The growth of the global bio-based economy and the desire for more sustainable products are driving increased interest in renewable chemicals derived from biomass as alternatives to petrochemicals. The polymer research community and commercial sector are actively seeking new bio-based polymers with novel functionality, enhanced performance, and low carbon footprints. Significant advances have been made in both bio-based thermosetting and thermoplastic polymers [1–5]. Lignin is the most abundant renewable aromatic polymer and so offers a widely available, low-cost source of ready-made aromatic polyols that can be used in making new, sustainable thermosets and thermoplastics.

A key challenge for a modern biorefinery is deriving maximum value from the biomass resource. Currently, the polysaccharide components of lignocellulosic biomass are valorized as fibers, fuels, and chemical products, while limited value is extracted from lignin. Valorization of the lignin component of lignocellulosic biomass into a suite of products, ranging in value and volume, can significantly improve the economics of biorefining operations [6–8]. The major opportunities for valorizing lignin, beyond energy products, are as a macromolecule and as a source of aromatic chemicals [9], both of which could be used in polymer applications. Technical or industrial lignins include lignosulfonates (production = c. 1.3 million tonnes/annum), Kraft lignin (c. 260,000 tonnes/annum), soda lignin (SL), and biorefinery (or enzymatic hydrolysis) lignins, and organosolv lignins [10]. The production of

Lignin Chemistry: Characterization, Isolation, and Valorization, First Edition.
Edited by Yuhe Liao and Bert F. Sels.

soda and biorefinery lignins combined has been estimated at c. 75,000 tonnes/annum [10]. Organosolv lignins are currently not produced commercially. In all cases, the structure of the original native lignin in the biomass has been modified with the level of modification depending on the severity of the processing conditions (e.g. pulping, extraction, biomass pretreatment, etc.) used to produce/extract the technical lignin. This is because the functional groups present in native lignin make it very reactive under most processing conditions.

Kraft, sulfite, and soda pulping lead to highly condensed and chemically modified lignins (Kraft, lignosulfonates, and SLs, respectively) due to base/acid-catalyzed condensation reactions that occur during pulping that create new stable C—C bonds [11]. These repolymerization reactions also occur, to a lesser extent, in organosolv processing (organosolv lignins) and biomass pretreatment (biorefinery/hydrolysis lignins), which use acid catalysts and/or thermomechanical processing to promote extraction and/or improve cellulose accessibility to enzymes. Biorefinery/hydrolysis lignins also tend to be low purity and contain undigested carbohydrates. These modifications can affect the properties of the lignins and their versatility as a starting material for new bio-based polymers. The key point here is that all lignins are not equal, with some lignins making better feedstocks for polymer synthesis.

More recently, Lignin-first biorefining, or more specifically reductive catalytic fractionation (RCF), is being used to produce low molecular weight lignin-derived polyols from native lignins [12–14]. These depolymerized native lignins can have high concentrations of monomers and dimers that can provide advantages over technical lignins in polymer synthesis. In lignin-first biorefining/RCF of lignocellulosic biomass, the native lignin is targeted first for valorization to avoid undesirable lignin repolymerization which limits further upgrading. Several active stabilization approaches have been used, including precious metal-catalyzed hydrogenolysis of native lignin in situ in lignocellulosic biomass with stabilization using pressurized hydrogen [15–17] (Figure 12.1) and acid-catalyzed depolymerization of native lignins with alcohol stabilization [18–20]. Another promising approach involves functionalizing and stabilizing the lignin, commonly using aldehydes, during its extraction from the biomass (Figure 12.1). The fractionated and functionalized lignin can then be catalytically depolymerized to monomers in high yields [20–22].

Novel thermoset and thermoplastic polymers made using both technical lignins and depolymerized native lignins are discussed in this chapter. The past literature in this area is comprehensive, and the reader is directed toward a number of excellent reviews on this topic [4, 23–35]. This book chapter is not a comprehensive review but rather a brief overview of recent advances in the space that the authors consider important and interesting. For the most part, the focus is on examples where lignin is used as a building block in polymer synthesis and is a major and integral part of the novel bio-based thermoset or thermoplastic materials that have been produced. Lignin/polymer blends, lignin-containing composite materials, and polymers based on lignin monomers not directly derived from lignin, lie outside the scope of this review.

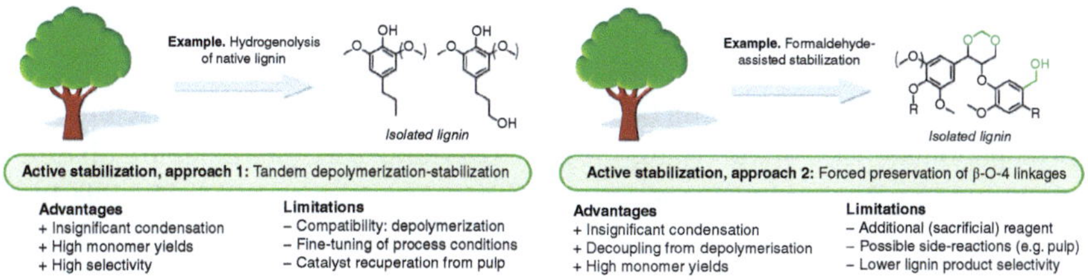

Figure 12.1 Two active stabilization approaches in lignin-first biorefining or reductive catalytic fractionation of lignocellulosic biomass with the aim of producing lignin-derived polyols for chemical and polymer applications. Source: Reproduced from Ref. [12] with permission from the Royal Society of Chemistry.

12.2 Lignin-Based Thermosets

A range of lignin-derived thermosetting polymers have been developed with the most widely studied being polyurethane, epoxy, and phenol-formaldehyde polymers. The complex cross-linked aromatic structure of lignin makes it inherently suitable for making thermosetting polymers. For thermosets, aromatic building blocks are favored due to their structural rigidity and stability, which gives the cured polymer desirable thermal and mechanical properties [1, 4, 5]. Most petrochemical-based aromatic thermoset polymers are derived from monomers with high toxicity or environmental concerns, such as bisphenol A, phenol, and methylene diphenyl isocyanate. Because thermosetting polymers are difficult to recycle, it is desirable to produce these polymers with a high renewable carbon content. Using lignin as a building block for thermosets offers advantages such as improved sustainability, enhanced performance, low toxicity, biodegradability, and potentially low cost [4, 32, 35–37].

12.2.1 Polyurethanes

The global polyurethane (PU) market was valued at US$ 72.8 billion in 2021 [38]. Over three-quarters of this PU is used in flexible and rigid foams and coatings. Other important applications include adhesives, sealants, and elastomers. Ninety percentage of PUs are based on polyethers, 9% on polyesters, and 1% on specialty polyols [39].

Lignin has been applied to PU materials because of its availability as a bio-based aromatic polyol and its high hydroxyl functionality. The lignin polyol alone, or in combination with other polyols, is reacted with isocyanates to produce the cross-linked polyurethane network. Using unmodified technical lignins as a polyol gives rise to stiffness and rigidity. Hence it is common to combine lignin polyols with long–chain aliphatic polyols such as poly(ethylene glycol) (PEG) and poly(propylene glycol) (PPG) to introduce flexibility, elasticity, and toughness to the PU material [24, 32]. Other challenges of using technical lignins in PUs include the lignin's complex nonlinear structure and high molecular weight, lower reactivity of aliphatic and phenolic hydroxyl groups, and brittleness of PU materials at high lignin loadings [4, 32, 39, 40].

Lignin can be used in producing both flexible and rigid foams. Flexible foams are block copolymers with hard (rigid, cross-linked) and soft (stretchable) phases. Lignin is used in the hard phase to provide firmness to the material. Flexible PU foams are commonly used in insulation, cushioning products (e.g. furniture, automotive interiors, and carpet underlay), and acoustic absorbing materials. Rigid foams are cross-linked polymer networks with a closed-cell structure. They typically have low density, low moisture permeability, low thermal conductivity, and good shock absorption properties [40, 41]. Rigid foams are mainly used in construction, insulation, and refrigeration applications [39].

Producing PU thermoset polymers from various lignins has been extensively studied with a number of recent reviews on the topic [24, 32, 35, 39]. In general, three broad approaches can be used to incorporate lignin into PU thermoset materials. These include: (i) direct use of technical lignin or solvent-fractionated lignins, (ii) use of chemical-modified technical lignins, and (iii) use of depolymerized native lignin produced by lignin-first biorefining/RCF approaches. The third approach is the most recent and least investigated to date. Some recent examples of these strategies are discussed in Sections 12.2.1.1–12.2.1.3.

12.2.1.1 Use of Technical or Fractionated Technical Lignins

A recent example of a unique use of technical and fractionated lignins in a PU coating is the work of de Haro et al. (2019) [42]. They cross-linked three technical lignins (mild acetone organosolv, THF-soluble fraction of kraft lignin, and SL) with a diisocyanate synthesized from vanillic acid (potentially derived from lignin) to make new bio-based polyurethane coatings with high lignin content. They showed the thermal and mechanical properties, and adhesion performance of the PU coatings correlated to the chemical–physical features of the parent lignins. This suggested the potential to tailor the coating characteristics by lignin selection. Liu et al. (2021) recently compared softwood kraft lignin (SKL), hydroxypropylated SKL, and organosolv pine and oak bark lignins as lignin-based polyols in the preparation of rigid PU foams with thermo-responsive shape-memory properties [43].

Incorporating technical lignins directly into PU foams is often limited by the high molecular weight and low solubility of the lignin, the low reactivity of hydroxyl groups, and the intrinsic brittleness of lignin. Solvent fractionation can be used to improve reactivity by producing lower molecular weight, more homogeneous lignin fractions with lower dispersity [32]. A variety of solvents have been investigated for fractionating lignin [32, 44, 45]. Wang et al. (2019) prepared Kraft lignin fractions ranging in molecular weight from 66,000 to 1000 $g\,mol^{-1}$ by sequential precipitation from an acetone–methanol–hexane cosolvent system [46]. PUs containing up to 40 wt% lignin were synthesized from the lignin fractions using PEG and poly(trimethylene glutarate) to form soft segments in the PU network. The mechanical properties of the PU materials could be tuned by the lignin molecular weight and the incorporation of the secondary polyol. Griffini et al. (2015) fractionated softwood kraft lignin by extracting with the bioderived solvent 2-methyl-tetrahydrofuran [47]. This fractionated lignin was cross-linked with a toluene diisocyanate-based polyisocyanate to produce PU films that were demonstrated to have potential as high-performance bio-based coatings and adhesives.

Despite advances in using technical lignin in PU polymers, the level of replacement of the petrochemical-based polyols is usually around 30 wt% [40]. Solvent fractionation can improve the level of lignin incorporation by providing lower molecular weight lignins; however, this adds complicated extraction steps, increases processing cost, and constrains scalability [32]. An alternative approach is to chemically modify the technical lignin.

12.2.1.2 Use of Chemically Modified Technical Lignins

The chemical modification of technical lignins for PU applications has a long history and is one of the most effective approaches to addressing the processability limitations of technical lignins [4, 24, 35, 40]. One popular approach to modifying technical lignins is oxyalkylation with alkylene oxides (e.g. propylene oxide, ethylene oxide) to produce a polyether polyol that can be used in PU synthesis. This approach etherifies the phenolic hydroxyl groups to give aliphatic hydroxyl groups, making them more available and reactive to isocyanate groups [32, 35, 39]. It also improves the miscibility of the lignin and enables a higher lignin content in the final PU polymer [32]. The original work dates back to the 1980s [48], with a number of studies subsequently using this approach to make lignin-based rigid PU foams [41, 49, 50]. Recently, Li et al. (2020) combined solvent fractionation and oxypropylation of an ethanol organosolv lignin from corn stover to produce lignin-based polyols for bio-based rigid polyurethane foams [51].

Oxyalkylated lignins can be combined with other polyols to modify chain flexibility and mechanical properties [32, 41]. Wang et al. (2019) grafted flexible polyethylene glycol (PEG) chains onto alkali lignin by first functionalizing PEG-2000 with epichlorohydrin and then reacting it with the lignin [52]. The PEG grafted alkali lignin was formulated with PEG-2000, hexamethylene diisocyanate, and foaming agents to produce highly resilient, flexible PU foams.

More recently, "greener" oxyalkylation methods have been achieved by replacing the volatile and toxic epoxides (propylene oxide or ethylene oxide) with corresponding cyclic carbonates [53, 54]. Zhang et al. (2019) oxyalkylated Kraft lignin by reacting with ethylene carbonate, in the presence of PEG-400 (nonreactive modifier), to produce biopolyols for producing rigid polyurethane foams [55]. The modification resulted in a 14-fold increase in reactivity toward isocyanates for the oxyalkylated lignin compared to the unmodified Kraft lignin. PU foams that were produced under optimized conditions had densities of 51–60 $kg\,m^{-3}$ and compressive strengths of 139–204 kPa. Sternberg and Pilla (2020), and Mimini et al. (2020), used green organic carbonates to synthesize cyclocarbonated Kraft lignin and lignosulfonate, respectively, and then cured these with diamines to prepare lignin-based isocyanate-free PU polymers [56, 57]. Sternberg and Pilla (2020) functionalized Kraft lignin with glycerol carbonate to make oxyalkylated lignin, which was in turn reacted with dimethyl carbonate to produce a cyclocarbonated lignin. This functionalized lignin was polymerized with a fatty acid-based diamine in the presence of a foaming agent to produce non-isocyanate polyurethane (NIPU) foams with 100% bio-based carbon (Figure 12.2) [57]. NIPU materials with shape memory capacity were also produced in this study.

An alternative strategy to chemical modification of technical lignins is depolymerization. Mahmood et al. (2016) recently reviewed the approach of depolymerizing technical lignins to produce reactive polyols for use in PU foams [40]. Cinelli et al. (2013) liquefied pine Kraft lignin with glycerol and PEG–400 by microwave treatment [58]. The liquefied lignin mixture was combined with either PPG triol or castor oil as a chain extender, polymeric diphenylmethane diisocyanate, and foaming agents to make flexible polyurethane foams. However, the lignin content in the final foams was only 6–13 wt%. Bernardini et al. (2015) extended this work by using liquefied SL and oxypropylated SL in similar PU formulations [59].

To produce rigid PU foams with high lignin contents, Mahmood et al. (2015) modified Kraft lignin by base-catalyzed depolymerization followed by oxypropylation with propylene oxide [60]. The depolymerized Kraft lignin (DKL) and oxypropylated DKL were used to replace PPG–400 or sucrose-based polyols in the preparation of rigid polyurethane foams with bio-based content up to 70 wt%. The lignin-based PU foams exhibit good thermal conductivity and compressive strengths compared with reference foams. In related work, Mahmood et al. (2013) depolymerized an enzymatic hydrolysis lignin by thermal treatment in ethanol:water 1 : 1 (v/v) and then oxypropylated the product by reacting with propylene oxide. This gave lignin-derived liquid polyols that were used to prepare rigid PU foams with compression properties that were superior to reference foams produced with a sucrose-based polyol [61].

Li et al. (2017) prepared lignin-based PU elastomers with high stiffness, strength, and toughness by reacting depolymerized alkali lignin, as hard segments, with PPG

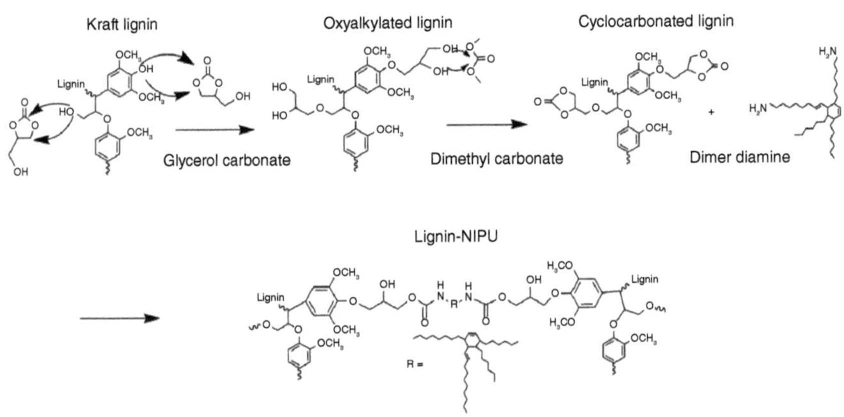

Figure 12.2 Reaction scheme for the preparation of lignin-based non-isocyanate polyurethanes by the functionalization of Kraft lignin with cyclic carbonates and curing with a fatty acid-based diamine. Source: Reproduced from Ref. [57] with permission from the Royal Society of Chemistry.

tolylene 2,4-diisocyanate terminated, as soft domains [62]. The depolymerized lignin ($M_n = 600\,g\,mol^{-1}$, 9.2 mmol OH/g) was prepared by base-catalyzed depolymerization of alkali lignin ($M_n = 3600\,g\,mol^{-1}$, 5.6 mmol OH/g). Increasing the lignin content (5–40 wt%) improved the thermal and mechanical properties of the lignin-based PU elastomers. In related work, Liu et al. (2019) also produced high-performance lignin-containing PU elastomers using depolymerized enzymatic hydrolysis lignin [63]. Polytetramethylene ether glycol was first reacted with hexamethylene diisocyanate to produce soft (stretchable) segments. This was then reacted with the depolymerized lignin to add the hard segments. The depolymerized lignin enhanced the tensile strength, toughness, and elasticity of the PU elastomers [63].

12.2.1.3 Use of Depolymerized Native Lignin Produced by Lignin-First Biorefining/RCF Approaches

Huang et al. (2018) investigated the lignin-first biorefinery concept with a focus on producing fuel and material products from the lignin component of the lignocellulosic biomass. They produced a lignin oil by Ru/SiC catalyzed depolymerization of native lignin in situ in hardwood biomass and then fractionated this oil by hexane extraction. The hexane-insoluble residue was used as bio-based polyols in up to 50 wt% for the preparation of rigid PU foams, while the hexane-soluble oil was further processed for use as a jet fuel additive [64]. Vendamme et al. (2020) used aldehyde-assisted fractionation (AAF) of lignocellulosic biomass to produce oligomers from native lignins from hardwood and softwood sources. These bioaromatic oligomers possessed purity, solubility, and functionality that facilitated their formulation into PU materials. The thermal and mechanical properties of the new AAF oligomer-based PU materials were compared to similar PU materials prepared from Kraft lignin-based oligomers. The results highlighted the versatility of AAF oligomers in designing tunable PU bioresins [65].

Quinsaat et al. (2021) employed a lignin-first biorefining approach by using lignin hydrogenolysis oil (LHO), obtained by depolymerizing native softwood lignin, to produce rigid PU foams [66]. The lignin oil, which was composed of a mixture of monomeric and oligomeric polyols [67], was fractionated or used directly to replace a sorbitol-based polyether polyol at 12.5–50 wt% in preparing mechanically robust PU foams with enhanced bio-based content. The lignin-based PU foams had similar densities and thermal properties, and substantially improved compression properties, compared to reference foams [66]. In subsequent work, Quinsaat et al. (2022) reported the preparation of isocyanate-free polyurethane/epoxy hybrid thermosets with a high bio-based content using the same depolymerized native softwood lignin oil [68]. The LHO was glycidylated with epichlorohydrin (LHO-GE) and then reacted with CO_2 to form cyclocarbonated LHO (LHO-CC). This was blended with glycerol diglycidyl ether (GDGE) and cured with two amine hardeners to produce novel hybrid bio-based PHU/epoxy thermosets materials (Figure 12.3) that outperformed the corresponding epoxy-only thermosets and reference hybrid PHU/epoxy materials produced using petrochemicals.

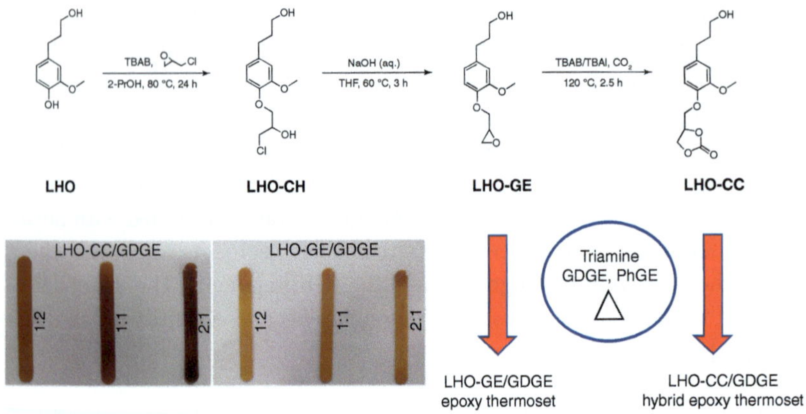

Figure 12.3 Reaction scheme for preparation of isocyanate-free lignin-based PU/epoxy hybrid thermoset material using a lignin-first approach. Source: Adapted from Ref. [68] with permission from the American Chemical Society.

12.2.2 Epoxies

In 2021, the global epoxy resin market was valued at US$ 22.9 billion [69]. Key applications for epoxy resins include adhesives, reinforcing resins (used in composites), coatings, electrical and electronic applications, polymer additives, and stabilizers [36]. The versatility of epoxy resins comes from their ability to be combined with a wide range of curing agents. The majority of commercial epoxy resins are based on bisphenol A diglycidyl ether (BADGE) made from bisphenol A [36], a known endocrine-disrupting compound [70], and epichlorohydrin. There is significant research interest in replacing bisphenol A with bio-based aromatic polyols, which are capable of delivering equivalent or better performance, on the grounds of improved sustainability, health, and environmental outcomes.

Similar to the PUs, a range of strategies have been used to produce high-performance lignin-based epoxy thermoset polymers. These include (i) Reactive blending of technical lignins with epoxys; (ii) Direct epoxidation of technical/fractionated technical lignins; (iii) Chemical modification of technical lignins prior to epoxidation; and (iv) Use of depolymerized native lignins produced by lignin-first biorefining/RCF approaches. The first three strategies are well established and have been effectively reviewed in several recent reviews on lignin thermosetting polymers [4, 24, 30, 35, 36, 71]. The fourth strategy is a new area that is beginning to receive attention [71]. Some recent examples of all these strategies are described in Sections 12.2.2.1–12.2.2.4.

12.2.2.1 Reactive Blending with Epoxy Resins

Engelmann and Ganster (2014) blended low molecular weight acetone-fractionated Kraft lignin with GDGE and cured with imidazole to prepare lignin-based epoxy polymers containing 20–50 wt% lignin [72]. The best-performing thermoset with

40 wt% lignin had a tensile strength of 37 MPa and Young's modulus of 2.2 GPa. Mechanical properties were further improved using pyrogallol as a copolyol and by introducing cellulose fiber reinforcement [72]. Li et al. (2018) expanded on using Kraft lignin reacted with GDGE to produce bio-based epoxy resins as a potential drop-in replacement for formaldehyde-based wood adhesives [73]. More examples of blending and reacting lignins with epoxies can be found in other recent reviews [24, 35, 36].

12.2.2.2 Direct Epoxidation of Technical or Fractionated Technical Lignins

Over et al. (2017) prepared a glycidylated organosolv lignin (GOL) by reaction with epichlorohydrin, blended with BADGE and then cured with isophorone diamine (IPDA) to produce epoxy thermosetting polymers with up to 42 wt% GOL content, leading to improvements in the glass-transition temperature (T_g) and stiffness of the material [74]. Gioia et al. (2018, 2020) extracted spruce and eucalyptus wood Kraft lignin by sequential solvent extraction to afford fractions of lower dispersity and increasing molecular weight [75, 76]. The fractions were glycidylated by reaction with epichlorohydrin and then cured with a flexible polyether diamine. The different lignin fractions resulted in epoxy polymers with different thermomechanical properties. Gioia et al. (2020) selected lignin fractions of comparable molecular weights to investigate the effect of differences in molecular structure on the thermomechanical properties of the final thermoset polymers [76].

12.2.2.3 Modification of Technical Lignins Prior to Epoxidation

Chemical functionalization of lignins to improve their solubility, hydroxyl content, and reactivity has been used in the preparation of lignin-based epoxy resins [4, 24, 31, 35]. Recently, Engelmann and Ganster (2016) reviewed a range of these functionalization methods [24]. Phenolation is the most well-known of these modifications for use in epoxy resin preparation and involves introducing new phenolic moieties, such as phenol or bisphenol A, into the lignin structure, usually at the α-positions of aliphatic side chains [4, 24, 77]. Other modifications include hydroxymethylation, oxyalkylation, demethylation, and oxidation with ozone [4, 24, 31, 77], although there have been few significant advances in recent years.

Depolymerization has also been used to improve the reactivity of technical lignins for use in epoxy resins. Ferdosian et al. (2015, 2016) studied the preparation and properties of lignin-derived epoxy thermoset polymers made from depolymerized Kraft, organosolv, and hydrolysis lignins [78–80]. The organosolv lignin was depolymerized using a Ru/C catalyzed hydrogenolysis process, while the Kraft and hydrolysis lignins were depolymerized by undisclosed proprietary processes. The depolymerized lignins were glycidylated with epichlorohydrin, blended with BADGE, and cured with an aromatic diamine to produce the epoxy thermoset polymers. Glass fiber-reinforced composites that were prepared using these lignin-derived epoxy resins showed tensile and flexural properties equivalent, or superior, to reference composites prepared from 100% BADGE-based epoxy resins [80].

12.2.2.4 Use of Depolymerized Native Lignins Produced by Lignin-First Biorefining/RCF Approaches

Lignin-first/RCF processing has been used to produce highly reactive lignin-based polyols as unique building blocks for epoxy thermoset polymers. van de Pas and Torr (2017) used Pd/C-catalyzed hydrogenolysis of native lignin in situ in *Pinus radiata* wood to produce a lignin oil containing mainly monomers, dimers, and oligomers of dihydroconiferyl alcohol and propyl guaiacol [81]. The whole LHO and an oligomer-rich fraction of the oil were glycidylated with epichlorohydrin, blended with BADGE or GDGE, and cured with diethylenetriamine (DETA) and IPDA to prepare high-performance lignin-based epoxy polymers. Using a combination of the epoxidized LHO and GDGE in a ratio of 2 : 1, 100% of the BADGE was replaced in the formulation, and an epoxy polymer was produced with superior flexural properties to the BADGE control resin [81]. Feghali et al. (2020) expanded on this research to: (i) show differences in mechanical properties of epoxy resins derived from LHOs from hardwood and softwood native lignins [82] and (ii) demonstrate this lignin-first biorefining concept at pilot scale for producing high-performance epoxy thermoset polymers from native lignin (Figure 12.4) [67]. This research also investigated process improvements aimed at making the overall process more scalable [67]. In related work, van Aelst et al. (2021) have recently reported using RCF to produce lignin oligomers as a replacement for bisphenol A in bio-based epoxy resins [83].

Kaiho et al. (2016) synthesized epoxy resins from depolymerized native lignin in situ in Eucalyptus wood [84]. An acid-catalyzed process with methanol stabilization was used to depolymerize the lignin to monomeric C_2-1,3-dioxolane phenols (i.e. C_2–acetals) in 33 wt% yield, based on the starting lignin. These products were then reacted to give flexible (by transacetalization) and rigid (by annulation)

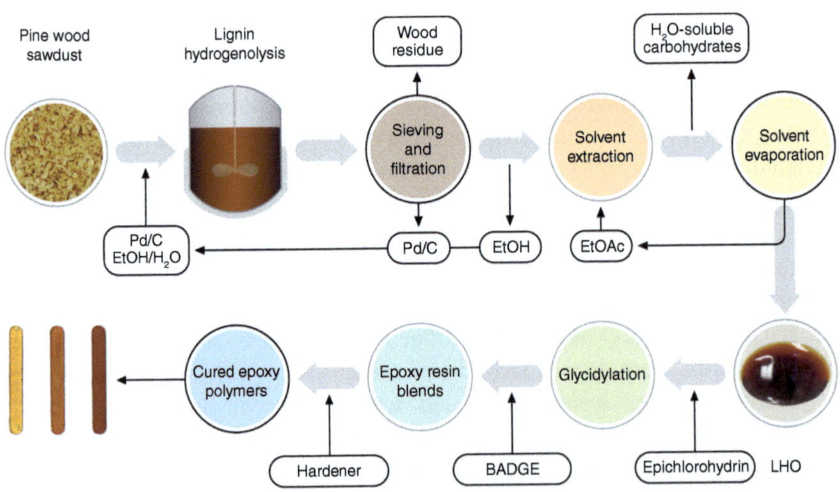

Figure 12.4 Schematic representation of a lignin-first approach to producing lignin-based epoxy thermosets from native pine wood lignin. Source: Reproduced from Ref. [67] with permission from the American Chemical Society.

difunctional intermediates that were subsequently glycidylated with epichlorohydrin. The resins were then cured with a phenol novolac curing agent. The epoxy thermoset based on the rigid annulated intermediate had a $T_g = 134\,°C$, which was superior to a BADGE-based resin ($T_g = 95\,°C$) [84].

Parsell et al. (2015) reported depolymerizing native lignin in situ in woody biomass using a bimetallic Zn/Pd/C catalyst to produce 4-propyl guaiacol (dihydroeugenol) and 4-propyl syringol (2,6-dimethoxy-4-propylphenol), in different proportions, in up to a combined yield of 54 wt%, depending on the percentage of G-type and S-type lignin in the wood [17]. Zhao and Abu-Omar (2016, 2017) subsequently used dihydroeugenol as a building block for epoxy thermoset polymers [85, 86]. In their 2016 paper, they report modification of dihydroeugenol through *o*-demethylation and reaction with formaldehyde, prior to glycidylation with epichlorohydrin. A range of dihydroeugenol-based epoxy networks is then produced by curing with DETA. The epoxy polymers with higher crosslinking densities were found to have superior thermal and mechanical properties [85]. In an extension of this work, Zhao and Abu-Omar (2017) combined dihydroeugenol with an organosolv lignin and used a sequence of modifications including demethylation, phenolation, and reaction with formaldehyde to produce a lignin-incorporated novolac polyphenol. This novolac polyphenol was glycidylated and then cured with DETA to produce a lignin-derived novolac epoxy polymer with improved thermal and mechanical performance, compared to the conventional route of blending an epoxidized lignin with comonomers [86].

These examples demonstrated how using lignin-first biorefining approaches can produce lignin-derived polyols that are more homogenous and less condensed than technical lignins, with lower molecular weights, narrower dispersity, and high monomer contents. These lignin-derived polyols are more structurally and functionally equivalent to petrochemically derived polyols and hence more amenable to further upgrading into high-performance thermoset polymers.

12.2.3 Phenol-Formaldehyde Resins

Phenol-formaldehyde resins are used in a variety of applications including as wood adhesives, as a binder in mineral fiber insulation, and in the manufacturing of laminates and moldings. They find wide applications because they have excellent thermal stability, mechanical strength, and moldability. The global phenolic resin market is estimated to be worth US$ 15 billion by 2024 [35]. The major application of phenol-formaldehyde resins is as adhesives for wood products, specifically plywood, particle board, medium-density fiberboard, oriented strand board, and laminates [31, 87]. As such, this section mainly covers aspects of research on phenol-formaldehyde wood adhesives.

Phenol-formaldehyde resins are prepared by condensing phenol or substituted phenols with formaldehyde. The two general types of phenol-formaldehyde resins are resoles (formed under alkaline conditions with excess formaldehyde) and novolacs (formed under acidic conditions with excess phenol). Lignin can be used to substitute the phenolic component of phenol-formaldehyde resins due to its phenolic structure [31, 35, 87–91]. There are many examples of lignin being used in resole resins, however, there has been little reported on the use of lignin in novolac resins [87, 88]. Recently in 2020, the Sumitomo Bakelite company developed novolac resins using lignin [92, 93].

The phenolic component in resole resins has been difficult to substitute with technical lignins such as Kraft, lignosulfonate, hydrolysis, and organosolv lignins because of their lower reactivity toward formaldehyde [31, 35, 88, 89]. Chemical modification has been used extensively to make lignins more reactive and has included methylolation (hydroxymethylation), phenolation, and demethylation [31, 35, 89, 90]. Despite using these chemical approaches, the corresponding lignin derivatives have only been used as a partial phenol replacement. These chemical approaches also come at an additional cost and suffer from the use of toxic chemicals.

Complete replacement of phenol has been achieved recently using enzymatic hydrolysis lignin from corn stover [94]. The lignin-formaldehyde resin was used for making plywood. The mechanical strength of the plywood panels was similar to those prepared using phenol resorcinol formaldehyde resin, even when cured at the same standard conditions used commercially. These were surprising results considering the enzymatic hydrolysis lignin was used without chemical modification. The high reactivity of the lignin was attributed to high levels of hydroxyl-bearing *p*-hydroxyphenyl phenolic units compared to less reactive guaiacyl and syringyl phenolic units. In a follow-up study, it was confirmed that lignins high in *p*-hydroxyphenyl units, such as from annual crops, were preferable as phenol replacements than those from softwood or hardwood sources [95].

Lignin can be depolymerized to increase the amount of reactive phenolic groups, rendering them more reactive toward formaldehyde [88]. Depolymerized technical lignins have been used as phenol replacements in wood adhesives, although only as partial replacements at levels up to 75% [96–99]. The individual phenols that constitute such depolymerized lignin oils can have either positive (e.g. guaiacol) or negative (e.g. syringol) impacts on phenol-formaldehyde resin performance, and therefore lignin type (e.g. softwood vs. hardwood) is important when considering feedstock source [100]. To our knowledge, there has been no research reported to date on the use of depolymerized lignins from lignin-first biorefining processes as phenol replacements in phenol-formaldehyde resins.

In a new strategy, Chen et al. (2019) partially replaced phenol with nanolignin particles and compared the adhesive properties to an unsubstituted control resin and to resins containing the unmodified alkali lignin that was used to prepare the nanolignin [101]. At 40% phenol substitution, the nanolignin-containing resin gave bond strengths that were superior to plywood specifications and formaldehyde emissions that met the stringent E0 grade ($\leq 0.5\,mg\,l^{-1}$ formaldehyde emission limit) for indoor applications. The nanolignin resins had much lower curing temperatures and improved thermal stability compared to the equivalent resins that incorporated the original alkali lignin. This study showed that the supramolecular assembly of lignin has a substantial influence on the performance of lignin-containing phenol-formaldehyde resins.

There is considerable interest in producing phenolic resins without using toxic formaldehyde. Common replacements for formaldehyde are glyoxal, polyethyleneimine, and bio-based reactants such as furfural [31, 88–90]. A 100% replacement of formaldehyde and phenol has been achieved using glyoxal and enzymatic hydrolysis lignin from corn stover [102]. The resin had satisfactory properties for application as an adhesive in interior-grade wood panels.

Lignin has also been used as a replacement for phenol in other applications besides wood adhesive applications such as in foams [103] and mineral wool insulation [104]. Li et al. (2017) have reported on the replacement of up to 50% of phenol in phenol-formaldehyde foams using depolymerized hydrolysis lignin [105]. Foams prepared at a phenol substitution level of 30% had comparable mechanical and thermal insulation properties compared to the phenol-formaldehyde foam, and foams prepared at a 50% substitution level had improved mechanical properties. Gao et al. (2021) have reported on the partial replacement of phenol in phenol-formaldehyde foams with SL treated with HBr under microwave irradiation [106]. This treatment in the presence of phenol led to the simultaneous depolymerization, demethylation, and phenolation of the lignin. Foams prepared using this treated lignin had better mechanical and thermal insulation properties compared to foams prepared using untreated lignin.

12.2.4 Other Polymers

This section briefly covers several polymer classes that have not been covered in the preceding sections. There are many examples that utilize technical lignins, and some of these utilize other bio-based reactants to give thermosets with a high bio-based content. A summary of selected recent examples is provided in Table 12.1.

Zhang et al. (2018) have reported on vitrimer materials from ozone-treated Kraft lignin [112]. The ozone treatment imparted carboxylic acid functionality to the lignin. These carboxylic groups along with the phenolic groups in the lignin were reacted with the diglycidyl ester of sebacic acid, a bio-based diacid, in the presence of a zinc catalyst. Increasing the content of ozonated lignin by a stoichiometric amount 1.5 times greater than the epoxy constituent resulted in increases in the tensile modulus and strength, up to 355 and 12.9 MPa, respectively. Transesterification exchange reactions occurred readily in the presence of the zinc catalyst at temperatures above 160 °C. The bio-based vitrimer exhibited excellent shape memory and repairing abilities at these elevated temperatures. It also showed promise as a recoverable adhesive, having good lap-shear strengths before and after a debond–rebond cycle of 6.3 and 5.0 MPa, respectively.

Di Francesco et al. (2022) have reported on novel polyadipate thermosets from Kraft lignin using an efficient two-step procedure [108]. The lignin was blended with an adipic acid-derived crosslinker that was prepared by reacting isopropenyl acetate and adipic acid in the presence of an acid catalyst. Partial esterification occurred when the blended heterogeneous mixture was heated briefly at 100 °C, resulting in a homogenous liquid prepolymer that could easily be poured into

Table 12.1 Minor classes of thermosetting polymers that incorporate technical lignins.

Polymer type	Lignin used[a]	Summary	References
Acrylate	Kraft (~21 wt%)	Lignin was methacrylated and incorporated into UV-curable formulations. These were cured to give coatings containing up to 31% methacrylated lignin.	[107]
Adipate	Kraft (61 wt%)	Partial esterification of the lignin with an adipic acid-derived crosslinker resulted in a pourable prepolymer that could be transformed into a thermoset with further heating.	[108]
Ester	Kraft (40 wt%)	Thermosets were prepared by crosslinking lignin and PEG with citric acid. The thermomechanical properties could be tailored by varying the lignin content between 20% and 40%. The thermosets also exhibited shape memory behavior.	[109]
Imine	Enzymatic hydrolysis (~56 wt%)	Esterification of the lignin with levulinic acid and subsequent crosslinking via the Schiff base reaction with diamines yielded thermosets with excellent properties.	[110]
Thiol-Ene	Kraft (~66 wt%)	The ethanol-soluble fraction of the lignin was allylated and subsequently crosslinked using thiol-ene chemistry. The resulting thermosets had glass-transition temperatures between 45 and 65 °C.	[111]
Vitrimer	Kraft (67 wt%)	Ozone-treated lignin was cured with epoxy-functionalized sebacic acid to give bio-based vitrimers that had excellent shape-memory, and shape changing and self-healing ability.	[112]

[a] Maximum lignin content used in parentheses. In some cases, lignin content was approximated (indicated by ~) based on factors including stoichiometry, conversion, and amounts and yields of lignin derivatives.

molds. Further heating of the material resulted in a crosslinked thermoset having a tensile modulus and strength of 1.4 GPa and 15 MPa, respectively. Model compounds were used to show that both the aliphatic and phenolic OH groups in the lignin participate in the esterification reactions.

Huang et al. (2019) have reported on polyimine thermosets from enzymatic hydrolysis lignin and levulinic acid [110]. The lignin was esterified with levulinic acid, a bio-based keto acid, in the presence of catalysts that were able to facilitate esterification of both the aliphatic and phenolic OH groups in the lignin. The active keto groups on the resulting lignin derivative were crosslinked via the Schiff base reaction with two diamines, hexamethylenediamine, and 1,12-diaminododecane, to give thermosets with excellent properties. The T_g values and tensile moduli and strengths of these thermosets were in the range 191–209 °C, 1.9–2.8 GPa, and 60.2–79.1 MPa, respectively. The thermosets also exhibited excellent UV resistance.

In contrast to using technical lignins, we have found only two recent examples of where depolymerized native lignins from lignin-first biorefining processes have been used to prepare thermoset polymers, other than polyurethanes or epoxy resins. Jedrzejczyk et al. (2022) have reported on anticorrosive protective films using a depolymerized lignin oil from a lignin-first process [113]. Birch wood was depolymerized using a nickel-catalyzed RCF process to give an oil containing lignin–derived monomers and oligomers. This oil along with fractions containing different proportions of monomers was used to produce films using thiol-yne chemistry. These films had excellent adhesive and barrier properties and excellent solvent resistance. Films prepared from the whole lignin oil performed just as well as the oil fractions. Bertella et al. (2022) have reported on multifunctional lignin from the AAF process and its incorporation in gelatin-based hydrogels for soft tissue engineering applications [114]. The AAF process was used to simultaneously extract and functionalize the lignin with two multifunctional aldehydes, glyoxylic acid, and terephthalic aldehyde. The resulting carboxyl functionality on the lignin imparted hydrophilic properties while the aldehyde functionality enabled crosslinking with gelatin. Hydrogels that contained lignin functionalized with equimolar amounts of the two multifunctional aldehydes were found to have superior mechanical and self-healing properties compared to hydrogels made from monofunctionalized lignin.

12.3 Lignin-Based Thermoplastics

There has been less research on lignin-based thermoplastic polymeric materials than thermosetting materials, with only a few recent reviews of the topic [27, 33, 34]. This is mainly due to the heterogeneous, complex branched structure of available lignins that, in many cases, have a functionality higher than two. This makes most technical lignins challenging as building blocks for linear thermoplastic polymers. Challenges such as high glass-transition temperatures (T_g) and poor flow properties hinder the melt and thermal processability of many technical lignins [27, 33]. The lack of thermoplasticity of technical lignins is due to crosslinking reactions that can occur during its isolation from lignocellulosic biomass and/or during subsequent thermal processing [33]. On the other hand, considering the extensive range of applications for thermoplastic polymers, both in industry and in our everyday lives, interest in using lignin for producing sustainable thermoplastic polymers is growing. Many publications featuring the keywords "lignin" and "thermoplastic" report on the use of lignins as fillers or blends in composite materials, rather than using the lignin as an integral building block for polymeric material, which is the topic of interest here.

More recently there has been increasing attention in using lignin as a macroinitiator core, for the graft polymerization, of various polymers as multi-armed side chains to form thermoplastic copolymer materials. Commonly, this is achieved via a required chemical modification of the lignin to improve the thermoplasticity, homogeneity, suitability, lower the hydroxyl group functionality, and allow for more predictable reactivity during polymer synthesis. These modifications can include strategies such as alkylation, acetylation, acrylation, esterification, and solvent fractionation [33]. This graft copolymerization is also a form of chemical modification of the lignin and the resulting thermoplastic copolymeric material may demonstrate improved miscibility with other polymers, that were previously incompatible with lignin, to form new composite materials with improved thermomechanical properties.

12.3.1 Graft Copolymers and Scope

Grafting enables the chemical incorporation of lignin with different types of polymers to form a random copolymeric material. It overcomes any immiscibility and incompatibility issues that can arise with the physical blending of two or more polymers. There are two grafting methods:

1) A "grafting from" approach where lignin behaves as a macroinitiator, via the hydroxyl groups, and monomer is directly polymerized from a lignin core as side chains (Figure 12.5). Ring-opening, free-radical, and controlled radical polymerizations have been used to make lignin-based thermoplastic polymers [115–120].
2) A "grafting onto" or "grafting to" approach where a polymer chain is separately synthesized and contains a functionalized end group that can react with the hydroxyl groups of the lignin to attach the polymer directly to the lignin core via covalent bond formation (Figure 12.5). Lignin-based thermoplastics have been produced by this approach using click chemistry (Huisgen cycloaddition) and boron-mediated or diisocyanate coupling [115, 121–125].

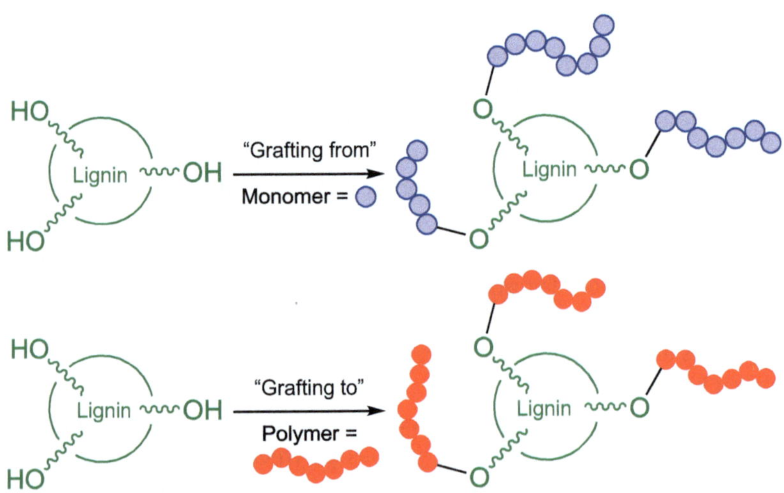

Figure 12.5 The two methods for graft polymerization from a lignin core: "grafting from" or "grafting to." Source: Oliver Driscoll.

Recently, since 2015, there has been significant and growing traction in two particular methods of graft polymerization using lignin as a core to graft from: (i) Ring-opening polymerization (ROP) using renewable cyclic esters such as lactide (LA) and ε-caprolactone (CL), and (ii) Controlled radical polymerization using vinylic monomers such as acrylates. In addition to this, a newer class of lignin copolymer, thermoplastic polyurethanes (TPUs), has been explored where diisocyanate substrates have been grafted to lignin. Therefore, our aim in the next section is to highlight some of the interesting, contemporary developments that have been made in these three significant areas of graft polymerization for their production of novel lignin-based thermoplastic polymers using both technical and lignin-first biorefining/RCF lignins. For further in-depth and broader reading, including other methods of polymerization, the reader is referred to the comprehensive reviews by Jiang, Venditti, and Bujanovic [27, 33, 34].

12.3.2 Lignin-Copolymer Thermoplastics Using Ring-Opening Polymerization

Poly(lactic acid) (PLA) and poly(ε–caprolactone) (PCL) are renewable, biocompatible, and biodegradable bioplastics [126–129]. These thermoplastic polyesters are commonly synthesized via ROP, a class of controlled living, chain–growth polymerization (Figure 12.6) [129]. They have received significant attention, both academically and industrially, but they suffer from unsuitable thermal properties, such as low T_g values, and poor mechanical properties such as brittleness [126, 129–132, 136, 137]. Therefore, recently there have been attempts to combine lignin

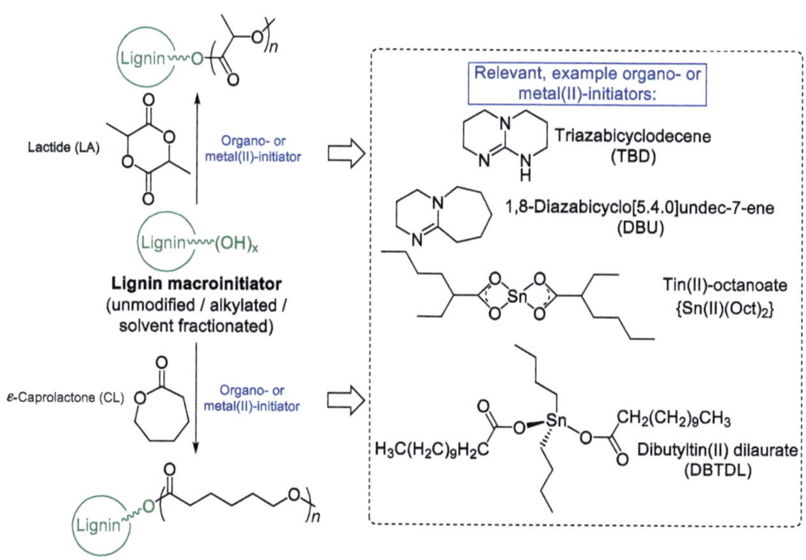

Figure 12.6 The general approach for the ring-opening polymerization of lactide and ε-caprolactone using lignin macroinitiator and organo-/metal-initiator; with example initiators relevant to this discussion shown. Source: Oliver Driscoll based on Ref. [130–135].

with LA and CL to make bio-based lignin–PLA and lignin–PCL copolymers with enhanced properties. These cyclic esters can be ring opened, polymerized, and subsequently grafted from the hydroxy groups or reactive sites of the lignin (Figure 12.6), generally with the aid of an external initiator, to form thermoplastic polyester materials (Table 12.2). Earlier methods have been attempted using tin(II)-octanoate, which is industrially used as the initiator to produce commercial poly(*L*-lactic acid) (PLLA) (Figure 12.6) [126]. However, alternative initiators are being investigated due to concerns over the toxicity of tin species [126, 128, 133]. The area of organo-initiation, using organic molecules (Figure 12.6), is very much in its infancy but has recently gained increasing attention as a metal- and halide-free approach to polymerization [138].

Ren et al. (2016) sought to develop a "greener" approach where lignin was added as a "toughening agent" to PLA to make copolymeric graft materials that were suitable for more durable applications [130]. Earlier reports suggested this was possible [131, 136, 137, 139]. Zong et al. (2016) had just developed lignin-g-PLLA copolymers using a purified lignin residue from biobutanol production from wheat straw/corn (a biorefinery lignin abbreviated as "BBL") as the macroinitiator ($M_n = 543\,g\,mol^{-1}$, $Đ = 1.61$, 5.62 mmol total OH/g). Triazabicyclodecene (TBD) was used as the organo-initiator for the ROP of *L*–LA (Table 12.2) under solvent-free conditions [131]. A significant challenge in this ROP approach, is the high hydroxyl content inherent in most lignins, causing poor molecular weights for the grafted PLA chains because of the high concentration of initiating molecules. Therefore, pre-chemical modification is typically essential before polymerization to decrease the hydroxyl content [130]. To achieve this, Ren et al. (2016) used aqueous dodecylation of 100% of the phenolic hydroxyl groups and 77% of the carboxylic acid hydroxyl groups of an alkaline lignin (AL) [130]. Racemic–LA was then polymerized from the unreacted aliphatic and remaining carboxylic hydroxyl groups at different feed ratios of the dodecylated lignin using 1,8-diazabicyclo[5.4.0]undec-7-ene (DBU) organo-initiator (Table 12.2). As the percentage in weight of dodecylated lignin increased from 2% (98 wt% LA, 1.4% lignin) to 40% (60 wt% LA, 19.6% lignin), the molecular weight decreased (from a M_n of 28200 to $1382\,g\,mol^{-1}$), as would be expected. The ROP remained well-controlled throughout ($Đ = 1.23–1.41$). These synthesized lignin-g-PLA copolymers were then physically blended at a range of different feed ratios with PLA. In comparison with pure PLA, the blends displayed similar thermal properties but improvements in tensile strength, elongation, Young's modulus, and higher hydrophobic contact angles and were more effective for blocking UV light [130].

Park et al. (2019) continued the work on grafting LA from lignin using organo-initiation [132]. SL was fractioned using a sequence of five organic solvents, consisting of ethyl acetate, 2–butanone, methanol, acetone, and water/dioxane, to give six lignin fractions with increasing molecular weights and decreasing hydroxyl contents. Here solvent fractionation was used as an alternative method to chemical modification for controlling the hydroxyl content in the lignin macroinitiator. Four fractions and the original SL were grafted with *L*-LA using TBD organo-initiator to

Table 12.2 A general overview of the discussed graft copolymers synthesized via ring-opening polymerization of lactide and ε-caprolactone using lignin macroinitiator and organo-/metal-initiator.

Author	Lignin source	Comonomer	Total OH content (mmol OH/g) before and after modification or without modification			Initiator	M_n (Da)	Đ	Lignin content (wt%)
			Before	After	Unmodified				
Ren (130)	AL	*racemic*–LA	6.9	1.5		DBU	1,382–28,200	1.23–1.41	1.4–19.6
Zong (131)	BBL	*L*–LA			5.6	TBD	2,544–7,033	1.49–2.39	1–40
Park (132)	SL	*L*–LA	5.2	3.9–6.5		TBD	880–16,140	—	—
Liu (133)	BBL	CL			5.6	TBD	367–8,167	—	1–40
Kai (134)	AL	*L*–LA and CL		—		$Sn(Oct)_2$	10,400–16,400	1.25–2.41	30–48
Park (135)	Kraft	CL			10.1	DBTDL	1,480–4,360	4.3–7.7	—

produce lignin-g-PLLA copolymers (Table 12.2). The *L*-LA conversion decreased (from 91% to 88% conversion) with the decreasing phenolic hydroxyl content of the sequential fractions and the molecular weights of the lignin-g-PLLA copolymers increased (M_n = 880 to 16,140 g mol^{-1}), because of the decrease in the overall concentration of initiating molecules. These copolymers were mixed with commercial PLA up to 15 wt% to form biocomposites. No improvement in tensile strength and modulus of the lignin-containing composites was found compared to the commercial PLA, although improved UV-blocking properties were noted [132].

ROP of CL from lignin has also been explored by a number of researchers (Figure 12.6). Liu et al. (2015) used a purified BBL as the macroinitiator and TBD organo-initiator to react with CL monomer (Table 12.2) to prepare BBL-g-PCL copolymers [133]. A range of copolymers with different molecular weights (M_n = 367–8167 g mol^{-1}) were produced by adjusting the feed ratios. As would be expected, the melting temperature (T_m = 47 to 55 °C) and semi-crystallinity were enhanced when the content of grafted PCL chains increased and overall lignin content in the material decreased (BBL = 1–40 wt%). Coated films of BBL-g-PCL observed 95% UV absorbance between 280 and 400 nm and UV-blocking properties, while transparency in the visible light region was maintained giving the possibility for their potential application specifically as transparent food packaging [133].

Building on the graft polymerization of LA and CL, Kai et al. (2017) went one step further and prepared lignin-g-PCL-*co*-PLLA copolymer using solvent-free ROP of both monomers (*L*-LA and CL) with an alkali lignin with an average M_w = 5000 g mol^{-1} (Table 12.2) [134]. The objective was to target more porous lignin-derived materials for biomedical applications, such as electrospun blended fibers, with properties such as antioxidant activity, biodegradability, and biocompatibility investigated. Tunable molecular weights could be controlled (M_n = 10,400–16,400 g mol^{-1}, Đ = 1.25–2.41), and the resulting T_g shifted from 34 to −37 °C with the lignin content ranging between 30% and 48%. These copolymers were blended, via electrospinning, with either pure PCL or PLLA polyester to produce ultrafine, uniform nanofibers (PCL or PLLA to lignin-g-PCL-*co*-PLLA = 9 : 1). The incorporation of lignin-g-PCL-*co*-PLLA into PCL in a blend greatly enhanced the mechanical properties of the nanofiber mats that were produced. The opposite was true for PLLA where the addition of lignin-g-PCL-*co*-PLLA had either an insignificant or negative influence on the mechanical properties. Good antioxidant activity and in vitro biocompatibility were observed for the blended lignin-containing nanofibers [134].

Park et al. (2019) used unmodified Kraft lignin (M_n = 970 g mol^{-1}, Đ = 1.57) as a macroinitiator for the limited ROP with CL using dibutyltin(II) dilaurate (DBTDL) as a co-initiator in the absence of solvent at 150 °C to form thermoplastic lignin-g-PCL polyesters (Table 12.2) [135]. Depending on the reaction conditions, the degree of polymerization of PCL on each lignin OH site ranged from 7 to 26, corresponding to 4780 and 32,600 g mol^{-1} of PCL chains per lignin macromolecule [135].

Although the aforementioned lignin-based ROP research is noteworthy, it does highlight some key challenges in produced lignin-based thermoplastic polymers employing a "grafting from" approach using a technical lignin as a macroinitiator. These include achieving a high lignin content in the thermoplastic polymer, achieving good control of the polymerization, and achieving suitable molecular weights for potential, future applications.

Recently, Wu et al. (2021) applied a RCF approach to native lignin in situ in hardwood in synthesizing a fully biomass-derived, thermoplastic polyester poly(ethylene terephthalate) (PET) analog [140]. PET is widely used globally and contributes significant ecological harm to the environment [127, 140]. Wu et al. (2021) used a two-step catalytic process, involving Cu_{20}-porous metal oxide catalyzed RCF of the poplar or beech wood followed by Raney-nickel mediated catalytic funneling, to produce lignin-derived 4-(3-hydroxypropyl)-cyclohexan-1-ol in 56% efficiency [140]. This diol monomer was then copolymerized via esterification with dimethyl-2,5-furandicarboxylate or dimethyl terephthalate, both of which can be sourced from the cellulosic woody residue remaining after RCF processing. Either zinc(II)-acetate or titanium(IV)-butoxide initiator was used to produce fully bio-based polyester with properties competitive with traditional PET (M_w = 19,274–50,676 g mol^{-1}, T_g = 71.4–86.9 °C compared to T_g = 72 °C for PET). The bio-based polyester could be chemically recycled via methanolysis back to the diol monomer, indicating promising potential for circular economy manufacturing [140]. This work is one of the few examples of using native lignin in a lignin-first/RCF biorefining approach to produce bio-based thermoplastic polymers from lignocellulosic biomass [140].

12.3.3 Lignin-Copolymer Thermoplastics Using Radical Polymerization

Atom transfer radical polymerization (ATRP) and reversible addition-fragmentation chain-transfer (RAFT) polymerization are powerful methods for synthesizing controlled, well-defined polymer architectures [141–147]. In particular, the former, developed by Matyjaszewski, has been finding growing success with lignin [33, 115, 148]. This is in contrast with classic free-radical polymerization using lignin where reports have been more limited [149–151] because of poor polymerization control, low reactivity, and low-molecular weights for the final polymer. Successful ATRP and RAFT are achieved by grafting and polymerizing vinylic monomers, such as acrylates and styrene, from a chemically modified lignin backbone behaving as a macroinitiator. For ATRP, this is achieved by introducing haloester functionality to the multiple hydroxyl groups in the lignin, via esterification, to act as initiation sites for the vinylic monomers.

Kim et al. (2015) built on previous work [136, 152, 153] by using ATRP to prepare lignin-*graft*-poly(styrene) (lignin-g-PS) copolymers using fully esterified Kraft lignin [154]. This Kraft lignin was fully esterified by introducing bromoisobutyryl moieties onto all the aliphatic and phenolic hydroxyl groups of the lignin (Figure 12.7). These functional groups acted as macroinitiators where styrene could be grafted from the lignin core. In DMF:water (3 : 2 v/v), the M_n of the PS side chains could be increased to a maximum of 206,000 g mol^{-1} via variation of the

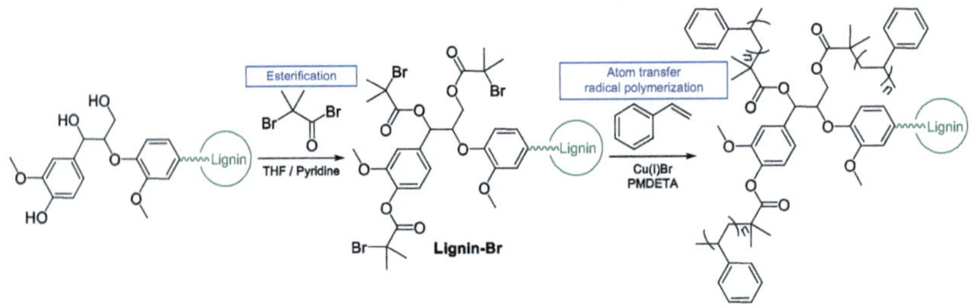

Figure 12.7 Kim's general reaction sequence for the preparation of the Kraft lignin-bromoester macroinitiators and lignin-*graft*-poly(styrene) copolymer. Source: Oliver Driscoll based on Ref. [154].

concentration of styrene monomer. However, the polymerization became less controlled as both the conversion and molecular weight increased. The thermal properties were enhanced at these high molecular weights as the copolymer became more comparable to pure PS homopolymer [154].

As alternatives to styrene, Yu et al. (2015) used methyl methacrylate (MMA) and butyl acrylate (BA) as monomers in the synthesis of multi-arm star, random lignin-*graft*-poly(methyl methacrylate-*co*-butyl acrylate) copolymers (lignin-g-poly(MMA-*co*-BA)) to be used as possible sustainable thermoplastic elastomers (Figure 12.8) [155]. Alkali lignin was modified with 2-bromoisobutyryl bromide to give a lignin-Br macroinitiator. The amount of both monomers was tuned with the overall molar ratio and reaction conditions remaining constant ([MMA + BA]:[lignin-Br]:[Cu(I)]:[PMDETA] = 1000 : 1 : 1 : 1) with a low lignin content of only 0.16–0.24 wt%. This low lignin content meant the thermal properties of the lignin-grafted copolymers were similar to that of the pure linear poly(MMA-*co*-BA). The mechanical properties were significantly improved in the lignin-grafted copolymers with higher tensile strengths, storage moduli (below the T_g), and elongation at break, and the stress at break was twice that when compared to the pure non-lignin copolymer. Despite the low lignin content, the lignin-g-poly(MMA-*co*-BA) copolymers exhibited enhanced UV-blocking properties indicating potential in UV-blocking thermoplastic elastomeric coating applications [155].

Lignin-based polymers with good UV absorption and antioxidant properties have the potential to serve as promising sunscreen agents. Kai et al. (2016) explored this potential by grafting poly(ethylene glycol) methacrylate (PEGMA), a main ingredient found in commercial sunscreens, onto an alkali lignin [156]. PEGMA was grafted onto the lignin-Br macroinitiator via ATRP using copper(I) bromide catalyst and 1,1,4,7,10,10-hexamethyltriethylenetetramine ligand (Figure 12.8) with the goal of enhancing the miscibility and dispersion efficiency of the lignin in the sunscreen. A range of lignin-g-PEGMA copolymers were synthesized by varying the lignin-Br initiator to PEGMA monomer molar ratio to tune both the molecular weight (M_n = 10,100–24,400 g mol^{-1}) and particle size (*Z*-average = 102–205 nm) of the copolymer. The lignin content in the copolymer ranged from low to moderate (mass% of lignin = 21–50%). The lignin copolymers showed good antioxidant

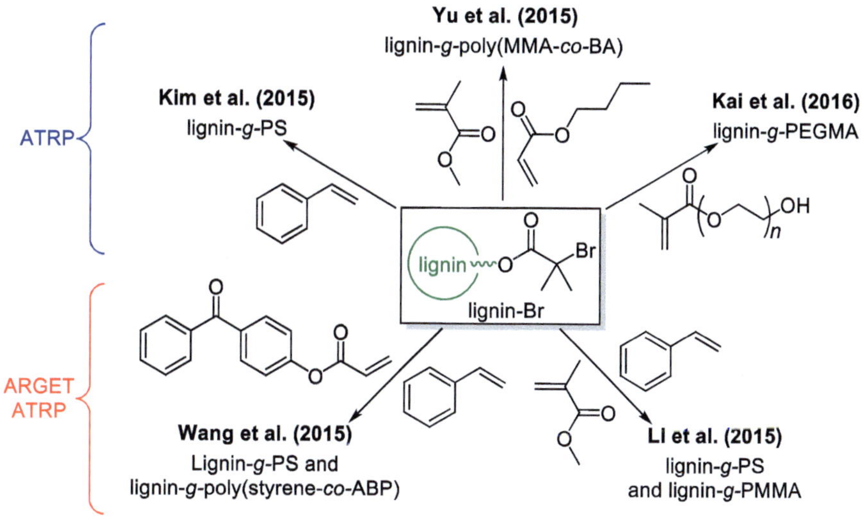

Figure 12.8 A general overview of the monomers employed with lignin macroinitiator for copolymer synthesis via atom transfer radical polymerization (ATRP) since 2015. Source: Oliver Driscoll, adapted from Ref. [148, 154–157] with permission from the Royal Society of Chemistry, Wiley and the American Chemical Society.

properties and blending the copolymers as an additive (10% with 5 wt% of lignin) with commercial sunblock cream enhanced the sun protection factor values from 15 to 38, without adversely affecting the color of the sunscreen [156].

A limitation of ATRP is that as the polymerization progresses, the rate of reaction slows because of increasing termination reactions and concentration of deactivator. Often, high amounts of catalyst are employed to counter this drawback [33, 148, 158]; however, removal of this catalyst is challenging making purification expensive [158]. Recently, an emerging method has been developed that can reduce the amount of catalyst required by regenerating the activator via electron transfer [148, 157–159]. This activator regenerated by electron transfer (ARGET) ATRP approach was employed by Wang and Venditti (2015) to produce UV-curable lignin graft copolymers from a lignin-Br macroinitiator (Figure 12.8) [157]. Purified hardwood Kraft lignin was first modified by bromoisobutyryl esterification and then used as a macroinitiator to synthesize lignin-g-PS and lignin-g-poly(styrene-*co*-acryloyl benzophenone) copolymers using a constant ratio of the lignin-Br macroinitiator, copper(II) bromide catalyst, 2,2′-bipyridine ligand and ascorbic acid of 1 : 0.1 : 0.1 : 0.1 (Figure 12.8). Ascorbic acid is required as a non-radical-forming reducing agent to regenerate the Cu(I) to Cu(II) for ARGET. Acryloyl benzophenone is a photosensitive compound and could be integrated as a comonomer, by tuning the feed ratios, to result in 6–9 mol% in the lignin-g-poly(styrene-*co*-acryloyl benzophenone) (47). Relatively low conversions (~20%) and molecular weights were observed (M_n = 6030–11,500 g mol^{-1}, Đ = 2.2–3.4) for all the resulting copolymers. The UV curing properties of the lignin-g-poly(styrene-*co*-acryloyl benzophenone) were demonstrated with UV irradiation resulting in the formation of an insoluble, gelled copolymer network with up to 30 wt% of the copolymer crosslinked, compared to <5 wt% crosslinking observed for the corresponding lignin-g-PS copolymer [157].

Li et al. (2015) also applied this ARGET ATRP approach, with the lignin-Br macroinitiator and ascorbic acid reducing agent, but instead of copper(II) bromide they used a more environmentally friendly iron(III) chloride catalyst ($FeCl_3{\cdot}6H_2O$) and triphenyl phosphine (PPh_3) as the ligand (Figure 12.8) [148]. Controlled graft polymerization was achieved using both styrene and MMA monomer at a constant molar ratio of [monomer]:[lignin-Br]:[$FeCl_3{\cdot}6H_2O$]:[PPh_3]:[ascorbic acid] = 200 : 1 : 1 : 3 : 0.5 in DMF for 24 hours. Both lignin-*graft*-poly(styrene) (lignin-g-PS) and lignin-*graft*-poly(methyl methacrylate) (lignin-g-PMMA) were obtained in higher molecular weights and reduced dispersities compared to the synthesized lignin-Br macroinitiator. However, low contents of lignin were present in the copolymers (9.5% in the lignin-g-PS and 6.2% in the lignin-g-PMMA by mass) [148].

In addition to ATRP polymerization, there have been interesting recent advances in using RAFT polymerization to synthesize lignin-containing thermoplastic polymers. Qu et al. (2019) employed RAFT polymerization in the direct synthesis of methacrylate polymers from pyrolytic lignin (PL) produced from pyrolysis of red oak wood [160]. The PL was first modified by full and partial esterification, using different amounts of methacryloyl chloride and acetyl chloride, to form PL-methacrylates. The RAFT polymerization was carried out using 2,2′-azobis (isobutyronitrile) (AIBN) initiator and 2-cyanopropan-2-yl methylcarbonotrithioate chain-transfer agent in 1,4-dioxane solvent. The fully esterified PL methacrylate experienced "gelation" to form an undesirable, extensively cross-linked polymer without a detectable T_g value. In contrast, partial esterification suppressed the "gel point" and a linear thermoplastic polymer was isolated in a low yield (19%) with a T_g of 161 °C, reasonable molecular weight (M_n = 7276 g mol^{-1}), narrow dispersity (Đ = 1.55) and T_d of 241 °C. It was believed that the yield and molecular weight were lowered due to radical inhibition by the residual, unmodified lignin phenol hydroxyl groups. In response, partial methacrylation followed by acetylation successfully eliminated these inhibition effects and increased the reactivity (yield = 43%, M_n = 10,420 g mol^{-1}) while dispersity remained narrow (Đ = 1.53). The potential of this lignin-based methacrylate polymer as a high-quality carbon fiber precursor was examined [160].

The process of producing pyrolytic lignin by fast pyrolysis severely modifies the lignin structure and this can limit its utility as a starting material for novel thermoplastic polymers. An alternative way of producing low molecular weight lignin-derived phenols is to use a lignin-first biorefining or RCF approach. Wang et al. (2018) depolymerized native lignin by RCF to produce a source of aromatic phenols for the development of high-performance, thermoplastic pressure-sensitive adhesives [161]. Hydrogenolysis of native lignin in situ in popular wood was achieved using a commercial Ru/C catalyst in methanol at 50 bar of H_2 at 250 °C for 15 hours (Figure 12.9). After filtration and cyclohexane extraction, 4-propylsyringol (4PS) and 4-propylguaiacol (4PG) were isolated in high yield. These lignin-derived phenols were functionalized with either acrylate (4PSA or 4PGA) or methacrylate

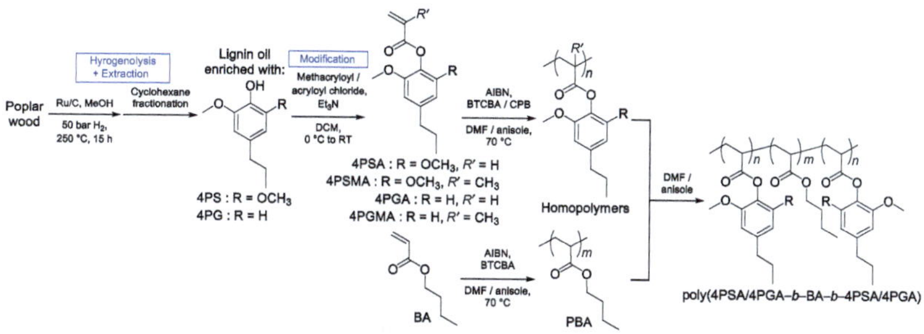

Figure 12.9 Wang's preparation of homopolymer and triblock copolymer derived from depolymerized, extracted, and modified lignin oil, sourced directly from poplar wood, using reversible addition-fragmentation chain-transfer (RAFT) polymerization. Source: Oliver Driscoll, adapted from Ref. [161] with permission from the American Chemical Society.

(4PSMA or 4PGMA) groups, using methacryloyl/acryloyl chloride and triethylamine, to produce monomers for RAFT polymerization (Figure 12.9) [161].

The RAFT polymerization was carried out with AIBN initiator, either 3,5-*bis*(2-dodecyl-thiocarbonothioylthio-1-oxopropoxy)benzoic acid (BTCBA) or 2-cyano-2-propyl benzodithioate (CPB) chain-transfer agent in anisole/DMF solvent (Figure 12.9) [161]. A well-controlled polymerization was demonstrated by the moderate to high molecular weights and low dispersities ($M_n = 12{,}400–30{,}400\,g\,mol^{-1}$, Đ = 1.16–1.44). Triblock polymers (poly(4PSA/4PGA-*b*-BA-*b*-4PSA/4PGA)) were designed and prepared, composed of poly(4PSA) or poly(4PGA) glassy end blocks and poly(*n*-butyl acrylate) (PBA) rubbery middle blocks, and were analogous of commercial block polymers composed of poly(methyl methacrylate) and poly(*n*-butyl acrylate). These block polymers were achieved via a two-step RAFT polymerization; the synthesis of the PBA and then chain extension using poly(4PSA)/poly(4PGA) end blocks (Figure 12.9). The lignin-derived triblock copolymers demonstrated significant potential for use as pressure-sensitive adhesives. The 180° peel forces were up to $4\,N\,cm^{-1}$ and tack forces were up to $2.5\,N\,cm^{-1}$, which were comparable with commercial adhesive tapes [161].

12.3.4 Lignin Thermoplastic Polyurethanes

In spite of the immense range of commercial applications for PUs; such as for foams, adhesives, coatings, and sealants, and their use in our everyday life, there has been a limited number of reports for lignin-based TPUs. Instead, significant research has been focused on lignin-derived thermoset PUs, as previously described. Despite high tensile strength and moduli, lignin-based thermoset PUs can suffer from brittleness with this problem escalated with an increasing lignin content. The thermally irreversible crosslinking also causes issues with material processing and end-of-life recycling. Therefore, the development of lignin-based TPUs can help to alleviate these issues. Initial, interesting progress has commenced in this contemporary area by employing the "grafting onto" approach, where technical lignins are used as a polyol source with the hydroxyl groups coupling to diisocyanate to form urethane linkages [123, 162, 163].

Jeong et al. (2013) prepared TPUs from partially acetylated (90%) softwood Kraft lignin and PEG reacted with 4,4-methylene diphenyl diisocyanate to produce a polymer composed of hard and soft segments [27, 164]. Controlled acetylation is required to decrease the high number of reactive hydroxyl groups in the lignin, reduce crosslinking, and enhance the thermoplasticity of the material. The PU properties could be tuned by the lignin content where high viscoplasticity was demonstrated when the lignin content was lower than 40 wt% and high viscoelasticity when the lignin content was more than 40 wt% [164]. These acetylated softwood Kraft lignin TPU materials exhibited mechanical properties comparable to other more conventional TPUs [165]. More recently the same research group used short jute fibers, modified by low temperature pyrolysis, for reinforcement of these acetylated lignin-based TPUs with the results indicating potential for their use as bio-based elastomers [165].

Gouveia et al. (2020) recently built on the work of Jeong et al. (2013) with the goal of developing a feasible commercial application for lignin-based TPUs [166]. Industrial-grade Kraft lignin was acetylated resulting in a reduction of 50% of the hydroxyl groups. This acetylated Kraft lignin and PEG were used as polyols, and 4,4-methylene diphenyl diisocyanate as the isocyanate in the synthesis of lignin TPU films. The effect of lignin content (31–50%), the molecular weight of the PEG (1000 or 1500 $g\,mol^{-1}$), and the NCO:OH ratio were studied. Increasing the lignin content improved the thermal and mechanical properties of the lignin TPUs. The lignin TPUs were then successfully applied as hot-melt adhesives for wood and evaluated by single-lap shear tests. Increasing lignin content to 50% increased average shear strength to around 5 MPa indicating potential for further development for industrial adhesive applications [166].

Recently, Quinsaat et al. (2022) used a lignin-first biorefining/RCF approach to producing native lignin-derived polyols for the preparation of robust lignin-derived TPU films [167]. The low molecular weight lignin polyols, termed LHO ($M_w = 395\,g\,mol^{-1}$, $Đ = 1.4$), were sourced using mild metal-catalyzed depolymerization of native lignin in situ *Pinus radiata* wood [67]. This LHO consists of monomeric and oligomeric fractions with low hydroxyl functionality ($f = 2.2$) relative to other bio-based polyols. The synthesis was conducted using a successive two-step polymerization process (Figure 12.10). First, poly(propylene glycol) (PPG) was selected as a co-polyol (M_n = 1000, 2000, and 4000 $g\,mol^{-1}$), to provide the needed flexible soft segments in the final material product, and this was reacted with 4,4-methylene diphenyl diisocyanate to form isocyanate end-capped PPG. Second, the LHO was then added to this prepolymer as a hard segment, crosslinking agent to produce randomly branched TPUs [167]. The isocyanate index was held at a low value of 0.5 (NCO:OH ratio = 1 : 1) to ensure a thermoplastic material was produced. The amounts of both the hard and soft segments were varied (LHO: 50 or 60 wt% and PPG: M_n = 1000, 2000, and 4000 $g\,mol^{-1}$), and the resulting polymers displayed low to respectable molecular weights (M_w = 4400–21,900 $g\,mol^{-1}$). At both higher lignin content and PPG molecular weight, the polymerization was less

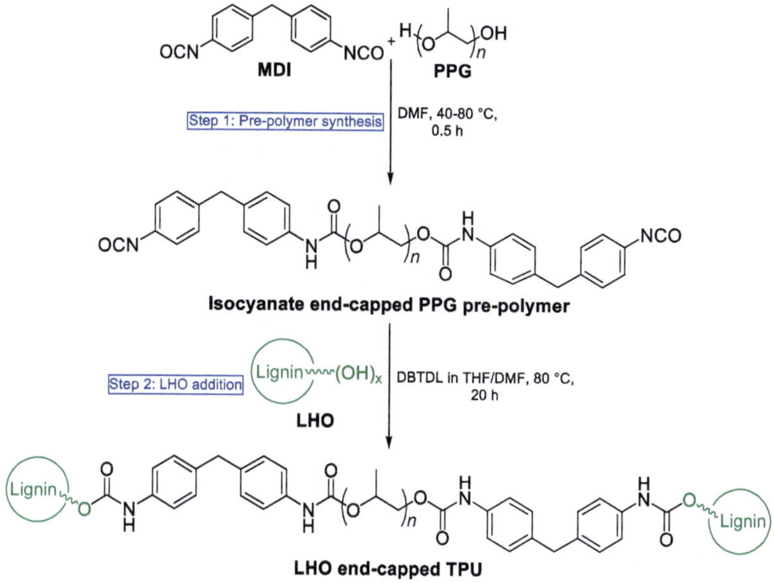

Figure 12.10 Quinsaat's two-step reaction sequence for the preparation of thermoplastic polyurethane (TPU) from lignin hydrogenolysis oil (LHO), 4,4-methylene diphenyl diisocyanate (MDI) and poly(propylene glycol) (PPG). Source: Oliver Driscoll, adapted from Ref. [167] with permission from the Elsevier.

controlled (Đ = 7.0–10.8) but the mechanical properties of the films were improved. The lignin-derived TPUs films were applied as hot-melt adhesives for wood substrates and their bonding strength was analyzed using single lap shear testing. The average lap shear strength increased upon decreasing the molecular weight of the PPG and increasing the LHO content with the best-performing adhesive having a shear strength of 6.5 MPa, which was competitive with other lignin-containing PU adhesives. The results demonstrated the potential of TPUs derived from depolymerized native lignin in future industrial adhesive applications [167].

12.4 Perspectives

Important advances have been made in the production of lignin-based thermoset and thermoplastic polymers in recent years as the research community and polymer industry seek more sustainable polymer feedstocks to traditional petrochemicals. Lignin sourced from lignocellulosic biomass is by far the most abundant natural source of renewable aromatic chemicals on the planet and as such has an important role to play in meeting our future material needs. However, valorizing lignin as a polymer feedstock has many technical and economic challenges that make the future of lignin in the emerging global bioeconomy far from certain. Traditionally, lignin has largely been seen as a byproduct of the pulp and paper industry with limited commercial value beyond its energy content, with the notable exception of a range of specialty products based on lignosulfonates and Kraft lignin. Traditional pulp and paper processing targets valorization of the cellulose and hemicellulose components of the lignocellulosic biomass ahead of the lignin. However, these industries are mature with limited growth potential and are in some cases waning. The emergence of new biorefinery operations that aim to produce a range of energy, chemical, and material products from lignocellulosic biomass could change this paradigm. These emerging biorefineries can no longer consider lignin as a low-value byproduct; instead, they must strive to extract maximum value from this resource to achieve economic competitiveness and sustainability in their operations. Lignin will be too valuable in the future as a source of aromatic chemicals and materials to be solely burnt for energy recovery. Of course, the most viable large-scale applications of lignin, in the short-term at least, will be strongly influenced by the availability and price of crude oil, clean energy/bioenergy prices, and geopolitical influences.

A key challenge of using lignin as a polymer feedstock is its complex chemical structure, which is only made more complicated and heterogeneous by the conventional methods used in isolating the lignin from lignocellulosic biomass. These approaches lead to rapid condensation reactions that create new C—C bonds in the lignin and produce a technical lignin that is more recalcitrant to downstream upgrading into chemicals and materials. Avoiding these condensation reactions and taking advantage of the inherent structure of native lignins to generate more uniform and versatile chemical feedstocks is a guiding principle behind recent developments in Lignin-first biorefining and RCF of lignocellulosic biomass. Although these approaches are still at a research and development stage, they offer an alternative way of processing lignocellulose biomass which seeks to preserve the value of the lignin as a source of low molecular weight aromatic chemicals. In this regard the future of Lignin-first biorefining is bright, notwithstanding the many technical, scaleup, and economic challenges that still need to be addressed.

Both traditional technical lignins produced through pulping processes and new Lignin-first biorefining lignins offer a source of renewable polyols that can be used to make novel high-performance bio-based thermoset and thermoplastic polymers. Technical lignins such as lignosulfonates and Kraft lignins are commercially available on kilotonne scales and are successfully used in a range of material applications such as concrete and bitumen additives, and miscellaneous dispersants and binders, whereas Lignin-first depolymerized native lignins are still very much at a developmental stage.

Promising large–scale thermoset polymer-based applications for these technical lignins include use as bio-based polyols in polyurethane foams and as a replacement of phenol in phenol-formaldehyde resins. In such cases, the modification of the lignin that occurs during its isolation, and the associated limitations that this brings to its downstream processing are less crucial and/or can be largely overcome by functionalizing or depolymerizing the lignin. Challenges still exist in commercially using technical lignins in polyurethane foams and phenol-formaldehyde resins, but extensive research and development in these areas would suggest many of these are not insurmountable given strong sustainability drivers and a competitive market price for the lignin. The caveat being that technical lignins, such as Kraft lignin, will likely be limited to partially polyol/phenol replacements (around 30-40 wt%), unless the lignin is significantly modified by further downstream processing. For many other applications, issues such as color and smell, incompatibility, low reactivity, and poor mechanical performance remain key barriers to the use of technical lignins. In contrast, Lignin-first/RCF processing can offer advantages that improve the versatility of the lignin. For example, native lignins are odorless and substantially less colored than technical lignins, and using reductive conditions to depolymerize native lignins can minimize chromophore formation reactions that readily occur in the more traditional lignin isolation processes, such as Kraft pulping.

The use of technical lignins as an integral building block for thermoplastic polymers is generally less developed than that for thermoset polymers. This is based on the heterogeneous, complex branched structure of technical lignins which severely limit their thermoplasticity and make them very challenging as building blocks for linear thermoplastic polymers. These inherent limitations constrain the level to which technical lignins can be incorporated into thermoplastic polymers. Here we have highlighted recent research on using Kraft, alkali, and SLs as macroinitiators or building blocks in the synthesis of thermoplastic polyesters, polyacrylates, and polyurethanes. Synthetic techniques such as ROP using renewable cyclic esters and controlled radical polymerization with vinylic monomers provide pathways to lignin-containing thermoplastic polymers with tunable properties. However, more research is needed as technical challenges and limitations remain, including:

1) the need to produce a lignin with a hydroxyl content that is conducive to producing linear polymers. In many cases, this requires the technical lignin to be functionalized to lower the hydroxyl content;
2) achieving effective polymerization of the lignin and copolymers;
3) limits to the level to which the lignin can be incorporated in the polymer;
4) problematic brittleness of the thermoplastic material at higher lignin loadings;
5) undesirable color and odor issues;
6) limited improvement in the thermal and mechanical properties of the material.

To date, the thermoplastic polymers produced by the presented approaches have been largely targeted toward niche applications such as specialty adhesives, carbon-fiber precursors, and sunscreen agents. Use in these applications is likely to only make a limited contribution to valorizing technical lignins at commercial scales.

Using technical lignins as feedstocks for many of our most widely used polymers is challenging, as well-defined, consistently uniform, low molecular weight building blocks are often desirable. This is what petrochemicals provide and what the polymer industry is accustomed to using. For example, over 90% of epoxy resins contain the monomeric building block bisphenol A. An ideal lignin-based polyol would be both structurally and functionally analogous to bisphenol A but without the same undesirable health and environmental concerns. Monomeric chemicals are difficult to produce in high yields from technical lignins due to the additional C—C bonds introduced into these lignins during their isolation. This is where Lignin-first biorefining/RCF of lignocellulosic biomass has a distinct advantage for producing polymer feedstock chemicals. These technologies can produce high yields of monomers, including such compounds as 4-propylguaiacol/syringol, dihydroconiferyl/sinapyl alcohol, and C_2-1,3-dioxolane phenols (C_2-acetal derivatives of homovanillin and homosyringaldehyde). High yields of monomers are possible as these technologies depolymerize the native lignin by cleaving the β-aryl ether linkages and stabilize the products against repolymerization by reducing or trapping the reactive functionalities. As highlighted in this chapter, monomeric lignin–derived building blocks have been used to make novel lignin-derived thermoset and thermoplastic polymers, including polyurethanes, polyepoxies, polyesters, polycarbonates, and polyacrylates. However, the versatility of these native lignin-derived monomers in the synthesis of novel bio-based thermoset and thermoplastic polymers has only started to be explored.

However, these lignin-derived monomers may be of higher value as chemical intermediates to fine chemicals, bioactives, and pharmaceuticals, rather than polymer feedstocks. Depending on the lignocellulosic feedstock, the use of Lignin-first/RCF processing of native lignin also produces good yields of oligomeric compounds, such as dimers, trimers, and larger oligomers. This is because in most cases C–C and aryl C–O interunit linkages in the lignin are not cleaved by the relatively mild processing conditions used. Only α- and β-aryl ether interunit linkages are cleaved in depolymerizing the native lignin. These oligomers are multifunctional and low molecular weight, and so are potentially good polymer feedstocks. On the downside, due to the structure complexity of native lignins, these oligomers are far from the uniform, homogeneous building blocks that petrochemicals have traditionally provided for the polymer industry. On the positive side, the depolymerized native lignin oils have been extensively characterized and the majority of the molecular structures of these oligomers have been elucidated and assigned. This characterization provides an excellent starting point for robust design and synthesis of new bio-based polymers and biomaterials. Despite the complexity of these lignin oligomers, they are still considerably better compared to technical lignins in terms of uniformity and homogeneity, with many technical lignins still relatively poorly characterized at a molecular level.

Another important consideration is the processing costs associated with these new Lignin-first/RCF technologies. The current state of the technology would suggest depolymerized native lignins are likely to come with higher manufacturing costs than technical lignins. The higher processing cost would need to be justified by enhancements in polymer synthesis, properties, and performance that cannot be achieved with technical lignins. Lignin-first/RCF biorefining to produce renewable biochemical feedstocks for the polymer industry is still very much in its infancy with research and development still at lab scale or early pilot scale. Research into technologies including the development of new catalysts, continuous processing, whole biomass valorization and downstream upgrading has seen these technologies advance significantly in recent years. With the advent of the global bioeconomy and the need for sustainable circular manufacturing, polyols derived from native lignin using Lignin-first/RCF technologies will have an important role to play, alongside more traditional technical lignins, in meeting the polymer and material needs of a more sustainable future.

References

1 Auvergne, R., Caillol, S., David, G. et al. (2014). Biobased thermosetting epoxy: present and future. *Chemical Reviews* 114 (2): 1082–1115.

2 Isikgor, F.H. and Becer, C.R. (2015). Lignocellulosic biomass: a sustainable platform for the production of bio-based chemicals and polymers. *Polymer Chemistry* 6 (25): 4497–4559.

3 Delidovich, I., Hausoul, P.J., Deng, L. et al. (2016). Alternative monomers based on lignocellulose and their use for polymer production. *Chemical Reviews* 116 (3): 1540–1599.

4 Upton, B.M. and Kasko, A.M. (2016). Strategies for the conversion of lignin to high-value polymeric materials: review and perspective. *Chemical Reviews* 116 (4): 2275–2306.

5 Llevot, A., Grau, E., Carlotti, S. et al. (2016). From lignin-derived aromatic compounds to novel biobased polymers. *Macromolecular Rapid Communications* 37 (1): 9–28.

6 Sethupathy, S., Murillo, M.G., Gao, L. et al. (2022). Lignin valorization: status, challenges and opportunities. *Bioresource Technology* 347: 126696.

7 Obydenkova, S.V., Kouris, P.D., Smeulders, D.M.J. et al. (2022). Evaluation of environmental and economic hotspots and value creation in multi-product lignocellulosic biorefinery. *Biomass and Bioenergy* 159: 106394.

8 Yuan, Z., Bals, B.D., Hegg, E.L., and Hodge, D.B. (2022). Technoeconomic evaluation of recent process improvements in production of sugar and high-value lignin co-products via two-stage Cu-catalyzed alkaline-oxidative pretreatment. *Biotechnology for Biofuels and Bioproducts* 15 (1): 45.

9 Holladay, J.E., White, J.F., Bozell, J.J., Johnson D. Top Value Added Chemicals from Biomass-Volume II, Results of Screening for Potential Candidates from Biorefinery Lignin. Pacific Northwest National Lab. (PNNL), Richland, WA (United States); National Renewable Energy Lab. (NREL), Golden, CO (United States); 2007. Contract No.: PNNL-16983.

10 Dessbesell, L., Paleologou, M., Leitch, M. et al. (2020). Global lignin supply overview and kraft lignin potential as an alternative for petroleum-based polymers. *Renewable and Sustainable Energy Reviews* 123: 109768.

11 Bertella, S. and Luterbacher, J.S. (2020). Lignin functionalization for the production of novel materials. *Trends in Chemistry* 2 (5): 440–453.

12 Renders, T., Van den Bosch, S., Koelewijn, S.F. et al. (2017). Lignin-first biomass fractionation: the advent of active stabilisation strategies. *Energy & Environmental Science* 10 (7): 1551–1557.

13 Korányi, T.I., Fridrich, B., Pineda, A., and Barta, K. (2020). Development of 'lignin-first' approaches for the valorization of lignocellulosic biomass. *Molecules* 25 (12): 2815.

14 Sun, Z., Cheng, J., Wang, D. et al. (2020). Downstream processing strategies for lignin-first biorefinery. *ChemSusChem* 13 (19): 5199–5212.

15 Torr, K.M., van de Pas, D.J., Cazeils, E., and Suckling, I.D. (2011). Mild hydrogenolysis of in-situ and isolated Pinus radiata lignins. *Bioresource Technology* 102 (16): 7608–7611.

16 Van den Bosch, S., Schutyser, W., Vanholme, R. et al. (2015). Reductive lignocellulose fractionation into soluble lignin-derived phenolic monomers and dimers and processable carbohydrate pulps. *Energy & Environmental Science* 8 (6): 1748–1763.

17 Parsell, T., Yohe, S., Degenstein, J. et al. (2015). A synergistic biorefinery based on catalytic conversion of lignin prior to cellulose starting from lignocellulosic biomass. *Green Chemistry* 17 (3): 1492–1499.

18 Deuss, P.J., Lancefield, C.S., Narani, A. et al. (2017). Phenolic acetals from lignins of varying compositions via iron(III) triflate catalysed depolymerisation. *Green Chemistry* 19 (12): 2774–2782.

19 De Santi, A., Galkin, M.V., Lahive, C.W. et al. (2020). Lignin-first fractionation of softwood lignocellulose using a mild dimethyl carbonate and ethylene glycol organosolv process. *ChemSusChem* 13 (17): 4468–4477.

20 Questell-Santiago, Y.M., Galkin, M.V., Barta, K., and Luterbacher, J.S. (2020). Stabilization strategies in biomass depolymerization using chemical functionalization. *Nature Reviews Chemistry* 4 (6): 311–330.

21 Shuai, L., Amiri, M.T., Questell-Santiago, Y.M. et al. (2016). Formaldehyde stabilization facilitates lignin monomer production during biomass depolymerization. *Science* 354 (6310): 329–333.

22 Lan, W., Amiri, M.T., Hunston, C.M., and Luterbacher, J.S. (2018). Protection group effects during α,γ-diol lignin stabilization promote high-selectivity monomer production. *Angewandte Chemie (International Ed. in English)* 57 (5): 1356–1360.

23 Chung, H. and Washburn, N.R. (2013). Chemistry of lignin-based materials. *Green Materials* 1 (3): 137–160.

24 Engelmann, G. and Ganster, J. (2016). Lignin reinforcement in thermosets composites. In: *Lignin in Polymer Composites* (ed. O. Faruk and M. Sain), 119–151. William Andrew.

25 Obaid, N., Kortschot, M.T., and Sain, M. (2016). Lignin-based foaming materials. In: *Lignin in Polymer Composites* (ed. O. Faruk and M. Sain), 217–232. William Andrew.

26 Westwood, N.J., Panovic, I., and Lancefield, C.S. (2016). Chemical modification of lignin for renewable polymers and chemicals. In: *Production of Biofuels and Chemicals from Lignin* (ed. Z. Fang and R.L. Smith), 183–216. Singapore: Springer Nature.

27 Wang, C., Kelley, S.S., and Venditti, R.A. (2016). Lignin-based thermoplastic materials. *ChemSusChem* 9 (8): 770–783.

28 Xu, C. and Ferdosian, F. (2017). *Conversion of Lignin into Bio-Based Chemicals and Materials*. Berlin: Springer-Verlag GmbH.

29 Feghali, E., Torr, K.M., van de Pas, D.J. et al. (2018). Thermosetting polymers from lignin model compounds and depolymerized lignins. *Topics in Current Chemistry* 376 (4): 32.

30 Huang, J., Fu, S., and Gan, L. (ed.) (2019). Lignin-modified thermosetting materials. In: *Lignin Chemistry and Applications*, 163–180. Elsevier.

31 Fulcrand, H., Rouméas, L., Billerach, G. et al. (2019). Advances in bio-based thermosetting polymers. *Recent Advances in Polyphenol Research* 6: 285–334.

32 Li, H., Liang, Y., Li, P., and He, C. (2020). Conversion of biomass lignin to high-value polyurethane: a review. *Journal of Bioresources and Bioproducts* 5 (3): 163–179.

33 Parit, M. and Jiang, Z. (2020). Towards lignin derived thermoplastic polymers. *International Journal of Biological Macromolecules* 165: 3180–3197.

34 Nagardeolekar, A., Ovadias, M., Dongre, P., and Bujanovic, B. (2021). Prospects and challenges of using lignin for thermoplastic materials. In: *Lignin Utilization Strategies: From Processing to Applications,* ACS Symposium Series, vol. 1377 (ed. C.G. Yoo and A.J. Ragauskas), 231–271. Washington, DC: American Chemical Society.

35 Lawoko, M. and Gioia, C. (2021). Recent strategies for lignin-based thermosets. In: *Lignin Utilization Strategies: From Processing to Applications*, ACS Symposium Series, vol. 1377 (ed. C.G. Yoo and A.J. Ragauskas), 175–206. Washington, DC: American Chemical Society.

36 Xu, C. and Ferdosian, F. (2017). Lignin-based epoxy resins. In: *Conversion of Lignin into Bio-Based Chemicals and Materials* (ed. C. Xu and F. Ferdosian), 111–131. Springer-Verlag GmbH.

37 Sternberg, J., Sequerth, O., and Pilla, S. (2021). Green chemistry design in polymers derived from lignin: review and perspective. *Progress in Polymer Science* 113: 101344.

38 Grand View Research (2022). *Polyurethane market size, share & trends analysis report by product (rigid foam, flexible foam), by application (construction, furniture & interiors), by region, and segment forecasts, 2022–2030.* Available at: https://www.researchandmarkets.com/reports/4118824/global-polyurethane-market-size-share-and-trends (Accessed: 8 May 2024).

39 Xu, C. and Ferdosian, F. (2017). Lignin-based polyurethane (PU) resins and foams. In: *Conversion of Lignin into Bio-Based Chemicals and Materials* (ed. C. Xu and F. Ferdosian), 133–153. Springer-Verlag GmbH.

40 Mahmood, N., Yuan, Z., Schmidt, J., and Xu, C. (2016). Depolymerization of lignins and their applications for the preparation of polyols and rigid polyurethane foams: a review. *Renewable and Sustainable Energy Reviews* 60: 317–329.

41 Li, Y. and Ragauskas, A.J. (2012). Kraft lignin-based rigid polyurethane foam. *Journal of Wood Chemistry and Technology* 32 (3): 210–224.

42 de Haro, J.C., Allegretti, C., Smit, A.T. et al. (2019). Biobased polyurethane coatings with high biomass content: tailored properties by lignin selection. *ACS Sustainable Chemistry & Engineering* 7 (13): 11700–11711.

43 Liu, L.-Y., Karaaslan, M.A., Hua, Q. et al. (2021). Thermo-responsive shape-memory polyurethane foams from renewable lignin resources with tunable structures–properties and enhanced temperature resistance. *Industrial and Engineering Chemistry Research* 60 (32): 11882–11892.

44 Park, S.Y., Kim, J.-Y., Youn, H.J., and Choi, J.W. (2018). Fractionation of lignin macromolecules by sequential organic solvents systems and their characterization for further valuable applications. *International Journal of Biological Macromolecules* 106: 793–802.

45 Wang, Y.-Y., Li, M., Wyman, C.E. et al. (2018). Fast fractionation of technical lignins by organic cosolvents. *ACS Sustainable Chemistry & Engineering* 6 (5): 6064–6072.

46 Wang, Y.-Y., Wyman, C.E., Cai, C.M., and Ragauskas, A.J. (2019). Lignin-based polyurethanes from unmodified kraft lignin fractionated by sequential precipitation. *ACS Applied Polymer Materials* 1 (7): 1672–1679.

47 Griffini, G., Passoni, V., Suriano, R. et al. (2015). Polyurethane coatings based on chemically unmodified fractionated lignin. *ACS Sustainable Chemistry & Engineering* 3 (6): 1145–1154.

48 Glasser, W.G., Barnett, C.A., Rials, T.G., and Saraf, V.P. (1984). Engineering plastics from lignin II. Characterization of hydroxyalkyl lignin derivatives. *Journal of Applied Polymer Science* 29 (5): 1815–1830.

49 Nadji, H., Bruzzèse, C., Belgacem, M.N. et al. (2005). Oxypropylation of lignins and preparation of rigid polyurethane foams from the ensuing polyols. *Macromolecular Materials and Engineering* 290 (10): 1009–1016.

50 Li, Y. and Ragauskas, A.J. (2012). Ethanol organosolv lignin-based rigid polyurethane foam reinforced with cellulose nanowhiskers. *RSC Advances* 2 (8): 3347–3351.

51 Li, B., Zhou, M., Huo, W. et al. (2020). Fractionation and oxypropylation of corn-stover lignin for the production of biobased rigid polyurethane foam. *Industrial Crops and Products* 143: 111887.

52 Wang, S., Liu, W., Yang, D., and Qiu, X. (2019). Highly resilient lignin-containing polyurethane foam. *Industrial and Engineering Chemistry Research* 58 (1): 496–504.

53 Kühnel, I., Podschun, J., Saake, B., and Lehnen, R. (2015). Synthesis of lignin polyols via oxyalkylation with propylene carbonate. *Holzforschung* 69 (5): 531–538.

54 Kühnel, I., Saake, B., and Lehnen, R. (2017). Comparison of different cyclic organic carbonates in the oxyalkylation of various types of lignin. *Reactive and Functional Polymers* 120: 83–91.

55 Zhang, X., Kim, Y., Elsayed, I. et al. (2019). Rigid polyurethane foams containing lignin oxyalkylated with ethylene carbonate and polyethylene glycol. *Industrial Crops and Products* 141: 111797.

56 Mimini, V., Amer, H., Hettegger, H. et al. (2020). Lignosulfonate-based polyurethane materials via cyclic carbonates: preparation and characterization. *Holzforschung* 74 (2): 203–211.

57 Sternberg, J. and Pilla, S. (2020). Materials for the biorefinery: high bio-content, shape memory Kraft lignin-derived non-isocyanate polyurethane foams using a non-toxic protocol. *Green Chemistry* 22 (20): 6922–6935

58 Cinelli, P., Anguillesi, I., and Lazzeri, A. (2013). Green synthesis of flexible polyurethane foams from liquefied lignin. *European Polymer Journal* 49 (6): 1174–1184.

59 Bernardini, J., Cinelli, P., Anguillesi, I. et al. (2015). Flexible polyurethane foams green production employing lignin or oxypropylated lignin. *European Polymer Journal* 64: 147–156.

60 Mahmood, N., Yuan, Z., Schmidt, J., and Xu, C. (2015). Preparation of bio-based rigid polyurethane foam using hydrolytically depolymerized Kraft lignin via direct replacement or oxypropylation. *European Polymer Journal* 68: 1–9.

61 Mahmood, N., Yuan, Z., Schmidt, J., and Xu, C. (2013). Valorization of hydrolysis lignin for polyols and rigid polyurethane foam. *Journal of Science & Technology for Forest Products and Processes.* 3 (5): 26–31.

62 Li, H., Sun, J.-T., Wang, C. et al. (2017). High modulus, strength, and toughness polyurethane elastomer based on unmodified lignin. *ACS Sustainable Chemistry & Engineering* 5 (9): 7942–7949.

63 Liu, W., Fang, C., Wang, S. et al. (2019). High-performance lignin-containing polyurethane elastomers with dynamic covalent polymer networks. *Macromolecules* 52 (17): 6474–6484.

64 Huang, Y., Duan, Y., Qiu, S. et al. (2018). Lignin-first biorefinery: a reusable catalyst for lignin depolymerization and application of lignin oil to jet fuel aromatics and polyurethane feedstock. *Sustainable Energy & Fuels* 2 (3): 637–647.

65 Vendamme, R., Behaghel de Bueren, J., Gracia-Vitoria, J. et al. (2020). Aldehyde-assisted lignocellulose fractionation provides unique lignin oligomers for the design of tunable polyurethane bioresins. *Biomacromolecules* 21 (10): 4135–4148.

66 Quinsaat, J.E.Q., Feghali, E., van de Pas, D.J. et al. (2021). Preparation of mechanically robust bio-based polyurethane foams using depolymerized native lignin. *ACS Applied Polymer Materials* 3 (11): 5845–5856.

67 Feghali, E., van de Pas, D.J., and Torr, K.M. (2020). Toward bio-based epoxy thermoset polymers from depolymerized native lignins produced at the pilot scale. *Biomacromolecules* 21 (4): 1548–1559.

68 Quinsaat, J.E.Q., Feghali, E., van de Pas, D.J. et al. (2022). Preparation of biobased non-isocyanate polyurethane/epoxy thermoset materials using depolymerized native lignin. *Biomacromolecules* 23 (11): 4562–4573.

69 Grand View Research (2022). *Epoxy resin market size, share & trends analysis report by application (adhesives, electrical & electronics, paints & coatings, wind turbines, composites, construction), by region, and segment forecasts, 2022–2030.* Available at: https://www.researchandmarkets.com/reports/3934213/epoxy-resin-market-size-share-and-trends-analysis (Accessed: 8 May 2024).

70 Vandenberg, L.N., Maffini, M.V., Sonnenschein, C. et al. (2009). Bisphenol-A and the great divide: a review of controversies in the field of endocrine disruption. *Endocrine Reviews* 30 (1): 75–95.

71 Pappa, C., Feghali, E., Vanbroekhoven, K., and Triantafyllidis, K.S. (2022). Recent advances in epoxy resins and composites derived from lignin and related bio-oils. *Current Opinion in Green and Sustainable Chemistry* 38: 100687.

72 Engelmann, G. and Ganster, J. (2014). Bio-based epoxy resins with low molecular weight kraft lignin and pyrogallol. *Holzforschung* 68 (4): 1–12.

73 Li, R.J., Gutierrez, J., Chung, Y.-L. et al. (2018). A lignin-epoxy resin derived from biomass as an alternative to formaldehyde-based wood adhesives. *Green Chemistry* 20 (7): 1459–1466.

74 Over, L.C., Grau, E., Grelier, S. et al. (2017). Synthesis and characterization of epoxy thermosetting polymers from glycidylated organosolv lignin and bisphenol A. *Macromolecular Chemistry and Physics* 218 (4): 1600411.

75 Gioia, C., Lo, R.G., Lawoko, M., and Berglund, L. (2018). Tunable thermosetting epoxies based on fractionated and well-characterized lignins. *Journal of the American Chemical Society* 140 (11): 4054–4061.

76 Gioia, C., Colonna, M., Tagami, A. et al. (2020). Lignin-based epoxy resins: unravelling the relationship between structure and material properties. *Biomacromolecules* 21 (5): 1920–1928.

77 Koike, T. (2012). Progress in development of epoxy resin systems based on wood biomass in Japan. *Polymer Engineering and Science* 52 (4): 701–717.

78 Ferdosian, F., Yuan, Z., Anderson, M., and Xu, C. (2015). Sustainable lignin-based epoxy resins cured with aromatic and aliphatic amine curing agents: curing kinetics and thermal properties. *Thermochimica Acta* 618: 48–55.

79 Ferdosian, F., Yuan, Z., Anderson, M., and Xu, C. (2016). Synthesis and characterization of hydrolysis lignin-based epoxy resins. *Industrial Crops and Products* 91: 295–301.

80 Ferdosian, F., Zhang, Y., Yuan, Z. et al. (2016). Curing kinetics and mechanical properties of bio-based epoxy composites comprising lignin-based epoxy resins. *European Polymer Journal* 82: 153–165.

81 van de Pas, D.J. and Torr, K.M. (2017). Biobased epoxy resins from deconstructed native softwood lignin. *Biomacromolecules* 18 (8): 2640–2648.

82 Feghali, E., van de Pas, D.J., Parrott, A.J., and Torr, K.M. (2020). Biobased epoxy thermoset polymers from depolymerized native hardwood lignin. *ACS Macro Letters* 9 (8): 1155–1160.

83 Van Aelst, K., Van Sinay, E., Vangeel, T. et al. (2021). Low molecular weight and highly functional RCF lignin products as a full bisphenol a replacer in bio-based epoxy resins. *Chemical Communications* 57 (46): 5642–5645.

84 Kaiho, A., Mazzarella, D., Satake, M. et al. (2016). Construction of the di(trimethylolpropane) cross linkage and the phenylnaphthalene structure coupled with selective β-O-4 bond cleavage for synthesizing lignin-based epoxy resins with a controlled glass transition temperature. *Green Chemistry* 18 (24): 6526–6535.

85 Zhao, S. and Abu-Omar, M.M. (2016). Renewable epoxy networks derived from lignin-based monomers: effect of cross-linking density. *ACS Sustainable Chemistry & Engineering* 4 (11): 6082–6089.

86 Zhao, S. and Abu-Omar, M.M. (2017). Synthesis of renewable thermoset polymers through successive lignin modification using lignin-derived phenols. *ACS Sustainable Chemistry & Engineering* 5 (6): 5059–5066.

87 Xu, C. and Ferdosian, F. (2017). Lignin-based phenol–formaldehyde (LPF) resins/adhesives. In: *Conversion of Lignin into Bio-Based Chemicals and Materials* (ed. C. Xu and F. Ferdosian), 91–109. Springer-Verlag GmbH.

88 Karthauser, J., Biziks, V., Mai, C., and Militz, H. (2021). Lignin and lignin-derived compounds for wood applications - a review. *Molecules* 26 (9): 2533.

89 Huang, C., Peng, Z., Li, J. et al. (2022). Unlocking the role of lignin for preparing the lignin-based wood adhesive: a review. *Industrial Crops and Products* 187: 115338.

90 Vásquez-Garay, F., Carrillo-Varela, I., Vidal, C. et al. (2021). A review on the lignin biopolymer and its integration in the elaboration of sustainable materials. *Sustainability* 13 (5): 2697.

91 Sarika, P.R., Nancarrow, P., Khansaheb, A., and Ibrahim, T. (2020). Bio-based alternatives to phenol and formaldehyde for the production of resins. *Polymers (Basel)* 12 (10): 2237.

92 Nakagawa, H., Matsumoto, M., Murai, T. (2018). *Lignin resin composition, cured product, and molded product* (U.S. Patent No. 10,059,793). U.S. Patent and Trade-mark Office. https://ppubs.uspto.gov/dirsearch-public/print/downloadPdf/10059793.

93 Sumitomo Bakelite Company Limited (2020, September 17). *Developed a plant-derived phenolic resins (lignin modified phenolic resins).* https://www.sumibe.co.jp/english/topics/2020/hpp/0917_01/index.html

94 Kalami, S., Arefmanesh, M., Master, E., and Nejad, M. (2017). Replacing 100% of phenol in phenolic adhesive formulations with lignin. *Journal of Applied Polymer Science* 134 (30): 45124.

95 Kalami, S., Chen, N., Borazjani, H., and Nejad, M. (2018). Comparative analysis of different lignins as phenol replacement in phenolic adhesive formulations. *Industrial Crops and Products* 125: 520–528.

96 Cheng, S., Yuan, Z., Leitch, M. et al. (2013). Highly efficient de-polymerization of organosolv lignin using a catalytic hydrothermal process and production of phenolic resins/adhesives with the depolymerized lignin as a substitute for phenol at a high substitution ratio. *Industrial Crops and Products* 44: 315–322.

97 Solt, P., Rossiger, B., Konnerth, J., and Van Herwijnen, H.W.G. (2018). Lignin phenol formaldehyde resoles using base-catalysed depolymerized kraft lignin. *Polymers* 10 (10): 1162.

98 Yan, L., Cui, Y., Gou, G. et al. (2017). Liquefaction of lignin in hot-compressed water to phenolic feedstock for the synthesis of phenol-formaldehyde resins. *Composites Part B: Engineering* 112: 8–14.

99 Di, B., Li, Z., Lei, Y. et al. (2021). Phenol-enriched hydroxy depolymerized lignin by microwave alkali catalysis to prepare high-adhesive biomass composites. *Polymer Engineering and Science* 61 (5): 1463–1475.

100 Biziks, V., Fleckenstein, M., Mai, C., and Militz, H. (2020). Suitability of a lignin-derived mono-phenol mimic to replace phenol in phenol-formaldehyde resin for use in wood treatment. *Holzforschung* 74 (4): 344–350.

101 Chen, Y., Gong, X., Yang, G. et al. (2019). Preparation and characterization of a nanolignin phenol formaldehyde resin by replacing phenol partially with lignin nanoparticles. *RSC Advances* 9 (50): 29255–29262.

102 Siahkamari, M., Emmanuel, S., Hodge, D.B., and Nejad, M. (2022). Lignin-glyoxal: a fully biobased formaldehyde-free wood adhesive for interior engineered wood products. *ACS Sustainable Chemistry & Engineering* 10 (11): 3430–3441.

103 Mimini, V., Kabrelian, V., Fackler, K. et al. (2019). Lignin-based foams as insulation materials: a review. *Holzforschung* 73 (1): 117–130.

104 Lucia, A., van Herwijnen, H.W.G., and Rosenau, T. (2020). Wood-based resins and other bio-based binders for the production of mineral wool. *Holzforschung* 74 (6): 539–550.

105 Li, B., Wang, Y., Mahmood, N. et al. (2017). Preparation of bio-based phenol formaldehyde foams using depolymerized hydrolysis lignin. *Industrial Crops and Products* 97: 409–416.

106 Gao, C., Li, M., Zhu, C. et al. (2021). One-pot depolymerization, demethylation and phenolation of lignin catalyzed by HBr under microwave irradiation for phenolic foam preparation. *Composites Part B: Engineering* 205: 108530.

107 Hajirahimkhan, S., Xu, C.C., and Ragogna, P.J. (2018). Ultraviolet curable coatings of modified lignin. *ACS Sustainable Chemistry & Engineering* 6 (11): 14685–14694.

108 Di Francesco, D., Rigo, D., Reddy, B.K. et al. (2022). A new family of renewable thermosets: kraft lignin poly-adipates. *ChemSusChem* 15 (11): e202200326.

109 Xu, Y., Odelius, K., and Hakkarainen, M. (2019). One-pot synthesis of lignin thermosets exhibiting widely tunable mechanical properties and shape memory behavior. *ACS Sustainable Chemistry & Engineering* 7 (15): 13456–13463.

110 Huang, K., Ma, S., Wang, S. et al. (2019). Sustainable valorization of lignin with levulinic acid and its application in polyimine thermosets. *Green Chemistry* 21 (18): 4964–4970.

111 Jawerth, M., Johansson, M., Lundmark, S. et al. (2017). Renewable thiol–ene thermosets based on refined and selectively allylated industrial lignin. *ACS Sustainable Chemistry & Engineering* 5 (11): 10918–10925.

112 Zhang, S., Liu, T., Hao, C. et al. (2018). Preparation of a lignin-based vitrimer material and its potential use for recoverable adhesives. *Green Chemistry* 20 (13): 2995–3000.

113 Jedrzejczyk, M.A., Madelat, N., Wouters, B. et al. (2022). Preparation of renewable thiol-yne “click” networks based on fractionated lignin for anticorrosive protective film applications. *Macromolecular Chemistry and Physics* 223 (13): 2100461.

114 Bertella, S., Rana, V.K., Peyman, K. et al. (2022). Controlled dual functionalization of lignin generates unique properties in gelatin-based hydrogels. *ChemRxiv Preprint*.

115 Chung, H., Al-Khouja, A., and Washburn, N.R. (2013). Lignin-based graft copolymers via ATRP and click chemistry. In: *Green Polymer Chemistry: Biocatalysis and Materials II*, ACS Symposium Series, vol. 1144 (ed. H.N. Cheng, R.A. Gross, and P.B. Smith), 373–391. Washington, DC: American Chemical Society.

116 Laurichesse, S. and Avérous, L. (2013). Synthesis, thermal properties, rheological and mechanical behaviors of lignins-grafted-poly(ε-caprolactone). *Polymer* 54 (15): 3882–3890.

117 Hatakeyama, T., Izuta, Y., Hirose, S., and Hatakeyama, H. (2002). Phase transitions of lignin-based polycaprolactones and their polyurethane derivatives. *Polymer* 43 (4): 1177–1182.

118 Matsushita, Y., Inomata, T., Takagi, Y. et al. (2011). Conversion of sulfuric acid lignin generated during bioethanol production from lignocellulosic materials into polyesters with ε-caprolactone. *Journal of Wood Science* 57: 214–218.

119 Phillips, R.B., Brown, W., and Stannett, V.T. (1972). The graft copolymerization of styrene and lignin. II. Kraft softwood lignin. *Journal of Applied Polymer Science* 16 (1): 1–14.

120 Ayoub, A., Venditti, R.A., Jameel, H., and Chang, H.-M. (2014). Effect of irradiation on the composition and thermal properties of softwood kraft lignin and styrene grafted lignin. *Journal of Applied Polymer Science* 131 (1): 39743.

121 Han, Y., Yuan, L., Li, G. et al. (2016). Renewable polymers from lignin via copper-free thermal click chemistry. *Polymer* 83: 92–100.

122 Korich, A.L., Fleming, A.B., Walker, A.R. et al. (2012). Chemical modification of organosolv lignin using boronic acid-containing reagents. *Polymer* 53 (1): 87–93.

123 Saito, T., Perkins, J.H., Jackson, D.C. et al. (2013). Development of lignin-based polyurethane thermoplastics. *RSC Advances* 3 (44): 21832–21840.

124 de Oliveira, W. and Glasser, W.G. (1994). Multiphase materials with lignin. 11. Starlike copolymers with caprolactone. *Macromolecules* 27 (1): 5–11.

125 de Oliveira, W. and Glasser, W.G. (1994). Multiphase materials with lignin. XIV. Star-like copolymers with styrene. *Journal of Wood Chemistry and Technology* 14 (1): 119–126.

126 Payne, J., McKeown, P., and Jones, M.D. (2019). A circular economy approach to plastic waste. *Polymer Degradation and Stability* 165: 170–181.

127 Payne, J. and Jones, M.D. (2021). The chemical recycling of polyesters for a circular plastics economy: challenges and emerging opportunities. *ChemSusChem* 14 (19): 4041–4070.

128 Byers, J.A., Biernesser, A.B., Delle Chiaie, K.R. et al. (2018). Catalytic systems for the production of poly(lactic acid). In: *Synthesis, Structure and Properties of Poly(lactic acid)*, Advances in Polymer Science, vol. 279 (ed. M.L. Di Lorenzo and R. Androsch), 67–118. Cham: Springer International Publishing.

129 Stanford, M.J. and Dove, A.P. (2010). Stereocontrolled ring-opening polymerisation of lactide. *Chemical Society Reviews* 39 (2): 486–494.

130 Ren, W., Pan, X., Wang, G. et al. (2016). Dodecylated lignin-g-PLA for effective toughening of PLA. *Green Chemistry* 18 (18): 5008–5014.

131 Zong, E., Jiang, J., Liu, X. et al. (2016). Combination of lignin and L-lactide towards grafted copolymers from lignocellulosic butanol residue. *International Journal of Biological Macromolecules* 86: 80–88.

132 Park, S.Y., Kim, J.-Y., Youn, H.J., and Choi, J.W. (2019). Utilization of lignin fractions in UV resistant lignin-PLA biocomposites via lignin-lactide grafting. *International Journal of Biological Macromolecules* 138: 1029–1034.

133 Liu, X., Zong, E., Jiang, J. et al. (2015). Preparation and characterization of lignin-graft-poly (ε-caprolactone) copolymers based on lignocellulosic butanol residue. *International Journal of Biological Macromolecules* 81: 521–529.

134 Kai, D., Zhang, K., Jiang, L. et al. (2017). Sustainable and antioxidant lignin–polyester copolymers and nanofibers for potential healthcare applications. *ACS Sustainable Chemistry & Engineering* 5 (7): 6016–6025.

135 Park, I.-K., Sun, H., Kim, S.-H. et al. (2019). Solvent-free bulk polymerization of lignin-polycaprolactone (PCL) copolymer and its thermoplastic characteristics. *Scientific Reports* 9 (1): 7033.

136 Chung, Y.-L., Olsson, J.V., Li, R.J. et al. (2013). A renewable lignin–lactide copolymer and application in biobased composites. *ACS Sustainable Chemistry & Engineering* 1 (10): 1231–1238.

137 Sun, Y., Yang, L., Lu, X., and He, C. (2015). Biodegradable and renewable poly(lactide)–lignin composites: synthesis, interface and toughening mechanism. *Journal of Materials Chemistry A* 3 (7): 3699–3709.

138 Ryzhakov, D., Printz, G., Jacques, B. et al. (2021). Organo-catalyzed/initiated ring opening copolymerization of cyclic anhydrides and epoxides: an emerging story. *Polymer Chemistry* 12 (20): 2932–2946.

139 Kim, S.J., Kim, Y.S., Lee, O.-K., and Ahn, B.-J. (2016). Synthesis and characterization of kraft lignin-graft-polylactide copolymers. *Wood Science and Technology* 50: 1293–1304.

140 Wu X., Galkin M.V., Sun Z., Barta K. (2022) Fully lignocellulose-based PET analogues for the circular economy, *Nature Communications* 13: 3376.

141 Coessens, V., Pintauer, T., and Matyjaszewski, K. (2001). Functional polymers by atom transfer radical polymerization. *Progress in Polymer Science* 26 (3): 337–377.

142 Matyjaszewski, K. and Tsarevsky, N.V. (2014). Macromolecular engineering by atom transfer radical polymerization. *Journal of the American Chemical Society* 136 (18): 6513–6533.

143 Matyjaszewski, K. and Xia, J. (2001). Atom transfer radical polymerization. *Chemical Reviews* 101 (9): 2921–2990.

144 Tsarevsky, N.V., Sarbu, T., Göbelt, B., and Matyjaszewski, K. (2002). Synthesis of styrene-acrylonitrile copolymers and related block copolymers by atom transfer radical polymerization. *Macromolecules* 35 (16): 6142–6148.

145 Tsarevsky, N.V. and Matyjaszewski, K. (2007). "Green" atom transfer radical polymerization: from process design to preparation of well-defined environmentally friendly polymeric materials. *Chemical Reviews* 107 (6): 2270–2299.

146 Perrier, S.b. (2017). 50th anniversary perspective: RAFT polymerization–a user guide. *Macromolecules* 50 (19): 7433–7447.

147 Semsarilar, M. and Abetz, V. (2021). Polymerizations by RAFT: developments of the technique and its application in the synthesis of tailored (co)polymers. *Macromolecular Chemistry and Physics* 222 (1): 2000311.

148 Li, H., Pang, Z., Gao, P., and Wang, L. (2015). Fe(III)-catalyzed grafting copolymerization of lignin with styrene and methyl methacrylate through AGET ATRP using triphenyl phosphine as a ligand. *RSC Advances* 5 (67): 54387–54394.

149 Liu, X., Xu, Y., Yu, J. et al. (2014). Integration of lignin and acrylic monomers towards grafted copolymers by free radical polymerization. *International Journal of Biological Macromolecules* 67: 483–489.

150 Zong, E., Liu, X., Liu, L. et al. (2018). Graft polymerization of acrylic monomers onto lignin with $CaCl_2–H_2O_2$ as initiator: preparation, mechanism, characterization, and application in poly(lactic acid). *ACS Sustainable Chemistry & Engineering* 6 (1): 337–348.

151 Ryan, J., Elsmore, M.T., Binner, E.R. et al. (2021). Solvent-free manufacture of methacrylate polymers from biomass pyrolysis products. *Reaction Chemistry & Engineering* 6 (2): 335–344.

152 Hilburg, S.L., Elder, A.N., Chung, H. et al. (2014). A universal route towards thermoplastic lignin composites with improved mechanical properties. *Polymer* 55 (4): 995–1003.

153 Liu, X., Wang, J., Yu, J. et al. (2013). Preparation and characterization of lignin based macromonomer and its copolymers with butyl methacrylate. *International Journal of Biological Macromolecules* 60: 309–315.

154 Kim, Y.S., Youe, W.-J., Kim, S.J. et al. (2015). Preparation of a thermoplastic lignin-based biomaterial through atom transfer radical polymerization. *Journal of Wood Chemistry and Technology* 35 (4): 251–259.

155 Yu, J., Wang, J., Wang, C. et al. (2015). UV-absorbent lignin-based multi-arm star thermoplastic elastomers. *Macromolecular Rapid Communications* 36 (4): 398–404.

156 Kai, D., Chua, Y.K., Jiang, L. et al. (2016). Dual functional anti-oxidant and SPF enhancing lignin-based copolymers as additives for personal and healthcare products. *RSC Advances* 6 (89): 86420–86427.

157 Wang, C. and Venditti, R.A. (2015). UV cross-linkable lignin thermoplastic graft copolymers. *ACS Sustainable Chemistry & Engineering* 3 (8): 1839–1845.

158 Jakubowski, W. and Matyjaszewski, K. (2006). Activators regenerated by electron transfer for atom-transfer radical polymerization of (meth)acrylates and related block copolymers. *Angewandte Chemie, International Edition* 45 (27): 4482–4486.

159 Jakubowski, W. and Matyjaszewski, K. (2005). Activator generated by electron transfer for atom transfer radical polymerization. *Macromolecules* 38 (10): 4139–4146.

160 Qu, W., Huang, Y., Luo, Y. et al. (2019). Controlled radical polymerization of crude lignin bio-oil containing multihydroxyl molecules for methacrylate polymers and the potential applications. *ACS Sustainable Chemistry & Engineering* 7 (9): 9050–9060.

161 Wang, S., Shuai, L., Saha, B. et al. (2018). From tree to tape: direct synthesis of pressure sensitive adhesives from depolymerized raw lignocellulosic biomass. *ACS Central Science* 4 (6): 701–708.

162 Duong, L.D., Nam, G.-Y., Oh, J.-S. et al. (2014). High molecular-weight thermoplastic polymerization of kraft lignin macromers with diisocyanate. *BioResources* 9 (2): 2359–2371.

163 Zhang, Y., Liao, J., Fang, X. et al. (2017). Renewable high-performance polyurethane bioplastics derived from lignin–poly(ε-caprolactone). *ACS Sustainable Chemistry & Engineering* 5 (5): 4276–4284.

164 Jeong, H., Park, J., Sunghoon, K. et al. (2013). Preparation and characterization of thermoplastic polyurethanes using partially acetylated kraft lignin. *Fibers and Polymers* 14: 1082–1093.

165 Roh, H.-G., Kim, S., Lee, J., and Park, J. (2018). Effect of low-temperature pyrolysis on the properties of jute fiber-reinforced acetylated softwood kraft lignin-based thermoplastic polyurethane. *Polymers* 10 (12): 1138.

166 Gouveia, J.R., de Sousa Júnior, R.R., Ribeiro, A.O. et al. (2020). Effect of soft segment molecular weight and NCO:OH ratio on thermomechanical properties of lignin-based thermoplastic polyurethane adhesive. *European Polymer Journal* 131: 109690.

167 Quinsaat, J.E.Q., Falireas, P.G., Feghali, E. et al. (2023). Depolymerized lignin oil: A promising building block towards thermoplasticity in polyurethanes. *Industrial Crops and Products* 194: 116305.

13

Advances in Preparation and Applications of Lignin Nanoparticles

Mika H. Sipponen and Liyang Liu

Stockholm University, Department of Materials and Environmental Chemistry, Svante Arrhenius väg 16C, Stockholm, SE-106 91, Sweden

13.1 Introduction

Research on lignin micro- and nanoparticles (here referred to LNPs for simplicity) has rapidly evolved during the past 10 years since the pioneering papers on the topic [1, 2]. During this period, the number of publications has increased from five papers in 2012 to 89 papers in 2022 (search terms: "lignin nanoparticle" in Web of Science), and it is safe to say that LNPs have become a mainstream research topic in materials science. Apart from the motivation to utilize lignin as an abundant and renewable resource, LNPs offer supplementary advantages, including precise control over particle shape and size. The simplicity of their preparation through a solvent exchange method has significantly contributed to the swift advancements and active participation of research teams worldwide. At present, several start-ups and industrial companies are dedicated to scaling up and commercializing LNPs for applications in bio-based materials.

Technological advances, application opportunities, and industrialization aspects (including some pertinent challenges) of LNPs have been covered in several reviews during the past few years [3–12]. In particular, Österberg and coworkers focused on sustainability and application aspects, highlighting the need for life cycle assessment and other impact studies [12]. In addition to technical lignins available from the pulp and paper industry today, there is a growing opportunity window in sourcing lignins from agro-industrial waste and residue streams [6].

This chapter provides an overview of recent developments in the field, focusing on publications from the past five years. However, due to the abundance of works, particularly in important replication and validation studies related to new biomass resources and biorefinery processes as sources of LNPs, our coverage is selective rather than comprehensive. The goal is to consolidate progress from various fronts

Lignin Chemistry: Characterization, Isolation, and Valorization, First Edition.
Edited by Yuhe Liao and Bert F. Sels.

and offer readers a clear understanding of the preparation methods, inherent and adjustable properties, and emerging applications of LNPs.

13.2 Preparation Methods

The common methods used in the preparation of LNPs have been reviewed elsewhere [3, 4, 8, 11–13]. Among the techniques, there appear three broadly used particle preparation methods to prepare aqueous colloidal dispersions and dry particle powders: solvent exchange, pH shifting, and aerosol evaporation. The working principle of these methods is shown in Figure 13.1 and will be discussed more in detail below. In addition, other preparation methods will be outlined with intention to identify opportunities for their further development.

13.2.1 Solvent Exchange

Solvent exchange and solvent polarity shifting methods are the general term used for the preparation of LNPs from organic solvents in dispersion state. The supramolecular aggregation of lignin is often brought upon by diluting the lignin solution by addition of water. Regardless of the order of mixing (lignin solution into water or vice versa) it is the change in the solubility of lignin in the solvent that changes

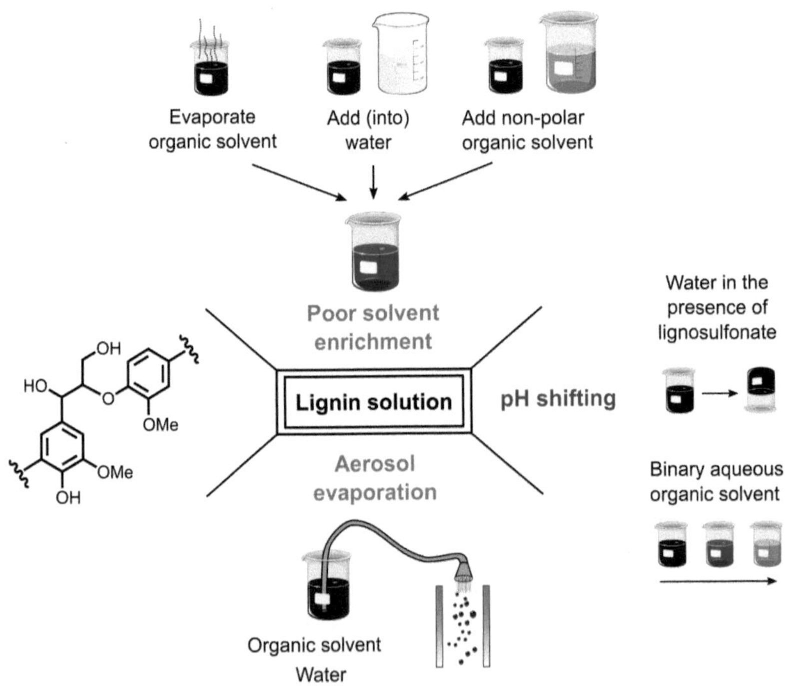

Figure 13.1 Common methods used for the preparation of LNPs.

upon enrichment of the poor solvent (water). Table 13.1 summarizes different solvent systems and lignin sources used in the preparation of LNPs using variations of the solvent exchange method. Such poor solvent enrichment can also be brought about by evaporation of volatile organic solvent from binary aqueous–organic solvent mixtures such as acetone–water or tetrahydrofuran–water [32, 33]. LNPs have also been formed by using a solvent system based on organic solvents only, with 1,4-dioxane as lignin solvent and cyclohexane as the diluting poor solvent [34]. Such a process may be energy efficient owing to facilitated evaporation and reuse of anhydrous organic solvents, but production of LNPs in aqueous dispersions is preferred in many water-based applications and preparation of nanocomposites with hydrophilic polymers.

The choice of lignin solvent plays a critical role in the solvent exchange process. Zou et al. compared aqueous tetrahydrofuran, acetone and dimethylsulfoxide solvent systems for the preparation of LNPs [35]. The particle size followed the above order of solvents: THF > acetone > DMSO, and the particles from aqueous acetone displayed a higher uniformity compared to those from aqueous THF. These authors also reported that the particles were compact and displayed low porosity at the resolution based on cryo-TEM imaging. Another comparison report came from Figueiredo et al., who used acetone–water and 70% ethanol solutions for the solvent-shifting method side by side with an acidification method [20]. Both of these studies reported that aqueous acetone and tetrahydrofuran solutions produce colloidally stable particles with better performance than the DMSO or ethanol-based solvent systems.

Table 13.1 Comparison of solvent systems used in the production of LNPs using the solvent exchange method and water as the diluting solvent.

Solvent system	Lignin type (conc.)[a]	Particle size (nm)[b]	References
Acetone[c]	Kraft lignin	140	[14]
Acetone–water 8 : 1 v/v	EHL ($10\,g\,l^{-1}$)	210	[15]
	EHL ($0.1\,g\,l^{-1}$)	50	[15]
Acetone–water 3 : 1 v/v	SKL ($10\,g\,l^{-1}$)	100	[16]
	SKL ($10\,g\,l^{-1}$)	100	[17]
	SKL ($10\,g\,l^{-1}$)	100	[18]
	SKL ($10\,g\,l^{-1}$)	70	[19]
	SKL, BB, PB ($10\,g\,l^{-1}$)	90–150	[20]
Acetone–water 7 : 3 v/v	Alkali lignin[d] ($5–100\,g\,l^{-1}$)	30–120	[21]
	EHL ($40\,g\,l^{-1}$)	130	[22]
THF	SKL	~300	[23]
—	EHL	100–500[d]	[24]
—	Kraft lignin	Bimodal: ~90; ~280	[25]
—	SKL ($20–60\,g\,l^{-1}$)	100–250	[26]

(Continued)

Table 13.1 (Continued)

Solvent system	Lignin type (conc.)[a]	Particle size (nm)[b]	References
THF–water 3 : 1 w/w	SKL	260	[27]
—	SKL	~200	[28]
Dioxane–water 7 : 3 v/v	KL ($1\,g\,l^{-1}$)	240	[29]
Ethanol	HKL ($2\,g\,l^{-1}$)	140	[20]
Ethanol–water 7 : 3 v/v	SL ($3\,g\,l^{-1}$)	70–300	[30]
DES (choline chloride ethylene glycol)	AL, SKL (3–9 wt%)	20–100	[31]

BB, soda-type lignin extracted from hydrothermally treated birch wood; PB, soda lignin from wheat straw; HKL, hardwood kraft lignin; SL, soda lignin from annual plants.
[a] Concentration of lignin in solution prior to enriching the volume fraction of the poor solvent.
[b] Average values or range reported based on dynamic light scattering and rounded to the nearest ten.
[c] Supercritical carbon dioxide used as a non-solvent.
[d] Two particle size populations were observed. SKL, Softwood kraft lignin. Alkali lignin from Sigma-Aldrich. EHL, enzymatic hydrolysis lignin.

In addition to the common organic solvents, ionic solvents such as deep eutectic solvents (DES) [31, 36, 37] and ionic liquids (IL) [38, 39] have been used. However, despite their apparent contributions to safety and practical handling in the laboratory, solvent systems such as DES and ILs pose challenges related to solvent recycling, which is central for scaling up. Furthermore, though hydrotropic liquids such as sodium *p*-toluenesulfonate that have been used for the production of lignin particles from initial lignin concentrations up to $200\,g\,l^{-1}$, downstream process considerations such as separation and redispersion of the particles in water need due attention [40].

Lignin nanospheres and nanorods have also been successfully produced using solvent-fractionated lignosulfonates [41]. This was achieved by dissolving the fractions in water and introducing ethanol as a poor solvent. To eliminate residual water, the solution underwent dialysis against ethanol. While these lignin nanomaterials are dispersed in ethanol, they possess unique properties that make them suitable for applications beyond typical aqueous dispersions of LNPs.

In addition to their effect on particle properties, it is important to take into consideration the concentration limitations and differences between the solvent systems. For example, aqueous solvent mixtures with tetrahydrofuran, acetone, or ethanol restrict dispersion concentrations of LNPs before removal of the solvent to approximate levels of 20, 10, and $1.5\,g\,l^{-1}$, respectively. Attempting to increase productivity by using higher initial lignin concentrations usually leads to agglomeration and reduced mass yields. Moreover, it is noted that particle size is also a function of initial lignin concentration and thus needs to be controlled to ensure consistent results from one batch to another. The various parameters that provide control over the properties of LNPs are presented more in detail in Section 13.3.2.

13.2.2 pH Shifting

Nanoparticle dispersions can be achieved through acid precipitation alone or with the assistance of ultrasonication in aqueous lignin solutions [20, 42, 43]. These particles can be either used as such or stabilized at low concentrations by adding surfactants or coating with cationic polymers such as polydiallyldimethylammonium chloride [44]. Irregularly shaped lignin particles have been formed from softwood kraft lignin [45] and alkali lignin [46] through acid-induced precipitation from an ethylene glycol solution. These particles were utilized for infusing silver ions, showcasing their potential as antimicrobial agents.

To ensure the absence of any residual organic solvents in the nanomaterials, it is crucial to use lignin that is free from such solvents. Pylypchuk and Sipponen devised a method to generate colloidal lignin micelles and high-consistency gels without the use of organic solvents [47]. In their approach, they neutralized alkaline solutions of poorly water-soluble lignins in the presence of sodium lignosulfonate. This resulted in the formation of micellar complexes, where lignosulfonates acted as amphiphilic molecules, offering stability due to their anionic charge and relatively higher molecular weight. Notably, the ability to produce LNPs at consistencies above 30 wt% is unprecedented compared to regular LNPs produced at lower concentrations. The method demonstrated versatility by utilizing various lignin sources, including kraft lignin, soda lignin, and organosolv lignin. The resulting flowable lignin dispersions exhibited shear-thinning behavior and underwent gelation within specific pH and concentration ranges, providing control over the process.

13.2.3 Aerosol Evaporation

The production of lignin particles utilizing aerosol drying techniques has gained significant attention in recent research [48–50]. Aerosol drying involves atomizing a lignin solution into fine droplets, which are then subjected to a controlled drying process to yield solid lignin particles. The drying can be conducted thermally, during an electrospray process [51], or by freeze-drying [52]. This approach offers several advantages, including the ability to tailor the morphology and particle size, while obtaining particles as dry powders, with the possibility to classify the particles according to their particle size [48]. Though usually produced in particle sizes in the micron range, recently it was achieved particles with sizes below 100 nm [50].

13.3 Properties

13.3.1 Inherent Properties of LNPs

LNPs are expected to exhibit surface properties that reflect the specific interface at which they are formed. When using the aqueous–organic solvent exchange method, it can be anticipated that LNPs will have hydrophilic surfaces. On the other hand, aerosol evaporation methods, where the interface is air, or solvent exchange utilizing organic solvent mixtures like cyclohexane in an anhydrous solvent exchange process, would likely result in relatively more hydrophobic surfaces. In the most common scenario, LNPs tend to assemble in a manner where hydrophobic segments preferentially gather at the particle's core, while hydrophilic components, such as carboxylic acid and phenolic hydroxyl groups, along with other hydrophilic units, decorate the particle surface.

Owing to their small size, one of the most attractive assets of LNPs is their surfaces which can adsorb and react with soluble and insoluble matter. Thus, the ability to study these interfaces with surface-sensitive methods are of key interest. Farooq et al. developed methods for production of model particle surfaces to allow characterization by quartz crystal microbalance, AFM, and other surface-sensitive techniques to better understand the interfacial properties of the particles [53]. Elsewhere, Pylypchuk et al. developed a dispersion state ^{1}H NMR spectroscopy analysis of LNPs and confirmed that their surfaces are enriched in polar functional groups, leading to hydrophilic amphiphilic surface [54]. Furthermore, Chen et al. revisited the insight of lignin aggregation through the molecular dynamic analysis, revealing that the driving force for lignin aggregation includes the intra-molecular interactions of lignin aromatic rings and the solvent during solvent shifting [55]. Deriving from these pioneering works and other contributions, we summarize in Figure 13.2 the common properties of LNPs. It is important to highlight that these properties can vary depending on the attributes of the source lignin, as well as the particle preparation methods and various process parameters.

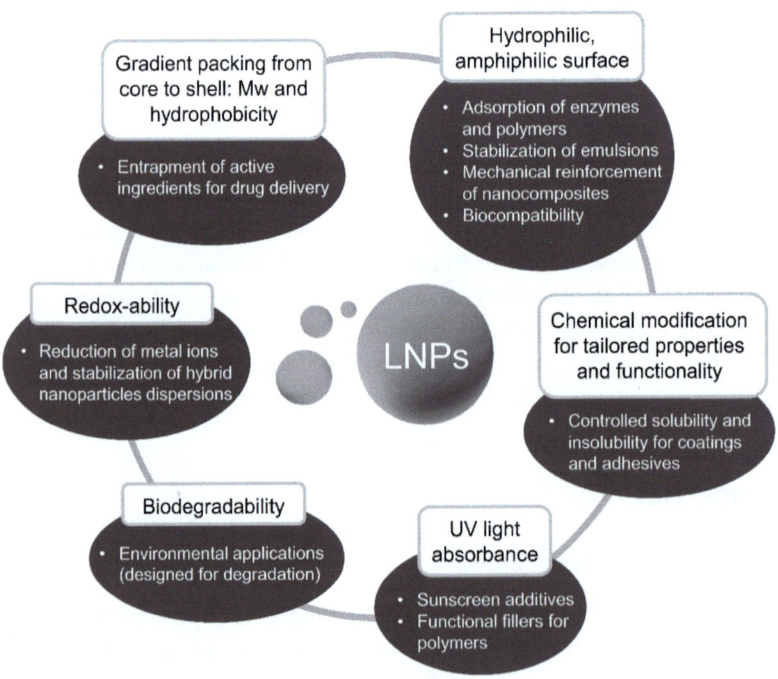

Figure 13.2 Properties and applications of LNPs.

The kinetic biodegradability of LNPs may exhibit variations based on the source of technical lignin. However, if no new covalent cross-linking or functionalization reactions are introduced, the ultimate biodegradability of LNPs is expected to be comparable to that of the original technical lignin. To their advantage, LNPs are often better dispersed in polymers compared to crude lignin, with superior UV shielding performance within the wavelength range of 250–400 nm. This performance can be ascribed to their chromophore groups, including (i) the conjugated electron system of the aromatic ring, (ii) conjugated aliphatic structures, and (iii) quinones [56]. Another benefit worth mentioning is that the LNPs show low cytoxicity, suggesting they are safe for potential applications in biomedicine and agriculture [57]. Furthermore, their surface functional groups, especially phenolic groups, provide new opportunities to tailor the structure-property of LNPs for enhanced performances of resulting materials in UV shielding, degradability, and particle morphology [58]. Noteworthy, the impacts of chemical modification on the cytotoxicity of resulting materials require comprehensive studies to identify and quantify the effects of different functional groups.

13.3.2 Size Control by Different Solvent Systems and Process Parameters

There are several parameters that offer control over the particle size of colloidal lignin particle dispersions. In this context, our focus will be on the aqueous–organic solvent exchange process. Table 13.2 presents a compilation of these key parameters, their typical impact on particle size, and important considerations and limitations that should be taken into account to achieve desired particle properties. The underlying intermolecular forces are discussed elsewhere [66]. Among the parameters that require careful attention, solution concentration and pH (in the case of aqueous–organic solvent systems) are the most prominent ones to begin with. It is evident from the literature that as the solution concentration increases, the particle size also increases, indicating the aggregation of nuclei during the initial stages of particle formation. In terms of the dependency on molecular weight, it has been observed that fractionated lignin with varying molecular weights generates particles in such a way that higher molecular weight (M_w) lignin yields smaller particles compared to the corresponding lower M_w fractions [59, 62]. The precise underlying reason is not entirely clear, whether it is associated with reduced electrostatic repulsion and, consequently, more efficient packing of high molecular weight (M_w) lignin fractions, or the increased capability of longer lignin chains to fold compared to the lower M_w counterparts. In addition, it has been postulated that the faster nucleation kinetics associated with the higher M_w lignins lead to the formation of smaller LNPs [57]. Solvent fractionation of lignin reduced dispersity and help to produce LNPs with narrow PDI [24, 54, 59, 67–69]. This foundation has been exploited for the production of photonic crystals by centrifugation-assisted assembly of LNPs [68–70].

Table 13.2 Effect of process parameters and lignin characteristics on particle size and stability of LNPs from the solvent exchange method.

Parameter	Effect on particle size[a]	Notes	References
Lignin concentration	Increase–increase	Consider concentration limit based on lignin type and solvent system; PDI optimum at intermediate lignin concentration	[28, 59–61]
pH of the solvent system	Increase–decrease	Maintain within pH 3–6 to avoid agglomeration	[28]
Solvent dilution rate	Increase–decrease	Nonlinear relation	[30]
Temperature during solvent-exchange	Increase–increase[b]	Nonvolatile solvents such as DMSO differ from the trend	[26]
Stirring rate	Increase–decrease	Avoid foaming and resulting templating	[26]
M_w of lignin	Increase–decrease	Differences in chemical structure	[59, 62]
Hydrophilicity of lignin	Increase–increase	Extensive carboxylation leads to water solubility that restricts particle formation	[63–65]

[a] Increase–increase: increase in the corresponding parameter causes increased particle size.
[b] For THF-based solvent system.

13.3.3 Versatile Particle Morphologies and Tailored Properties

LNPs with tunable morphology (e.g. compact, compact with single hole, core–shell, capsule, and micellar) have been fabricated to tailor their properties for different applications over last several years [25] (Figure 13.3). Previous studies have investigated and discussed the mechanism affecting the LNPs morphology [59, 62, 71]. These studies harnessed solvent-fractionated lignin to acquire more uniform fractions from different types of lignin resources. Their results suggested the structural features of technical lignin, including the molecular weight, surface functional groups, and inter-unit linkages, all play pivotal roles in the controlled design and synthesis of LNPs. Simultaneously, the process parameters, particularly the initial

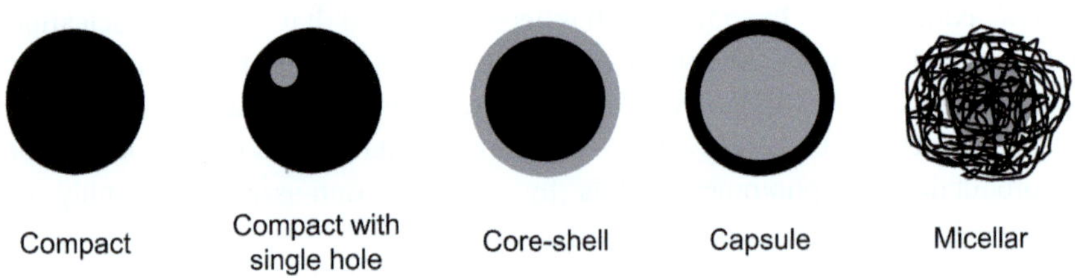

Figure 13.3 Diversity in the appearance of LNPs prepared by different methods.

concentration and dilution rate of the aqueous–organic solvent mixture, not only impact the particle size but also provide the ability to adjust the morphology of LNPs.

Moreno and Sipponen have thoroughly reviewed previous chemical modification methods, including esterification, etherification, and amination, highlighting their positive impacts on the properties and morphologies of LNPs [58]. In this chapter, we will introduce several innovative approaches to acquire LNPs or hybrid LNPs with enhanced performances, such as reactivity, pH-responsive, solvent resistance, radical scavenger, and UV shielding [72]. Most of these chemical approaches can convert phenolic hydroxyl groups in the lignin to desired functional groups from simple structural units to sophisticated structures. The carboxylation of lignin prior to the formation of LNPs has been demonstrated to prepare particles for enhanced reactivity (increasing anionic groups) toward amide conjugation for drug delivery [63], nano-pesticide [73], and iron absorption [39, 74]. The surface amination reaction is capable of enhancing the radical scavenging activity and UV resistance of LNPs for antioxidant applications [75]. The mannish reaction can functionalize the lignin surface with histidine-containing imidazole for the pH-responsive drug release [76]. Additionally, the substitution of phenolic hydroxyl groups with hydrophobic benzene ring [77], alkane [78], oleate [79], or cholesterol [80] can tune the hydrophobicity or amphiphilic interfaces to enhance the stability of LNPs, trap the pesticide or drugs, and realize the controllable release. The surface hydroxyl groups on the lignin can also react with poly(lactic-*co*-glycolic) acid for the synthesis of nanoparticles with a controllable particle size (100–200 nm) [81]. The possible applications of the newly developed particles remain to be demonstrated, but they potentially can be used for the biomedicine and agriculture sectors.

Preparation of LNPs hybrid particles containing organic fillers is another promising approach to tune the morphology of LNPs and bring synergetic impacts for more advanced applications. For example, the hollow lignin sphere can be formed by adding butylated hydroxytoluene (BHT) as an additive. The lignin molecules would aggregate through the aromatic conjugate interaction on the internal surface of the membrane layers between the BHT–THF and water. With the evaporation of THF, the BHT can deposit on the vessel wall leading to the formation of a cavity in the resulting spheres (Figure 13.3) [71]. Sipponen et al. developed the hybrid LNPs, containing 40% weight of fatty acid (lauric acid and oleic acid) and 20 nm thick lignin shells (stabilizer), that can be potentially applied as an environmentally benign and scalable phase-change material for the thermal energy storage at the dry and wet state [82]. Jesionowski and his collaborators used the cationic surfactant (hexadecyltrimethylammonium bromide, CTAB) to tailor the chemical affinity and change the electrokinetic properties of LNPs. The resulting LNPs with CTAB can serve as a functional sorbent of harmful ions for the removal of metal ions, e.g., the vanadium (V) from an aqueous system [83]. Jiang et al. adopted renewable xylan to tune the morphology of lignin macromolecular in producing hybrid nanospheres as pesticide carriers. The interaction of polysaccharides for the hybrid LNPs can lead to the enzyme-mediated release property [84].

The presence of surface phenolic hydroxyl groups on lignin can enable the synthesis of lignin–inorganic hybrid materials for various novel applications. These include adsorbents, antimicrobial materials, catalysts, and energy storage materials, as reviewed by Budnyak et al. [85] We will highlight a few recent studies in this area, demonstrating how lignin and inorganic materials can bring synergic effects for new functional materials. For example, Lintinen et al. harnessed lignin to integrate with iron isopropoxide to form a series of lignin/iron oxide hybrid nanoparticles. Adjusting the reaction can tune the morphology of LNPs from a fishnet-type open network structure to particles with various diameters (10–400 nm) to hollow spheres [86]. The potential applications of these hybrid particles can be used as recyclable magnetic fertilizers. In the same line of research, Li et al. integrated iron oxide nanoparticles with aminated LNPs to create magnetic LNPs as phosphorous adsorbents and reusable slow-release fertilizers [87]. Moreover, LNPs/inorganic composite hybrid nanoparticles can contribute to composite fabrication. LNPs can contribute to more evenly distributed metal oxide (e.g. TiO_2) particles and consequently enhance mechanical performances, UV absorption, and antibacterial properties, and create new opportunities for the utilization of lignin-based composites [88].

13.3.4 Stabilized and Solvent-Resistant Lignin Particles

Instability and dissolution in solvents and alkali of non-covalently assembled LNPs limit their functionalization for chemical formulations and applications under harsh conditions. Different strategies have been employed to overcome these limitations. Leskinen et al. [28] and later Wang et al. [89] subjected LNPs to hydrothermal treatment in temperatures reaching 190 °C. Hydrothermal treatment presumably causes intermolecular cross-linking within the individual particles, effectively improving stability, with lower solubility in organic solvents, while the particle size was also reduced compared to the original LNPs. Another approach to stabilize the particles was presented by Morsali et al. who synthesized hydroxymethylated LNPs and stabilized them through hydrothermal curing and cross-linking reactions [90]. The resulting particles were colloidally stable, containing 97% bio-based content. The particles exhibited structural stability in aqueous media (pH 3–12) and polar organic solvents. Earlier, Zou et al. endowed LNPs with high solvent resistance by the fabrication of lignin-epoxy hybrid nanoparticles following with hydrothermal curing, producing particles that can be surface-modified toward desired applications [19].

13.4 Emerging Applications

LNPs are central building blocks for lignin-based smart materials as reviewed by Moreno and Sipponen [91]. This section intends to give an overview of the emerging applications and determine areas to focus on future research. Design for circularity will likely be one of the key drivers in this field [10, 92]. It is also important to recognize that innovations are driven by private companies that contribute to applied research and market penetration of lignin-based materials.

13.4.1 Polymer Composites, Gels, Adhesives, and Coatings

LNPs have well-defined surfaces and are easy to disperse in an aqueous media that can serve as fillers to blend with hydrophilic polysaccharide matrix, such as cellulose nanofibrils [16], and chitosan [93]. These renewable biopolymers own high strength and stiffness combined with low weight and biodegradability. However, these materials face challenges regarding their elastances and moisture strength. Due to inherent properties (Figure 13.2), the integration of LNPs with these polysaccharide leads to synergetic impacts on resulting nanocomposites in terms of the reinforced impacts, UV shielding, antibacterial, enzyme retarded degradability, and water resistance (Table 13.3). For instance, Liu et al. embedded 20 wt% of LNPs into a chitosan matrix to make a film, demonstrating an efficient UV blocking of 97% at 400 nm. Coating the film on the glass can absorb solar energy to manage the temperature under the glass and provide a sustainable solution to facilitate indoor thermal management and improve the energy efficiency of construction [93]. The LNP-chitosan composites also demonstrated as a new sorbent for enhanced removal of dye pollution from aqueous solutions, as Sohni and his collaborators presented [101]. Tian and his collaborators blend LNPs with bacterial nano cellulose-lignin resulting in composite films and investigate their retarded biodegradability [94]. As such, LNPs can open new avenues for applying these renewable hydrophilic polysaccharide polymers in food packaging, water purification, energy-efficient buildings, and biomedicine.

When blending with hydrophilic synthetic polymers such as poly(vinyl alcohol) [102] and polyacrylates [103], LNPs demonstrated superior compatibility with this polymer matrix. For instance, poly(vinyl alcohol) is a water-soluble polymer with abundant surface hydroxyl groups demonstrating superior compatibility with LNPs in an aqueous solution. Yang et al. blend different amounts of LNPs with a PVA matrix via a facile solvent casting method. Results suggest that additional LNPs marked enhanced water barrier ability, UV shielding impacts, and radical scavenging activity for this composite that can significantly broaden their applications [96]. Besides the solvent casting, another approach is to apply the DMSO/H_2O binary solvent, leading to the self-assembly of lignin macromolecules into nanoparticles in the PVA matrix [97]. LNP-based hydrogel can also be produced through an in situ free radical polymerization of monomers, e.g., acrylamide, with the phenolic hydroxyl groups on lignin surfaces. The resulting nanocomposite hydrogels demonstrated highly mechanical properties and noncytotoxicity compared to pure PAM hydrogel [98]. These works provide innovative solutions for the design of stretchable hydrogels for flexible wearable electronics and biomedical materials such as artificial muscles and strong underwater antifouling materials [104].

Blending LNPs with thermoplastics like poly(lactic acid) faces challenges due to their low compatibility and low-temperature resistances, because the melting temperature of PLA (180 °C) is higher than the T_g of technical lignin (100–160 °C) [99]. Long et al. created LNPs by solvent shifting and blended them with the PLA matrix through melt processing. Their results suggested the formation of lignin microparticles in the resulting composite, but it is unclear about the particle size and

Table 13.3 LNPs-based composites and their applications.

Polymer	Lignin	Lignin wt%	Nanoparticles	Composite preparation	Additional properties	Applications	References
Bacterial cellulose	Soda/DES/ Organosolv lignin	42–44	Yes	Solvent casting	Retarded biodegradability, decreased DP, cellulose crystallinity, and water-holding capacity	Food packaging	[94]
Cellulose nanofibrils	Kraft lignin	15	Yes	Solvent casting & hot pressing	Enhanced mechanical, UV shielding, hydrophobic, and antioxidative	Food packaging, water purification, and biomedicine	[95]
Chitosan	Alkali lignin	20	Yes	Solvent casting	UV shielding, hydrophobicity, and mechanical performances	Thermal regulation materials	[93]
Polyvinyl alcohol	Enzymatic hydrolysis lignin	1–2	Yes	Solvent casting	Enhanced hydrophobicity, barrier ability to water vapor, UV shielding, radical scavenging activity	Food package	[96]
Polyvinyl alcohol	Alkali lignin	1–3	Yes	In situ nanoparticle formation	Reinforcing impacts	Wearable electronics and healthcare monitoring	[97]
Polyacrylamide	Alkali lignin	1–8	Yes	Radical polymerization	Enhanced toughness and slightly decreased cell viability	Biomedical materials	[98]
Polylactic acid	Lignin	0.25–1.25	Yes	Plastic extrusion	Enhanced mechanical properties	Packaging materials	[99]
Polylactic acid	Modified organosolv lignin	1–5	Yes	Melt mixing and hot compression	Enhanced UV shielding, but lower mechanical performances	Packaging materials	[72]
Thermoset	Epoxidized lignin	20–80	Yes	Anionic polymerization	Stabilizers, solvent resistance, and flame retardancy	Wood adhesives	[100]

morphology of these particles in the polymer matrix [99]. Chemical functionalization is another approach to enhance the dispersibility of LNPs in the PLA and their inherited properties, such as UV-shielding for advanced PLA composites [72]. Leveraging on chemical modification, LNPs have recently found utilizations in making thermoset materials for adhesives and coatings. For example, the nanoprecipitation of lignin and BADGE mixture could lead to the formation of hybrid particles that can be used as wood adhesives or solvent-stable colloidal particles [19]. Henn et al., prepared wood coating formulations by simply mixing LNPs with glycerol diglycidyl ether. The particulate coatings enabled protection against water while retaining the moisture uptake of wood [105]. Later, the same group demonstrated LNPs as curing agents for epoxidized lignin prepared in Pickering emulsion phase toward wood coatings with lignin contents above 80% [106].

Elsewhere, Ferruti et al. employed an alternative approach to produce epoxy-functionalized LNPs. They initiated the process by reacting soluble or insoluble lignin obtained through solvent fractionation with epichlorohydrin. Subsequently, they combined this epoxidized fraction with the remaining non-derivatized fraction during the formation of LNPs through the solvent exchange method. This resource-efficient process harnessed all the lignin and allowed controlling the location of the epoxy group predominantly either in the core of the particles or deposited on the outer surface. Consequently, the particles with epoxidized lignin at the surfaces were demonstrated as adhesives by curing the LNPs on amine-functionalized glass at a high temperature, resulting in the high-performance lignin-epoxy composite [100].

13.4.2 Carrier Systems for Agriculture and Biomedicine

LNPs as the carrier system for agriculture and biomedicine have been the subject of a few recent reviews [3, 107–110], and thus we focus here on the recent results and critical perspectives for promising paths forward. Frequently, there are three ways, including adsorption, entrapment, and encapsulation, to store active ingredients in the LNPs-based carrier system (Figure 13.4). In doing so, the system can realize controlled release, protection from extreme conditions such as UV and pH, and preservation of activity. This provides innovative solutions to utilize the active ingredients for agriculture and biomedicine more efficiently.

Nanotechnology applications in agriculture have received a lot of attention to increase crop production [111]. The valorization of LNPs from plant resources for

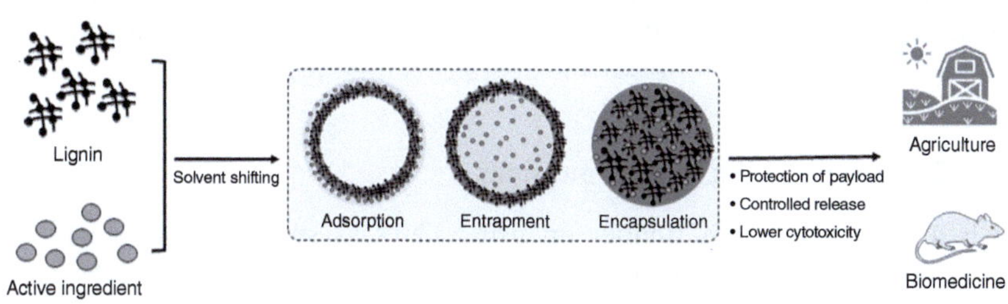

Figure 13.4 LNPs carrier system for agriculture and biomedicine.

nano-pesticides and nano-fertilizers can achieve more efficient use of resources by controlling the release rate and decreasing the consumption of activated ingredients [112, 113]. Compared to conventional carrier systems, the inherent properties of LNPs can bring synergetic impacts to these applications: (i) lignin, one of the most abundant biopolymers from plant resources, can return to the soil, promoting carbon sequestration [112]; (ii) the UV shielding and radical scavenger properties allow them to protect the activated ingredients from photodegradation; Yiamsawas et al. found LNPs demonstrates excellent UV-light absorption with a peak around 210–280 nm, providing potential as UV protection materials [114]; (iii) the enriched surface functional groups, including hydroxyl groups, ionic groups, and aromatic ring, make it possible to trape ingredients such as abscisic acid [115], fungicides [116, 117], essential oil [118], and pesticides such as avermectin [73] and pyraclostrobin [119]. Sipponen and his collaborators thoroughly summarized these different types of activated ingredients that can be trapped by LNPs-based carrier systems [107]; (iv) as explained above, surface functional groups, such as reactive hydroxyl groups, endows for tailoring surface functionality for different types of applications. The etherification of lignin with benzyl groups, for example, can enhance their hydrophobicity to trap hydrophobic pesticides that can slow down the release rate for using the ingredients more efficiently [77].

One of the stunning performances of LNP-based carrier systems is the controlled releasing performances depending on the pH [76, 120–122] and temperature [123], which are crucial for biomedicine applications. Siddiqui et al. studied the cytotoxicity and genotoxicity, suggesting that LNPs are compatible with normal and cancer cells even at high concentration, indicating they are safe to be used as drug carriers [124]. Therefore, there has been an increased attention on the valorization of LNPs for biomedicine over the last decade. Chen et al. prepared the lignin/chitosan oligosaccharide nanoparticles that can trap the hydrophilic cytarabine efficiently, which can avoid the risk of drug–drug interactions. Their results suggested that LNPs/chitosan could provide excellent UV, thermal, and oxidation protection to anticancer drugs and efficiently kill leukemia cells compared with conventional single drug-loaded systems [125]. Similar to this strategy, a series of LNPs-based drug carriers containing, for instance, doxorubicin [126], peptide-guided resiquimod [127], β-cyclodextrin [128], doxorubicin [129], curcumin [130–132], and other anticancer drugs [80] have been developed to realize the utilization of drug more efficiently. LNPs can scavenge reactive oxygen species deemed beneficial for the health of the gastrointestinal tract [30]. Overall, we envision that LNPs offer a promising platform to facilitate the efficient utilization of active ingredients in agriculture and biomedicine.

13.4.3 Emulsifiers for Pickering Emulsions

There is an active research frontier on using LNPs as surfactants to stabilize Pickering emulsions. Owing to their assembly according to polarity LNPs are effective particulate emulsifiers (Figure 13.5). In addition to oil-in-water (o/w)

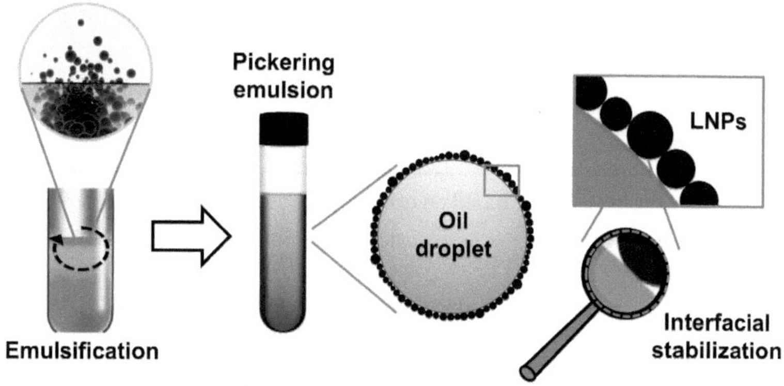

Figure 13.5 Stabilization of Pickering emulsions with LNPs.

emulsions [21, 27, 48, 133] also reverse emulsions (w/o) have been reported recently [134]. In case of o/w emulsions, a variety of inorganic and organic liquids have been used as oil phases, including silicone oil, toluene, styrene, methyl methacrylate, n-tetradecanol, n-hexadecane, dichloromethane, fruit and vegetable oils such as orange and coconut oils, olive oil, and rapeseed oil. Cationic LNPs achieved with coating quaternary ammonium derivatives of lignin or chitosan have in cases of vegetable oils proved more efficient emulsion stabilizers than the corresponding anionic LNPs [17, 27]. Moreno et al. used chitosan-coated LNPs and later carbohydrate-containing LNPs for the stabilization of Pickering emulsions with styrene and acrylic monomers as oil phase for free radical polymerization [18, 103, 135]. The subsequent fabrication of polymeric nanocomposites, with particles uniformly dispersed throughout the matrix, was successfully achieved, overcoming the challenges typically encountered with conventional melt processing techniques. Lignin-based Pickering emulsions can also be employed to produce engineered materials such as phase-change materials for thermal energy storage and management [136]. In addition to application oriented research future work in this area should focus on mapping surface tension of LNPs at interfaces and relating this data to particle size and surface chemistry.

13.4.4 Sensors

LNPs are versatile components of materials that can be used for sensing biosensing, environmental monitoring, and mechanical sensing applications. LNPs can be modified with different functional groups to enhance their sensing capabilities and selectivity. Table 13.4 gives an overview of the sensor applications along with some notes by the authors of the present chapter. Lignin-capped silver nanoparticles [137] and more recently silver nanoparticles grown on surfaces of stabilized LNPs [90] have been used for the colorimetric sensing of hydrogen peroxide via redox chemistry.

Table 13.4 Effect of process parameters and lignin characteristics on particle size and stability of LNPs from the solvent exchange method.

Sensed analyte of stimulus	Detection method	Notes	References
Hydrogen peroxide	Colorimetric	Based on oxidation of metal NPs	[90, 137]
Pressure and strain	Electrical resistance	Resistance dependent on strain angle	[138]
Metal ions	Colorimetric	Lignin as stabilizer for AuNPs for detection of 0.01–0.1 mM Pb^{2+}	[139]
Glucose	Colorimetric	Glucose oxidase–peroxidase enzyme cascade	[140]
Formaldehyde	Fluorescence	Augmented fluorescence in the presence of formaldehyde	[141]

13.4.5 Hybrid Particles with Metal and Inorganic Nanoparticles

Morsali et al. developed silver@HLNPs by employing alkali-stable particles derived from hydroxymethylated lignin as reductive templates [90]. The cured particles effectively reduce silver ions, leading to the creation of lignin-silver hybrid nanoparticles suitable for hydrogen peroxide sensing. Additionally, the particles can be modified through base-catalyzed epoxy ring-opening reactions, providing them with pH-dependent agglomeration and redispersion properties. With their scalability and solvent stability, these LNPs hold potential as circular biobased nanomaterials for sensor and catalyst applications. In an alternative approach, Moreno et al. conducted the synthesis of gold nanoparticles on the surfaces of LNPs, which were prepared from softwood kraft lignin that had undergone partial derivatization using fluorinated oleate esters [142]. In addition to being stabilized in alkaline media, the particles displayed reversible swelling in response to oxygen gas, a property related to migration or orientation of the fluorinated oleate chains. These gas-responsive nanoparticles were also used to accelerate the synthesis of gold nanoparticles, which deposited as particle clusters on their surfaces, giving rise to stable colloidal dispersion with dark color originating from the surface plasmon resonance peak of gold nanoparticles (Figure 13.6). Elsewhere, LNPs have also been used as supports for catalytic materials such as cuprous oxide for Huisgen [3+2] cycloadditions under relatively mild conditions [143].

13.4.6 Sunscreens

Lignin possesses natural UV-absorbing properties and antioxidant activity due to its conjugated aromatic structures and the presence of phenolic hydroxyl groups. This foundation has inspired researchers around the world to harness lignin as an eco-friendly and renewable substitute for conventional mineral and synthetic components found in sunscreens, specifically those designed for UV protection. Most of

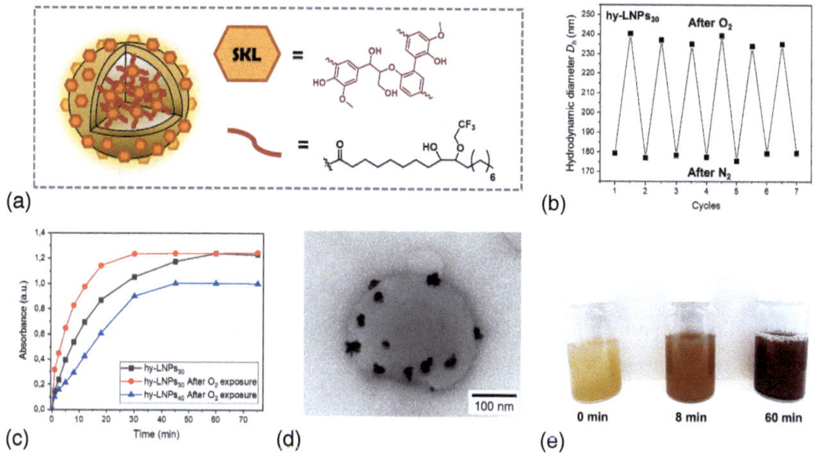

Figure 13.6 Gas-responsive LNPs. (a) Schematic structure of physically mixed (co-aggregated) softwood kraft lignin (SKL) and SKL-lignin fluorinated oleate particles. (b) Change in particle diameter as a response to oxygen and nitrogen gas bubbling cycles. (c) Kinetics of formation of gold nanoparticles in the dispersion of the hybrid lignin fluorinated oleate nanoparticles as such and after oxygen gas bubbling. (d) TEM micrograph of a single particle containing gold nanoparticle clusters on the surface. (e) Appearance in digital photographs of the dispersion after different reaction times in the presence of the gold precursor. Source: Copyright Wiley 2023: https://doi.org/10.1002/smll.202205672.

these formulations are o/w emulsions which benefit from surface-active properties of lignin originating from its amphiphilic structure. It is worth to note that when comparing the results, one should keep in mind that in vitro and in vivo results may deviate considerably from each other. In addition to the ability to absorb light energy and transmit it photothermally, the LNPs are considered important for their radical scavenging ability, especially in the case of complex formulations involving photoactive inorganic sunscreens such as titanium dioxide (TiO2) particles [144]. Table 13.5 gives an overview of sunscreens that contain LNPs as a booster of the sun protection factor (SPF). From this data, it is clear that typically 5–10% of LNPs are incorporated either in base lotion or as additives in commercial sunscreens. Furthermore, it is worth mentioning that the presence of free phenolic hydroxyl groups in lignin is beneficial for enhancing the SPF of sunscreens.

Conventional sunscreen formulations come with white or whiteish tone, but lignin-based sunscreens are gifted with different tones of brown originating from its natural color. Despite the fact that natural colors derived from lignin tend to complement human skin tones more effectively, endeavors have been undertaken to mitigate the brown coloration of lignin-based formulations through various modifications of lignin. For instance, alkaline oxidation of lignin with urea and hydrogen peroxide was reported to increase whiteness of lignin [145]. Another way to increase whiteness of LNPs is via esterification with long-chain fatty acids, e.g., lignin oleate nanoparticles [79]. However, the consumer preference for color tones of sunscreen formulations may differ regionally and evolve over time toward color tones that are

Table 13.5 Sunscreen formulations with LNPs.

Sunscreen formulation	SPF	LNPs (wt%)	References
LNPs	15	10	[15]
LNPs added to commercial sunscreens (SPF 19 and 55)	38, 86	5	[145]
LNPs (acetylated AL) blended into a base cream	1–2	1–5	[146]
LNPs, natural additives required to act in synergy	5–61	1–5	[147]
LNPs (OSL)	3–5	5	[148]
LNPs (AL) and polydopamine-grafted LNPs (AL)	77–223	10	[149]
LNPs (regular and oxidized SKL and HKL)	>21	10	[150]

OSL, organosolv lignin; AL, alkali lignin, SKL, softwood kraft lignin; HKL, hardwood kraft lignin.

closer to that of skin. To explore the subject of lignin modification for sunscreen applications, refer to the recent review by Tran et al. [151].

13.4.7 Biologically Active Lignin Nanomaterials

As reviewed, LNPs encapsulated with active agents have been used as biologically active agents. Maldonado-Carmona et al. encapsulated 5,10,15,20-tetrakis(4-hydroxyphenyl)-21H,23H-porphyrin into acetylated lignin water-dispersible nanoparticles as photosensitizers against several gram-negative and gram-positive bacteria [152]. LNPs have also been used as contrast agents for fluorescence imaging of *Escherichia coli* cells [153]. Moreover, lignin has the ability to reduce metal ions and particularly silver into nanoparticles [85]. The combination of metal nanoparticles and lignin has been approached in versatile ways to produce antimicrobial materials with increased stability and demonstrated antimicrobial activity against a plethora of bacteria [44, 45, 154, 155].

Compared to antimicrobial agents biomedical applications require materials to meet much more stringent requirements. Different methods of administration (intravenous, pulmonary, transdermal, oral) pose unique challenges. LNPs used in biomedicine must be biocompatible, exhibit appropriate mechanical properties, balance stability and biodegradation, and have surface properties supporting cell adhesion, proliferation, tissue integration, and sterilization. Research on the biomedical applications of LNPs is still in its early stages, with the majority of efforts focused on drug delivery systems [142] and, to a lesser extent, wound healing materials. LNPs derive several inherent properties of lignin, such as antioxidant activity, radical scavenging ability, absorbance of UV radiation, and surface activity, which are attractive toward biomedical materials [156, 157]. Fully biobased hydrogels with cellulose nanofibrils, calcium alginate, and LNPs were three-dimensionally printed and used as cell culture media with tailorable mechanical properties and no negative effects on proliferation of hepatocellular carcinoma cell line HepG2 [158]. Figueiredo and Santos have edited a book that explores various applications of LNPs in the field of biomedicine, including drug delivery systems. Readers are encouraged to explore the book for additional examples of the utilization of LNPs in this context [159]. The authors of this chapter hold the perspective that lignin and LNPs can be strategically employed to leverage the inherent functionalities of lignin. This approach includes utilizing lignin's radical scavenging ability in orally administered drug delivery systems for intestinal diseases, as well as incorporating its anti-inflammatory and UV-protective properties into wound dressings, among other potential applications.

13.5 Future Perspectives

Lignin nanoparticles have emerged as a research area that connects and intersects multiple fields, including bio-based materials and polymer chemistry. In recent years, there has been rapid technological advancement in the preparation of these colloidal particles, primarily through solvent-based and aerosol evaporation processes. The principles of green chemistry serve as essential guidelines for surpassing the current state of the art, either through process optimization and intensification or through the development of innovative processes. These green and energy-efficient approaches aim to minimize or eliminate the use of organic solvents and generate minimal waste. Consequently, the production of colloidal lignin particles without organic solvents, achieved through aerosol processes or neutralization of aqueous alkaline solutions of lignin mixtures, shows promising potential. Furthermore, there is a need to standardize particle properties and establish correlations between lignin structure, particle properties, and material performance in various applications. Additionally, as the scaling up and testing of lignin particles in relevant environments progresses, new opportunities for their introduction in novel applications emerge. This advancement pushes the technology readiness level of these unique bio-based materials, derived from renewable lignocellulosic resources.

References

1 Stewart, H., Golding, M., Matia-Merino, L. et al. (2014). Manufacture of lignin microparticles by anti-solvent precipitation: effect of preparation temperature and presence of sodium dodecyl sulfate. *Food Research International* 66: 93–99. https://doi.org/10.1016/j.foodres.2014.08.046.

2 Frangville, C., Rutkevičius, M., Richter, A.P. et al. (2012). Fabrication of environmentally biodegradable lignin nanoparticles. *ChemPhysChem* 13 (18): 4235–4243. https://doi.org/10.1002/cphc.201200537.

3 Liu, K., Zhuang, Y., Chen, J. et al. (2022). Research progress on the preparation and high-value utilization of lignin nanoparticles. *International Journal of Molecular Sciences* 23: 7254. https://doi.org/10.13386/j.issn1002-0306.2021010260.

4 Iravani, S. and Varma, R.S. (2020). Greener synthesis of lignin nanoparticles and their applications. *Green Chemistry* 22 (3): 612–636. https://doi.org/10.1039/c9gc02835h.

5 Chauhan, P.S. (2020). Lignin nanoparticles: eco-friendly and versatile tool for new era. *Bioresource Technology Reports* 9: https://doi.org/10.1016/j.biteb.2019.100374.

6 Duarah, P., Haldar, D., and Purkait, M.K. (2020). Technological advancement in the synthesis and applications of lignin-based nanoparticles derived from agro-industrial waste residues: a review. *International Journal of Biological Macromolecules* 163: 1828–1843. https://doi.org/10.1016/j.ijbiomac.2020.09.076.

7 Chen, K., Wang, S., Qi, Y. et al. (2021). State-of-the-art: applications and industrialization of lignin micro/nano particles. *ChemSusChem* 14 (5): 1284–1294. https://doi.org/10.1002/cssc.202002441.

8 Deng, J., Sun, S.F., Zhu, E.Q. et al. (2021). Sub-micro and nano-lignin materials: small size and rapid progress. *Industrial Crops and Products* 164 (March): 113412. https://doi.org/10.1016/j.indcrop.2021.113412.

9 Gao, W. and Fatehi, P. (2019). Lignin for polymer and nanoparticle production: current status and challenges. *The Canadian Journal of Chemical Engineering* https://doi.org/10.1002/cjce.23620.

10 Lizundia, E., Sipponen, M.H., Greca, L.G. et al. (2021). Multifunctional lignin-based nanocomposites and nanohybrids. *Green Chemistry* 23: 6698–6760. https://doi.org/10.1039/d1gc01684a.

11 Daniel, W., Schneider, H., Jos, A. et al. (2021). Lignin nanoparticles enter the scene: a promising versatile green tool for multiple applications. *Biotechnology Advances* 47: 107685. https://doi.org/10.1016/j.biotechadv.2020.107685.

12 Österberg, M., Sipponen, M.H., Mattos, B.D., and Rojas, O.J. (2020). Spherical lignin particles: a review on their sustainability and applications. *Green Chemistry* 22 (9): 2712–2733. https://doi.org/10.1039/d0gc00096e.

13 Beisl, S., Miltner, A., and Friedl, A. (2017). Lignin from micro- to nanosize: production methods. *International Journal of Molecular Sciences* 18 (6): https://doi.org/10.3390/ijms18061244.

14 Lu, Q., Zhu, M., Zu, Y. et al. (2012). Comparative antioxidant activity of nanoscale lignin prepared by a supercritical antisolvent (SAS) process with non-nanoscale lignin. *Food Chemistry* 135 (1): 63–67. https://doi.org/10.1016/j.foodchem.2012.04.070.

15 Qian, Y., Zhong, X., Li, Y., and Qiu, X. (2017). Fabrication of uniform lignin colloidal spheres for developing natural broad-spectrum sunscreens with high sun protection factor. *Industrial Crops and Products* 101: 54–60. https://doi.org/10.1016/j.indcrop.2017.03.001.

16 Farooq, M., Zou, T., Riviere, G. et al. (2019). Strong, ductile, and waterproof cellulose nanofibril composite films with colloidal lignin particles. *Biomacromolecules* 20 (2): https://doi.org/10.1021/acs.biomac.8b01364.

17 Zou, T., Sipponen, M.H., and Österberg, M. (2019). Natural shape-retaining microcapsules with shells made of chitosan-coated colloidal lignin particles. *Frontiers in Chemistry* 7 (May): 370. https://doi.org/10.3389/fchem.2019.00370.

18 Moreno, A. and Sipponen, M.H. (2020). Biocatalytic nanoparticles for the stabilization of degassed single electron transfer-living radical Pickering emulsion polymerizations. *Nature Communications* 11: 5599. https://doi.org/10.1038/s41467-020-19407-3.

19 Zou, T., Sipponen, M.H., Henn, A., and Österberg, M. (2021). Solvent-resistant lignin-epoxy hybrid nanoparticles for covalent surface modification and high-strength particulate adhesives. *ACS Nano* https://doi.org/10.1021/acsnano.0c09500.

20 Figueiredo, P., Lahtinen, M.H., Agustin, M.B. et al. (2021). Green fabrication approaches of lignin nanoparticles from different technical lignins: a comparison study. *ChemSusChem* 14 (21): 4718–4730. https://doi.org/10.1002/cssc.202101356.

21 Li, X., Shen, J., Wang, B. et al. (2021). Acetone/water cosolvent approach to lignin nanoparticles with controllable size and their applications for Pickering emulsions. *ACS Sustainable Chemistry and Engineering* 9 (15): 5470–5480. https://doi.org/10.1021/acssuschemeng.1c01021.

22 Gao, K., Liu, J., Li, X. et al. (2022). Lignin nanoparticles as highly efficient, recyclable emulsifiers for enhanced oil recovery. *ACS Sustainable Chemistry and Engineering* https://doi.org/10.1021/acssuschemeng.2c01101.

23 Lievonen, M., Valle-Delgado, J.J., Mattinen, M.-L. et al. (2016). A simple process for lignin nanoparticle preparation. *Green Chemistry* 18 (5): 1416–1422. https://doi.org/10.1039/C5GC01436K.

24 Pang, T., Wang, G., Sun, H. et al. (2020). Lignin fractionation for reduced heterogeneity in self-assembly nanosizing: toward targeted preparation of uniform lignin nanoparticles with small size. *ACS Sustainable Chemistry and Engineering* 8 (24): 9174–9183. https://doi.org/10.1021/acssuschemeng.0c02967.

25 Li, H., Deng, Y., Liang, J. et al. (2016). Direct preparation of hollow nanospheres with Kraft lignin: a facile strategy for effective utilization of biomass waste. *BioResources* 11 (2): 3073–3083. https://doi.org/10.15376/biores.11.2.3073-3083.

26 Manisekaran, A., Grysan, P., Duez, B. et al. (2022). Solvents drive self-assembly mechanisms and inherent properties of Kraft lignin nanoparticles (<50 Nm). *Journal of Colloid and Interface Science* 626: 178–192. https://doi.org/10.1016/j.jcis.2022.06.089.

27 Sipponen, M.H., Smyth, M., Leskinen, T. et al. (2017). All-lignin approach to prepare cationic colloidal lignin particles: stabilization of durable Pickering emulsions. *Green Chemistry* 19 (24): 5831–5840. https://doi.org/10.1039/C7GC02900D.

28 Leskinen, T., Smyth, M., Xiao, Y. et al. (2017). Scaling Up production of colloidal lignin particles. *Nordic Pulp & Paper Research Journal* 32 (4): 586–596. https://doi.org/10.3183/npprj-2017-32-04_p586-596_leskinen.

29 Li, H., Deng, Y., Wu, H. et al. (2016). Self-assembly of Kraft lignin into nanospheres in dioxane–water mixtures. *Holzforschung* 70 (8): 725–731. https://doi.org/10.1515/hf-2015-0238.

30 Sipponen, M.H., Lange, H., Ago, M., and Crestini, C. (2018). Understanding lignin aggregation processes. a case study: budesonide entrapment and stimuli controlled release from lignin nanoparticles. *ACS Sustainable Chemistry and Engineering* 6 (7): 9342–9351. https://doi.org/10.1021/acssuschemeng.8b01652.

31 Luo, T., Wang, C., Ji, X. et al. (2021). Preparation and characterization of size-controlled lignin nanoparticles with deep eutectic solvents by nanoprecipitation. *Molecules* 26 (1): https://doi.org/10.3390/molecules26010218.

32 Österberg, M., Sipponen, M., Kostiainen, M., Äkräs, L., Rivière, G., Zhang, X. (2022) Lignin particle based hydrogel and the method for preparation of lignin colloidal particles by solvent evaporation process. US 2022/0010077 A1.

33 Wang, M., Yang, D., Xu, Q. et al. (2021). Highly efficient evaporation method to prepare PH-responsive lignin-hollow-nanosphere with controllable size and its application in oral drug delivery. *Industrial Crops and Products* 162: 113230. https://doi.org/10.1016/j.indcrop.2020.113230.

34 Qian, Y., Qiu, X., Zhong, X. et al. (2015). Lignin reverse micelles for UV-absorbing and high mechanical performance thermoplastics. *Industrial and Engineering Chemistry Research* 54 (48): 12025–12030. https://doi.org/10.1021/acs.iecr.5b03360.

35 Zou, T., Nonappa, N., Khavani, M. et al. (2021). Experimental and simulation study of the solvent effects on the intrinsic properties of spherical lignin nanoparticles. *Journal of Physical Chemistry B* 125 (44): 12315–12328. https://doi.org/10.1021/acs.jpcb.1c05319.

36 Luo, T., Wang, C., Ji, X. et al. (2021). Innovative production of lignin nanoparticles using deep eutectic solvents for multifunctional nanocomposites. *International Journal of Biological Macromolecules* 183: 781–789. https://doi.org/10.1016/j.ijbiomac.2021.05.005.

37 Guo, Y., Xu, L., Shen, F. et al. (2022). Insights into lignocellulosic waste fractionation for lignin nanospheres fabrication using acidic/alkaline deep eutectic solvents. *Chemosphere* 286: https://doi.org/10.1016/j.chemosphere.2021.131798.

38 Liu, C., Li, Y., and Hou, Y. (2019). A simple environment-friendly process for preparing high-concentration alkali lignin nanospheres. *European Polymer Journal* 112: 15–23. https://doi.org/10.1016/j.eurpolymj.2018.12.031.

39 Xie, B., Hou, Y., and Li, Y. (2020). Modified lignin nanosphere adsorbent for lead and copper ions. *BioResources* 16 (1): 249–262. https://doi.org/10.15376/BIORES.16.1.249-262.

40 Chen, L., Zhou, X., Shi, Y. et al. (2018). Green synthesis of lignin nanoparticle in aqueous hydrotropic solution toward broadening the window for its processing and application. *Chemical Engineering Journal* 346 (March): 217–225. https://doi.org/10.1016/j.cej.2018.04.020.

41 Jiang, W., Liu, S., Wu, C. et al. (2020). Super-stable, solvent-resistant and uniform lignin nanorods and nanospheres with a high yield in a mild and facile process. *Green Chemistry* 22 (24): 8734–8744. https://doi.org/10.1039/d0gc02887h.

42 Padilha, C.E.D.A., Nogueira, C.D.C., Matias, S.C.B. et al. (2020). Fabrication of hollow polymer microcapsules and removal of emulsified oil from aqueous environment using soda lignin nanoparticles. *Colloids and Surfaces A: Physicochemical and Engineering Aspects* 603 (July): 125260. https://doi.org/10.1016/j.colsurfa.2020.125260.

43 Agustin, M.B., Penttilä, P.A., Lahtinen, M., and Mikkonen, K.S. (2019). Rapid and direct preparation of lignin nanoparticles from alkaline pulping liquor by mild ultrasonication. *ACS Sustainable Chemistry & Engineering* 7: 19925–19934. https://doi.org/10.1021/acssuschemeng.9b05445.

44 Richter, A.P., Bharti, B., Armstrong, H.B. et al. (2016). Synthesis and characterization of biodegradable lignin nanoparticles with tunable surface properties. *Langmuir* 32 (25): 6468–6477. https://doi.org/10.1021/acs.langmuir.6b01088.

45 Richter, A.P., Brown, J.S., Bharti, B. et al. (2015). An environmentally benign antimicrobial nanoparticle based on a silver-infused lignin core. *Nature Nanotechnology* 10 (9): 817–823. https://doi.org/10.1038/nnano.2015.141.

46 Yang, W., Fortunati, E., Gao, D. et al. (2018). Valorization of acid isolated high yield lignin nanoparticles as innovative antioxidant/antimicrobial organic materials. *ACS Sustainable Chemistry and Engineering* 6 (3): 3502–3514. https://doi.org/10.1021/acssuschemeng.7b03782.

47 Pylypchuk, I. and Sipponen, M.H. (2022). Organic solvent-free production of colloidally stable spherical lignin nanoparticles at high mass concentrations. *Green Chemistry* 24 (22): 8705–8715. https://doi.org/10.1039/D2GC02316D.

48 Ago, M., Huan, S., Borghei, M. et al. (2016). High-throughput synthesis of lignin particles (~30 Nm to ~2 Mm) via aerosol flow reactor: size fractionation and utilization in Pickering emulsions. *ACS Applied Materials & Interfaces* 8 (35): 23302–23310. https://doi.org/10.1021/acsami.6b07900.

49 Lourencon, T.V., Greca, L.G., Tarasov, D. et al. (2020). Lignin-first integrated hydrothermal treatment (HTT) and synthesis of low-cost biorefinery particles. *ACS Sustainable Chemistry & Engineering* 8 (2): 1230–1239. https://doi.org/10.1021/acssuschemeng.9b06511.

50 Modi, S., Foston, M.B., and Biswas, P. (2023). Controlled synthesis of smaller than 100 Nm lignin nanoparticles in a furnace aerosol reactor. *ACS ES&T Engineering* 3 (5): 671–681. https://doi.org/10.1021/acsestengg.2c00386.

51 Zhou, H., Modi, S., and Biswas, P. (2022). Controlled synthesis of charged lignin nanocarriers by electrospray. *Colloids and Surfaces A: Physicochemical and Engineering Aspects* 648: 129314. https://doi.org/10.1016/j.colsurfa.2022.129314.

52 Mishra, P.K. and Wimmer, R. (2017). Aerosol assisted self-assembly as a route to synthesize solid and hollow spherical lignin colloids and its utilization in layer by layer deposition. *Ultrasonics Sonochemistry* 35: 45–50. https://doi.org/10.1016/j.ultsonch.2016.09.001.

53 Farooq, M., Tao, Z., Valle-Delgado, J.J. et al. (2020). Well-defined lignin model films from colloidal lignin particles. *Langmuir* 36 (51): 15592–15602. https://doi.org/10.1021/acs.langmuir.0c02970.

54 Pylypchuk, I.V., Lindén, P.A., Lindström, M.E., and Sevastyanova, O. (2020). New insight into the surface structure of lignin nanoparticles revealed by 1H liquid-state NMR spectroscopy. *ACS Sustainable Chemistry & Engineering* 8 (36): 13805–13812. https://doi.org/10.1021/acssuschemeng.0c05119.

55 Chen, L., Luo, S.-M., Huo, C.-M. et al. (2022). New insight into lignin aggregation guiding efficient synthesis and functionalization of a lignin nanosphere with excellent performance. *Green Chemistry* 24: 285–294. https://doi.org/10.1039/d1gc03651c.

56 Zhang, Y. and Naebe, M. (2021). Lignin: a review on structure, properties, and applications as a light-colored UV absorber. *ACS Sustainable Chemistry & Engineering* 9 (4): 1427–1442. https://doi.org/10.1021/acssuschemeng.0c06998.

57 Freitas, F.M.C., Cerqueira, M.A., Gonçalves, C. et al. (2020). Green synthesis of lignin nano- and micro-particles: physicochemical characterization, bioactive properties and cytotoxicity assessment. *International Journal of Biological Macromolecules* 163: 1798–1809. https://doi.org/10.1016/j.ijbiomac.2020.09.110.

58 Moreno, A., Liu, J., Morsali, M., and Sipponen, M.H. (2021). Chemical modification and functionalization of lignin nanoparticles. In: *Micro and Nanolignin in Aqueous Dispersions and Polymers* (ed. D. Puglia, C. Santulli, and F. Sarasini), 385–424. Amsterdam: Elsevier.

59 Pylypchuk, I.V., Riazanova, A., Lindström, M.E., and Sevastyanova, O. (2021). Structural and molecular-weight-dependency in the formation of lignin nanoparticles from fractionated soft- and hardwood lignins. *Green Chemistry* 23 (8): 3061–3072. https://doi.org/10.1039/d0gc04058d.

60 Conner, C.G., Veleva, A.N., Paunov, V.N. et al. (2020). Scalable formation of concentrated monodisperse lignin nanoparticles by recirculation-enhanced flash nanoprecipitation. *Particle and Particle Systems Characterization* 37 (7): 1–10. https://doi.org/10.1002/ppsc.202000122.

61 Lu, X., Zhang, Y., Chen, J. et al. (2021). Preparation of tailored lignin nanospheres with specific size and tunable structure by self-assembly of lignin from corn straw. *Industrial Crops and Products* 172 (August): 113993. https://doi.org/10.1016/j.indcrop.2021.113993.

62 Ma, M., Dai, L., Xu, J. et al. (2020). A simple and effective approach to fabricate lignin nanoparticles with tunable sizes based on lignin fractionation. *Green Chemistry* 22 (6): 2011–2017. https://doi.org/10.1039/D0GC00377H.

63 Figueiredo, P., Sipponen, M.H., Lintinen, K. et al. (2019). Preparation and characterization of dentin phosphophoryn-derived peptide-functionalized lignin nanoparticles for enhanced cellular uptake. *Small* 15 (24): 1901427. https://doi.org/10.1002/smll.201901427.

64 Cao, Q., Wu, Q., Dai, L. et al. (2021). Size-controlled lignin nanoparticles for tuning the mechanical properties of poly(vinyl alcohol). *Industrial Crops and Products* 172 (July): https://doi.org/10.1016/j.indcrop.2021.114012.

65 Liu, Z.-H., Hao, N., Shinde, S. et al. (2019). Codesign of combinatorial organosolv pretreatment (COP) and lignin nanoparticles (LNPs) in biorefineries. *ACS Sustainable Chemistry & Engineering* 7 (2): 2634–2647. https://doi.org/10.1021/acssuschemeng.8b05715.

66 Mishra, P.K. and Ekielski, A. (2019). The self-assembly of lignin and its application in nanoparticle synthesis: a short review. *Nanomaterials* 9 (2): 243. https://doi.org/10.3390/nano9020243.

67 Zhang, J., Tian, Z., Ji, X.-X., and Zhang, F. (2023). Light-colored lignin extraction by ultrafiltration membrane fractionation for lignin nanoparticles preparation as

UV-blocking sunscreen. *International Journal of Biological Macromolecules* 231: 123244. https://doi.org/10.1016/j.ijbiomac.2023.123244.

68 Liu, J., Nero, M., Jansson, K. et al. (2023). Photonic crystals with rainbow colors by centrifugation-assisted assembly of colloidal lignin nanoparticles. *Nature Communications* 14 (1): 3099. https://doi.org/10.1038/s41467-023-38819-5.

69 Wang, J., Chen, W., Yang, D. et al. (2022). Monodispersed lignin colloidal spheres with tailorable sizes for bio-photonic materials. *Small* 18 (19): 1–11. https://doi.org/10.1002/smll.202200671.

70 Wang, J., Chen, W., Yang, D. et al. (2022). Photonic lignin with tunable and stimuli-responsive structural color. *ACS Nano* 16 (12): 20705–20713. https://doi.org/10.1021/acsnano.2c07756.

71 Xiong, F., Wang, H., Xu, H. et al. (2020). Self-assembled lignin nanospheres with solid and hollow tunable structures. *Industrial Crops and Products* 144: 112063. https://doi.org/10.1016/j.indcrop.2019.112063.

72 Cao, X., Huang, J., He, Y. et al. (2021). Biodegradable and renewable UV-shielding polylactide composites containing hierarchical structured POSS functionalized lignin. *International Journal of Biological Macromolecules* 188 (August): 323–332. https://doi.org/10.1016/j.ijbiomac.2021.08.033.

73 Li, Y., Yang, D., Lu, S. et al. (2018). Modified lignin with anionic surfactant and its application in controlled release of avermectin. *Journal of Agricultural and Food Chemistry* 66 (13): 3457–3464. https://doi.org/10.1021/acs.jafc.8b00393.

74 Liu, C., Li, Y., and Hou, Y. (2019). Preparation of a novel lignin nanosphere adsorbent for enhancing adsorption of lead. *Molecules* 24 (15): https://doi.org/10.3390/molecules24152704.

75 Yang, W., Ding, H., Qi, G. et al. (2021). Enhancing the radical scavenging activity and UV resistance of lignin nanoparticles via surface Mannich amination toward a biobased antioxidant. *Biomacromolecules* 22 (6): 2693–2701. https://doi.org/10.1021/acs.biomac.1c00387.

76 Zhao, J., Zheng, D., Tao, Y. et al. (2020). Self-assembled pH-responsive polymeric nanoparticles based on lignin-histidine conjugate with small particle size for efficient delivery of anti-tumor drugs. *Biochemical Engineering Journal* 156: 107526. https://doi.org/10.1016/j.bej.2020.107526.

77 Zhou, M., Xiong, Z., Yang, D. et al. (2018). Preparation of slow release nanopesticide microspheres from benzoyl lignin. *Holzforschung* 72: 599–607.

78 Li, B., You, S., Qi, W. et al. (2020). Structure-tunable assembly of lignin sub-micro spheres by modifying the amphiphilic interfaces of lignin via n-alkane. *European Polymer Journal* 126 (January): 109539. https://doi.org/10.1016/j.eurpolymj.2020.109539.

79 Moreno, A., Liu, J., Gueret, R. et al. (2021). Unravelling the hydration barrier of lignin oleate nanoparticles for acid- and base-catalyzed functionalization in dispersion state. *Angewandte Chemie International Edition* 60 (38): 20897–20905. https://doi.org/10.1002/anie.202106743.

80 Pourmoazzen, Z., Sadeghifar, H., Yang, G., and Lucia, L. (2020). Cholesterol-modified lignin: a new avenue for green nanoparticles, meltable materials, and drug delivery. *Colloids and Surfaces B: Biointerfaces* 186: 110685. https://doi.org/10.1016/j.colsurfb.2019.110685.

81 Astete, C.E., De Mel, J.U., Gupta, S. et al. (2020). Lignin-graft-poly(lactic-co-glycolic) acid biopolymers for polymeric nanoparticle synthesis. *ACS Omega* 5 (17): 9892–9902. https://doi.org/10.1021/acsomega.0c00168.

82 Sipponen, M.H., Henn, A., Penttilä, P., and Österberg, M. (2020). Lignin-fatty acid hybrid nanocapsules for scalable thermal energy storage in phase-change materials. *Chemical Engineering Journal* 393: 124711. https://doi.org/10.1016/j.cej.2020.124711.

83 Stanisz, M., Klapiszewski, Ł., Kołodyńska, D., and Jesionowski, T. (2021). Development of functional lignin-based spherical particles for the removal of Vanadium(V) from an aqueous system. *International Journal of Biological Macromolecules* 186 (July): 181–193. https://doi.org/10.1016/j.ijbiomac.2021.07.046.

84 Jiang, Y., Chen, Y., Tian, D. et al. (2020). Fabrication and characterization of lignin-xylan hybrid nanospheres as pesticide carriers with enzyme-mediated release property. *Soft Matter* 16 (39): 9083–9093. https://doi.org/10.1039/d0sm01402h.

85 Budnyak, T.M., Slabon, A., and Sipponen, M.H. (2020). Lignin–inorganic interfaces: chemistry and applications from adsorbents to catalysts and energy storage materials. *ChemSusChem* 4344–4355. https://doi.org/10.1002/cssc.202000216.

86 Lintinen, K., Latikka, M., Sipponen, M.H. et al. (2016). Structural diversity in metal–organic nanoparticles based on iron isopropoxide treated lignin. *RSC Advances* 6 (38): 31790–31796. https://doi.org/10.1039/C6RA03865D.

87 Li, T., Lü, S., Wang, Z. et al. (2020). Lignin-based nanoparticles for recovery and separation of phosphate and reused as renewable magnetic fertilizers. *Science of the Total Environment* 142745. https://doi.org/10.1016/j.scitotenv.2020.142745.

88 Yang, J., Dai, J., Liu, X. et al. (2022). A lignin-based epoxy/TiO2 hybrid nanoparticle for multifunctional bio-based epoxy with improved mechanical, UV absorption and antibacterial properties. *International Journal of Biological Macromolecules* 210 (March): 85–93. https://doi.org/10.1016/j.ijbiomac.2022.04.229.

89 Wang, H., Xiong, F., Tan, Y. et al. (2021). Preparation and formation mechanism of covalent-noncovalent forces stabilizing lignin nanospheres and their application in superhydrophobic and carbon materials. *ACS Sustainable Chemistry and Engineering* 9 (10): 3811–3820. https://doi.org/10.1021/acssuschemeng.0c08780.

90 Morsali, M., Moreno, A., Loukovitou, A. et al. (2022). Stabilized lignin nanoparticles for versatile hybrid and functional nanomaterials. *Biomacromolecules* 23 (11): 4597–4606. https://doi.org/10.1021/acs.biomac.2c00840.

91 Moreno, A. and Sipponen, M.H. (2020). Lignin-based smart materials: a roadmap to processing and synthesis for current and future applications. *Materials Horizons* 7 (9): 2237–2257. https://doi.org/10.1039/d0mh00798f.

92 Tardy, B.L., Lizundia, E., Guizani, C. et al. (2023). Prospects for the integration of lignin materials into the circular economy. *Materials Today* 65: 122–132. https://doi.org/10.1016/j.mattod.2023.04.001.

93 Liu, J., Moreno, A., Chang, J. et al. (2022). Fully biobased photothermal films and coatings for indoor ultraviolet radiation and heat management. *ACS Applied Materials and Interfaces* 14 (10): 12693–12702. https://doi.org/10.1021/acsami.2c00718.

94 Tian, D., Guo, Y., Huang, M. et al. (2021). Bacterial cellulose/lignin nanoparticles composite films with retarded biodegradability. *Carbohydrate Polymers* 274 (August): 118656. https://doi.org/10.1016/j.carbpol.2021.118656.

95 Wang, L., Tan, L., Hu, L. et al. (2021). On laccase-catalyzed polymerization of biorefinery lignin fractions and alignment of lignin nanoparticles on the nanocellulose surface via one-pot water-phase synthesis. *ACS Sustainable Chemistry and Engineering* 9 (26): 8770–8782. https://doi.org/10.1021/acssuschemeng.1c01576.

96 Yang, W., Qi, G., Kenny, J.M. et al. (2020). Effect of cellulose nanocrystals and lignin nanoparticles on mechanical, antioxidant and water vapour barrier properties of glutaraldehyde crosslinked PVA films. *Polymers* 12 (6): https://doi.org/10.3390/POLYM12061364.

97 Feng, Y., Yu, J., Sun, D. et al. (2022). Solvent-induced in-situ self-assembly lignin nanoparticles to reinforce conductive nanocomposite organogels as anti-freezing and anti-dehydration flexible strain sensors. *Chemical Engineering Journal* 433 (P2): 133202. https://doi.org/10.1016/j.cej.2021.133202.

98 Chen, Y., Zheng, K., Niu, L. et al. (2019). Highly mechanical properties nanocomposite hydrogels with biorenewable lignin nanoparticles. *International Journal of Biological Macromolecules* 128: 414–420. https://doi.org/10.1016/j.ijbiomac.2019.01.099.

99 Long, H., Hu, L., Yang, F. et al. (2022). Enhancing the performance of polylactic acid composites through self-assembly lignin nanospheres for fused deposition modeling. *Composites Part B: Engineering* 239: 109968. https://doi.org/10.1016/j.compositesb.2022.109968.

100 Ferruti, F., Pylypchuk, I., Zoia, L. et al. (2023). Recombinatorial approach for the formation of surface-functionalised alkaline-stable lignin nanoparticles and adhesives. *Green Chemistry* 25 (2): 639–649. https://doi.org/10.1039/D2GC03406A.

101 Sohni, S., Hashim, R., Nidaullah, H. et al. (2019). Chitosan/nano-lignin based composite as a new sorbent for enhanced removal of dye pollution from aqueous solutions. *International Journal of Biological Macromolecules* 132: 1304–1317. https://doi.org/10.1016/j.ijbiomac.2019.03.151.

102 Xiong, F., Wu, Y., Li, G. et al. (2018). Transparent nanocomposite films of lignin nanospheres and poly(vinyl alcohol) for UV-absorbing. *Industrial and Engineering Chemistry Research* 57 (4): 1207–1212. https://doi.org/10.1021/acs.iecr.7b04108.

103 Moreno, A., Morsali, M., Liu, J., and Sipponen, M.H. (2021). Access to tough and transparent nanocomposites via Pickering emulsion polymerization using biocatalytic hybrid lignin nanoparticles as functional surfactants. *Green Chemistry* 23: 3001–3014.

104 Yang, W., Fortunati, E., Bertoglio, F. et al. (2018). Polyvinyl alcohol/chitosan hydrogels with enhanced antioxidant and antibacterial properties induced by lignin nanoparticles. *Carbohydrate Polymers* 181: 275–284. https://doi.org/10.1016/j.carbpol.2017.10.084.

105 Henn, K.A., Forsman, N., Zou, T., and Österberg, M. (2021). Colloidal lignin particles and epoxies for bio-based, durable, and multiresistant nanostructured coatings. *ACS Applied Materials & Interfaces* 13 (29): 34793–34806. https://doi.org/10.1021/acsami.1c06087.

106 Henn, K.A., Forssell, S., Pietiläinen, A. et al. (2022). Interfacial catalysis and lignin nanoparticles for strong fire- and water-resistant composite adhesives. *Green Chemistry* 24 (17): 6487–6500. https://doi.org/10.1039/D2GC01637K.

107 Sipponen, M.H., Lange, H., Crestini, C. et al. (2019). Lignin for nano- and microscaled carrier systems: applications, trends, and challenges. *ChemSusChem* 12 (10): 2039–2054. https://doi.org/10.1002/cssc.201900480.

108 Pereira, A.D.E.S., Luiz de Oliveira, J., Maira Savassa, S. et al. (2022). Lignin nanoparticles: new insights for a sustainable agriculture. *Journal of Cleaner Production* 345: 131145. https://doi.org/10.1016/j.jclepro.2022.131145.

109 Wijaya, C.J., Ismadji, S., and Gunawan, S. (2021). A review of lignocellulosic-derived nanoparticles for drug delivery applications: lignin nanoparticles, xylan nanoparticles, and cellulose nanocrystals. *Molecules* 26: 676. https://www.mdpi.com/1420-3049/26/3/676 (accessed 16 June 2023.

110 Kumar, R., Butreddy, A., Kommineni, N. et al. (2021). Lignin: drug/gene delivery and tissue engineering applications. *International Journal of Nanomedicine* 16: 2419–2441. https://doi.org/10.2147/IJN.S303462.

111 Kah, M., Kookana, R.S., Gogos, A., and Bucheli, T.D. (2018). A critical evaluation of nanopesticides and nanofertilizers against their conventional analogues. *Nature Nanotechnology* 13 (8): 677–684. https://doi.org/10.1038/s41565-018-0131-1.

112 Zhou, M., Wang, D., Yang, D. et al. (2019). Avermectin loaded nanosphere prepared from acylated alkali lignin showed anti-photolysis property and controlled release performance. *Industrial Crops and Products* 137 (May): 453–459. https://doi.org/10.1016/j.indcrop.2019.04.037.

113 Fu, X., Zheng, Z., Sha, Z. et al. (2022). Biorefining waste into nanobiotechnologies can revolutionize sustainable agriculture. *Trends in Biotechnology* 40 (12): 1503–1518. https://doi.org/10.1016/j.tibtech.2022.09.013.

114 Yiamsawas, D., Kangwansupamonkon, W., and Kiatkamjornwong, S. (2021). Lignin-based nanogels for the release of payloads in alkaline conditions. *European Polymer Journal* 145: 110241. https://doi.org/10.1016/j.eurpolymj.2020.110241.

115 Yin, J.M., Wang, H.L., Yang, Z.K. et al. (2020). Engineering lignin nanomicroparticles for the antiphotolysis and controlled release of the plant growth regulator abscisic acid. *Journal of Agricultural and Food Chemistry* 68 (28): 7360–7368. https://doi.org/10.1021/acs.jafc.0c02835.

116 Andeme Ela, R.C., Tajiri, M., Newberry, N.K. et al. (2020). Double-shell lignin nanocapsules are a stable vehicle for fungicide encapsulation and release. *ACS Sustainable Chemistry and Engineering* 8 (46): 17299–17306. https://doi.org/10.1021/acssuschemeng.0c06686.

117 MacHado, T.O., Beckers, S.J., Fischer, J. et al. (2020). Bio-based lignin nanocarriers loaded with fungicides as a versatile platform for drug delivery in plants.

Biomacromolecules 21 (7): 2755–2763. https://doi.org/10.1021/acs.biomac.0c00487.

118 Zikeli, F., Vinciguerra, V., Sennato, S. et al. (2020). Preparation of lignin nanoparticles with entrapped essential oil as a bio-based biocide delivery system. *ACS Omega* 5 (1): 358–368. https://doi.org/10.1021/acsomega.9b02793.

119 Luo, J., Zhang, D.X., Jing, T. et al. (2020). Pyraclostrobin loaded lignin-modified nanocapsules: delivery efficiency enhancement in soil improved control efficacy on tomato fusarium crown and root rot. *Chemical Engineering Journal* 394 (March): https://doi.org/10.1016/j.cej.2020.124854.

120 Chandna, S., Thakur, N.S., Kaur, R., and Bhaumik, J. (2020). Lignin-bimetallic nanoconjugate doped pH-responsive hydrogels for laser-assisted antimicrobial photodynamic therapy. *Biomacromolecules* 21 (8): 3216–3230. https://doi.org/10.1021/acs.biomac.0c00695.

121 Cheng, L., Deng, B., Luo, W. et al. (2020). pH-responsive lignin-based nanomicelles for oral drug delivery. *Journal of Agricultural and Food Chemistry* 68 (18): 5249–5258. https://doi.org/10.1021/acs.jafc.9b08171.

122 Sgarzi, M., Gigli, M., Giuriato, C., and Crestini, C. (2022). Simple strategies to modulate the pH-responsiveness of lignosulfonate-based delivery systems. *Materials* 15 (5): https://doi.org/10.3390/ma15051857.

123 Dai, L., Li, Y., Kong, F. et al. (2019). Lignin-based nanoparticles stabilized Pickering emulsion for stability improvement and thermal-controlled release of trans-resveratrol. *ACS Sustainable Chemistry and Engineering* 7 (15): 13497–13504. https://doi.org/10.1021/acssuschemeng.9b02966.

124 Siddiqui, L., Bag, J., Seetha et al. (2020). Assessing the potential of lignin nanoparticles as drug carrier: synthesis, cytotoxicity and genotoxicity studies. *International Journal of Biological Macromolecules* 152: 786–802. https://doi.org/10.1016/j.ijbiomac.2020.02.311.

125 Chen, K., Qian, Y., Wang, C. et al. (2021). Tumor microenvironment-responsive, high internal phase Pickering emulsions stabilized by lignin/chitosan oligosaccharide particles for synergistic cancer therapy. *Journal of Colloid and Interface Science* 591: 352–362. https://doi.org/10.1016/j.jcis.2021.02.012.

126 Zhou, Y., Han, Y., Li, G., and Chu, F. (2019). Effects of lignin-based hollow nanoparticle structure on the loading and release behavior of doxorubicin. *Materials* 12 (10): https://doi.org/10.3390/MA12101694.

127 Figueiredo, P., Lepland, A., Scodeller, P. et al. (2020). Peptide-guided resiquimod-loaded lignin nanoparticles convert tumor-associated macrophages from M2 to M1 phenotype for enhanced chemotherapy. *Acta Biomaterialia* https://doi.org/10.1016/j.actbio.2020.09.038.

128 Zhou, Y., Han, Y., Li, G. et al. (2019). Lignin-based hollow nanoparticles for controlled drug delivery: grafting preparation using β-cyclodextrin/enzymatic-hydrolysis lignin. *Nanomaterials* 9 (7): https://doi.org/10.3390/nano9070997.

129 Zhou, Y., Han, Y., Li, G. et al. (2019). Preparation of targeted lignin-based hollow nanoparticles for the delivery of doxorubicin. *Nanomaterials* 9 (2): https://doi.org/10.3390/nano9020188.

130 Alqahtani, M.S., Alqahtani, A., Al-Thabit, A. et al. (2019). Novel lignin nanoparticles for oral drug delivery. *Journal of Materials Chemistry B* 7 (28): 4461–4473. https://doi.org/10.1039/c9tb00594c.

131 Alqahtani, M.S., Alqahtani, A., Kazi, M. et al. (2020). Wound-healing potential of curcumin loaded lignin nanoparticles. *Journal of Drug Delivery Science and Technology* 60 (August): 102020. https://doi.org/10.1016/j.jddst.2020.102020.

132 Dinari, A., Abdollahi, M., and Sadeghizadeh, M. (2021). Design and fabrication of dual responsive lignin-based nanogel via "grafting from" atom transfer radical polymerization for curcumin loading and release. *Scientific Reports* 11 (1): 1–16. https://doi.org/10.1038/s41598-021-81393-3.

133 Gordobil, O., Blažević, N., Simonič, M., and Sandak, A. (2023). Potential of lignin multifunctionality for a sustainable skincare: impact of emulsification process parameters and oil-phase on the characteristics of O/W Pickering emulsions. *International Journal of Biological Macromolecules* 233: 123561. https://doi.org/10.1016/j.ijbiomac.2023.123561.

134 Borrero-López, A.M., Wang, L., Li, H. et al. (2023). Oleogels and reverse emulsions stabilized by acetylated Kraft lignins. *International Journal of Biological Macromolecules* 242: 124941. https://doi.org/10.1016/j.ijbiomac.2023.124941.

135 Tarasov, D., Schlee, P., Pranovich, A. et al. (2022). AqSO biorefinery: a green and parameter-controlled process for the production of lignin–carbohydrate hybrid materials. *Green Chemistry* 24 (17): 6639–6656. https://doi.org/10.1039/D2GC02171D.

136 Wang, Y., Li, X., Shen, C. et al. (2020). Lignin assisted Pickering emulsion polymerization to microencapsulate 1-tetradecanol for thermal management. *International Journal of Biological Macromolecules* 146: 1–8. https://doi.org/10.1016/j.ijbiomac.2019.12.175.

137 Saratale, R.G., Saratale, G.D., Ghodake, G. et al. (2019). Wheat straw extracted lignin in silver nanoparticles synthesis: expanding its prophecy towards antineoplastic potency and hydrogen peroxide sensing ability. *International Journal of Biological Macromolecules* 128: 391–400. https://doi.org/10.1016/j.ijbiomac.2019.01.120.

138 Han, X., Zhang, Y., Ran, F. et al. (2022). Lignin nanoparticles for hydrogel-based pressure sensor. *Industrial Crops and Products* 176: 114366. https://doi.org/10.1016/j.indcrop.2021.114366.

139 Li, S., Xue, Y., Mai, Y. et al. (2022). Light-induced facile and efficient synthesis of color-variable lignin-based gold nanoparticles and its application as Pb2+ sensor. *International Journal of Biological Macromolecules* 211: 26–34. https://doi.org/10.1016/j.ijbiomac.2022.05.029.

140 Capecchi, E., Piccinino, D., Tomaino, E. et al. (2020). Lignin nanoparticles are renewable and functional platforms for the concanavalin a oriented immobilization of glucose oxidase–peroxidase in cascade bio-sensing. *RSC Advances* 10 (48): 29031–29042. https://doi.org/10.1039/D0RA04485G.

141 Ma, Z., Liu, C., Niu, N. et al. (2018). Seeking brightness from nature: J-aggregation-induced emission in cellulolytic enzyme lignin nanoparticles. *ACS*

Sustainable Chemistry & Engineering 6 (3): 3169–3175. https://doi.org/10.1021/acssuschemeng.7b03265.

142 Moreno, A., Delgado-Lijarcio, J., Ronda, J.C. et al. (2022). Breathable lignin nanoparticles as reversible gas swellable nanoreactors. *Small* 2205672. https://doi.org/10.1002/smll.202205672.

143 Zhou, Z., Peng, X., Zhong, L. et al. (2018). Lignin nanosphere-supported cuprous oxide as an efficient catalyst for Huisgen [3+2] cycloadditions under relatively mild conditions. *Polymers* 10 (7): 1–16. https://doi.org/10.3390/polym10070724.

144 Li, Y., Yang, D., Lu, S. et al. (2019). Encapsulating TiO_2 in lignin-based colloidal spheres for high sunscreen performance and weak photocatalytic activity. *ACS Sustainable Chemistry & Engineering* 7 (6): 6234–6242. https://doi.org/10.1021/acssuschemeng.8b06607.

145 Yu, H., Wang, B., Wang, Y. et al. (2023). Lignin nanoparticles with high phenolic content as efficient antioxidant and sun-blocker for food and cosmetics. *ACS Sustainable Chemistry & Engineering* 11 (10): 4082–4092. https://doi.org/10.1021/acssuschemeng.2c06070.

146 Wang, B., Sun, D., Wang, H.-M. et al. (2019). Green and facile preparation of regular lignin nanoparticles with high yield and their natural broad-spectrum sunscreens. *ACS Sustainable Chemistry & Engineering* 7: acssuschemeng.8b05735. https://doi.org/10.1021/acssuschemeng.8b05735.

147 Piccinino, D., Capecchi, E., Delfino, I. et al. (2021). Green and scalable preparation of colloidal suspension of lignin nanoparticles and its application in eco-friendly sunscreen formulations. *ACS Omega* 6 (33): 21444–21456. https://doi.org/10.1021/acsomega.1c02268.

148 Li, S.X., Li, M.F., Bian, J. et al. (2019). Preparation of organic acid lignin submicrometer particle as a natural broad-spectrum photo-protection agent. *International Journal of Biological Macromolecules* 132: 836–843. https://doi.org/10.1016/j.ijbiomac.2019.03.177.

149 Zhou, Y., Qian, Y., Wang, J. et al. (2020). Bioinspired lignin-polydopamine nanocapsules with strong bioadhesion for long-acting and high-performance natural sunscreens. *Biomacromolecules* https://doi.org/10.1021/acs.biomac.0c00696.

150 Widsten, P., Tamminen, T., and Liitiä, T. (2020). Natural sunscreens based on nanoparticles of modified Kraft lignin (CatLignin). *ACS Omega* 5 (22): 13438–13446. https://doi.org/10.1021/acsomega.0c01742.

151 Tran, M.H., Phan, D., and Lee, E.Y. (2021). Review on lignin modifications toward natural UV protection ingredient for lignin-based sunscreens. *Green Chemistry* 23: 4633–4646. https://doi.org/10.1039/d1gc01139a.

152 Maldonado-Carmona, N., Marchand, G., Villandier, N. et al. (2020). Porphyrin-loaded lignin nanoparticles against bacteria: a photodynamic antimicrobial chemotherapy application. *Frontiers in Microbiology* 11 (November): 1–17. https://doi.org/10.3389/fmicb.2020.606185.

153 Paul, S., Thakur, N.S., Chandna, S. et al. (2021). Development of a light activatable lignin nanosphere based spray coating for bioimaging and antimicrobial

photodynamic therapy. *Journal of Materials Chemistry B* 9 (6): 1592–1603. https://doi.org/10.1039/d0tb02643c.

154 Lintinen, K., Luiro, S., Figueiredo, P. et al. (2019). Antimicrobial colloidal silver-lignin particles via ion and solvent exchange. *ACS Sustainable Chemistry and Engineering* 7 (18): 15297–15303. https://doi.org/10.1021/acssuschemeng.9b02498.

155 Morena, A.G., Bassegoda, A., Hoyo, J., and Tzanov, T. (2021). Hybrid tellurium–lignin nanoparticles with enhanced antibacterial properties. *ACS Applied Materials & Interfaces* 13 (13): 14885–14893. https://doi.org/10.1021/acsami.0c22301.

156 Sugiarto, S., Leow, Y., Tan, C.L. et al. (2022). How far is lignin from being a biomedical material? *Bioactive Materials* 8: 71–94. https://doi.org/10.1016/j.bioactmat.2021.06.023.

157 Pei, W., Deng, J., Wang, P. et al. (2022). Sustainable lignin and lignin-derived compounds as potential therapeutic agents for degenerative orthopaedic diseases: a systemic review. *International Journal of Biological Macromolecules* 212 (May): 547–560. https://doi.org/10.1016/j.ijbiomac.2022.05.152.

158 Zhang, X., Morits, M., Jonkergouw, C. et al. (2020). Three-dimensional printed cell culture model based on spherical colloidal lignin particles and cellulose nanofibril-alginate hydrogel. *Biomacromolecules* 21 (5): 1875–1885. https://doi.org/10.1021/acs.biomac.9b01745.

159 Figueiredo, P. and Santos, H.A. (2021). *Lignin-Based Materials for Biomedical Applications: Preparation, Characterization, and Implementation*. Elsevier.

14

Carbon (Nano)Fibers and Carbon Materials from Lignin and Their Applications

Muhammad Muddasar[1], Anne Beaucamp[1], Mark Vaughan[1], Misbah Mushtaq[3], Mario Culebras[4], Marina M. Leite[3], Tadhg Kennedy[2,3] and Maurice N. Collins[1,2]

[1] *University of Limerick, Bernal Institute, School of Engineering, Stokes Laboratories, Limerick, Ireland*
[2] *University of Limerick, SFI AMBER Centre, Ireland*
[3] *University of Limerick, Department of Chemical Sciences, Bernal Institute, Limerick, Ireland*
[4] *University of Valencia, Institute of Material Science, 46980, Paterna, Valencia, Spain*

14.1 Introduction

Lignin, an abundant natural biopolymer, is one of the three basic components of wood and other lignocellulosic biomasses. Woody biomasses generally contain a large amount of lignin, with quantities ranging between 15% and 40% of plant weight. Currently, lignin is not directly extracted as a raw material for particular applications but is produced from the paper and pulp industries and biorefineries. Around 60–70 million tons of lignin are produced worldwide annually. Additionally, lignocellulosic biorefineries will continuously increase the availability of lignin in the future, making it increasingly important to find viable solutions for its commercialization [1].

Paper and pulp industries use three major separation processes (Kraft, soda, and organosolv) to remove natural lignin from cellulose and hemicellulose. The polymeric structure of lignin is degraded in each case to varying degrees owing to harsh chemical conditions, resulting in modified or so-called technical lignins (organosolv, kraft, and lignosulphonates) [2]. While technically lignins have undergone several modifications, they still retain three important properties that may make them useful. One distinguishing feature is their high carbon content in comparison to other abundant biopolymers (cellulose and hemicellulose). It is also worth noting that all types of lignin possess abundant oxygen functional groups, which permits the chemical structure to be modulated. Furthermore, they maintain a chemical structure rich in aromatic compounds [3]. The above-mentioned characteristics indicate that the conversion of lignin into carbon

Lignin Chemistry: Characterization, Isolation, and Valorization, First Edition.
Edited by Yuhe Liao and Bert F. Sels.

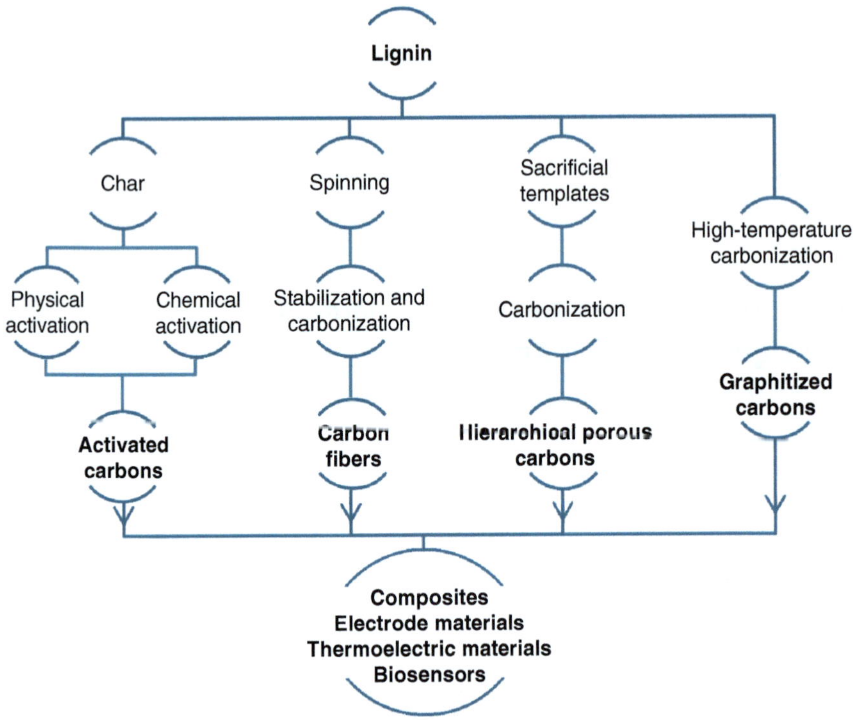

Figure 14.1 Synthesis routes of various carbon materials from lignin and their potential applications. Source: Muhammad Muddasar.

materials/fibers is a promising approach toward lignin valorization. All the major synthesis routes of lignin-derived carbon materials and fibers have been illustrated in Figure 14.1.

14.2 Lignin-Derived Carbon Materials

Carbon materials are utilized in numerous energy and sustainability applications because of their outstanding conductivity, tuneability, mechanical and chemical stability, remarkable stiffness, and significant tensile strength. Currently, carbon materials are mostly produced from petroleum-associated materials such as polyacrylonitrile (PAN), coke, and pitch. Nevertheless, these precursor materials remain largely untapped commercially since their conversion requires risky, expensive, and sophisticated techniques [4]. It is, therefore, crucial to find sustainable, low-cost precursors that can be converted into carbon materials in order to ensure an effective, efficient, and environmentally friendly production of carbon materials. The abundance, aromaticity, and high carbon content of lignin have prompted scientists to synthesize lignin-derived carbon materials and fibers in order to replace petroleum-based chemicals.

Generally, the carbonization of lignin precursors at elevated temperatures under an inert environment produces a solid carbon material known as char. Carbonization temperature, sample retention time, and the type of lignin precursor significantly affect the carbon yield and morphology of the final product. The obtained carbonaceous material is activated physically or chemically to produce lignin-derived activated carbons. The physical activation process involves partial gasification of char at a temperature between 700 and 900 °C in the presence of an oxidative gaseous agent (air, steam, CO_2, etc.). Lignin-derived char can also be activated by chemical means using various activation agents, such as potassium hydroxide, phosphoric acid, bicarbonates, metal thiosulfates, metal nitrates, and metal carbonates. Chemically activated carbon materials produce micropores and mesopores with surfaces between 1000 and 3000 $m^2 g^{-1}$ [5]. It is important to note that activated carbon with a well-developed porosity is a value-added material that has many applications such as energy storage, water treatment, air purification, and catalysis.

More advanced carbon materials have also been prepared from lignin, including hierarchical porous carbons and highly ordered graphitized carbons. The templating strategy is generally used for synthesizing porous carbon material with hierarchical structures. When carbonization is carried out, templates serve as sacrificial porogens for generating tunable hierarchical structures. These hierarchical porous carbons are particularly suitable as electrode materials in energy storage applications especially supercapacitors [6]. The carbonization process requires high temperatures (over 2000 °C) in order to produce highly graphitized carbon. Such carbon material has ordered crystalline structure and is suitable for the synthesis of anode material for lithium-ion (LIBs), sodium-ion (SIBs), and potassium-ion batteries (KIBs) [7].

14.3 Lignin-Based Carbon Fibers

Lignin has also been researched as a potential replacement for petroleum-based carbon fibers (CFs) owing to its high carbon content (30–40 wt%), low environmental impact, and cost-effectiveness ($3.08/kg) [8]. This path adds value to a resource normally burned as low-value fuel. Currently, over 90% of commercial CFs are made from petroleum-based PAN due to their high strength (4.15–7 GPa) and processability [9, 10]. However, the use of PAN-based CFs is limited because of increasing production costs and their negative environmental impact.

Other alternatives are available besides PAN, such as rayon or cellulose fibers, but they are constrained to low-strength applications. The three major lignin categories are hardwood, softwood, and grass [11]. Each one of these sources possesses different structures and molecular weights. To produce CFs, hardwood and softwood lignins have been studied in depth.

CFs are of interest for structural and functional applications. Structural applications depend on mechanical properties, whereas functional applications rely more on microstructure, electrical and thermal conductivity. Structural CFs are used as reinforcing agents in composite materials and are commonly found in automotive, aerospace, energy, construction, and sporting sectors. On the contrary, functional CFs are utilized for energy conversion and storage, biomaterials, and so on [12, 13]. The production of CFs from lignin involves multiple steps, but the most essential are lignin selection, spinning, stabilization, and carbonization. Other processes may be involved such as surface treatments, which are necessary for composite manufacture or graphitization, which is used to develop high-tensile CF for aerospace applications.

Despite the environmental and economic advantages that lignin has for CF production, results so far have produced CFs that are considerably weaker than their PAN-based counterparts. As a result, the types of lignin and spinning methods implemented have been a primary focus to eliminate defects that compromise mechanical properties during stabilization and carbonization for structural components. Fiber diameter is of utmost importance for structural CFs, as it has been reported that tensile strength increases at finer diameters due to molecular order [14].

14.3.1 Spinning of Lignin-Based Precursor Fibers

Several methods can be employed to produce lignin precursors. The most widely researched techniques involve solution spinning, electrospinning, and melt spinning (Figure 14.2). Each of these techniques generally produces different morphologies, diameters, and mechanical properties to each other. Melt and solution spinning are applied to structural CF applications, whereas electrospinning caters to functional applications. The resultant fibers from melt or solution spinning usually

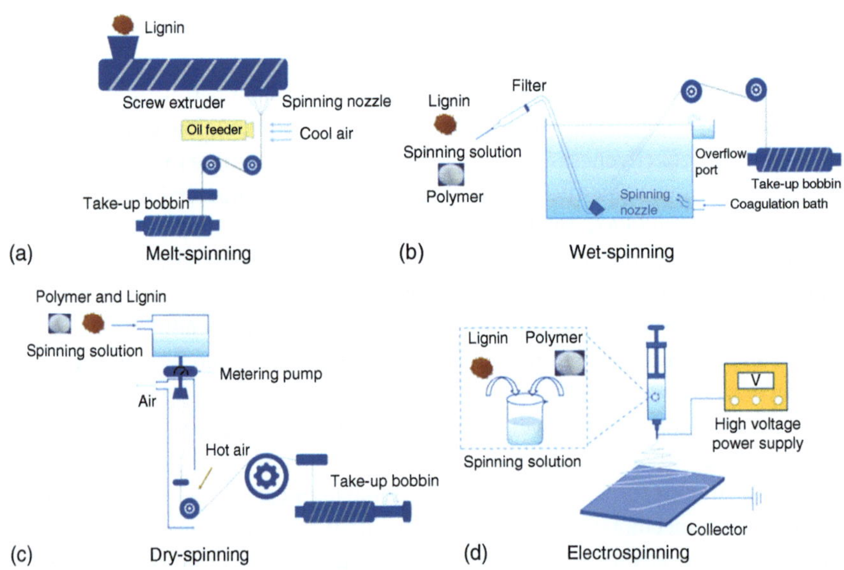

Figure 14.2 Schematics of (a) melt-spinning, (b) wet-spinning, (c) dry-spinning, and (d) electrospinning methods. Source: [16] Reproduced with permission of RSC Publishing.

exhibit different mechanical properties, with melt spinning showing the most favorable so far. There is often a disparity in fiber diameter, leading to some results looking stronger or weaker than others.

The first patents and reports on the manufacture of carbon materials from alkali-modified lignin emerged during the late 1960s and early 1970s [15]. It was concluded from these that existing methods used to spin materials of high molecular weight could be applied to spinning lignin, that is, melt spinning and solution spinning. Utilizing melt or solution spinning is generally desired, as these production methods share similarities with industrial PAN or pitch-based manufacturing practices while also allowing micron-scale fibers to be produced. Melt spinning is particularly interesting because of its production speed, low cost, and minimal environmental impact compared to other techniques.

Hardwood lignins have shown the most promise due to their lower glass transition (T_g) of 110–130 °C, allowing easier processing through techniques such as melt spinning without chemical modification or special blends [17]. The ratio of S and G units can affect the melting efficiency, needing to be tuned for melt spinning and stabilization.

14.3.1.1 Spinning of Lignin-Based Precursor Fibers

Melt spinning is primarily used to turn thermoplastics into fibers. The polymer is melted and extruded through a spinneret to give it its shape. As it exits the spinneret, it starts to solidify either via air or water cooling. As it is cooling, the polymer undergoes a drawing, or stretching process, which orients the molecular chains to improve mechanical properties and reduce fiber cross-section. Despite the relative ease of manufacturing precursors in comparison to other methods, there are limitations that dictate the behavior during spinning. Some improvements include chemical modification, lignin purification, thermal treatment, and plasticization with another polymer.

Melt Spinning of Lignin

Purification is an essential step as a result of impurities introduced during the pulping process. These impurities vary depending on the technique used, and some lignins may not need to undergo this step, such as organosolv lignin, which is typically used without modification [11]. Washing is the simplest form of purification, allowing ash and water-soluble impurities to be removed with distilled water. Kadla et al. [14] reported purifying hardwood and softwood kraft lignin through repeated washing cycles, maintaining a pH of less than 5 with HCl. This allowed for kraft lignin to be blended with polyethylene oxide (PEO) and melt spun. Washing with acids has been shown to reduce –OH groups, increase aliphatic chains, and reduce crosslinking during the spinning process [11].

Sudo and Shimizu [18] prepared CFs from steam-exploded lignin modified through hydrogenolysis and phenolysis that exhibited good melt spinning and mechanical properties, achieving fiber diameters of 7.6 μm and tensile strength of 660 MPa. Further work focused on adapting the pulping process, with melt spinning kraft and organosolv lignin, and lignin blends. Adapting the pulping process was found to produce CFs that were comparable in mechanical performance [19]. Uraki et al. [19] used melt spinning with organosolv lignin obtained through aqueous acetic acid pulping. Partial acetylation of hydroxyl groups was observed to prevent condensation reactions from occurring during heating. Thermal treatment aided spinnability, allowing continuous melt spinning at 400 m min^{-1}.

Chemical modification is another option that may be used to improve lignin processability. Similarly to some washing techniques, the aim is to block phenolic –OH groups [11]. Although this major functional group can benefit the oxidization process and provide thermal stability, it is prone to crosslink during spinning. It has also been shown to negatively impact melt stability with other polymers. Esterification approaches such as acetylation, butylation, and methylation have been studied. A patent by Eckert and Abdullah [20] describes using the acetylation method with chemicals such as acetyl chloride, acetic anhydride, and acetic acid for improving melt-spinnability of softwood kraft lignin. However, it was noted that stabilization times were prolonged.

Melt Spinning of Lignin-polymer Blends

Plasticization with a polymer aids spinnability by increasing its melt-flow index. This prevents crosslinking occurring during extrusion, while also keeping the glass transition (T_g) temperature relatively high, which is necessary for thermostabilization. Blending kraft or Organosolv lignins with a thermoplastic polymer to produce a melt-spinnable composition has been reported multiple times. In comparison to other solutions that have been proposed, this method has been deemed a relatively easy way to improve the spinnability of lignin fibers [11]. However, it can decrease the overall carbon yield. Current research has attempted blending either PEO, PET, PP, PLA, TPU, and polyamide with lignin to improve melt-spinnability, each with varying degrees of success [21–23]. Spinnability and the resulting precursor properties are dictated by the processing parameters and polymer selected for use with lignin.

Kadla et al. [14] first reported using PEO as a plasticizer to reduce the T_g of kraft lignin, which enabled the spinning of finer lignin precursor fibers due to its miscibility. However, stabilization was impaired at high concentrations of PEO (<5 wt% was considered optimal). The resultant CFs exhibited tensile strength and modulus of 400–550 MPa and 30–60 Gpa, respectively, attributed to PEO's high miscibility with lignin. Ebers et al. [23] described the preparation of PFs from beech Organosolv lignin combined with epoxidized PEO and CNTs. In this study, it was shown that epoxidized PEO had an ability to covalently bind to lignin and increase the degree of crystallinity. It was shown that the addition of CNT did not influence the properties, however. Further studies by Kubo and Kadla [24] examined the effect on fiber properties when blending PET and PP with hardwood Kraft lignin (HKL). Their research determined that the spinnability of lignin is heavily dependent on the polymer it is blended with and its thermal properties. PET and PP were discovered to be spun with relative ease, but PET was the only miscible composition of the two. The immiscible PP fibers produced had hollow or porous morphologies, rendering them unusable for CF production.

Wang et al. [25] attempted to blend HKL with PLA due to its high miscibility. In this study, it was shown that it was possible to produce continuous precursor fibers as the addition of PLA increased overall spinnability. Despite having good spinnability, the blend was compromised owing to the presence of voids caused by the volatilization of PLA in the subsequent stabilization and carbonization steps. More recently, Luo et al. [26] investigated synthesizing an orientable polymer from lignin bio-oil that could be melt spun. Initially, this bio-oil is transformed into a new lightly branched, acrylate polymer. This produced a melt-spinnable thermoplastic that could result in high yields during the spinning process. Aside from this, the mechanical properties are the greatest so far for lignin-based CFs, with tensile strengths of 1.7 GPa, and a tensile modulus of 182 GPa.

Culebras et al. [27] explored blending Alcell organosolv and hydroxypropyl-modified kraft lignin with TPU. These studies indicated a high degree of melt processability with compositions containing more than 30 wt% TPU and carbon yields up to 35 wt%. TPU, in comparison to other polymer blends, showed excellent miscibility without sacrifices in other areas, such as the morphological structure. This approach achieved CF diameters of 25 μm, with tensile strength and moduli of 1.1 and 80 GPa, respectively. Nordström et al. [28] successfully produced CFs from SKL and HKL by using HKL permeate (HKLP) as a softening agent. This technique allowed for the creation of continuously spun fibers to be produced that were also defect free. Muthuraj et al. [22] reported blending biobased polyamides with hydroxypropyl-modified lignin as potential precursors. Lignin particles had to be treated with compatibilizers to enhance homogeneity of blends. It was found during TGA analysis that there were significant char residues, suggesting that bio-polyamide acted as a promoting agent for lignin, which in turn could improve carbon yields.

14.3.1.2 Precursors via Solution Spinning

While melt spinning is the most studied form of producing lignin-based CFs, solution spinning has also been investigated. Unlike melt spinning, solution spinning applies to polymers that cannot be melted or softened whilst maintaining high mechanical properties [29]. Solution spinning covers dry, wet, and dry-wet spinning methods, and their use depends on the solvent evaporation behavior of compositions. However, a shortcoming of this technique is a much lower production speed and the necessity of a solvent.

Zhang and Ogale [30] used partial acetylation and acetone as a solvent to enable dry spinning of precursors, while reducing stabilization times. This technique created CFs around 7 μm. A thermally fusible material was produced which could be melt spun, yet stabilization was impractical due to heating rates as low as $0.01\,°C\,min^{-1}$. Nevertheless, a tensile moduli and strength of 52 and 1.04 GPa, respectively, were achieved. Jiang et al. [31] were able to produce CFs from lignin/PAN compositions that had tensile strength between 300 and 500 MPa. Tensile moduli were stated to be below 100 GPa. Bengtsson et al. [32] researched lignin-cellulose blends with reduced stabilization times via dry-jet spinning. These two macromolecules synergize well with each other, combining cellulose's molecular order and lignin's high carbon content. It was possible to achieve a maximum tensile strength of 1 GPa and modulus of 74 GPa. In a subsequent study, significantly increased carbon yield with impregnation of ammonium dihydrogen phosphate (ADHP) was achieved, increasing from 31 to 49 wt% [33]. However, CF mechanical properties suffered, particularly, the tensile modulus. Jin et al. [34] produced CFs from SKL that were <7 μm in diameter, with tensile strength and moduli values of 1.39 and 98 GPa, respectively, via dry-spinning. The mechanical properties achieved in this study are the greatest lignin-based CFs produced using solution spinning to date.

14.3.2 Stabilization and Carbonization of Lignin Fibers

Stabilization is a key step that is performed before carbonization due to the low glass transition temperature of lignin. During stabilization, lignin crosslinks without softening, allowing for carbonization to be performed. Thermostabilization is the most used method, but other means have been experimented with. Carbonization is the final step in producing CFs from lignin. It is performed at temperature ranges of 600–1000 °C in an inert gas. Heating rates are higher relative to those applied at stabilization, generally ranging from 5 to 10 °C min^{-1} [35].

Braun et al. [36] were one of the first in this field to examine the thermostabilization process in detail using HKL. An increase in oxygen was observed at a temperature range between 200 and 250 °C. Temperatures below 250 °C allowed oxidation reactions to take place. At higher temperatures, crosslinking of the lignin molecule occurs due to carbonyl and carboxyl structures becoming incorporated into ester and anhydride linkages. They also noted that the fraction of carbonyls increasing during stabilization is likely to contribute to the increase in T_g. Despite this, a heating rate at or lower than 0.06 °C min^{-1} was necessary to maintain T_g during thermostabilization. In comparison, Bengtsson et al. [33] investigated precursor thermostabilization strategies and found it was possible to reduce oxidative stabilization times to under two hours, or omit the step entirely without fibers fusing together. They concluded that holding a temperature of 250 °C took precedence over a slow heating rate. Previously to the aforementioned study, Norberg et al. [37] were able to thermostabilize softwood kraft lignin permeate (SKLP) fibers at a rate of 15 °C min^{-1} up to 250 °C in air. This allowed for stabilization to be completed within 45 minutes. It also allowed for stabilization and carbonization to be performed as a single-step process, reducing the complexity and cost of using lignin in CF production and resulting flawless lignin CFs, as shown in Figure 14.3.

Zhang et al. [39] utilized ultraviolet (UV) light in conjunction with thermostabilization. It was found that this approach could lead to a 90% reduction in stabilization time, as a rapid heating rate can be used without fibers fusing together. A contributing factor to this rapid stabilization is the increased T_g of approximately 15 °C compared to untreated fibers.

At carbonization, a turbostratic structure is formed [40]. The turbostratic carbon layers are affected by the maximum temperature during carbonization and polymer selection. This structure is described as layers of bent carbon. While it is less ordered than its counterpart graphitic carbon, it is key to CF tensile strength [40, 41]. Bengtsson et al. [42] investigated the effect on carbonization conditions for

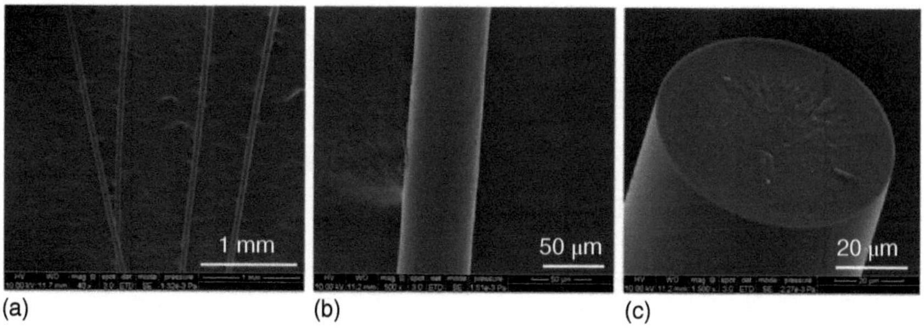

Figure 14.3 (a) SEM image of several CFs. (b) SEM image of the surface for a single CF. (c) Cross-sectional view of CF. Source: [38] Reproduced with permission of Elsevier.

lignin-cellulose precursors at heating rates ranging from 1 to 40 °C min^{-1}. It showed the influence on carbonization temperature from 600 to 1600 °C. It was shown that carbonization temperatures of 1600 °C increased tensile modulus to 77 GPa. Raman spectroscopy observed a change in structure, forming nanocrystalline graphite. Tensile strength was reduced at higher temperatures (Table 14.1), with the maximum strength of 1050 MPa being recorded at 1000 °C.

14.4 Applications of Lignin-Derived Carbon Materials – Supercapacitors

Electric double-layer capacitors (EDLCs), commonly known as supercapacitors, store charge by the adsorption of ions at interfaces between porous activated carbon electrodes and organic electrolytes. The high power density of these devices, their fast charging and discharging characteristics, the wide operating temperature range, and their long cycle life have made them widely used in electric and hybrid automobiles. Electrostatic adsorption of ions onto the electrode–electrolyte interface or a faradaic reaction is the primary mechanisms for supercapacitors to store energy [43]. The following characteristics should be present in an ideal carbon electrode material:

- Extremely porous structure, providing a high specific surface area (SSA) for charge accumulation.
- Reasonable pore structure distribution, promoting electrolyte ion transport.
- High electrical conductivity, providing a low-resistance electron conduction path.
- Excellent wettability, ensuring that the electrolyte fully wets the pores.

Lignin is a highly complex aromatic polymer with a carbon content of up to 60%. This makes it an ideal precursor for carbon materials. Additionally, lignin has unique advantages over traditional activated carbon due to its 3D polymeric structure. The morphology of lignin-based carbon materials, along with surface wettability, pore distribution, pore size, degree of graphitization, etc., also have a significant impact on the electrochemical performance of supercapacitors. The use of lignin as

Table 14.1 Lignin CFs properties from melt and solution spinning.

Lignin source	Modifications	Spinning type	Stabilization and carbonization	Diameter	Mechanical properties	Source
Alcell Organosolv, SWKL, HWKL	SWKL and HWKL repeated washing with distilled water	Melt spinning	12–180 °C h–^{1} up to 250 °C 180 °C h–^{1} up to 1000 °C nitrogen	30–55 μm (HWKL and Alcell) 7–14 μm (acetic and phenolated steam-exploded)	TS – 400–550 MPa TM – 30–60 GPa	[14]
Steam-exploded birch lignin	Hydrogenated in 2% NaOH solution at 250 °C with Raney-Ni as catalyst	Melt spinning	1–2 °C min^{-1} up to 210 °C 5 °C min^{-1} up to 1000 °C	7.6 ± 2.7 μm	TS – 660 ± 230 MPa TM – 40.7 ± 6.3 GPa	[18]
Organosolv lignin (AWL)	N/A	Melt spinning	0.5 °C min^{-1} up to 250 °C 5 °C min^{-1} up to 1000 °C	14 ± 1 μm	TS – 355 ± 53 MPa TM – 39.1 ± 13.3 GPa	[19]
HKL	Blended with PET and PP	Melt spinning	12–180 °C h up to 250 °C 180 °C h^{-1} up to 1000 °C	PET – 31 – 45 μm PP – 44–76 μm	TS PET – 669–703 MPa TM PET – 84–94 GPa TS PP – 167–437 MPa TM PP – 28–57 GPa	[24]
HKL	Blended with PLA (MW – 1.77 * 10^5)	Melt spinning	0.25 °C min^{-1} 60–280 °C 3 °C min^{-1} up to 1000 °C	30–60 μm	TS – 159.2–258.6 MPa TM – 1.7–11.6 GPa	[25]
Lignin bio-oil	Acrylation, Acetylation, RAFT polymerization	Melt spinning	0.1–10 °C min^{-1} 25–280 °C 7 °C min^{-1} 25–500, 3 °C min^{-1} 500–1000 °C	5.1 μm	TS – 1.7 GPa TM – 182 GPa	[26]

TcC and TcA	Blended with TPU	Melt spinning	0.1 °C min^{-1} 250 °C 10 °C min^{-1} 1000 °C	23–32 μm	TS – 1.1 GPa TM – 80 GPa	[27]
SKL	Acetylation (0.55 ml Ac2O/g SKL)	Dry spinning	0.2 °C min^{-1} 220 °C 4.5 °C min^{-1} 1000 °C	7 μm	TS – 1.04 ± 0.1 GPa TM – 52 ± 2 GPa	[30]
SKL	Ultrafiltration, precipitation, and washing. Kraft pulp dissolved for cellulose	Dry-jet wet spinning	1 °C min^{-1} 200–250 °C 10 °C min^{-1} 1400 °C	9–15 μm	TS – 720–1000 MPa TM – 64–74 GPa	[32]
SKL	SKL and KP dissolved Emlm-Oac at 70 °C for 1 h	Dry-jet wet spinning	0.2 °C min^{-1} to 200 °C, 1 °C min^{-1} 200–250 °C 1 °C min^{-1} 600, 3 °C min^{-1} 600–1000 °C	6–8 μm	TS – 1070 MPa TM – 76 GPa	[33]
SKL	SKL purified by ALPHA process	Dry spinning	20 °C min^{-1} 165–250 °C 7 °C min^{-1} 1000 °C	<7 μm	TS – 1.39 GPa TM – 98 GPa	[34]

Source: Mark Vaughan.

a carbon precursor material for high-performance supercapacitors has therefore been extensively studied in recent years. Lignin-based carbon materials are typically synthesized by various methods including self-templating, soft-templating, hard-templating, aerogels, physical or chemical activation techniques, and electrospinning [44]. All these synthesizing techniques have their own advantages. However, free-standing electrodes simplify the supercapacitor further by eliminating the current collector and reducing environmental and commercial costs.

14.4.1 Lignin-Derived Activated Carbon Materials

The electrostatic double layer interface between the electrodes and electrolyte contributes significantly to the performance of supercapacitors, so electrodes with a higher SSA will likely have a higher capacity. Consequently, lignin-derived activated carbon electrode materials have widely been reported for supercapacitor applications owing to their high SSAs (500–2000 $m^2 g^{-1}$), moderate electrical conductivity, and low cost.

The first lignin-derived carbon material was derived from steam pyrolysis of apricot and cherry stones, almond shells, and grape seeds in the 1990s for adsorption applications [45]. However, the use of lignin-derived electrode material for supercapacitors did not appear until 2010. Jianhui and team tested KOH-activated lignin-derived carbon material in EDLCs for the first time in 2011 [46]. Lignin was extracted from strong black liquor and heated in an inert environment at 900 °C for 1.5 hours, and the effects of the activation agent were analyzed by adjusting the KOH:Lignin ratio from 1 : 1 to 3 : 1. It was observed that SSA increased from 718 to 1506 $m^2 g^{-1}$ as the KOH ratio increased from 1 to 3. In addition, the specific capacitance also increased from 196.1 to 267.8 $F g^{-1}$ at a current density of 50 $mA g^{-1}$. A capacitance retention ratio of 83.9% was also noted after 2000 cycles. These excellent results are attributable to the large SSA, suitable pore dispersion, and low electrical resistance of synthesized material.

Many researchers aimed to enhance the capacitance of lignin-based supercapacitors by improving the surface area. Adopting a two-stage activation method with a pre-carbonization step is a modern way to obtain the largest surface areas and pore dispersion. Guo et al. utilized enzymatic hydrolysis lignin, derived from corn straw, and hydrothermally carbonized it at 180 °C in 5 wt% H_2SO_4 for 18 hours to obtain hydrochar [47]. The acquired hydrochar was chemically activated with different KOH/hydrochar weight ratios to evaluate the influence of these ratios on its electrochemical properties. It was noted that LHC-3K with SSA of 1660 $m^2 g^{-1}$ is mainly made up of abundant micropores (0.66 cm^3 g) and small mesopores (0.10 cm^3 g). Three-dimensional hierarchical texture of LHC-3K attained an excellent specific capacitance of 420 $F g^{-1}$ in 6 M KOH electrolyte solution with superior energy density of 46.8 $Wh kg^{-1}$ (Figure 14.4a).

Lignin-derived activated carbons have excellent performance in supercapacitors but the use of highly corrosive activation agents (e.g. KOH, NaOH, H_3PO_4, $ZnCl_2$, etc.) for pore formation is one of the primary challenges to environmental sustainability. Thankfully, attempts are being undertaken to switch to more environmentally friendly synthesis techniques. A green, facile synthesis route was reported by Beaucamp et al. to prepare lignin-derived tailored porous carbon from carbon aerogel [48]. The synthesis method combines chemical crosslinking and lyophilization, followed by carbonization of lignin aerogel. Without the use of any physical or chemical activation agents, a carbon material with a highly porous tailored structure was accomplished. Additionally, the electrode made from aerogel performed better electrochemically, having a specific capacitance of $18.7\,F\,g^{-1}$ at $1\,A\,g^{-1}$. Such novel synthesis techniques open up new avenues for the development of green and environmentally friendly lignin-derived carbon electrodes for future supercapacitors (Figure 14.3).

Most of the primary studies focused on increasing the surface area of activated carbons to enhance the performance of supercapacitors. However, carbon nanostructure and pore size are more important to specific capacitance than the raw surface area in increasing specific capacitance. It is hard to tailor morphology of activated carbons to attain predefined porous structures. Therefore, researchers are now focusing on creating tailored carbon materials containing specific nanostructures using templates to increase the energy density of supercapacitors based on lignin (Figure 14.4).

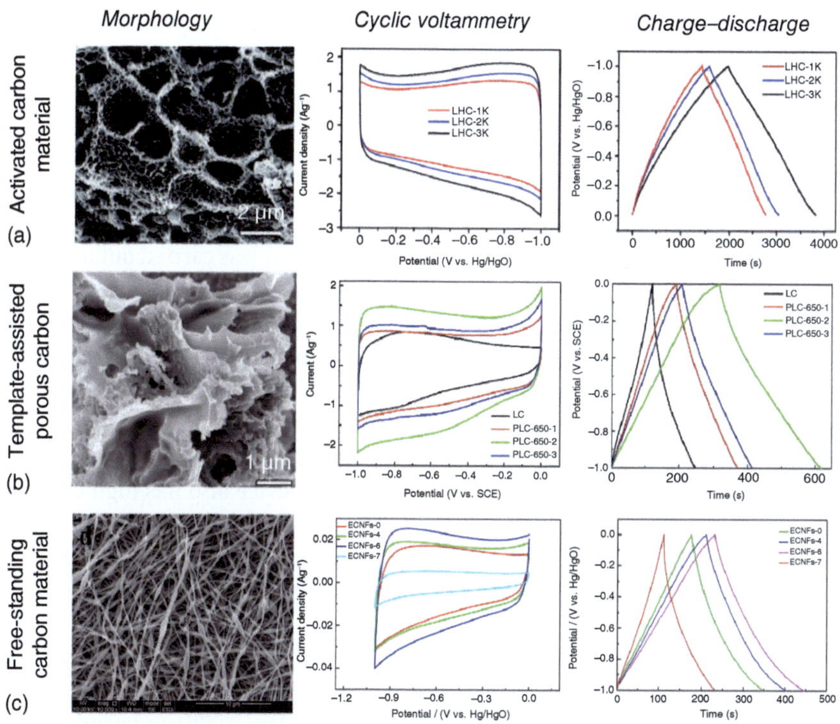

Figure 14.4 (a) Porous activated carbon. Source: [47]. Reproduced with permission of RSC publication. (b) template-assisted porous carbon material. Source: [49]. Reproduced with permission of Elsevier. (c) free-standing carbon material. Source: [50]. Reproduced with permission of MDPI publisher.

14.4.2 Template-Assisted Porous Carbon Materials

Template method is the most mature methodology for the synthesis of porous carbon with tunable pore size distribution. The templates are the porogen for generating pores in the carbonization process. Hard and soft templates are two strategies to obtain tunable pore size. The process of "hard templating" involves introducing inorganic substances into the lignin matrix, such as MgO, ZnO, Montmorillonite, Halloysite, $MnCO_3$, silica, and zeolites. Generally, the removal of hard templates from carbonized samples yields macro and mesopores in the carbon matrix. Fu et al. used sodium lignosulfonate as carbon source and synthesized a high-performing porous carbon electrode for supercapacitors via zinc oxalate-assisted *in-situ* templating [49]. The ZnO template was removed with 1 M HCl, resulting in a SSA of 1300 $m^2\,g^{-1}$ and possessing a hierarchical porous structure with a micropore distribution (0.5–2 nm) and a broad mesopore distribution (2–20 nm). Synthesized material exhibits an excellent specific capacitance of 320 $F\,g^{-1}$ at 1 $A\,g^{-1}$ and maintains a superior capacity retention of 93.5% at 5 $A\,g^{-1}$ after 10,000 cycles (Figure 14.4b). These excellent electrochemical performances indicate that the as-prepared carbon material should hold great promise for the energy storage devices. On the other hand, removing ZnO with HCl produces aqueous $ZnCl_2$, which is harmful to plants and aquatic life.

Soft template, on the other hand, decomposed fully during the carbonization process, eliminating the need of post template removal step. Soft templates include fully decomposable Pluronic P123, Pluronic F127, and partially decomposable polymers, such as polyvinyl chloride. Besides the above-mentioned hard and soft templates. Furthermore, some decomposable salts, such as $ZnCO_3$ and $MgCO_3$, can also be used as novel templates for tunable porosity. Scientists from Huaqiao University, China, recently reported the use of F127 (soft template) to synthesize lignin-derived mesoporous carbon [51]. A deep eutectic salt pretreatment was carried out on pine alkali lignin prior to soft templating with F127. A reduction in pretreatment time with increased phenolic hydroxyl content of modified lignin was found to be associated with water addition to deep eutectic solvent. In comparison to untreated lignin, pretreated lignin with F127 templates results in an ordered mesoporous carbon structure with an increased SSA (1195 $m^2\,g^{-1}$) and pore volume (0.7 $cm^3\,g^{-1}$). This resulted in a specific capacitance improvement of 197.32 $F\,g^{-1}$ in the pretreated sample compared to 72 $F\,g^{-1}$ in the untreated sample. The device also has high energy density and excellent cyclic stability with 99.09% capacitance retention after 10,000 cycles. It is typically necessary to employ activation agents in order to increase microporosity within the generally mesoporous structure created by the template for optimum performance.

Porosity, porosity distribution, and hierarchical structure of lignin-derived carbon materials can be controlled using a templating strategy. However, electrodes can be formed by crushing and mixing these materials with binder (PTFE) and conductive agents (carbon black). These extra components increase the size of the device as well as create a resistive interface between the active material and the current collector. It may therefore be possible to reduce complexity and increase volumetric capacity by eliminating these components with electrically conductive free-standing electrodes instead of powdered materials.

14.4.3 Lignin-derived Free Standing Carbon Materials

The development of flexible supercapacitor electrodes based on lignin is one of the latest major advancements in energy storage technology. Melt spinning, centrifugal spinning, dry spinning, gel spinning, wet spinning, or electrospinning are the most common techniques for the synthesis of lignin-based flexible CF mats. In each of these methods, precursor polymers are converted into (nano) fibers and carbonized at high temperatures (>600 °C) under inert atmospheres to synthesize free-standing carbon materials. Since high-temperature carbonization changes the morphology of fibers, stabilizing them is essential. Lignin polymers are typically stabilized to form thermoset polymers for further treatment. Fibers are stabilized by heating them at 200–300 °C for several hours under an oxidative atmosphere.

CFs can be made from pure lignin by dissolving it in appropriate solvents [52]. However, the lower molecular weight of lignin makes this difficult to electrospun. Therefore, lignin is typically mixed with polymers, such as polyvinyl alcohol (PVA), PEO, and PAN to increase the spinnability and enhance the properties of the fibers. Electrospinning of enzymatically hydrolyzed lignin with PAN polymer was reported by Wang et al. in order to produce binder-free electrodes for supercapacitors [50]. Mats prepared from precursors were stabilized at 250 °C prior to being carbonized at 800 °C under N_2. Fiber mats produced by 40 : 60 PAN:lignin solution had the highest SSA of $675\,m^2\,g^{-1}$. When tested in 6 M KOH solution, these lignin-derived electrodes demonstrated $216.8\,F\,g^{-1}$ specific capacitance (Figure 14.4c). Park et al. conducted an innovative research, creating both electrode and electrolyte of supercapacitor using lignin [53]. Various lignin/PAN solutions were mixed together and electrospun at 15 kV high voltage. These electrospun fibers were then carbonized at 900 °C for one hour in inert environment. A flexible electrolyte was produced by crosslinking 33 wt% alkali lignin solution with poly(ethylene glycol) digycidyl ether crosslinker at $1.2\,mol\,g^{-1}$ lignin concentration. Combining these two materials resulted in all-lignin-based flexible supercapacitor with a superior specific capacitance of $129.23\,F\,g^{-1}$ at $0.5\,A\,g^{-1}$.

Lignin-spun fibers are prone to beading, a phenomenon in which tiny beads of solution are deposited along the fiber length. This problem can be tackled by optimizing the electrospinning parameters, i.e. voltage, humidity, temperature, etc. However, the sustainability of lignin CFs will continue to be hampered by the use of fossil-derived polymers (PAN, PEO, PVA, etc.). Consequently, researchers have shifted their attention to the use of biomass-based precursors containing lignin for electrospinning. Cao et al. reported the synthesis of lignin-based nanofibers by covalently bonding lignin with cellulose acetate in their recent work [54]. Epichlorohydrin is used to link lignin and cellulose acetate covalently through epoxy ring reaction, eliminating the use of fossil-derived polymers. Lignin provides the thermal stability in the synthesized freestanding mats, whilst cellulose provides the flexibility. Prior to high-temperature carbonization, the fibers underwent thermal stabilization at a very low heating rate of 0.4 °C min^{-1} till 220 °C. These biobased carbon nanofibers exhibited a superior specific capacitance of 320.3 F g^{-1} in 6 M KOH solution owing to their outstanding SSA of 1061.7 m^2 g^{-1}. This innovative method shows potential for producing high-quality and environmentally friendly CNFs.

14.5 Applications of Lignin-Derived Carbon Materials – Batteries

Another most crucial field for sustainable growth in modern society is energy storage via rechargeable batteries. It is imperative to develop more effective and affordable rechargeable batteries for electric cars and energy-hungry multipurpose portable electronic devices such as smartphones, laptops, tablets, and biomedical sensors. Among all the reported rechargeable batteries, LIBs have achieved great success during the last decade due to their high energy density and long cycle life. Furthermore, SIBs and KIBs are also gaining prominence due to their low cost and enhanced sustainability.

Typically, a battery comprises two electrodes (cathode, anode), a separator, electrolyte, binders, and active materials. Current research in this area mainly focuses on the development of electrode materials with enhanced rate capability, longer cycling lifetimes, and higher energy densities. Furthermore, sustainable and low-cost materials are a research focus in the field of rechargeable batteries [55]. Carbon-based materials are well-suited for use as an active material due to their versatility, high electrical conductivity, and general stability throughout cycling; however, common carbonaceous active materials such as synthetic graphite are not sustainable as they are usually obtained from nonrenewable sources such as fossil fuels. As these resources dwindle and new regulations come into effect to make the environment cleaner and safer, there is an interest in developing more sustainable feedstocks. Lignin is an excellent candidate in this regard as it has the potential to be transformed into compounds that serve various purposes in the battery because it contains many functional groups and modification pathways. As a result, lignin not only satisfies the criteria for enhanced electrochemical performance but also reduces costs and provides a substantial opportunity to produce an active material with enhanced sustainability [44].

To promote and emphasize these value-added applications of lignin, this section provides a succinct description of the most recent advancements in lignin-derived electrode materials, specifically as an anode material for rechargeable batteries. Graphite is the standard anode material in LIBs due to its good electronic conductivity and stable Li^+ storage properties. However, there are sustainability and cost issues, in particular, related to synthetic graphite which is derived from fossil fuels. For SIBs, graphite displays limited performance due to the larger ionic radius of Na^+; therefore, hard carbon is the anode material of choice for SIBs due to its hierarchical pore structure and enlarged interlayer distance. Given the importance of carbon-based materials for both LIBs and SIBs anodes, lignin-derived graphitic and hard carbon anode materials are promising routes to enhance the sustainability of both technologies. The following sections provide a comprehensive summary of the recent development of lignin-derived anode materials in rechargeable batteries. The section begins with a brief introduction of lignin, its physical and chemical characteristics, and a comparison of many popular lignin types used in batteries. Lignin-derived porous carbon and carbon nanofibers anodes are summarized.

14.5.1 Lignin-derived Anode Materials

Carbon-based materials such as graphite and hard carbon are commonly used as the anode active materials in LIBs and SIBs, respectively, where intercalation and deintercalation of Li^+ or Na^+ ions occur during charging and discharging [56]. In SIBs, ion adsorption and pore filling also have significant contributions to the anode capacity [57]. While graphite is used as the anode material in conventional LIBs, it has disadvantages such as slow diffusion of Li^+, relatively low capacity, and poor rate performance. In this respect, porous carbon is frequently regarded as a good substitute due to the numerous ion diffusion pathways and enhanced conductivity network. These materials are distinguished by their multi-scale pore structure, which typically includes micropore (up to 2 nm), mesopore (2–50 nm), and macropore (>50 nm) structures [58]. The linked microstructure of porous carbon allows for shortened pathways, easier diffusion kinetics, and enhanced conductivity. Furthermore, the electrode's porous structure can store more active material by accommodating the change in electrode volume [59]. Additionally, carbon-based materials can offer an important conductive network for electron transport in electrodes, facilitating the electrochemical processes that occur during charging and discharging [60, 61].

Porous carbon materials derived from lignin can be employed as an anode material in both LIBs and SIBs. To provide lignin-derived carbon with a customized and pliable microstructure, many manufacturing techniques have been devised. A chemical activator can be added to change the porous structure of lignin-derived carbon and enhance the electrochemical performance. By impregnating potassium carbonate (K_2CO_3) as a chemical activator during carbonization, Xi et al. fabricated highly graphitized lignin-based mesoporous and microporous carbon with a SAA of $2300\,m^2\,g^{-1}$ which provided a high-volume reservoir for the rapid transportation of Li^+ and achieved enhanced reversible capacity of $520\,mAh\,g^{-1}$ at a current density of $200\,mA\,g^{-1}$ over 200 cycles [62]. Zinc carbonate ($ZnCO_3$) has also recently been proposed by Xi et al. as a noncorrosive and recyclable activator to manufacture hierarchical lignin-derived porous carbon (HLPC) [63]. The lignin particles were gas exfoliated by the carbon dioxide generated from $ZnCO_3$, which stopped the lignin structure from shrinking due to dehydration during pyrolysis. Additionally, nano-sized zinc oxide particles directly generated from $ZnCO_3$ were used to construct hierarchical pores. The resultant HLPC LIBs retained capacity of $550\,mAh\,g^{-1}$ after 200 cycles at $0.2\,A\,g^{-1}$. Furthermore, lignin-derived carbon materials' microstructure can also be affected by carbonization parameters. Du et al. established a pre-oxidation approach to increase the reversible capacity of lignin-derived carbon produced by direct pyrolysis (Figure 14.5a) [64]. It is thought that the pre-oxidation mechanism impacted the microstructure of the resulting carbon, leading to an increased layer spacing and a more porous structure that provides better Li^+ storage properties.

Graphite is not generally applied directly as the active material in SIB anodes as the larger ionic radius (102 pm) of the Na^+ compared to the Li^+ (76 pm) hinders intercalation between the graphitic layers. Furthermore, sodium-graphite intercalation compounds have a low thermodynamic stability [67]. Hard carbons, however, are electrochemically active due to their disordered structure that cannot be completely graphitized. This allows for reversible Na^+ charging and discharging via intercalation, pore-filling, and adsorption, making it more suitable as an SIB anode material, with capacities that range from 150 to 350 mAh g^{-1}. Hard carbon is

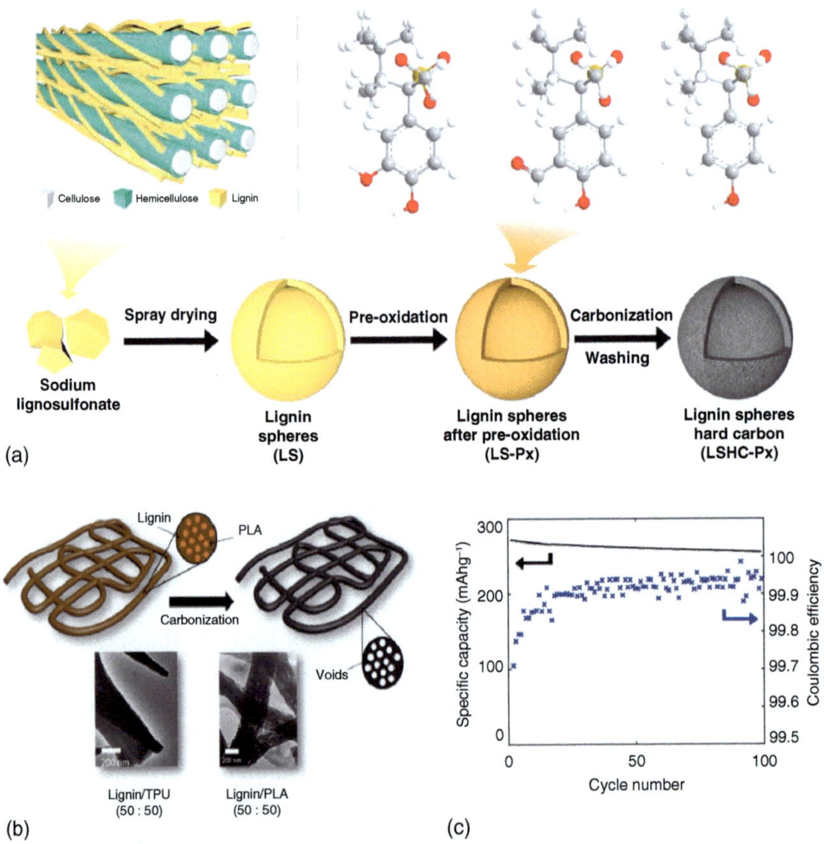

Figure 14.5 (a) Schematic illustration of hard carbon preparation process. Source: [64]. Reproduced with permission of Elsevier. (b) Schematic of phase separation for lignin/PLA fibers and TEM images of lignin/TPU-50:50 and lignin/PLA-50:50. Source: [65]. Reproduced with permission of Wiley. (c) Specific capacity (black) and coulombic efficiency (blue) of the LCFs carbonized at 1200°C cycled at 100 mA g$-^{1}$ after being cycled at 30 mA g$-^{1}$ for one cycle. Source: [66]. Reproduced with permission of IOP Publishing.

frequently derived from biomass, and several investigations have demonstrated that lignin-derived carbon has enormous promise as a SIB anode material. Zhao et al. improved the electrochemical performance by using a low-temperature pre-oxidation process before pyrolysis [68]. In order to increase the interlayer distance for the insertion of Na^+, pre-oxidation can be conducted to reinforce the lignin's intrinsic crosslinking structure and prevent the development and orientation of graphitic layers during carbonization. For example, Li et al. used sodium lignosulfonate as a precursor in a simple spray drying technique followed by carbonization to create hard carbon microspheres (HCMs) as SIBs anode material [57]. Despite having a relatively low SSA of $11.89\,m^2\,g^{-1}$, the HCM structure exhibited micro- and mesopores. The HCM provided an excellent initial coulombic efficiency (CE) of 88.3% and a high reversible capacity of $339\,mAh\,g^{-1}$. In addition, the higher capacity of $150\,mAh\,g^{-1}$ at 2.0 C can be attributed to the fact that the interlayer distance was able to be increased.

Porous carbon materials derived from lignin are also desirable for use as an electrode's conductive framework because they can provide an electron channel to facilitate electrochemical reactions occurring in the active material. Zhong et al. fabricated a hierarchical porous structure of molybdenum disulfide-coated porous carbon nanospheres (MoS_2@PCN) as an LIB anode material with high SSA, excellent conductivity, and stable reversible electrochemical processes [69]. Furthermore, the carbon skeleton of PCN buffered the volume expansion of MoS_2 while charging and discharging, leading to more consistent cycle performance of $519\,mAh\,g^{-1}$ at $0.1\,A\,g^{-1}$ after 50 cycles. Similarly, lignin-derived porous carbons are suitable as anode materials for sustainable battery technologies such as KIBs in which the production of hard carbon from low-cost bio-sources may be essential to bringing the technology to a competitive level. For Instance, Zhenrui Wu et al. studied a lignin-derived HC KIB anode which showed a high reversible capacity of 300 at $50\,mA\,g^{-1}$ [70].

Like LIBs, SIBs can also operate with alloying anodes, in which case a reversible reaction occurs by alloying and de-alloying Na in the negative electrode as opposed to Na^+ adsorption and intercalation. Even though alloying produces high capacities, it also causes significant volume variations that can quickly lead to electrode degradation. Wang et al. used lignin as a precursor to create Sn/C particles as anodes for SIB and achieved good cyclability with a capacity of $374\,mAh\,g^{-1}$ at $20\,mA\,g^{-1}$, corresponding to 91% capacity retention after 1000 cycles. In the future, similar strategies may be looked at to enable various alloying metal anodes for high-energy density SIBs and KIBs [71].

The aromatic structure of lignin and its thermal characteristics make it a suitable resource for CFs/CNFs synthesis and therefore a replacement for PAN, at least partially. LIB or SIB anodes have been constructed from monolithic fused fibrous mats derived from lignin. High molecular weight polymers are blended with lignin to increase their viscosity for fiber spinning such as poly(vinyl alcohol) (PVA), PAN, PEO, poly polyurethane (TPU), and polylactic acid (PLA). For example, Culebras et al. fabricated CNFs from lignin/PLA and lignin/TPU blends exhibited a high capacity values of 611 $mAh\,g^{-1}$ after 500 cycles as anode material for LIBs (Figure 14.5b) [65]. To further enhance the electrochemical performance of lignin-derived CNFs, Culebras et al. embedded Si particles in lignin blends and achieved the capacity of 921 $mAh\,g^{-1}$ for LIBs anode material [72]. To improve the electrochemical performance of lignin-derived CFs and CNFs, it is helpful to include a chemical or physical activation procedure that allows for the rational tailoring of the porous structure or the increase of porosity. For example, Peuvot et al. investigated various carbonization temperatures for stabilized lignin fibers and achieved a specific capacity of 310 $mAh\ g^{-1}$ with CE of 89% when carbonization is 1200 °C (Figure 14.5c) [66].

14.6 Applications of Lignin Derived-Carbon Materials – Sensing

Sensors are devices that respond to a physical stimulus (such as pH, heat, light, sound, pressure, magnetism, or a particular motion) and transmit a resulting, measurable impulse [73]. Carbon materials derived from lignin present sensing capabilities, which are currently investigated [74]. Table 14.2 presents the most relevant sensing devices reported in literature of the past two years. Lignin-based carbons are responsive to various stimuli such as light, strain, and chemical compounds. The nature of the stimuli sensed is dependent on the preparation of the lignin-based carbon.

14.6.1 Light Stimuli Carbon Quantum Dots for Sensing Applications

Carbon quantum dots (CQDs) are carbon nanomaterials with diameters below 10 nm, with amorphous or graphene-like carbon structures. These materials have excellent water dispersibility, biocompatibility, light stability, non-flickering photoluminescence, and adjustable excitation and emission wavelengths [83]. Their synthesis can be achieved through various techniques such as hydrothermal, electrochemical, sonochemical, solvothermal, laser ablation, or microwave cutting [84]. The aromatic nature of lignin renders it an ideal candidate for CQDs synthesis. O-related functional groups in the CQD from the lignin substructures induce chromophoric properties [76]. The processing is based on the depolymerization of lignin by acidic cleavage of ether β aryl bonds, before repolymerization in CQDs at elevated temperature and pressure [85]. Parameters such as temperature, solvent, or acidic conditions were found to influence the excitation and emission wavelengths of the CQD. The doping of the CQD with nitrogen or boron was found to influence the quantum yield and fluorescence. Zhu et al. induced triple-emission fluorescence at 300, 330, and 490 nm by doping the CQD with boron and nitrogen functional groups, enhancing intramolecular charge transfer and improving emission efficiency. The resulting sensor detected hexavalent chromium with a limit of detection (LOD) of 0.077 μM and a linear range up to 100 μM [76].

Table 14.2 Relevant lignin-based carbon materials for sensing applications from the past 2 years.

Type of material	Application	Preparation	Further properties	Refs.
Carbon dots	Formaldehyde sensor	Ag^+-mediated nitrogen–sulfur co-doped carbon dots (Ag-M-NSCDs)	Fluorescent sensing platform	[75]
	Hexavalent chromium sensor	Nitrogen and boron co-doped carbon quantum dots	Fluorescent sensing platform	[76]
	Formaldehyde sensor	Nitrogen-doped carbon quantum dots	Fluorescent sensing platform	[77]
Activated carbon	Strain and pressure sensor	Loaded in a polyvinyl alcohol, carboxymethyl chitosan, cellulose nanofibrils hydrogel		[78]
Carbon nanofibers	Uric acid sensor	Electrospun phosphorylated lignin – Cellulose acetate dope	Electrochemical platform	[79]
	flexible strain sensor	lignocellulose nanofiber-decorated graphite-flake nanocomposites	Embedded in an hydrogel	[80]
	Glucose detection	Electrospun 50% cell lignin and 50% polylactic acid dope	Hierarchical porous structure; electrochemical platform	[81]
Aerogel	Shape memory flexible strain sensor	Lignin/single-walled carbon nanotubes/ graphene oxide aerogel	Pressure sensor and supercapacitor	[82]

Source: Anne Beauchamp.

14.6.2 Strain and Pressure Sensors

The conductivity of nanoscale carbon materials fillers is used to enhance the inherent piezoelectricity effect within strain-sensitive materials such as cement or zinc oxide. [78, 86] such materials convert a mechanical deformation into measurable electric signals. Li et al. used activated carbon from cornhob lignin to induce conductivity within PVA hydrogels. The particles also acted as a filler that reinforced the mechanical properties of the hydrogels [78]. Peng et al. showed that channels aligned within a conductive lignosulfonate-based aerogel allowed for wide detection region, high sensitivity, and shape memory [82].

14.6.3 Electrochemical Sensors

The conductivity and nanostructured morphology of lignin-based carbon materials render these good candidates for electrochemical sensing. In particular, lignin-based carbon nanofibers present high surface area and graphitic structures for high conductivity. However, lignin alone cannot be electrospun in a single axial set-up, due to its low molecular weight which limits chain entanglement [87]. A binder polymer is used to improve the processability of lignin, such as PEO [88], PAN [89], or PVA [87]. Dimethylformamide (DMF) is generally the solvent of choice for electrospinning, at concentrations varying between 10 and 50 wt% of polymer. The electrospun precursor fibers are stabilized at 200 to 300 °C in the presence of oxygen to avoid fusion of the fibers before carbonization in inert atmosphere at 800 to 1200 °C [35]. Reportedly, the stabilization phase can be omitted for KOH-impregnated lignin-PEO nanofibers [90] or when adding nanoparticles of cellulose [91]. Finally, the addition of alkali lignin into PVA-based CNF reportedly created a porous structure (average pores size of 3.5 nm, BET surface area 583 $m^2 g^{-1}$ for 70% lignin blend) [87] The porosity of lignin-based CNF can be attributed to volatile material removed during the carbonization process. This porosity, inherent to lignin carbonization, renders this precursor attractive for the manufacturing of high surface area carbonaceous material. Zheng et al. developed phosphate lignin which, when electrospun with cellulose acetate as a binder, presents improved spinnability. When mounted as a working electrode, the cyclovoltammetry graph of the carbonized nanofibers presented a peak at 0.55 V in the presence of uric acid. However, their work did not determine limit of detection or linear range for the sensing system. Beaucamp et al. developed a glucose detection platform by tailoring the porosity of lignin-based CNF through polymer blending. The use of PLA allowed for increased porosity due to its sacrificial nature during carbonization. The resulting hierarchical porous structure of the CNF immobilizes the redox centers of glucose oxidase for selectivity, high sensitivity, a linear range, and a detection limit that are suitable for detecting glucose in human saliva samples. Lignin is useful as a precursor to carbon materials because of its polymeric structure based on functionalized aromatic units.

14.7 Applications of Lignin Derived-Carbon Materials – Thermoelectric Devices

Thermoelectric effect represents a huge opportunity to decrease carbon footprint of our society due to the fact that is possible to transform temperature gradients into electricity thanks to the Seebeck effect present in semiconducting materials (voltage generated due to a thermal gradient). Traditionally, inorganic compounds such as Bi_2Te_3, PbTe, and SiGe have dominated thermoelectric applications for energy harvesting devices. However, serious drawbacks such as toxicity, scarcity of raw materials, and high cost have limited their application and pushed research to newer alternative materials which are highly abundant, low cost, and nontoxic. The basic unit of a thermoelectric generator consists in n-type and p-type semiconductors that are connected electrically in series and thermally in parallel. As a consequence when heat is applied through the thermoelectric junction a current flow is generated due to the Seebeck effect.Thermoelectric efficiency is measured by the dimensionless figure of merit *ZT*:

$$ZT = \frac{S^2 \sigma T}{\kappa} \tag{14.1}$$

where *S, T,* σ, and κ are the Seebeck coefficient, absolute temperature, electrical conductivity, and thermal conductivity, respectively. The power factor is calculated as follows:

$$PF = S^2 \sigma \tag{14.2}$$

and is used to compare the thermoelectric efficiency of samples with similar thermal conductivities [92–98]. Currently, there is no viable sustainable alternative in the market to be used as thermoelectric material compared to traditional inorganic semiconductors. The most promising one consist in layered materials based on carbon nanostructures which have reached efficiency values close to Bi_2Te_3 [99].

Looking at this scenario, lignin has an enormous potential for a promising sustainable source to produce thermoelectric materials due to its ideal molecular structure to produce carbon nanostructures with semiconducting properties. This is an emerging field of application for lignin and there are not many studies published until now. One study published by the University of Limerick [100] showed enormous potential of lignin as precursor for carbon nanostructures with TE properties. Carbon nanofibers (CNFs) from Lignin/PAN blends produced by electrospinning and their TE properties as a function of lignin content and processing conductions. The addition of lignin (up to 70%) reduces the diameter of CNFs from 450 to 250 nm, increases sample flexibility, and promotes inter-fiber fusion. The results showed the possibility of a conversion of p-type to n-type semiconducting behavior through doping with hydrazine vapor which allows the production of TEGs utilizing both types of semiconductors based on lignin. CNFs depicted a maximum p-type power factor of 9.27 $\mu W\,cm^{-1}\,K^{-2}$ for CNFs carbonized at 900 °C with 70% lignin which is a 34.5-fold increase to the CNFs with 0% lignin. For the hydrazine-treated samples, the results showed a maximum n-type power factor of 10.2 $\mu W\,cm^{-1}\,K^{-2}$ for the CNFs produced in the same way.

Another strategy to use lignin as part of thermoelectric materials is lignin can be used as a dopant for carbon-based nanostructures to tune their electric properties due to its aromatic chemical structure that can tailor thermoelectric properties of multi-walled carbon nanotubes (MWCNTs) [101]. The addition of lignin to Carbon Nanotube Yarns (CNTYs) improves their TE performance by one order of magnitude, showing for the first time that lignin can influence the transport properties of TE materials such as carbon nanotubes. In this case, lignin increases electrical conductivity and Seebeck coefficients simultaneously (Figure 14.6). Moreover, these materials show the possibility to manufacture sustainable TE devices with relatively good power values of 3.5 μW, which represent one of the highest values reported in literature for fully organic TE generators.

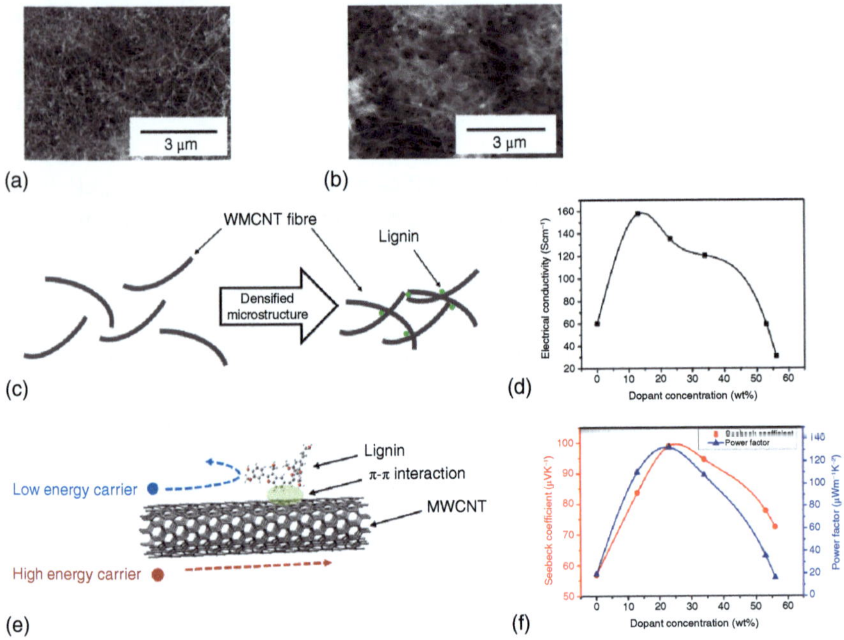

Figure 14.6 SEM images of (a) pristine CNTY and (b) CNTY/lignin nanocomposite doped with 34 wt% lignin; schematic diagrams of (c) densified MWCNT fiber microstructure by incorporating lignin, (e) charge carrier filtering mechanism by introducing lignin; thermoelectric properties of CNTY/lignin nanocomposites with varying dopant levels: (d) electrical conductivity, (f) Seebeck coefficient and power factor. Source: Adapted with permission from ref. (101). Copyright 2020, Wiley.

Another way to use lignin for thermal energy harvesting is the use of lignin as a photothermal material which can be combined with conventional TEGs to produce energy due to the temperature gradient generated due to light absorption of lignin [102]. This study demonstrated that lignin nanoparticles (L-NPs) can carry out photothermal conversion, which was attributed to $\pi-\pi$ stacking of lignin molecules L-NPs showed a stable photothermal effect (22%) generating around 0.12 V under irradiation of 100 mW cm^{-2}. Other studies have shown this strategy for photoresponsive actuator devices where lignin is blended with castor oil-derived polyamide elastomers to develop the photoactive part [103].

Acknowledgments

This research was supported by Irish Government funding via the DAFM NXTGENWOOD research program 2019PROG704. M.V., A.B., and M.N.C. acknowledge support from VIBES which has received funding from the Biobased Industries Joint Undertaking (JU) under the European Union's Horizon 2020 research and innovation program under grant agreement No 101023190. The JU receives support from the European Union's Horizon 2020 research and innovation program and the Bio-based Industries Consortium.

References

1 Jędrzejczak, P., Collins, M.N., Jesionowski, T., and Klapiszewski, Ł. (2021). The role of lignin and lignin-based materials in sustainable construction – a comprehensive review. *International Journal of Biological Macromolecules* 187: 624–650.

2 Beaucamp, A., Muddasar, M., Amiinu, I.S. et al. (2022). Lignin for energy applications – state of the art, life cycle, technoeconomic analysis and future trends. *Green Chemistry* 24 (21): 8193–8226.

3 Collins, M.N., Culebras, M., and Ren, G. (2022). Chapter 8 – The use of lignin as a precursor for carbon fiber–reinforced composites. In: *Micro and Nanolignin in Aqueous Dispersions and Polymers* (ed. D. Puglia, C. Santulli, and F. Sarasini), 237–250. Elsevier.

4 Muddasar, M., Beaucamp, A., Culebras, M., and Collins, M.N. (2022). Cellulose: characteristics and applications for rechargeable batteries. *International Journal of Biological Macromolecules* 219: 788–803.

5 Rodríguez, J.J., Cordero, T., and Rodríguez-Mirasol, J. (2016). Carbon materials from lignin and their applications. In: *Production of Biofuels and Chemicals from Lignin* (ed. Z. Fang and J.R.L. Smith), 217–262. Singapore: Springer Singapore.

6 Zhang, W., Lin, H., Lin, Z. et al. (2015). 3D hierarchical porous carbon for supercapacitors prepared from lignin through a facile template-free method. *ChemSusChem* 8 (12): 2114–2122.

7 Chatterjee, S. and Saito, T. (2015). Lignin-derived advanced carbon materials. *ChemSusChem* 8 (23): 3941–3958.

8 Jin, Y., Lin, J., Cheng, Y., and Lu, C. (2021). Lignin-based high-performance fibers by textile spinning techniques. *Materials* 14 (12): 3378.

9 Jang, D., Lee, M.E., Choi, J. et al. (2022). Strategies for the production of PAN-Based carbon fibers with high tensile strength. *Carbon* 186: 644–677.

10 Qian, X., Wang, X., Zhong, J. et al. (2019). Effect of fiber microstructure studied by Raman spectroscopy upon the mechanical properties of carbon fibers. *Journal of Raman Spectroscopy* 50 (5): 665–673.

11 Fang, W., Yang, S., Wang, X.-L. et al. (2017). Manufacture and application of lignin-based carbon fibers (LCFs) and lignin-based carbon nanofibers (LCNFs). *Green Chemistry* 19 (8): 1794–1827.

12 Saito, N., Aoki, K., Usui, Y. et al. (2011). Application of carbon fibers to biomaterials: a new era of nano-level control of carbon fibers after 30-years of development. *Chemical Society Reviews* 40 (7): 3824–3834.

13 Liu, H., Xu, T., Liu, K. et al. (2021). Lignin-based electrodes for energy storage application. *Industrial Crops and Products* 165: 113425.

14 Kadla, J.F., Kubo, S., Venditti, R.A. et al. (2002). Lignin-based carbon fibers for composite fiber applications. *Carbon* 40 (15): 2913–2920.

15 Otani, S., Fukuoka, Y., Igarashi, B., and Sasaki, K. (1969). Method for producing carbonized lignin fiber. Google Patents.

16 Sun, S.-C., Xu, Y., Wen, J.-L. et al. (2022). Recent advances in lignin-based carbon fibers (LCFs): precursors, fabrications, properties, and applications. *Green Chemistry* 24 (15): 5709–5738.

17 Choi, D., Kil, H.-S., and Lee, S. (2019). Fabrication of low-cost carbon fibers using economical precursors and advanced processing technologies. *Carbon* 142: 610–649.

18 Sudo, K. and Shimizu, K. (1992). A new carbon fiber from lignin. *Journal of Applied Polymer Science* 44 (1): 127–134.

19 Uraki, Y., Kubo, S., Nigo, N. et al. (1995). Preparation of carbon fibers from organosolv lignin obtained by aqueous acetic acid pulping. *Holzforschung* .

20 Eckert, R.C. and Abdullah, Z. (2010). Carbon fibers from kraft softwood lignin. Google Patents.

21 Qu, W., Yang, J., Sun, X. et al. (2021). Towards producing high-quality lignin-based carbon fibers: a review of crucial factors affecting lignin properties and conversion techniques. *International Journal of Biological Macromolecules* 189: 768–784.

22 Muthuraj, R., Horrocks, A., and Kandola, B.K. (2020). Hydroxypropyl-modified and organosolv lignin/bio-based polyamide blend filaments as carbon fibre precursors'. *Journal of Materials Science* 55 (16): 7066–7083.

23 Ebers, L.-S., Auvergne, R., Boutevin, B., and Laborie, M.-P. (2020). Impact of PEO structure and formulation on the properties of a Lignin/PEO blend. *Industrial Crops and Products* 143: 111883.

24 Kubo, S. and Kadla, J.F. (2005). Lignin-based carbon fibers: effect of synthetic polymer blending on fiber properties. *Journal of Polymers and the Environment* 13 (2): 97–105.

25 Wang, S., Li, Y., Xiang, H. et al. (2015). Low cost carbon fibers from bio-renewable lignin/poly(lactic acid) (pla) blends. *Composites Science and Technology* 119: 20–25.

26 Luo, Y., Qu, W., Cochran, E., and Bai, X. (2021). Enabling high-quality carbon fiber through transforming lignin into an orientable and melt-spinnable polymer. *Journal of Cleaner Production* 307: 127252.

27 Culebras, M., Beaucamp, A., Wang, Y. et al. (2018). Biobased structurally compatible polymer blends based on lignin and thermoplastic elastomer polyurethane as carbon fiber precursors. *ACS Sustainable Chemistry & Engineering* 6 (7): 8816–8825.

28 Nordström, Y., Norberg, I., Sjöholm, E., and Drougge, R. (2013). A new softening agent for melt spinning of softwood kraft lignin. *Journal of Applied Polymer Science* 129 (3): 1274–1279.

29 Wang, S., Bai, J., Innocent, M.T. et al. (2022). Lignin-based carbon fibers: formation, modification and potential applications. *Green Energy & Environment* 7 (4): 578–605.

30 Zhang, M. and Ogale, A.A. (2014). Carbon fibers from dry-spinning of acetylated softwood kraft lignin. *Carbon* 69: 626–629.

31 Jiang, X., Ouyang, Q., Liu, D. et al. (2018). Preparation of low-cost carbon fiber precursors from blends of wheat straw lignin and commercial textile-grade polyacrylonitrile (PAN). *Holzforschung* 72 (9): 727–734.

32 Bengtsson, A., Bengtsson, J., Olsson, C. et al. (2018). Improved yield of carbon fibres from cellulose and kraft lignin. *Holzforschung* 72 (12): 1007–1016.
33 Bengtsson, A., Bengtsson, J., Sedin, M., and Sjöholm, E. (2019). Carbon fibers from lignin-cellulose precursors: effect of stabilization conditions. *ACS Sustainable Chemistry & Engineering* 7 (9): 8440–8448.
34 Jin, J., Ding, J., Klett, A. et al. (2018). Carbon fibers derived from fractionated–solvated lignin precursors for enhanced mechanical performance. *ACS Sustainable Chemistry & Engineering* 6 (11): 14135–14142.
35 Poursorkhabi, V., Abdelwahab, M.A., Misra, M. et al. (2020). Processing, carbonization, and characterization of lignin based electrospun carbon fibers: a review. *Frontiers in Energy Research* 8: 208.
36 Braun, J.L., Holtman, K.M., and Kadla, J.F. (2005). Lignin-based carbon fibers: oxidative thermostabilization of kraft lignin. *Carbon* 43 (2): 385–394.
37 Norberg, I., Nordström, Y., Drougge, R. et al. (2013). A new method for stabilizing softwood kraft lignin fibers for carbon fiber production. *Journal of Applied Polymer Science* 128 (6): 3824–3830.
38 Wang, R., Zobeiri, H., Lin, H. et al. (2019). Anisotropic thermal conductivities and structure in lignin-based microscale carbon fibers. *Carbon* 147: 58–69.
39 Zhang, M., Jin, J., and Ogale, A.A. (2015). Carbon fibers from UV-assisted stabilization of lignin-based precursors. *Fibers* 3 (2): 184–196.
40 Frank, E., Steudle, L.M., Ingildeev, D. et al. (2014). Carbon fibers: precursor systems, processing, structure, and properties. *Angewandte Chemie, International Edition* 53 (21): 5262–5298.
41 Rodríguez-Mirasol, J., Cordero, T., and Rodríguez, J.J. (1996). High-temperature carbons from kraft lignin. *Carbon* 34 (1): 43–52.
42 Bengtsson, A., Hecht, P., Sommertune, J. et al. (2020). Carbon fibers from lignin–cellulose precursors: effect of carbonization conditions. *ACS Sustainable Chemistry & Engineering* 8 (17): 6826–6833.
43 Ji, S.M. and Kumar, A. (2022). Cellulose-derived nanostructures as sustainable biomass for supercapacitors: a review. *Polymers* 14 (1): 169.
44 Wu, X., Jiang, J., Wang, C. et al. (2020). Lignin-derived electrochemical energy materials and systems. *Biofuels, Bioproducts and Biorefining* 14 (3): 650–672.
45 Gergova, K., Petrov, N., and Eser, S. (1994). Adsorption properties and microstructure of activated carbons produced from agricultural by-products by steam pyrolysis. *Carbon* 32 (4): 693–702.
46 Fang J, Yao B, Nan H, He X, Wang T, editors. Effect of KOH char ratio on the properties of lignin-derived activated carbons for electric double layer capacitors. 2011 International Conference on Materials for Renewable Energy & Environment; 2011 20-22 May 2011.
47 Guo, N., Li, M., Sun, X. et al. (2017). Enzymatic hydrolysis lignin derived hierarchical porous carbon for supercapacitors in ionic liquids with high power and energy densities. *Green Chemistry* 19 (11): 2595–2602.
48 Beaucamp, A., Muddasar, M., Crawford, T. et al. (2022). Sustainable lignin precursors for tailored porous carbon-based supercapacitor electrodes. *International Journal of Biological Macromolecules* 221: 1142–1149.

49 Fu, F., Yang, D., Zhang, W. et al. (2020). Green self-assembly synthesis of porous lignin-derived carbon quasi-nanosheets for high-performance supercapacitors. *Chemical Engineering Journal* 392: 123721.

50 Wang, X., Zhang, W., Chen, M., and Zhou, X. (2018). Electrospun enzymatic hydrolysis lignin-based carbon nanofibers as binder-free supercapacitor electrodes with high performance. *Polymers* 10 (12).

51 Sima, G., Gan, L., Chang, L. et al. (2021). Efficient fabrication of ordered mesoporous carbon derived from lignin via deep eutectic solvent pretreatment for supercapacitors. *Microporous and Mesoporous Materials* 323.

52 Khan, I., Hararak, B., and Fernando, G.F. (2021). Improved procedure for electrospinning and carbonisation of neat solvent-fractionated softwood Kraft lignin. *Scientific Reports* 11 (1): 16237.

53 Park, J.H., Rana, H.H., Lee, J.Y., and Park, H.S. (2019). Renewable flexible supercapacitors based on all-lignin-based hydrogel electrolytes and nanofiber electrodes. *Journal of Materials Chemistry A* 7 (28): 16962–16968.

54 Cao, Q., Zhang, Y., Chen, J. et al. (2020). Electrospun biomass based carbon nanofibers as high-performance supercapacitors. *Industrial Crops and Products* 148: 112181.

55 Duehnen, S., Betz, J., Kolek, M. et al. (2020). Toward green battery cells: perspective on materials and technologies. *Small Methods* 4 (7): 2000039.

56 Xiao, B., Rojo, T., and Li, X. (2019). Hard carbon as sodium-ion battery anodes: progress and challenges. *ChemSusChem* 12 (1): 133–144.

57 Li, C., Sun, Y., Wu, Q. et al. (2020). A novel design strategy of a practical carbon anode material from a single lignin-based surfactant source for sodium-ion batteries. *Chemical Communications* 56 (45): 6078–6081.

58 Li, R., Zhou, Y., Li, W. et al. (2020). Structure engineering in biomass-derived carbon materials for electrochemical energy storage. *Research* 2020.

59 Ding, Z., Li, F., Wen, J. et al. (2018). Gram-scale synthesis of single-crystalline graphene quantum dots derived from lignin biomass. *Green Chemistry* 20 (6): 1383–1390.

60 Wang, K.X., Zhu, Q.C., and Chen, J.S. (2018). Strategies toward high-performance cathode materials for lithium–oxygen batteries. *Small* 14 (27): 1800078.

61 Liu, P., Wang, Y., and Liu, J. (2019). Biomass-derived porous carbon materials for advanced lithium sulfur batteries. *Journal of Energy Chemistry* 34: 171–185.

62 Xi, Y., Wang, Y., Yang, D. et al. (2019). K_2CO_3 activation enhancing the graphitization of porous lignin carbon derived from enzymatic hydrolysis lignin for high performance lithium-ion storage. *Journal of Alloys and Compounds* 785: 706–714.

63 Xi, Y., Huang, S., Yang, D. et al. (2020). Hierarchical porous carbon derived from the gas-exfoliation activation of lignin for high-energy lithium-ion batteries. *Green Chemistry* 22 (13): 4321–4330.

64 Du, Y.-F., Sun, G.-H., Li, Y. et al. (2021). Pre-oxidation of lignin precursors for hard carbon anode with boosted lithium-ion storage capacity. *Carbon* 178: 243–255.

65 Culebras, M., Geaney, H., Beaucamp, A. et al. (2019). Bio-derived carbon nanofibres from lignin as high-performance Li-Ion anode materials. *ChemSusChem* 12 (19): 4516–4521.

66 Peuvot, K., Hosseinaei, O., Tomani, P. et al. (2019). Lignin based electrospun carbon fiber anode for sodium ion batteries. *Journal of The Electrochemical Society* 166 (10): A1984–A1990.

67 Dou, X., Hasa, I., Saurel, D. et al. (2019). Hard carbons for sodium-ion batteries: structure, analysis, sustainability, and electrochemistry. *Materials Today* 23: 87–104.

68 Zhao, N., Wang, C., Li, B. et al. (2022). Construction of flexible lignin/polyacrylonitrile-based carbon nanofibers for dual-carbon sodium-ion capacitors. *Journal of Materials Science* 57 (25): 11809–11823.

69 Chen, F., Wu, L., Zhou, Z. et al. (2019). MoS_2 decorated lignin-derived hierarchical mesoporous carbon hybrid nanospheres with exceptional Li-ion battery cycle stability. *Chinese Chemical Letters* 30 (1): 197–202.

70 Wu, Z., Zou, J., Zhang, Y. et al. (2022). Lignin-derived hard carbon anode for potassium-ion batteries: interplay among lignin molecular weight, material structures, and storage mechanisms. *Chemical Engineering Journal* 427: 131547.

71 Wang, J., Yin, H., Wang, Z. et al. (2022). High-performance Sn-based anode with robust lignin-derived hard carbon support for sodium-ion batteries. *Asia-Pacific Journal of Chemical Engineering* e2768.

72 Culebras, M., Collins, G.A., Beaucamp, A. et al. (2022). Lignin/Si hybrid carbon nanofibers towards highly efficient sustainable li-ion anode materials. *Engineered Science* .

73 Merriam-Webster (2022). Sensor: Merriam-Webster.

74 Moreno, A. and Sipponen, M.H. (2020). Lignin-based smart materials: a roadmap to processing and synthesis for current and future applications. *Materials Horizons* 7 (9): 2237–2257.

75 Zhou, X., Hu, Y., Cao, Y. et al. (2022). A novel signal-on fluorometric sensor based on metal ion-mediated carbon dots for formaldehyde determination and lysosome-targeted bioimaging. *New Journal of Chemistry* 46 (36): 17540–17547.

76 Zhu, L., Shen, D., and Hong, L.K. (2022). Triple-emission nitrogen and boron co-doped carbon quantum dots from lignin: highly fluorescent sensing platform for detection of hexavalent chromium ions. *Journal of Colloid and Interface Science* 617: 557–567.

77 Wang, Y., Liu, Y., Zhou, J. et al. (2021). Hydrothermal synthesis of nitrogen-doped carbon quantum dots from lignin for formaldehyde determination. *RSC Advances* 11 (47): 29178–29185.

78 Li, M., Tu, Q., Long, X. et al. (2021). Flexible conductive hydrogel fabricated with polyvinyl alcohol, carboxymethyl chitosan, cellulose nanofibrils, and lignin-based carbon applied as strain and pressure sensor. *International Journal of Biological Macromolecules* 166: 1526–1534.

79 Zheng, H., Han, X., Wei, Q. et al. (2022). A green flexible and wearable biosensor based on carbon nanofibers for sensitive detection of uric acid in artificial urine. *Journal of Materials Chemistry B* 10 (41): 8450–8461.

80 Lin, F., Qiu, Y., Zheng, X. et al. (2022). One-pot mechanochemical assembly of lignocellulose nanofiber/graphite nanocomposites for wearable electronic devices. *Chemical Engineering Journal* 437: 135286.

81 Beaucamp, A., Culebras, M., and Collins, M.N. (2021). Sustainable mesoporous carbon nanostructures derived from lignin for early detection of glucose. *Green Chemistry* 23 (15): 5696–5705.

82 Peng, Z., Yu, C., and Zhong, W. (2022). Facile preparation of a 3D porous aligned graphene-based wall network architecture by confined self-assembly with shape memory for artificial muscle, pressure sensor, and flexible supercapacitor. *ACS Applied Materials & Interfaces* 14 (15): 17739–17753.

83 Sun, L., Mo, Z., Li, Q. et al. (2021). Facile synthesis and performance of pH/temperature dual-response hydrogel containing lignin-based carbon dots. *International Journal of Biological Macromolecules* 175: 516–525.

84 Abbas, A., Tabish, T.A., Bull, S.J. et al. (2020). High yield synthesis of graphene quantum dots from biomass waste as a highly selective probe for Fe^{3+} sensing. *Scientific Reports* 10 (1): 21262.

85 Zhao, S., Chen, X., Su, J. et al. (2022). Interaction of lignin and xylan in the hydrothermal synthesis of lignocellulose-based carbon quantum dots and their application in in-vivo bioimaging. *International Journal of Biological Macromolecules* 222: 1876–1887.

86 Chen, C., Bai, X., Ding, Y., and Lee, I.S. (2019). Electrical stimulation as a novel tool for regulating cell behavior in tissue engineering. *Biomaterials Research* 23: 25.

87 Lai, C., Zhou, Z., Zhang, L. et al. (2014). Free-standing and mechanically flexible mats consisting of electrospun carbon nanofibers made from a natural product of alkali lignin as binder-free electrodes for high-performance supercapacitors. *Journal of Power Sources* 247: 134–141.

88 Dallmeyer, I., Ko, F., and Kadla, J.F. (2010). Electrospinning of technical lignins for the production of fibrous networks. *Journal of Wood Chemistry and Technology* 30 (4): 315–329.

89 Seo, D.K., Jeun, J.P., Kim, H.B., and Kang, P.H. (2011). Preparation and characterization of the carbon nanofiber mat produced from electrospun PAN/lignin precursors by electron beam irradiation. *Reviews on Advanced Materials Science* 28 (1): 31–34.

90 Hu, S. and Hsieh, Y.-L. (2013). Ultrafine microporous and mesoporous activated carbon fibers from alkali lignin. *Journal of Materials Chemistry A* 1 (37): 11279–11288.

91 Cho, M., Karaaslan, M., Chowdhury, S. et al. (2018). Skipping oxidative thermal stabilization for lignin-based carbon nanofibers. *ACS Sustainable Chemistry & Engineering* 6 (5): 6434–6444.

92 Culebras, M., Gómez, C.M., and Cantarero, A. (2014). Enhanced thermoelectric performance of PEDOT with different counter-ions optimized by chemical reduction. *Journal of Materials Chemistry A* 2 (26): 10109.

93 Culebras, M., Gómez, C., and Cantarero, A. (2014). Review on polymers for thermoelectric applications. *Materials* 7 (9): 6701–6732.

94 Culebras, M., Uriol, B., Gomez, C.M., and Cantarero, A. (2015). Controlling the thermoelectric properties of polymers: application to PEDOT and polypyrrole. *Physical Chemistry Chemical Physics* 17 (23): 15140–15145.

95 Culebras, M., de Lima, M.M., Gómez, C., and Cantarero, A. (2016). Organic thermoelectric modules produced by electrochemical polymerization. *Journal of Applied Polymer Science* .

96 Gao, J., Liu, C., Miao, L. et al. (2015). Power factor enhancement via simultaneous improvement of electrical conductivity and Seebeck coefficient in tellurium nanowires/reduced graphene oxide flexible thermoelectric films. *Synthetic Metals* 210: 342–351.

97 Shi, H., Liu, C., Xu, J. et al. (2013). Facile fabrication of PEDOT:PSS/ polythiophenes bilayered nanofilms on pure organic electrodes and their thermoelectric performance. *ACS Applied Materials & Interfaces* 5 (24): 12811–12819.

98 Culebras, M., Igual-Muñoz, A.M., Rodríguez-Fernández, C. et al. (2017). Manufacturing Te/PEDOT films for thermoelectric applications. *ACS Applied Materials & Interfaces* 9 (24): 20826–20832.

99 Culebras, M., Cho, C., Krecker, M. et al. (2017). High thermoelectric power factor organic thin films through combination of nanotube multilayer assembly and electrochemical polymerization. *ACS Applied Materials & Interfaces* 9 (7): 6306–6313.

100 Dalton, N., Lynch, R.P., Collins, M.N., and Culebras, M. (2019). Thermoelectric properties of electrospun carbon nanofibres derived from lignin. *International Journal of Biological Macromolecules* 121: 472–479.

101 Culebras, M., Ren, G., O'Connell, S. et al. (2020). Lignin doped carbon nanotube yarns for improved thermoelectric efficiency. *Advanced Sustainable Systems* 4 (11): 2000147.

102 Zhao, X., Huang, C., Xiao, D. et al. (2021). Melanin-inspired design: preparing sustainable photothermal materials from lignin for energy generation. *ACS Applied Materials & Interfaces* 13 (6): 7600–7607.

103 Chen, J., Qi, J., He, J. et al. (2022). Biobased composites with high lignin content and excellent mechanical properties toward the ingenious photoresponsive actuator. *ACS Applied Materials & Interfaces* 14 (10): 12748–12757.

15

Lignin-based Hydrogel: Mechanism, Properties, and Applications

Qiang Wang[1], *Baobin Wang*[1], *Jiachuan Chen*[1], *Guihua Yang*[1], *Lei Zhang*[1], *Kefeng Liu*[1], *Qimeng Jiang*[1] *and Pedram Fatehi*[1,2]

[1] *Qilu University of Technology (Shandong Academy of Sciences), Faculty of Light Industry, State Key Laboratory of Biobased Material and Green Papermaking, Jinan, Shandong, 250353, China*
[2] *Lakehead University, Biorefining Research Institute and Chemical Engineering Department, Thunder Bay, ON, P7B 5E1, Canada*

15.1 Introduction of Hydrogels

Hydrogels are soft and elastic three-dimensional (3D) polymer networks with a strong affinity toward the water. Hydrogels are typically made from chemicals that can absorb water or aqueous solutions several hundred times their dry weight [1, 2]. Due to the highly tailorable structure of hydrogel, it has shown versatile properties, including transparency, biocompatibility, conductivity, stimuli responsiveness, and self-heling property, which enabled promising applications in the fields of water treatment, drug delivery, wound dressing, cosmetics, agriculture, and energy storage [3–5]. Figure 15.1 displays the classification, physicochemical properties, and applications of hydrogels made of synthetic chemicals (physical or chemical crosslinking).

According to their crosslinking strategies, hydrogels can be classified into two categories: chemical hydrogels and physical hydrogels. Table 15.1 shows the primary crosslinking approach for the fabrication of hydrogels. Chemical hydrogels are fabricated through different covalent bonds of the polymer networks [24]. The common covalent chemistry introduced to hydrogel networks includes radical polymerization, aldehyde-mediated crosslinking, stereo-complex formation, and click-chemistry-derived crosslinking [25–29]. Radical polymerization can be initiated by light, heat, radiation, and other methods, such as plasmonic activation. The polymerization process could be affected by monomer type and composition, polymerization method, crosslinking reagent, and type. The dominant monomers for hydrogel are monomers of acrylic acid (AA), acrylamide, and vinyl acetate [27]. In addition, monomers could be grafted onto biological polymers to form a hydrogel

Lignin Chemistry: Characterization, Isolation, and Valorization, First Edition.
Edited by Yuhe Liao and Bert F. Sels.

matrix. Click chemistry demonstrates the advantages of high yields, high specificity, and high selectivity, and the main reactions associated with click chemistry-derived hydrogel are Diels–Alder, oxime, and Michael addition (Figure 15.1). Diels–Alder reaction occurs between furan and maleimide, and their high selectivity enabled environmentally friendly preparation of hydrogels [10, 11]. Oxime reactions occur between aminooxy/hydroxylamine group and aldehyde/ketone. In this respect, aldehyde, aminooxy, and catechol functionalized polyethylene glycol were proposed to prepare oxime crosslinked poly(ethylene glycol) (PEG) hydrogel [12]. The reaction between nucleophiles across the carbon–carbon bond and activated electrophilic olefins or alkynes is the Michael reaction [30]. The Michael donors include enolates as well as nonenolate nucleophiles, such as amines, thiols, and phosphines. Generally, the acceptors of the Michael reaction are mainly acrylate esters, acrylonitrile, maleimides, alkyl methacrylates, cyanoacrylates, and vinyl sulfone [13]. The chemical crosslinking of polymer networks endows hydrogel with a covalent bond, which introduces chemical and mechanical stability to hydrogels.

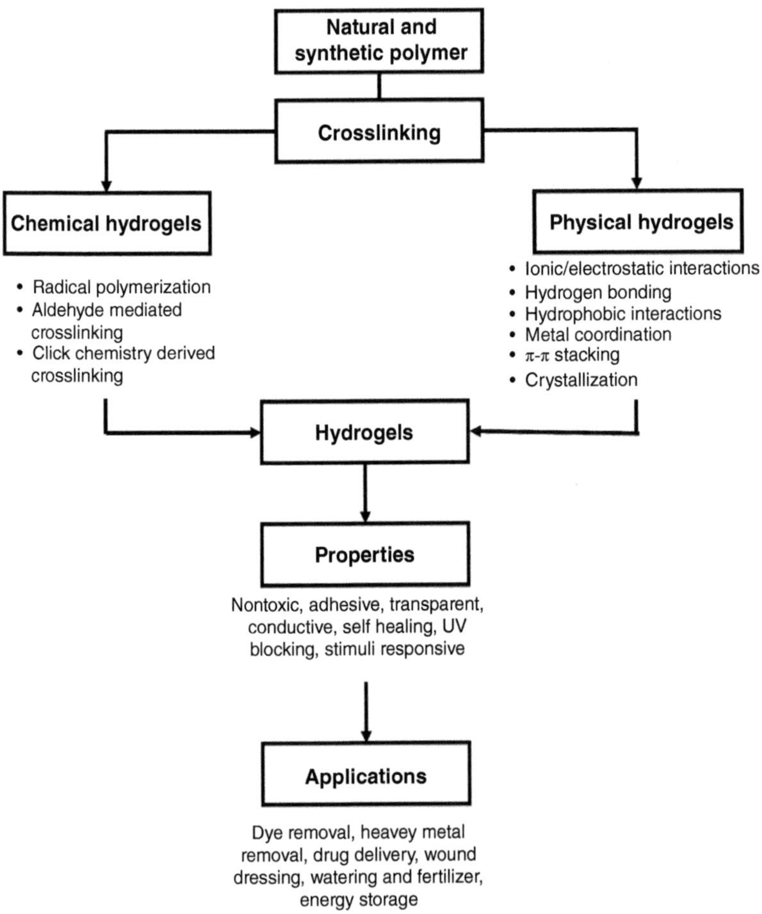

Figure 15.1 Classification, significant properties, and applications of hydrogels.

Table 15.1 Classifications and mechanisms of hydrogel preparation.

Type of hydrogels	Sub classifications	Reaction media	Reaction approach	Crosslinking mechanism	References
Chemical hydrogels	Radical polymerization	Free-radical monomers (acrylic acid, acrylamide and vinyl acetate)	Photo/initiators induced	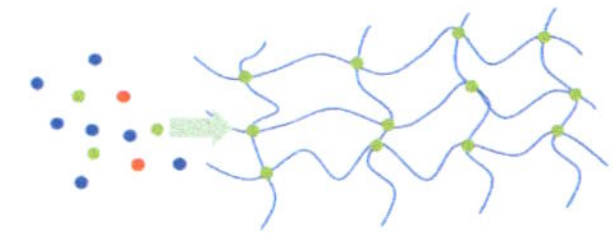	[6]
	Aldehyde-mediated crosslinking	Polymers with side chains of alcohol, amine, or hydrazide groups	Self-crosslinking	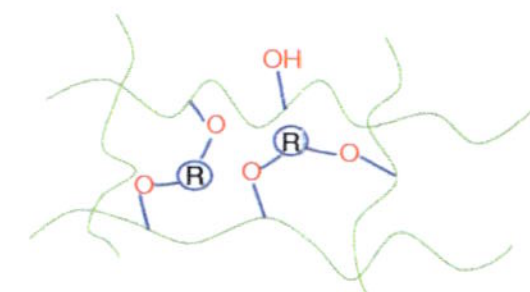	[7–9]
	Click-chemistry derived crosslinking	Diels–Alder Oxime Michael addition reactions		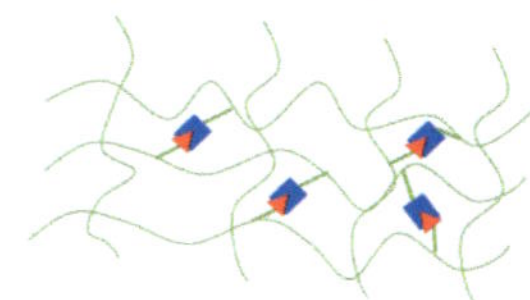	[10, 11] [12] [13]
	Ionic/electrostatic interactions	Molecules of opposite electric charges		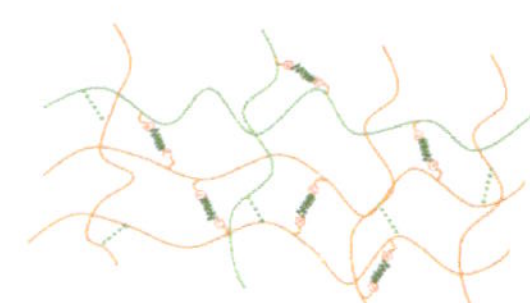	[14, 15]
	Hydrogen bonding	Hydrogen atom with electronegative molecules (O, S, etc.)		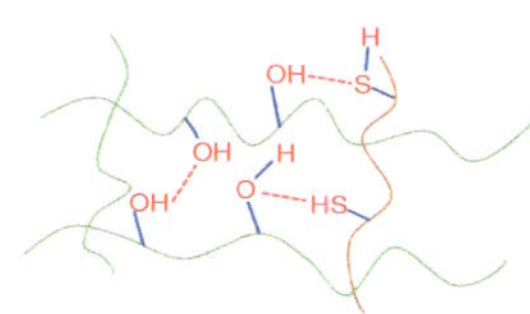	[16]

(Continued)

Table 15.1 (Continued)

Type of hydrogels	Sub classifications	Reaction media	Reaction approach	Crosslinking mechanism	References
Physical hydrogels	Hydrophobic interactions,	Polymers with hydrophobic end groups, side chains, or monomers		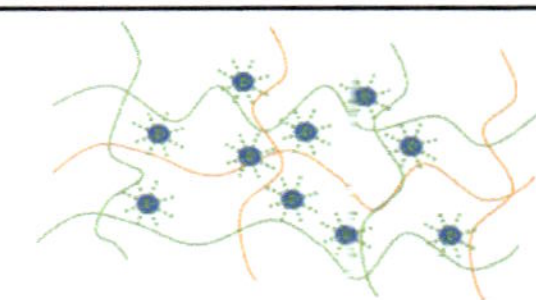	[17]
	Metal coordination	Polymers with groups that could be bonded with metal ions			[18, 19]
	π–π stacking	Polymers with π-bonds/ aromatic rings		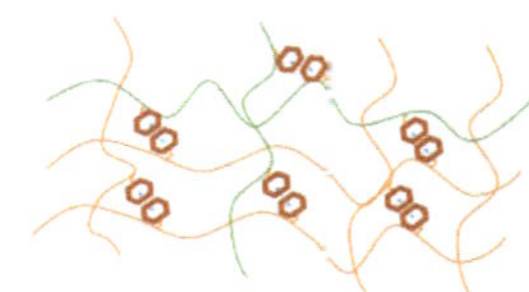	[20]
	Crystallization	All kinds of synthetic and biological hydrogels	Freeze thawing, freeze drying	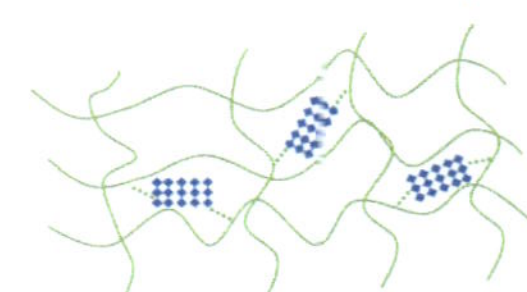	[21]
	Stereo complex Formation	Complexation of PDLA and PLLA		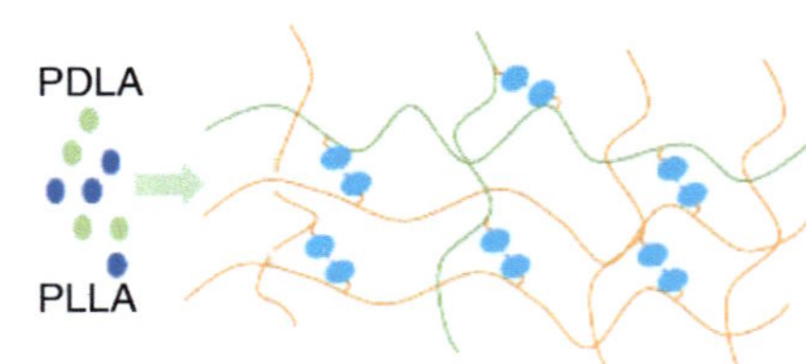	[22, 23]

In comparison with chemical pathways to produce hydrogels, physically fabricated hydrogels have networks created by molecular entanglement and/or physical interaction of macromolecules [31]. The main strategies in this approach include the utilization of molecular interaction, such as crystallization, ionic/electrostatic interactions, hydrogen bonding, hydrophobic interactions, metal coordination, π–π stacking, and stereo-complex formation. Physical networking is normally weak and even reversible under external stimuli effect. Generally, combining chemical crosslinking and physical networking is necessary to achieve multifunctional hydrogels. Crystallization is an effective physical crosslinking route for the high elastic strength of physically produced hydrogels. The most common way to realize crystallization is through a cyclic freeze-thawing process. The freeze-thawing process would show more porous, spongy, and higher elastic performance [32]. Crystallized endow the hydrogel with enhanced elastic modulus via different factors, such as molecular weight, concentration, and the number of cycles.

Electrostatic/ionic interactions are another physical crosslinking strategy that occurs with the charged functional groups of biological macromolecules or synthetic polymers. The basic principle of this method is the complexation of polyelectrolyte solutions with multivalent ions of opposite charge. Ionic crosslinked hydrogels synthesized by the introduction of divalent (e.g. Ca^{2+}, Ba^{2+}, and Mg^{2+}) are considered to be of this type [33, 34]. In addition, hydrogen bonding is ubiquitous noncovalent interaction in various hydrogels. The functional groups, e.g. amide, phosphate, carboxylic acid, urea, carbazole, and hydroxyl, could interact with themselves or with electron donor groups, forming hydrogen bonds. Hydrophobic interaction occurs in polymer networks with hydrophobic pendants. The hydrophobic hydrogel could be prepared through the sol–gel phase transition induced by thermal variation based on lower critical solution temperature (LCST) or upper critical solution temperature (UCST) polymers [17]. Metal coordination is referred to as metal–ligand interaction, a Lewis-acid-derived interaction. The relatively weak interaction makes it reversible upon external stimuli [18]. For instance, transient network hydrogels crosslinked through histidine–divalent cation coordination bonds were studied by conventional rheologic methods using histidine-modified star PEG polymers. The viscoelastic properties of such hydrogels were studied as a function of the metal, pH, concentration, and ionic strength [19]. The π–π stacking refers to the interactions with polymers containing π conjugate structures, such as aromatic conductive polymers (e.g. polypyrrole, polyaniline, polythiophene, and polyethyldioxythiophene) and inorganic materials (e.g. graphene, carbon nanotubes) incorporating in the hydrogel matrix. The reduced graphene oxide (GO)-based hydrogels could be prepared through strong van der Waals and π–π interactions [20]. In addition, the stereoisomer poly lactic acid (PLA) exhibited a melting temperature of 170 °C. However, the mixture of high molecular weight PDLA and PLLA would induce a phase with a higher melting temperature (230 °C), which is ascribed to stereo-complex formation. The PDLA and PLLA could form hydrogels with temperature-induced sol–gel transformation [35]. However, one notable drawback of stereo complexation is the relatively restricted range of polymer compositions that can be used [22]. Compared with the chemical crosslinking approach, physical crosslinking achieved via a noncovalent bond demonstrates reversible interactions, which could endow hydrogel with interesting properties, such as stretchability, self-healing, and mechanical robustness. Also, compared with the chemical crosslinking method, physical crosslinking method has the advantages of faster formation and less environmental impact, but the hydrogel prepared by this method has poor stability under different conditions (such as strong acid or strong base).

15.2 Introduction of Lignin

Lignin is one of the major components that constitute the cell walls of lignocellulosic plants and the second most abundant renewable natural resource on the planet. It was first discovered and described by Payen in 1838 and was granted its modern name by Schulze in 1857 [36]. Since then, the physicochemical aspects of lignin have been widely studied. Lignin is a complex natural aromatic polymer with a highly branched three-dimensional structure, made up of three phenylpropane units (monolignols), namely, coniferyl, sinapyl, and *p*-coumaryl alcohol (Figure 15.2), which bond together by a series of linkages, such as C—C and C—O—C bonds [37]. Generally, lignin units derived from three monolignols are called guaiacyl (G), syringyl (S), and *p*-hydroxyphenyl (H) units. These units contain many different functional groups, including hydroxyl (OH), carboxyl (COOH), carbonyl (C═O), and methoxyl, which are active sites for further chemical modification and lignin utilization [38].

15.2.1 Sources of Lignin

Lignin can be extracted from various renewable raw materials, such as sugarcane, bagasse, pine or cedar trees, and wood. The proportions of three phenylpropanoid alcohols differ for lignin of different origins. Table 15.2 lists the proportions of different monolignols in lignin of different plant sources. The alcohol units in softwood consist of mainly G-units. The lignin of hardwoods, however, contains both G

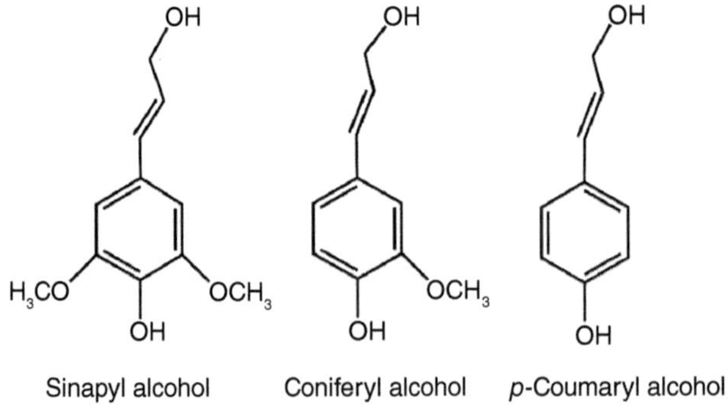

Figure 15.2 Chemical structure of three monolignols.

Table 15.2 Cellulose, hemicellulose, and lignin contents of different plant types as well as the proportion of three monolignols in lignin of different plant sources.

Type	Cellulose (%)	Hemicellulose (%)	Lignin (%)	Coniferyl alcohol (%)	Sinapyl alcohol (%)	*p*-Coumaryl alcohol (%)
Softwood	45–50	25–35	25–35	>95	0	<5
Hardwood	40–55	24–40	18–25	25–50	45–75	0–8
Grass	25–40	35–50	10–30	35–80	20–55	5–35

Source: Adapted from [39, 40].

and S units as the main building blocks, while all three alcohols are the lignin precursors in grass-derived lignin (monocotyledonous tissue).

15.2.2 Characteristics of Lignin

Natural lignin is colorless or light yellow, but when treated with acid or alkali, its color changes to light or dark brown. The variation in the monolignol content in plant sources yields multiplicity in the physicochemical properties of the resulting lignin materials. The molecular masses of isolated lignin are in the range of 1000–20,000 $g\,mol{-}^{1}$, but the degree of polymerization of lignin in nature is difficult to estimate [41]. Lignin has excellent properties, such as high thermal stability, antioxidant, biodegradability, and antimicrobial behavior, which are favorable for exploiting diverse applications, such as hydrogels [24, 42].

15.3 Lignin-Based Hydrogel Preparation

As the most abundant aromatic compounds in nature, lignin contains a variety of active groups, such as phenolic hydroxyl, alcohol hydroxyl, methoxy, and aldehyde, which make it an excellent potential for the fabrication of functional hydrogels through diverse strategies. The basic polymer matrix of hydrogel is formed from monomers or polymers through crosslinking with water infiltration. The functional groups of polymeric backbones induce unique properties and structures in the hydrogels. Basically, hydrogels can have single, semi-interpenetrating, and interpenetrating networks (IPNs) via various crosslinking strategies (Figure 15.3). The single-component hydrogel network is derived from one material that normally cannot meet the requirement of multi-functionalities required in advanced applications, such as adsorption, drug delivery, wound dressing, and energy storage (I). Therefore, other strategies such as IPN, semi-IPN, and nanocomposite incorporation were introduced to increase the multi-properties of the hydrogel. The IPN has two polymeric networks, one of which is pre-crosslinked. In this system, two polymer networks are interconnected with each other, and the interlocked structure of the IPN ensures its excellent mechanical properties due to the presence of a dense

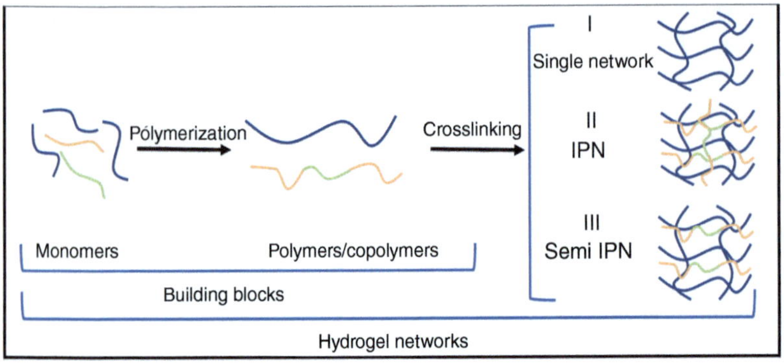

Figure 15.3 Methods for generating building blocks of hydrogel network.

and compact matrix (II). Similarly, constructing a (semi-)IPN is a unique strategy for utilizing a combination of crosslinked polymers, which endow hydrogel with superior properties to single network hydrogel (III). An IPN polymer is a class of polymers with two or more interlaced networks that are not covalently linked [43]. Semi-IPN polymers are characterized by the fact that one of the polymer components in the IPN has a linear structure rather than a network structure [44].

The synthesis approaches of interpenetrating polymer network hydrogels can be divided into three types according to the chemical mechanism of preparation. (i) The combination of two or more monomers is polymerized separately with an initiator and activator to simultaneously synthesize two or more network structures interlaced [45, 46]. (ii) A single network polymer hydrogel is prepared first and is then swelled to a hybrid solution containing another monomer, initiator, and activator. In this process, the monomer is crosslinked and polymerized on the formed network substrate, thus forming an IPN structure [45, 47]. (iii) The monomer is crosslinked and polymerized into a network structure, while another linear polymer is trapped in the network substrate to form a semi-IPN structure of polymers. Then, the IPN polymer is prepared by selective crosslinking of linear polymer chains [48, 49]. In the following section, each method of preparing lignin-derived hydrogel is discussed.

15.3.1 Blending of Lignin in Hydrogel Networks

Generally, the blending/complexing of lignin in hydrogels via physical interaction, including hydrogen bonding, electrostatic interaction, and van der Waals force, is reversible. The physical mixture consists of the simple dispersion or blending of lignin with other polymers. This approach benefits from the inherent properties of lignin, such as three-dimensional structure and large functional groups, that improve the mechanical properties of formed hydrogels. In addition, lignin incorporation endows hydrogel with unique properties of lignin, including antibacterial and antioxidant.

Bio-based composite hydrogels can also be prepared by blending lignin with other biomass, such as cellulose and chitosan (Table 15.3). In one study, the hydrogel was synthesized by blending polyvinyl alcohol (PVA) and hydroxyethyl cellulose with

Table 15.3 Preparation of hydrogel by blending lignin into the hydrogel matrix.

Lignin source	Polymer network	Crosslinking mechanism	Preparation method	Properties	Applications	References
Lignosulfonate	Graphene oxide	Hydrogen bonds, π–π interaction, and electrostatic interactions	Hydrothermal treatment	Multiporosity ($459.3\,m^2\,g^{-1}$)	Water purification	[50]
Alkali lignin	Chitosan	Electrostatic interaction	Blending	Non-toxicity, adhesion	Tissue Engineering Wound healing; wastewater treatment	[22]
Alkali lignin	Cellulose	Physical interactions	Blending	Biocompatibility, biodegradability, controllable properties	Biocatalytic, biomedical, and bioelectronic fields	[51]
Lignosulfonate	PVA, chitosan	Physical interactions	Blending	Robust, protein adsorption capacity, wound environmental regulation ability	Wound healing	[52]

lignin in the presence of borax as a crosslinker [53]. Uniformly dispersed lignin in this network acts as a plasticizer and increases the mobility of the polymer chain, resulting in high tensile properties. The excellent self-healing properties of the hydrogel are attributed to the dynamic hydrogen bonds formed between lignin, hydroxyethyl cellulose, PVA, and reversible diol-borax complexation, which makes these hydrogels useful for flexible devices. Ravishankar et al. prepared chitosan-alkali lignin-based hydrogels by means of solution blending [22]. Due to the electrostatic interaction between the benzene oxide group in lignin and the ammonium group on the main chain of chitosan, a novel lignin–chitosan–PVA composite hydrogel was prepared by introducing lignin [52]. It was confirmed that hydrogen bonds existed between hydroxyl groups of lignin and PVA, while electrostatic interactions existed between the sulfonic moiety of lignin and the cationic amino group of chitosan [22]. Furthermore, lignin enhanced the mechanical strength, protein adsorption capacity, and wound environmental regulation of hydrogel. In another work, alkali lignin was blended with chitosan by electrostatic interactions between the ammonium groups of chitosan and the phenoxide groups of alkali lignin to synthesize biocompatible hydrogel [22]. Cellulose/lignin composite hydrogel could also be prepared through the co-dissolution of cellulose and lignin in ionic liquid (1-ethyl-3-methylimidazolium acetate) and then reconstituted with distilled water [51]. The blending of lignin into the hydrogel matrix could enhance the mechanical properties of the hydrogel. Jaiswal et al. utilized lignin to prepare silver nanoparticles (Ag NPs) and then incorporated the composites into the Carrageenan matrix through crosslinking with divalent cations [54].

Lignin can also be physically linked with hydrophilic polymers using hydrogen bonds due to the abundance of polar sites on the lignin skeleton. Oveissi et al. prepared polyurethane hydrogel in the presence of lignin, and the results demonstrated that the hydrogen bonds formed between lignin and polyurethane improved the mechanical strength and the processability of the composite hydrogel [55]. In the same vein, Li et al. prepared lignosulfonate/GO hydrogel (LS-GH) by the hydrothermal treatment of lignosulfonate and GO [50], and the results confirmed that diverse oxygen-containing groups of lignosulfonate helped with the improvement of specific surface area, multiple porosities and active sites for the adsorption of metal ions. The simple blending of the lignin enabled the easy fabrication of the hydrogel, however, the amount of lignin that can be added to hydrogel is rather limited due to the issue of compatibility and solubility of lignin in the polymer network.

15.3.2 Polymerization of Lignin with Bio-Based/Synthetic Chemicals for Hydrogels

Lignin can be polymerized with other materials to produce hydrogels. In the reaction process, a stable covalent bond is formed between the crosslinking reagent and lignin. The formation process of the chemical crosslinking method is irreversible, and the product has the advantage of a stable structure. Generally, the polymerization of lignin with chemicals for hydrogels mainly includes crosslinking copolymerization, crosslinking grafted lignin, atom transfer radical polymerization (ATRP), reversible addition-fragmentation chain transfer polymerization (RAFT), and interpenetrating network polymer, which will be discussed in the following sections.

15.3.2.1 Crosslinking

Lignin hydrogels can be directly synthesized by crosslinking lignin and crosslinking reagents [39]. As shown in Table 15.4, lignin could act as a crosslinker or is directly crosslinked with a polymer matrix with the aid of an external crosslinking reagent during the preparation of hydrogel. Various sources of lignin were utilized with different polymers, such as synthetic polymers or biopolymers, to improve composite hydrogels' properties.

The hydroxyl groups of lignin are usually crosslinked by methylolation or epoxidation with formaldehyde or epichlorohydrin. Alternatively, lignin was phenolated, which increased the active sites of industrial lignin (corn cob lignin, poplar wood lignins, and wheat straw lignin) and improved the properties of induced hydrogels [62].

Epichlorohydrin (ECH) is a common crosslinker, which could react with OH groups on the backbone of lignin. Generally, ECH facilitates the enhancement of pore size distribution, chemical stability, mechanical properties, and adsorption capacity of hydrogels [63]. The addition of ECH endows hydrogel with increasing pore size, which leads to the increased swelling capacity of the hydrogel network. As an example, lignin was crosslinked with PVA by ECH [56, 58]. Lignin–PVA super-absorbent hydrogels exhibited excellent swelling ratio, excellent water retention, and biodegradability (Table 15.4).

Lignin can also be crosslinked with polyethylene glycol diglycidyl ether (PEGDGE) [64]. Generally, the etherification reaction occurs between PEGDGE and the phenolic hydroxyl group of lignin. Passauer et al. prepared lignin-based hydrogels by crosslinking technical lignin with PEGDGE [65]. Alkali lignin was also used to crosslink with PEGDGE to prepare hydrogels chemically. In these cases, the phenolic functional group of lignin acted as the polymer backbone, while two epoxy rings of PEGDGE were used as the chemical crosslinker [59].

Bio-based materials, such as cellulose, can be bonded with lignin to prepare bio-based hydrogels with specific properties. Dai et al. proposed a facile synthetic approach to crosslink lignin and cellulose with epichlorohydrin, in which cellulose and alkali lignin were used as the polymer backbone ECH was used to bind lignin and cellulose chains [66]. The modification of lignin could also be a strategy to prepare hydrogel. Li et al. modified lignosulfonate through the Mannich reaction to obtain lignin amine (LA), which was further crosslinked by the reaction between amine groups of LA and carbonyl groups of PVA to form hydrogel [67].

15.3.2.2 Free Radical Polymerization of Lignin and Monomers

Lignin can be grafted with others chemicals via a free radical polymerization mechanism to generate hydrogels. Radical polymerization includes initiation, propagation, chain transfer, and termination. The initiator-derived free radicals are generated by thermal or photochemical homolytic cleavage of covalent bonds or by a redox process [68]. Table 15.5 summarizes the free radical polymerization

Table 15.4 Lignin-based hydrogels prepared by chemical crosslinking.

Lignin source	Preparation method	Polymer network	Crosslinker/initiator	Crosslinking mechanism	Properties	Application	References
Steam explosion lignin of aspen	crosslinking copolymerization	Polyvinyl alcohol	Epichlorohydrin	Etherification	Porous Structure	—	[56]
Steam explosion lignin		Cellulose and lignin	Epichlorohydrin	Etherification	—	Polyphenyl release	[57]
Alkaline lignin		Polyvinyl alcohol and lignin or lignin-epoxy	Epichlorohydrin	Etherification	Super swelling capacity ($456\,g\,g^{-1}$)	Dye pollutant removal	[58]
Alkaline		Poly(ethyleneglycol) diglycidylether	Lignin	Etherification	Mechanical integrity, ionic conductivity	Energy storage	[59]
Kraft and organosolv lignin		Poly(ethylene glycol) diglycidyl ether	Kraft and organosolv lignin	Etherification	Swelling capacity	—	[60]
—		Poly(ethylene glycol) and poly(methyl vinyl ether-*co*-maleic acid)	lignin	Etherification	Swelling capacity	Drug delivery	[61]

strategies of lignin to produce hydrogels. Various types of hydrogels can be obtained by copolymerizing lignin and hydrophilic monomers.

For example, a hydrogel was prepared by polymerizing alkaline or kraft lignin [69] with acrylamide (AM) and PVA, and lignin contributed to the mechanical properties of hydrogels. Waleed K. El-Zawawy reported the generation of hydrogel from PVA, acrylamide (PAAm), and lignin of the black liquor procured from the alkaline pulping of rice straw [70].

Lignin can be incorporated into other polymers by means of interpenetration to synthesize hydrogels while maintaining their intrinsic properties [77–79]. In this method, lignin-based polymers are formed by radical reaction with monomers, while the monomer polymerizes into a copolymer network [80]. Jiang et al. synthesized a super-adsorbent bentonite-doped lignin hydrogel via reacting AA, 2-acrylamido-2-methyl-propanesulfonic acid (AMPS), and lignosulfonate (LHS-BT) [72].

Temperature-sensitive lignin-based hydrogel was also prepared by the graft copolymerization of acetic acid lignin [81]. In this reaction, lignin acetate was polymerized by *N*-isopropyl acrylamide (NIPAAm) and *N,N′*-methylene bisacrylamide (MBBAm) using H_2O_2 as initiator. In this system, lignin contributed to the porous structure and active sites of hydrogels. In the same vein, lignin-based pH/temperature dual-response hydrogel was prepared by free radical polymerization of PVA and poly(*N*-isopropylacrylamide) in the presence of lignin-based carbon dots [71]. In this hydrogel, lignin carbon dots linked to PVA and PNIPAM via hydrogen bonds were responsible for pH stimulation, and the interpenetrating polymer network structure formed by PVA and PNIPAM provided porous honeycomb structure and strong mechanical strength [70]. Sabino et al. synthesized IPN hydrogels by free radical reaction of acrylamide and *N,N′*-methylene-bis-acrylamide as the crosslinking agent and lignin as the interpenetrated phase [70]. The lignin incorporation modified the morphological property and active sites, which facilitated the lignin-based hydrogel's adsorption capacities.

15.3.2.3 Atom Transfer Radical Polymerization (ATRP) of Lignin for Hydrogel Production

ATRP is an effective method to realize activity-controlled radical polymerization in mild conditions. The mechanism of ATRP is shown in Figure 15.4 [82]. The first step of ATRP is to synthesize lignin-based organic halide as a precursor. The second step is to copolymerize the precursor and monomers. The lignin-based organic halide was prepared as an initiator, on which the radical reaction with vinyl monomers was processed. The advantages of this strategy include precise control of polymer composition, topology, and functionality, enabling the synthesis of well-defined polymers with various structures, which is beneficial for the preparation of hydrogel with a well-defined structure [83]. However, the complex process and relatively low efficiency limited its application.

ATRP was used for preparing lignin-based hydrogel. In this system, *N*-isopropylacrylamide (NIPAM) and *N,N*-dimethylaminoethylmethacrylate (DMAEMA) comonomers were copolymerized and the fully brominated lignin

Table 15.5 Lignin-based hydrogels prepared by polymerizing lignin and monomers.

Lignin source	Conditions	Chemicals	Crosslinker/initiator	Properties	Application	References
Alkaline or kraft lignin	40 °C, 6 h	Acrylamide and polyvinyl alcohol	Potassium persulfate and sodium metabisulfite	Swelling capacity	Adsorption	[69]
Black liquor	25 °C, 24 h	Poly(vinyl alcohol) and polyacrylamide	Ammonium cerium sulfate and *N*,*N*′-methylene bisacrylamide	Swelling capacity	Adsorption	[70]
Sodium lignosulfonate	0 °C, 24 h	Acrylic acid and 2-acrylamido-2-methyl-propanesulfonic acid	Potassium persulfate	Dual responsive and mechanical property	Adsorption	[71]
Sodium lignosulfonate	90 °C, 1 h	Polyvinyl alcohol	Aminated lignin	Antibacterial and biocompatibility	Antibacterial application	[67]
Sodium lignosulfonate	70 °C, 10 min	Acrylic acid and 2-acrylamido-2-methyl-propanesulfonic acid	Grafted lignin	Adsorption capacity	Wastewater remediation	[72]
Sodium lignosulphonate	70 °C, 12 h	Sulfobetaine methacrylate and methacrylated lignin	—	—	Antimicrobial and antifouling	[73]
Sodium lignosulphonate	60 °C, 1 h	Poly(acrylic acid-*co*-poly(vinyl pyrrolidone)) hydrogel SLS-g-P(AA-*co*-PVP)	Sodium lignosulphonate	Swelling and pH responsive	Drug delivery	[74]
—	50 °C, 1 h	Acrylamide and *N*-isopropyl acrylamide	Lignin	Thermal stability and mechanical properties	Dye removal	[75]
Alkaline lignin	60 °C, 15 min	*N*,*N*′-methylene-bisacrylamide (LM) and acrylic acid (AA)	Lignin	—	Metal ions removal	[76]

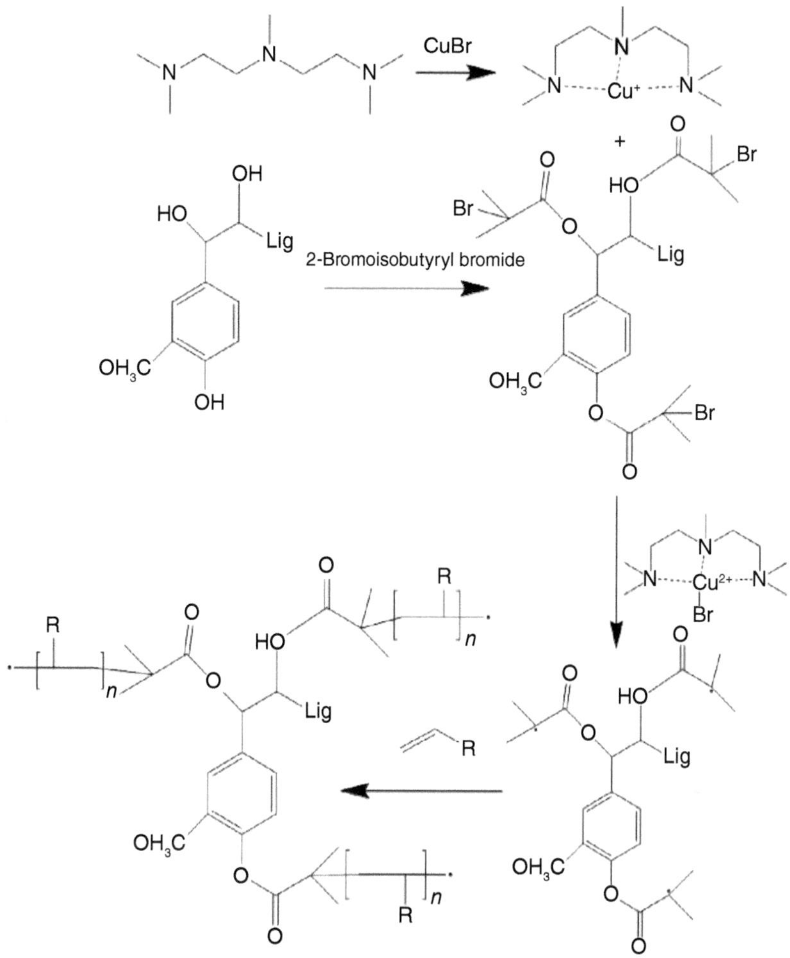

Figure 15.4 The mechanism of ATRP.

acted as macroinitiator to prepare dual responsive hydrogel [84]. Kai et al. synthesized lignin-based supramolecular hydrogel (PEGMA-grafted lignin hyperbranched copolymer) by ATRP [85]. The PEG methyl ether methacrylate (PEGMA)-grafted lignin hyperbranched copolymers prepared by ATRP exhibited molecular weight ranging between 38.7 and 65.0 kDa by adjusting the PEGMA-to-lignin weight ratio. Due to the hyperbranched architecture, their aqueous solutions were complexed with α-cyclodextrin (α-CD) to prepare hydrogel with a tunable mechanical response and excellent self-healing capability.

15.3.2.4 Reversible Addition-Fragmentation Chain Transfer (RAFT) Polymerization

Similar to ATRP, reversible addition-fragmentation chain transfer (RAFT) has revolutionized the polymer synthesis field. The RAFT meets the principles of green chemistry involving the design and application of chemical production and processes to reduce environmental impact. To generate lignin-based hydrogel following the RAFT method, the lignin-RAFT was prepared and served as a reaction core first, followed by the polymerization of the monomer onto lignin RAFT core through azobisisobutyronitrile (AIBN) initiation and radical reaction (Figure 15.5). Gao et al. synthesized lignin-based networks by using RAFT polymerization [86]. In this process, vanillin methacrylate (VMA) was synthesized by methacrylic anhydride and vanillin with 4-dimethylaminopyridine (DMAP) serving as the catalyst. Then, the lignin-4-cyano-4-(phenylcarbonothioylthio) pentanoic acid (CTA) macroinitiator was prepared via the esterification reaction between the hydroxyl groups of lignin and CTA in the presence of DMAP. Next, the lignin–CTA acted as the chain transfer agent to induce the copolymerization of VMA and lauryl methacrylate via RAFT polymerization. The main synthetic strategies for generating hydrogels following the RAFT system are shown in Table 15.6.

15.4 Applications

15.4.1 Dye Decontamination

Since the continuous development of chemical manufacturing industries, such as textile industries, wastewater containing organic dyes is a severe threat to human health and water safety. Therefore, it has great practical significance to treat dye-containing wastewater to alleviate the water resource crisis. At present, some methods, such as biotransformation, photodegradation, membrane technology, adsorption, and advanced oxidation, are applied to treat wastewater effluents containing dyes. Adsorption is considered to be the most promising method to remov pollutants, attributed to its high efficiency, simple operation, and low secondary pollution. Efforts to develop adsorbents for capturing pollution have been intensified. Although various adsorbents have been designed to remove these organic dyes from wastewater effluents, most of the existing materials suffer from complex preparation processes and poor adsorption capacity, which are unfavorable to practical applications. Consequently, it is crucial to develop an adsorbent for efficiently

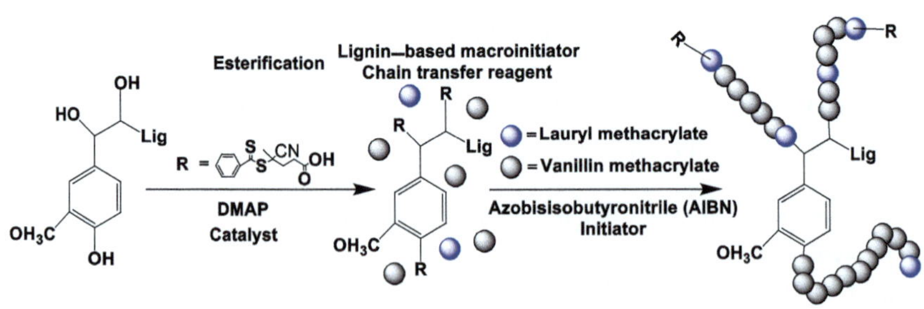

Figure 15.5 Mechanism for lignin-RAFT polymerization [86].

Table 15.6 Lignin-based hydrogels are prepared via ATRP and RAFT mechanisms.

Lignin source	Preparation method	Chemicals	Crosslinker/initiator	Application	References
Kraft lignin	Atom transfers radical polymerization (ATRP) reversible addition-fragmentation chain transfer polymerization (RAFT).	*N*-isopropylacrylamide and *N*,*N*-dimethylaminoethylmethacrylate	Brominated lignin as ATRP macroinitiator	Drug delivery	[84]
Alkali lignin		Methyl methacrylate-*co*-butyl acrylate and lignin-g-P(MMA-*co*-BA)	Lignin-g-P(MMA-*co*-BA)	UV blocking	[87]
DES-lignin		*N*-vinyl caprolactam and DES-lignin-graft thermo-responsive polymer	DES-lignin-Br	—	[88]
Kraft lignin		Poly(*N*-isopropylacrylamide) (PNIPAAm) and poly(ethylene glycol) (PEG) and poly(propylene glycol) (PPG)	Brominated lignin	—	[89]
Organosolv lignin		Lignin-graft-poly(*N*,*N*-diethylaminoethyl methacrylate) (PDEAEMA) copolymer	Organosolv lignin	UV resistance	[90]
—		Lignin-derived vanillin as building block and fatty acid derivative as the comonomer	Diamines	Adhesives	[86]
Organosolv lignin		Oil-derived methacrylate monomers	Lignin-based chain transfer agent (Lignin-RAFT)	—	[91]
Organosolv lignin		Lignin-graft-poly(*n*-butyl acrylate-*co*-1-vinylimidazole) copolymers	Lignin-based macro-RAFT agent	Water-resistant adhesive	[92]
Kraft lignin		Azobis-(isobutyronitrile)	Lignin RAFT Macroinitiator	—	[93]

Table 15.7 Application of lignin-based hydrogels for dye elimination.

Lignin source	Preparation method	Polymer network	Properties	Performance	References
Sodium lignosulfonate	Free radical graft copolymerization	Acrylamide and acryloyloxyethyltrimethyl ammonium chloride	Recyclability (10 times)	409.84 $mg\,g^{-1}$	[94]
Alkali lignin	Hydrogen bonding	Polyvinyl alcohol	Super absorbent (500 $g\,g^{-1}$)	Rhodamine 6G, 196 $mg\,g^{-1}$	[58]
Alkali lignin	Free radical graft copolymerization	Acrylic acid- 2-acrylamido-2-methyl-propanesulfonic acid, and lignin	Fast adsorption rate, recyclability	2541.76 $mg\,g^{-1}$ for MG	[72]
—	Free radical polymerization-graft copolymerization	Acrylamide	Swelling capacity	9646.92 $mg\,g^{-1}$	[75]
Alkali lignin	Covalent crosslinking	Long-chain hydroxyethyl celluloses (HECs) as framework backbones, short-chain polyvinyl alcohols (PVAs) as branched chains	Swelling ratio (1220 $g\,g^{-1}$), water retention, biodegradability	Crystal violet 184 $mg\,g^{-1}$	[95]
Alkali lignin	Hydrogen bonding	Polyvinyl alcohol	Mechanical robustness	MB 91.8% in 120 min	[96]
Alkali lignin	Sol–gelt transition (hydrogen bonding)	Polyethylene glycol	—	Malachite green (MG), 155 $mg\,g^{-1}$	[97]

removing dyes from wastewater. Interestingly, lignin-based hydrogels have been utilized to serve this purpose, and the results are tabulated in Table 15.7.

The physicochemical properties of the hydrogel, such as porous structure, swelling property, and functional groups of the polymer backbone, allow for the excellent adsorption of various compounds in wastewater [98]. The active groups, such as the phenolic, carboxyl, and carbonyl groups, facilitated dye removal via electrostatic interactions. Lignin-containing hydrogels have high water absorption, thermal stability, and biodegradability, which also favors the adsorption of various compounds [99]. These hydrogels are used for purifying the dye pollution from wastewater treatment. Wei et al. reported a lignosulfonate-based hydrogel for cleaning wastewater containing Acid Red dyes [94]. Alternatively, Wu et al. articulated a lignin–PVA hydrogel with excellent adsorption capacities for rhodamine, crystal violet, and methylene blue dyes [58].

However, the common lignin-derived hydrogel preparation process is not attractive due to the long preparation time, long adsorption equilibrium, and complex post-treatment process. To fabricate more efficient lignin-derived hydrogels as adsorbents for dyes, a more facile strategy to prepare hydrogel would be needed to scale up the production of such materials. Jiang et al. synthesized a super-adsorbent bentonite-doped lignin-containing hydrogel (LHS-BT) via a simple template method to remove cationic dyes from wastewater [72]. Adsorption experiments showed that LHS-BT had ultra-high adsorption capacity (2541.76 $mg\,g^{-1}$ for malachite green dyes). Additionally, LHS-BT exhibited excellent recyclability without a significant loss for at least five cycles, which depicted great potential for use in wastewater remediation processes. Furthermore, lignin-containing hydrogels, such as lignin–PVA, lignin-poly acrylamide/*N*-isopropyl acrylamide, and lignin-hydroxyethyl cellulose-PVA, were prepared with excellent adsorption capacities owing to the swelling property, porous structure, and abundant functional groups [58, 75, 95, 96].

The recyclability of the hydrogel is essential for its practical application as an adsorbent. Meng et al. prepared lignin-derived magnetic hydrogel microspheres via blending with Fe_3O_4 [97]. The adsorbents could be recycled by magnetic separation, regenerating from the acid condition, and reused for multiple cycles. In another study, a functional lignin-based hydrogel as an adsorbent achieved more than 90% adsorption efficiency after recycling the adsorption–desorption of dyes 10 times by using 0.1 $mol\,l^{-1}$ nitric acid solution as a desorption medium, which indicated that the lignin-containing hydrogel had effective recycling performance [94]. Still, the precise control of the properties and recycling performance of the lignin-based hydrogel limits its practical application.

15.4.2 Heavy Metal and Other Ion Elimination

Heavy metal contamination of water resources has become an increasing environmental concern due to the highly toxic, nondegradable, and carcinogenic properties of heavy metals. The porous structure and adjustable active sites of the lignin-based hydrogels allow for their potential application for heavy metal removals. Lignin-derived hydrogels demonstrated high adsorption activity for heavy metal ions due to

Table 15.8 The application of lignin-based hydrogels for heavy metal removal.

Lignin source	Preparation method	Chemicals	Properties	Performance	References
Sodium lignosulfonate	Free radical graft copolymerisation; in situ synthesis reaction	Acrylamide	Mechanical stability	nFeS@lh reduced the total, DTPA-soluble, and fixed Cd in the heavily polluted soil by 22.4%, 13.5%, and 40.1%	[100]
Sodium lignosulfonate	The free radical graft copolymerization	Acrylic acid	Absorbing capacity	Co^{2-} 385 mg g^{-1}	[101]
—	The free radical graft copolymerization	Acrylic acid	Recyclability, adsorption property	1.17 mmol g^{-1}	[102]
Alkali lignin	Grafting	Acrylic acid	Porous structure, selectivity toward different heavy metals	Pb^{2+} 1.076 mmol g^{-1}	[103]
Sodium lignosulfonate	Crosslinking polymerization	Acrylamide; sodium lignosulfonate	Swelling property, thermal property		[104]
Lignosulfonate	Hydrogen bonding and π–π conjugation	Reduced graphene oxide (rGO)	Recyclability	1210 mg g^{-1} Pb(II)	[50]

their highly porous structures and abundant functional groups. The application of lignin-based hydrogels for heavy metal decontamination is listed in Table 15.8.

In one study, nanoparticles incorporated-hydrogel was developed and used for the removal of heavy metals from sewage. Liu et al. developed a FeS nanoparticle loading lignin hydrogel for efficient Cd^{2+} removal [100]. In this system, lignin-based hydrogels inherited lignin's natural mechanical and environmental stability, and the FeS nanoparticles efficiently adsorbed Cd^{2+} and enhanced Cd^{2+} desorption from soils by producing H^+. Lignin-derived hydrogels enabled the efficient dispersion and recovery of FeS nanoparticles in the soil. In this system, the hydrogel's high sorption capacity ($833.3\,g\,kg^{-1}$) was attributed to four mechanisms including cadmium sulfide precipitation via chemical reaction, lignin complexation, hydrogel swelling, and nanoparticle sorption. Lignin-based hydrogel systems, including lignin-chitosan-poly-AA, lignin-montmorillonite, and lignin-poly-AA, were also used for the adsorption of heavy metal ions [102, 103]. The results showed the prepared hydrogels' excellent desorption performance and reusability for reuse.

Nanomaterials incorporated in the lignin-based hydrogel have been extensively applied for environmental remediations. The size and content of nanoparticles directly affected the removal of contaminants in such systems. The adsorption capacity of heavy metal ions was enhanced via nanoparticle incorporation [50, 104]. The adsorbents demonstrated high capacity, low cost, ecofriendliness, and recyclability, which is beneficial for the scale production of relative absorbents. Though lignin-based hydrogel exhibited excellent adsorption performance, the recyclability of lignin-based hydrogel still restricted its practical applications.

15.4.3 Drug Delivery Application

Lignin-based hydrogels have been studied for pharmaceutical applications owing to their high hydrophilicity, thermal stability, and biocompatibility. The application of lignin-based hydrogels for drug delivery systems is listed in Table 15.9.

The thermo-responsive properties of hydrogels will facilitate their use in delivering a drug to the human body as they function differently at altered temperatures. Currently, lignin-based thermo-response polymers have attracted extensive attention in drug sustained-release. Liu et al. synthesized lignin-grafted poly(*N*-Vinyl caprolactam), which can be regulated with the lowest critical solution temperature [88]. It provided the foundation for regulating lignin-based thermo-responsive materials to adapt to human temperature in drug loading. In the past, studies focused on the fabrication of lignin-based hydrogel following different strategies, such as supramolecular chemistry, reaction extrusion, and polymer complexation,

Table 15.9 The application of lignin-based hydrogel for drug delivery applications.

Lignin source	Preparation method	Chemicals	Properties	Performance	References
Kraft lignin	Chemical crosslinking	Polysaccharides and lignin	Mucoadhesiveness, bioadhesiveness, mechanical resistance, release kinetic, and antimicrobial activity	Strong swelling ability dependent on pH	[105]
Alkali lignin	Electrostatically complex	Chitosan-alkali lignin	Biocompatibility	Nontoxic to Mesenchymal stem cells, in vitro, and to zebrafish up to $100\,\mu g\,ml^{-1}$, *in vivo*	[22]
—	Chemical crosslinking	Collagen/polyurethane	—	Elastic modulus : 1.928 MPa	[106]
Sulfonated lignin	Graft copolymerization	Lignin-graft-Polyoxazoline	Anti-inflammatory	Anti-biofilm properties	[107]
Kraft lignin	Atom transfer radical polymerization (ATRP) reversible addition-fragmentation chain transfer polymerization (RAFT).	*N*-isopropylacrylamide and *N*,*N*-dimethylaminoethylmethacrylate	Dual temperature and pH responsive	Drug delivery	[84]
Kraft lignin	Radical polymerization	Acrylic acid	Antimicrobial property, pH responsive	Very low IC50 value ($0.1\,\mu g\,ml^{-1}$)	[108]

for use in drug delivery applications [22, 84, 85, 105, 106, 108]. The active groups, such as the carboxyl group, together with the polymer matrix, induced the pH responsiveness to the hydrogels. However, the cytotoxicity of industrial lignin still needs to be systematically investigated to facilitate the use of lignin in hydrogels intended for drug delivery systems.

15.4.4 Lignin-Based Hydrogel for Wound Dressing

Wound healing is a complicated and dynamic process that includes sequential molecular, physiological, biochemical, and cellular functions of replacing damaged structures with new cells and tissues [84]. Skin acts as a barrier to prevent the body from microorganism invasions. However, skin can be damaged due to external factors, such as burns, scalding, laceration, and sharp cut, which make skin wound healing essential. In general, wound healing consists of three phases: inflammatory, proliferative, and remodeling phases. Wound dressing needs to reduce the probability of potential bacterial infection and maintain a moist environment with the ability of antioxidant and free radical scavenging. The hydrogel could satisfy the function of a physical barrier, bactericidal material, adsorption medium, and environment conditioner, which makes it an ideal candidate for wound dressing. Therefore, lignin would endow hydrogels with multi-functionalities of high biocompatibility, low immunogenicity, antimicrobial activity, and biodegradability [52]. The status of previous attempts to use lignin-based hydrogels for wound dressing is listed in Table 15.10. In one study, Zhang et al. synthesized a novel lignin–chitosan–PVA composite hydrogel, and the introduction of lignin effectively improved the mechanical strength (tensile stress is up to 46.87 MPa), protein adsorption capacity, and wound environmental regulation of the hydrogel. The lignin–chitosan–PVA composite hydrogels significantly accelerated wound healing in a murine wound model. The addition of lignin endowed the hydrogel with excellent antioxidant activity, impact resistance, and tensile properties because of the presence of these functional structures and groups. In addition, lignin enhanced the adsorption capacity of the protein and the mechanical strength of the lignin–chitosan–PVA composite hydrogel, resulting in better removal of the metabolites and resistance to external reinjury [52]. In another study, a multi-functional hydrogel was fabricated with self-healing, shape adaptability, biocompatibility, blood compatibility, antioxidant excellent broad-spectrum antimicrobial good tissue adhesion, and electrical conductivity properties. The in vivo experiment results showed excellent hemostasis properties in a rat liver injury model [109]. Similarly, a novel lignin/poly(ionic liquids) composite hydrogel with excellent mechanical strength, self-healing properties, bactericidal activity, and antioxidant activity was fabricated for wound dressing [110]. Another study reported a lignin-derived oligomer incorporated bacterial-based cellulose hydrogel for wound dressing [111], which acted as an antibacterial agent against selected pathogenic bacterial and accelerated skin repair and decreased pain for patients.

Table 15.10 The application of lignin-based hydrogel for wound dressing.

Lignin source	Preparation method	Chemicals	Properties	Performance	References
Lignin with low sulfonate content	Chemical crosslinking	Carrageenan	Strength, flexibility, swelling ratio	Antibacterial activity; reduce wound area <3%	[54]
Kraft lignin	Atom transfers radical polymerization	*N*-isopropylacrylamide (NIPAM) and *N*,*N*-dimethylaminoethylmethacrylate (DMAEMA)	—	High drug loading capacity	[84]
—	Freeze-thaw cycle	Lignin–chitosan–PVA	Biocompatibility, Mechanical strength	Accelerated wound healing	[52]
—	Covalent bonding	Hydroxypropyl cellulose	Antioxidant, self-healing, shape adaptability	Antibacterial activity; speed up the healing process	[109]
—	Covalent bonding	Lignin/poly(ionic liquids) composite	Self-healing, bactericidal activity	Recovery antibacterial properties	[110]
—	—	Bacterial cellulose-lignin composite	Antimicrobial	Inhibitory/bactericidal effect; high swellability	[111]

Due to the polyphenol structure, lignin can also be used as a reductant for the preparation of metal nanoparticles. Then, the metal nanoparticles loaded with lignin could be composited with other polymers, such as Carrageenan and PVA, to form hydrogels with antibacterial and biocompatibility properties, which are suitable for wound healing applications [54, 67]. However, the potential toxicity to the human body needs to be extensively studied with respect to in vivo and in vitro experiments to develop lignin-containing hydrogels for commercial wound dressing applications.

15.4.5 Lignin-Derived Hydrogel for Agriculture of Watering Plants and Fertilizers

The increasing human population and shrinking agricultural soils have become critical issues in meeting the food supply demand. Watering and fertilizer are significant in maintaining soil fertility and crop production since the macronutrients, especially nitrogen (N), potassium (K), and phosphorus (P), are key elements to boost crop yields. Conventional fertilizer leads to vaporization, leaching, and surface runoff. Excessive fertilizer causes severe environmental issues, such as compaction and acidity or alkalization of soil, or even a decline in soil fertility after long-term use [112]. Hydrogel-based fertilizers have drawn attention to increase the efficiency in controlled release and decrease negative environmental impacts. Lignin-based hydrogel is used to improve the controlled release of fertilizers and the water-holding capacity of soil [113]. The addition of lignin to hydrogels has improved the ecofriendliness, costs, and biodegradability of hydrogels. Also, lignin-based hydrogels have been known to be utilized as fertilizers and soil decontamination materials to resist drought stress [114, 115]. The application of lignin-based hydrogels in agriculture for watering plants and fertilizer is listed in Table 15.11. Lignin-incorporated hydrogels have demonstrated excellent properties for biodegradability and the water-holding ability for plants under drowning conditions [116, 117], which are beneficial for the handling and further utilization of the soil [118]. Moreover, the incorporation of lignin in the hydrogel enabled the slow release of fertilizers [119].

15.4.6 Lignin-Derived Hydrogels for Energy Storage Applications

With the rapid development of artificial intelligence (AI) and the Internet of Things (IoT), electrochemical energy storage (EES) devices have drawn much attention [120]. Owing to the tailorable physiochemical and biological properties, the hydrogels with outstanding physicochemical properties, such as mechanical strength, flexibility, conductivity, and electron transport rate, offer potential applications as electrodes and electrolytes. For flexible energy storage applications, lignin-based hydrogels would increase ionic conductivity, which is favorable to the migration of electrolyte ions, and lignin-induced redox reaction would facilitate the increase in the pseudocapacitance of supercapacitors. The application of lignin-based hydrogels in energy storage materials is listed in Table 15.12. Liu et al. reported

Table 15.11 The application of lignin-based hydrogels as fertilizing and controlled watering materials.

Lignin source	Crosslinking mechanism	Polymer network	Properties	Performance	References
—	Chemical crosslinking	Lignin	Mechanical stability, Swelling ratio ($50\,g\,g^{-1}$)	The water content of sandy soil increased by 70%	[115]
Alkaline lignin	Chemical crosslinking	Lignin/sodium polyacrylate	Biodegradability, water retaining ability	Alleviate water scarcity in agricultural soils	[116]
Ligninsulfonate	Hydrogen bonding	Lignin/sodium alginate	Water holding capacity, saturated hydraulic conductivity, water/nutrient retention	Extends tobacco growth time for 9–14 days	[117]
Calcium lignosulfonate	Hydrogen bonding	Lignosulfonate/synthetic zeolite/polyvinylpyrrolidone	Biodegradability, water uptake capability	Satisfactory water solubility and uptake	[118]
Organosolv lignin	Chemical	Acrylic acid	Water holding property	Achieve controllable and slow nitrogen fertilizer release; improve the water-holding property	[119]

Table 15.12 The application of lignin-based hydrogel for energy storage.

Lignin source	Preparation method	Polymer network	Properties	Performance	References
—	Radical polymerization	Poly(ethylene glycol) methyl ether methacrylate and a methacrylamide catechol derivative monomer	Superior mechanical strength (the fracture compression stress of 4.74 MPa), good ionic conductivity (80 ms cm^{-1})	High water-swelling capability, ionic conductivity, etc.	[121]
Lignosulfonate	—	Polyvinyl alcohol	Mechanical stability, ionic conductivity	Good rate capability, long cyclic stability, high specific capacitance	[122]
Alkali lignin	Ring-opening polymerization	Polyvinylalcohol and poly(ethylene glycol) diglycidyl ether	Ionic conductivity, mechanical integrity	Outstanding charge storage capability, flexibility, and durability	[59]

a double-crosslinked lignin hydrogel via a two-step network formation, which showed superior mechanical strength (the fracture compression stress of 4.74 MPa), the good ionic conductivity of $80\,mS\,cm^{-1}$, and excellent shape recovery property [121]. The constructed flexible supercapacitor with lignin-derived hydrogels as electrolyte and polyaniline (PANI) deposited carbon cloth as electrodes demonstrated a high specific capacitance of $190\,F\,g^{-1}$ and excellent flexibility. Moreover, the phenolic groups of lignin that can undergo redox reactions through the quinone/hydroquinone (Q/QH2) structure would store electrons and protons in Q/QH2 structure. The redox charge transfer in the quinone/hydroquinone (Q/QH2) structure would enhance the charge storage capacity by simply incorporating lignin in the electrode. Based on the reversible redox charge transfer of quinone groups in lignin, lignin-derived hydrogels were utilized as electrodes to prepare flexible supercapacitors, which exhibit high specific capacitance, good retention of rate capability, and cycling stability [59, 122].

15.5 Perspectives and Future Work

The new era has witnessed the rocket growth of natural polymer-based functional materials, such as cellulose, chitin, silk, and starch. Lignin, the most abundant aromatic polymer, distinguished from other natural polymers due to its intrinsic structure, has piqued much attention for diverse applications. Due to its antibacterial activity, UV blocking, and oxidation resistance endowed by the polyphenol structure of lignin, research and development of lignin-derived hydrogels have demonstrated diverse applications in various fields.

In general, lignin incorporation into the hydrogel matrix through diverse strategies would demonstrate the intrinsic properties of lignin. In contrast, the high polydispersity and extremely complex structures of lignin hindered the preparation process of lignin-based hydrogels.

In spite of the intrinsic advantages of lignin, the wide molecular weight distribution and structural heterogeneity of lignin have limited its inclusion in high-performance hydrogels as they induced issues including compatibility and water resistance. Chemical modifications via various emerging techniques, such as ATRP and RAFT, represent a promising way to tune the properties of lignin-derived hydrogels. However, the complex process and precise control of the polymer network need to be further evaluated. Most of the research on lignin-derived hydrogel focused solely on the gelation chemistry and functions for specific applications. Nonetheless, a systematic study to reveal the structure–property function of the lignin-derived hydrogels would contribute to the versatile requirements of applications.

The investigation of some properties of lignin-based hydrogels, including toxicity, and compatibility needs to be carried out to accelerate their use in different application fields, especially for biomedical applications. Thus, extensive characterization concerning in vitro/in vivo experiments should be necessary for commercializing such products.

References

1 Zhou, Y., Wan, C., Yang, Y. et al. (2019). Highly stretchable, elastic, and ionic conductive hydrogel for artificial soft electronics. *Advanced Functional Materials* 29 (1).

2 Han, J.Q., Lei, T.Z., and Wu, Q.L. (2014). High-water-content mouldable polyvinyl alcohol-borax hydrogels reinforced by well-dispersed cellulose nanoparticles: dynamic rheological properties and hydrogel formation mechanism. *Carbohydrate Polymers* 102: 306–316.

3 Ahmed, E.M. (2015). Hydrogel: Preparation, characterization, and applications: a review. *Journal of Advanced Research* 6 (2): 105–121.

4 Kamoun, E.A., Kenawy, E.-R.S., and Chen, X. (2017). A review on polymeric hydrogel membranes for wound dressing applications: PVA-based hydrogel dressings. *Journal of Advanced Research* 8 (3): 217–233.

5 Parhi, R. (2017). Cross-linked hydrogel for pharmaceutical applications: a review. *Advanced Pharmaceutical Bulletin* 7 (4): 515–530.

6 Athawale, V. and Lele, V. (1998). Graft copolymerization onto starch. II. Grafting of acrylic acid and preparation of it's hydrogels. *Carbohydrate Polymers* 35 (1-2): 21–27.

7 Mansur, H.S., Oréfice, R.L., and Mansur, A.A. (2004). Characterization of poly(vinyl alcohol)/poly(ethylene glycol) hydrogels and PVA-derived hybrids by small-angle X-ray scattering and FTIR spectroscopy. *Polymer* 45 (21): 7193–7202.

8 Zhu, L., Liu, Y., Jiang, Z. et al. (2019). Highly temperature resistant cellulose nanofiber/polyvinyl alcohol hydrogel using aldehyde cellulose nanofiber as cross-linker. *Cellulose* 26 (9): 5291–5303.

9 Crescenzi, V., Francescangeli, A., Capitani, D. et al. (2003). Hyaluronan networking via Ugi's condensation using lysine as cross-linker diamine. *Carbohydrate Polymers* 53 (3): 311–316.

10 Yu, F., Cao, X., Du, J. et al. (2015). Multifunctional hydrogel with good structure integrity, self-healing, and tissue-adhesive property formed by combining Diels–Alder click reaction and acylhydrazone bond. *ACS Applied Materials & Interfaces* 7 (43): 24023–24031.

11 Zhang, W., Zhao, J., Cai, C. et al. (2022). Gas-sensitive cellulosic triboelectric materials for self-powered ammonia sensing. *Advanced Science* 2203428.

12 Fujita, M., Policastro, G.M., Burdick, A. et al. (2021). Preventing post-surgical cardiac adhesions with a catechol-functionalized oxime hydrogel. *Nature Communications* 12 (1): 3764.

13 Liu, Z.Q., Wei, Z., Zhu, X.L. et al. (2015). Dextran-based hydrogel formed by thiol-Michael addition reaction for 3D cell encapsulation. *Colloids and Surfaces B: Biointerfaces* 128: 140–148.

14 Zhao, Q.S., Ji, Q.X., Xing, K. et al. (2009). Preparation and characteristics of novel porous hydrogel films based on chitosan and glycerophosphate. *Carbohydrate Polymers* 76 (3): 410–416.

15 Ebara, M., Kotsuchibashi, Y., Narain, R. et al. (2014). *Smart Biomaterials Preface. Smart Biomaterials*. Tokyo: Springer-Verlag Tokyo. V-VI.

16 Takigami, M., Amada, H., Nagasawa, N. et al. (2007). Preparation and properties of CMC gel. *Transactions of the Materials Research Society of Japan* 32 (3): 713–716.

17 Washington, R.P. and Steinbock, O. (2001). Frontal polymerization synthesis of temperature-sensitive hydrogels. *Journal of the American Chemical Society* 123 (32): 7933–7934.

18 Zheng, S.Y., Ding, H., Qian, J. et al. (2016). Metal-coordination complexes mediated physical hydrogels with high toughness, stick–slip tearing behavior, and good processability. *Macromolecules* 49 (24): 9637–9646.

19 Fullenkamp, D.E., He, L., Barrett, D.G. et al. (2013). Mussel-inspired histidine-based transient network metal coordination hydrogels. *Macromolecules* 46 (3): 1167–1174.

20 Tiwari, J.N., Mahesh, K., Le, N.H. et al. (2013). Reduced graphene oxide-based hydrogels for the efficient capture of dye pollutants from aqueous solutions. *Carbon* 56: 173–182.

21 Holloway, J.L., Lowman, A.M., and Palmese, G.R. (2013). The role of crystallization and phase separation in the formation of physically cross-linked PVA hydrogels. *Soft Matter* 9 (3): 826–833.

22 Ravishankar, K., Venkatesan, M., Desingh, R.P. et al. (2019). Biocompatible hydrogels of chitosan-alkali lignin for potential wound healing applications. *Materials Science and Engineering: C* 102: 447–457.

23 Hennink, W.E. and van Nostrum, C.F. (2012). Novel crosslinking methods to design hydrogels. *Advanced Drug Delivery Reviews* 64: 223–236.

24 Thakur, V.K. and Thakur, M.K. (2015). Recent advances in green hydrogels from lignin: a review. *International Journal of Biological Macromolecules* 72: 834–847.

25 Ghobril, C. and Grinstaff, M. (2015). The chemistry and engineering of polymeric hydrogel adhesives for wound closure: a tutorial. *Chemical Society Reviews* 44 (7): 1820–1835.

26 Hu, W., Wang, Z., Xiao, Y. et al. (2019). Advances in crosslinking strategies of biomedical hydrogels. *Biomaterials Science* 7 (3): 843–855.

27 Laftah, W.A., Hashim, S., and Ibrahim, A.N. (2011). Polymer hydrogels: a review. *Polymer-Plastics Technology and Engineering* 50 (14): 1475–1486.

28 Malkoch, M., Vestberg, R., Gupta, N. et al. (2006). Synthesis of well-defined hydrogel networks using Click chemistry. *Chemical Communications* 26: 2774–2776.

29 Tillet, G., Boutevin, B., and Ameduri, B. (2011). Chemical reactions of polymer crosslinking and post-crosslinking at room and medium temperature. *Progress in Polymer Science (Oxford)* 36 (2): 191–217.

30 Nising, C.F. and Braese, S. (2012). Recent developments in the field of oxa-Michael reactions. *Chemical Society Reviews* 41 (3): 988–999.

31 Passauer, L., Hallas, T., Bäucker, E. et al. (2015). Biodegradation of hydrogels from oxyethylated lignins in model soils. *ACS Sustainable Chemistry & Engineering* 3 (9): 1955–1964.

32 Jayaramudu, T., Li, Y., Ko, H.U. et al. (2016). Poly(acrylic acid)-poly(vinyl alcohol) hydrogels for reconfigurable lens actuators. *International Journal of Precision Engineering and Manufacturing - Green Technology* 3 (4): 375–379.

33 Wang, J., Li, T., Chen, F. et al. (2018). Softening and shape morphing of stiff tough hydrogels by localized unlocking of the trivalent ionically cross-linked centers. *Macromolecular Rapid Communications* 39 (12): 1800143.

34 Yang, C.H., Wang, M.X., Haider, H. et al. (2013). Strengthening alginate/polyacrylamide hydrogels using various multivalent cations. *ACS Applied Materials & Interfaces* 5 (21): 10418–10422.

35 Fujiwara, T., Mukose, T., Yamaoka, T. et al. (2001). Novel thermo-responsive formation of a hydrogel by stereo-complexation between PLLA-PEG-PLLA and PDLA-PEG-PDLA block copolymers. *Macromolecular Bioscience* 1 (5): 204–208.

36 Low, L.E., Teh, K.C., Siva, S.P. et al. (2021). Lignin nanoparticles: the next green nanoreinforcer with wide opportunity. *Environmental Nanotechnology, Monitoring & Management* 15.

37 Lupoi, J.S., Singh, S., Parthasarathi, R. et al. (2015). Recent innovations in analytical methods for the qualitative and quantitative assessment of lignin. *Renewable & Sustainable Energy Reviews* 49: 871–906.

38 Thakur, V.K., Thakur, M.K., Raghavan, P. et al. (2014). Progress in green polymer composites from lignin for multifunctional applications: a review. *ACS Sustainable Chemistry & Engineering* 2 (5): 1072–1092.

39 Jin, Z., Yu, Y., Shao, S. et al. (2010). Lignin as a cross-linker of acrylic acid-grafted carboxymethyl lignocellulose. *Journal of Wood Science* 56 (6): 470–476.

40 Raschip, I.E., Hitruc, E.G., and Vasile, C. (2011). Semi-interpenetrating polymer networks containing polysaccharides. II. Xanthan/lignin networks: a spectral and thermal characterization. *High Performance Polymers* 23 (3): 219–229.

41 Xu, J., Li, C., Dai, L. et al. (2020). Biomass fractionation and lignin fractionation towards lignin valorization. *ChemSusChem* 13 (17): 4284–4295.

42 Argyropoulos, D.S., Sadeghifar, H., Cui, C. et al. (2014). Synthesis and characterization of poly(arylene ether sulfone) kraft lignin heat stable copolymers. *ACS Sustainable Chemistry & Engineering* 2 (2): 264–271.

43 Lohani, A., Singh, G., Bhattacharya, S.S. et al. (2014). Interpenetrating polymer networks as innovative drug delivery systems. *Journal of Drug Delivery* 2014: 1–11.

44 Wang, W., Hou, G., Zheng, Z. et al. (2017). Designing polymer nanocomposites with a semi-interpenetrating or interpenetrating network structure: toward enhanced mechanical properties. *Physical Chemistry Chemical Physics* 19 (24): 15808–15820.

45 Priya, P., Raja, A., and Raj, V. (2015). Interpenetrating polymeric networks of chitosan and egg white with dual crosslinking agents polyethylene glycol/polyvinylpyrrolidone as a novel drug carrier. *Cellulose* 23 (1): 699–712.

46 Wang, J.J. and Liu, F. (2013). Enhanced adsorption of heavy metal ions onto simultaneous interpenetrating polymer network hydrogels synthesized by UV irradiation. *Polymer Bulletin* 70 (4): 1415–1430.

47 Hoare, T.R. and Kohane, D.S. (2008). Hydrogels in drug delivery: progress and challenges. *Polymer* 49 (8): 1993–2007.

48 Dragan, E.S., Lazar, M.M., Dinu, M.V. et al. (2012). Macroporous composite IPN hydrogels based on poly(acrylamide) and chitosan with tuned swelling and sorption of cationic dyes. *Chemical Engineering Journal* 204-206: 198–209.

49 Dragan, E.S., Perju, M.M., and Dinu, M.V. (2012). Preparation and characterization of IPN composite hydrogels based on polyacrylamide and chitosan and their interaction with ionic dyes. *Carbohydrate Polymers* 88 (1): 270–281.

50 Li, F., Wang, X., Yuan, T. et al. (2016). A lignosulfonate-modified graphene hydrogel with ultrahigh adsorption capacity for Pb(II) removal. *Journal of Materials Chemistry A* https://doi.org/10.1039/C6TA03779H.

51 Park, S., Kim, S.H., Kim, J.H. et al. (2015). Application of cellulose/lignin hydrogel beads as novel supports for immobilizing lipase. *Journal of Molecular Catalysis B: Enzymatic* 119: 33–39.

52 Zhang, Y., Jiang, M., Zhang, Y. et al. (2019). Novel lignin–chitosan–PVA composite hydrogel for wound dressing. *Materials Science and Engineering: C* 104: 110002.

53 Huang, S., Shuyi, S., Gan, H. et al. (2019). Facile fabrication and characterization of highly stretchable lignin-based hydroxyethyl cellulose self-healing hydrogel. *Carbohydrate Polymers* 223: 115080.

54 Jaiswal, L., Shankar, S., Rhim, J.W. et al. (2020). Lignin-mediated green synthesis of AgNPs in carrageenan matrix for wound dressing applications. *International Journal of Biological Macromolecules* 159: 859–869.

55 Oveissi, F., Naficy, S., Le, T.Y.L. et al. (2018). Tough and processable hydrogels based on lignin and hydrophilic polyurethane. *ACS Applied Bio Materials* 1 (6): 2073–2081.

56 Ciolacu, D. and Cazacu, G. (2018). New green hydrogels based on lignin. *Journal of Nanoscience and Nanotechnology* 18 (4): 2811–2822.

57 Ciolacu, D., Oprea, A.M., Anghel, N. et al. (2012). New cellulose–lignin hydrogels and their application in controlled release of polyphenols. *Materials Science and Engineering: C* 32 (3): 452–463.

58 Wu, L., Huang, S., Zheng, J. et al. (2019). Synthesis and characterization of biomass lignin-based PVA super-absorbent hydrogel. *International Journal of Biological Macromolecules* 140: 538–545.

59 Park, J.H., Rana, H.H., Lee, J.Y. et al. (2019). Renewable flexible supercapacitors based on all-lignin-based hydrogel electrolytes and nanofiber electrodes. *Journal of Materials Chemistry A* 7 (28): 16962–16968.

60 Passauer, L. (2012). Highly swellable lignin hydrogels: novel materials with interesting properties. In: *Functional Materials from Renewable Sources*, 211–228.

61 Larraneta, E., Imízcoz, M., Toh, J.X. et al. (2018). Synthesis and characterization of lignin hydrogels for potential applications as drug eluting antimicrobial coatings for medical materials. *ACS Sustainable Chemistry & Engineering* 6 (7): 9037–9046.

62 Yang, S., Zhang, Y., Yuan, T.Q. et al. (2015). Lignin-phenol-formaldehyde resin adhesives prepared with biorefinery technical lignins. *Journal of Applied Polymer Science* 132 (36).

63 Zainal, S.H., Mohd, N.H., Suhaili, N. et al. (2021). Preparation of cellulose-based hydrogel: a review. *Journal of Materials Research and Technology* 10: 935–952.

64 Xu, L., Gao, S., Guo, Q. et al. (2020). A solvent-exchange strategy to regulate noncovalent interactions for strong and antiswelling hydrogels. *Advanced Materials* 32 (52): 2004579.

65 Passauer, L., Struch, M., Schuldt, S. et al. (2012). Dynamic moisture sorption characteristics of xerogels from water-swellable oligo(oxyethylene) lignin derivatives. *ACS Applied Materials & Interfaces* 4 (11): 5852–5862.

66 Dai, L., Zhu, W., Lu, J. et al. (2019). A lignin-containing cellulose hydrogel for lignin fractionation. *Green Chemistry* 21 (19): 5222–5230.

67 Li, M., Jiang, X., Wang, D. et al. (2019). In situ reduction of silver nanoparticles in the lignin based hydrogel for enhanced antibacterial application. *Colloids and Surfaces B: Biointerfaces* 177: 370–376.

68 Yamada, B. and Zetterlund, P.B. (2002). General chemistry of radical polymerization. In: *Handbook of Radical Polymerization*, 117–186.

69 El-Zawawy, W.K. (2005). Preparation of hydrogel from green polymer. *Polymers for Advanced Technologies* 16 (1): 48–54.

70 El-Zawawy, W.K. and Ibrahim, M.M. (2012). Preparation and characterization of novel polymer hydrogel from industrial waste and copolymerization of poly(vinyl alcohol) and polyacrylamide. *Journal of Applied Polymer Science* 124 (5): 4362–4370.

71 Sun, L., Mo, Z., Li, Q. et al. (2021). Facile synthesis and performance of pH/temperature dual-response hydrogel containing lignin-based carbon dots. *International Journal of Biological Macromolecules* 175: 516–525.

72 Jiang, M.W., Niu, N., and Chen, L.G. (2022). A template synthesized strategy on bentonite-doped lignin hydrogel spheres for organic dyes removal. *Separation and Purification Technology* 285.

73 Xu, C., Liu, L., Renneckar, S. et al. (2021). Chemically and physically crosslinked lignin hydrogels with antifouling and antimicrobial properties. *Industrial Crops and Products* 170.

74 Wang, X., Zhou, Z., Guo, X. et al. (2016). Ultrasonic-assisted synthesis of sodium lignosulfonate-grafted poly(acrylic acid-co-poly(vinyl pyrrolidone)) hydrogel for drug delivery. *RSC Advances* 6 (42): 35550–35558.

75 Wang, Y., Xiong, Y., Wang, J. et al. (2017). Ultrasonic-assisted fabrication of montmorillonite-lignin hybrid hydrogel: highly efficient swelling behaviors and super-sorbent for dye removal from wastewater. *Colloids and Surfaces A: Physicochemical and Engineering Aspects* 520: 903–913.

76 Ma, Y., Guo, Y., Zhou, J. et al. (2017). Porous lignin based poly(acrylic acid)/organo-montmorillonite nanocomposites: Swelling behaviors and rapid removal of Pb (II) ions. *Polymer* 128: 12–23.

77 Peñaranda, A., J.E. and Sabino, M.A. (2010). Effect of the presence of lignin or peat in IPN hydrogels on the sorption of heavy metals. *Polymer Bulletin* 65 (5): 495–508.

78 Jiang, P., Cheng, Y., Yu, S. et al. (2018). Study on the effect of 1-butanol soluble lignin on temperature-sensitive gel. *Polymers (Basel)* 10 (10).

79 Zerpa, A., Pakzad, L., and Fatehi, P. (2018). Hardwood kraft lignin-based hydrogels: production and performance. *ACS Omega* 3 (7): 8233–8242.

80 Gao, G., Xu, W.Z., and Kadla, J.F. (2014). Reversible pH-responsive hydrogels of softwood kraft lignin and poly[(2-dimethylamino)ethyl methacrylate]-based polymers. *Journal of Wood Chemistry and Technology* 35 (1): 73–90.

81 Liu, R., Dai, L., Xu, C. et al. (2020). Lignin-based micro-and nanomaterials and their composites in biomedical applications. *ChemSusChem* 13 (17): 4266–4283.

82 Matyjaszewski, K. (2018). Advanced materials by atom transfer radical polymerization. *Advanced Materials* 30 (23): e1706441.

83 Bortolamei, N., Isse, A.A., Di Marco, V.B. et al. (2010). Thermodynamic properties of copper complexes used as catalysts in atom transfer radical polymerization. *Macromolecules* 43 (22): 9257–9267.

84 Dinari, A., Abdollahi, M., and Sadeghizadeh, M. (2021). Design and fabrication of dual responsive lignin-based nanogel via "grafting from" atom transfer radical polymerization for curcumin loading and release. *Scientific Reports* 11 (1): 1962.

85 Kai, D., Low, Z.W., Liow, S.S. et al. (2015). Development of lignin supramolecular hydrogels with mechanically responsive and self-healing properties. *ACS Sustainable Chemistry & Engineering* 3 (9): 2160–2169.

86 Gao, S., Cheng, Z., Zhou, X. et al. (2020). Fabrication of lignin based renewable dynamic networks and its applications as self-healing, antifungal and conductive adhesives. *Chemical Engineering Journal* 394.

87 Yu, J., Wang, J., Wang, C. et al. (2015). UV-absorbent lignin-based multi-arm star thermoplastic elastomers. *Macromolecular Rapid Communications* 36 (4): 398–404.

88 Liu, R., Ding, T., Deng, P. et al. (2022). Preparation of LCST regulable DES-lignin-g-PNVCL thermo-responsive polymer by ARGET-ATRP. *International Journal of Biological Macromolecules* 194: 358–365.

89 Diao, B., Zhang, Z., Zhu, J. et al. (2014). Biomass-based thermogelling copolymers consisting of lignin and grafted poly(N-isopropylacrylamide), poly(ethylene glycol), and poly(propylene glycol). *RSC Advances* 4 (81): 42996–43003.

90 Huang, W., Wu, M., Liu, W. et al. (2019). Value-adding of organosolv lignin: designing mechanically robust UV-resistant polymeric glass via ARGET ATRP. *Applied Surface Science* 475: 302–311.

91 Xu, Y., Yuan, L., Wang, Z. et al. (2016). Lignin and soy oil-derived polymeric biocomposites by "grafting from" RAFT polymerization. *Green Chemistry* 18 (18): 4974–4981.

92 Wang, W., Wang, F., Zhang, C. et al. (2021). Versatile value-added application of hyperbranched lignin derivatives: water-resistance adhesive, UV protection coating, self-healing and skin-adhesive sensing. *Chemical Engineering Journal* 404.

93 Gupta, C. and Washburn, N.R. (2014). Polymer-grafted lignin surfactants prepared via reversible addition-fragmentation chain-transfer polymerization. *Langmuir* 30 (31): 9303–9312.

94 Wei, S.X., Chen, W., Tong, Z. et al. (2021). Synthesis of a functional biomass lignin-based hydrogel with high swelling and adsorption capability towards Acid Red 73. *Environmental Science and Pollution Research* 28 (37): 51306–51320.

95 Huang, S.Q., Wu, L., Li, T. et al. (2019). Facile preparation of biomass lignin-based hydroxyethyl cellulose super-absorbent hydrogel for dye pollutant removal. *International Journal of Biological Macromolecules* 137: 939–947.